Fitting Statistical Distributions

The Generalized Lambda Distribution and Generalized Bootstrap Methods

Fitting Statistical Distributions

Distributions

The Generalized Lambda Distribution and Generalized Bootstrap Methods

Zaven A. Karian and Edward J. Dudewicz

CRC Press
Taylor & Francis Group
Boca Raton London New York

CRC Press is an imprint of the
Taylor & Francis Group, an **informa** business

CRC Press
Taylor & Francis Group
6000 Broken Sound Parkway NW, Suite 300
Boca Raton, FL 33487-2742

First issued in paperback 2019

© 2000 by Taylor & Francis Group, LLC
CRC Press is an imprint of Taylor & Francis Group, an Informa business

No claim to original U.S. Government works

ISBN-13: 978-1-58488-069-1 (hbk)
ISBN-13: 978-0-367-39861-3 (pbk)

Library of Congress Cataloging-in-Publication Data
Karian, Zaven A.
Fitting statistical distributions : the Generalized Lambda Distribution and Generalized Bootstrap methods / Zaven A. Karian, Edward J. Dudewicz.
p. cm.
Includes bibliographical references and index.
ISBN 1-58488-069-4
1. Distribution (Probability theory) I.. Dudewicz, Edward J. II. Title.
QA273.6.K37 2000
519.2′4—dc21   00-036098

Library of Congress Card Number 00-036098

**Visit the Taylor & Francis Web site at
http://www.taylorandfrancis.com**

**and the CRC Press Web site at
http://www.crcpress.com**

Preface

Human endeavor in the hundreds of fields that comprise human knowledge (for example, modern universities may offer graduate degrees in over 100 programs, some of them having sub-areas—see Syracuse University (1999, pp. 7–9) for an example with 116 graduate programs) involves the construction of statistical models to describe fundamental variables. Of the variety of possible models, probability distributions are the most widely used ones. This book deals with how to model (or, how to fit) continuous probability distribution models to data.

The area of fitting distributions to data has seen explosive growth in recent years. In view of this and the widespread need for fitting methods in virtually all areas of endeavor, we believe there is a need for a text that can be used in applied programs (including graduate programs in the mathematical sciences, where many research topics in this area will be explored), and as a reference work by researchers and practitioners. This book therefore includes the following: a detailed exposition of the Generalized Lambda Distribution (GLD), relevant results related to the Generalized Bootstrap (GB) and Monte Carlo Simulation (MC), and the necessary tables, algorithms, and computer programs for fitting continuous probability distributions to data.

We restrict our coverage to the GLD/GB/MC methods that have been used successfully in many fields, where our own recent work has dealt with the refinement and completion of these methods, whose proper exposition for the audiences intended fills a volume. A reviewer of some of that work stated "I don't see why the authors [are] wait[ing] to write a $\cdots$ book about univariate and multivariate GLD applications" (Omey (1998)), and we appreciate this encouragement.

The plan of the book is detailed in the introduction to Chapter 1. A wide-ranging variety of applications is included to aid the practitioner, as well as proofs of key results to aid the person involved in theoretical development, teaching a course, or studying these topics. Various programs are provided to help with the computational difficulties associated with fitting distributions. These programs are available from the CRC Press website http://www.crcpress.com and are listed in Appendix A.

We thank our students who often have analyzed data from a broad range of fields with these methods, or developed extensions of the results in their research work (often in a published paper). Some examples include Nelson B. Beckwith (whose work on the GLD–2 of Chapter 5 was published in Beckwith and Dudewicz (1996)), and Mr. Fasheng Li and other students in Statistical Simulation and Nonstandard Data Analysis (MAT 621) at Syracuse University whose valuable applied and theoretical insights have aided us in our own development and understanding of the modeling process in both its applied and theoretical aspects. We also wish to express our gratitude to Susan Karian, Jarold Tawney, and Helen Viles for their considerable typesetting assistance.

We welcome communications from those who use this book and will try to include additional applications and references to new work in subsequent editions.

Zaven A. Karian
Denison University
Granville, Ohio

Edward J. Dudewicz
Syracuse University
Syracuse, New York

March 2000

About the Authors

Dr. Zaven A. Karian holds the Benjamin Barney Chair of Mathematics, and is Professor of Mathematics and Computer Science at Denison University in Ohio. He has been active as instructor, researcher and consultant in mathematics, computer science, statistics, and simulation for over thirty years. He has taught workshops in these areas for a dozen educational institutions and national and international conferences (International Conference on Teaching Mathematics, Greece; Asian Technology Conference in Mathematics, Japan; Joint Meetings of the American Mathematical Society/Mathematical Association of America).

Dr. Karian has taught short courses of varying lengths for colleges and universities (Howard University, Washington, D.C.; The Ohio State University; State University of New York; and Lyndon State College, Vermont), for professional societies (Society for Computer Simulation, American Statistical Association, Mathematical Association of America (MAA), the Ohio Section of the MAA), and private and public foundations (Alfred P. Sloane Foundation, National Science Foundation). His consulting activities include Cooper Tire and Rubber Company, Computer Task Group, and Edward Kelcey and Associates (New Jersey), as well as over forty colleges and universities.

Dr. Karian is the author and co-author of nine texts, reference works, and book chapters and he has published over thirty articles. He serves as editor for computer simulation of the *Journal of Applied Mathematics and Stochastic Analysis*. Dr. Karian holds the bachelor's degree from American International College in Massachusetts, master's degrees from the University of Illinois (Urbana-Champaign) and The Ohio State University, and his doctoral degree from The Ohio State University. He has been a Plenary Speaker on two occasions at the Asian Conference on Technology in Mathematics (Singapore, and Penang, Malaysia).

Dr. Karian has served on the International Program Committees of conferences (in Greece, Japan, and People's Republic of China), the Board of Governors of the MAA, and the governing board of the Consortium of Mathematics and its Applications. He was a member of the Joint MAA/Association for Computing Machinery (ACM) Committee on Retraining in Computer Science and he chaired the Task Force (of the MAA, ACM and IEEE Computer Society) on Teaching Computer Science, the Subcommittee on Symbolic Computation of the MAA, and the Committee on Computing of the Ohio Section of the MAA.

Dr. Karian has been the Acting Director of the Computer Center, Denison University; Visiting Professor of Statistics, Ohio State University; Chair of the Department of Mathematical Sciences, Denison University. He has been the recipient of the R.C. Good Fellowship of Denison University on three occasions and has been cited in *Who's Who in America* (Marquis Who's Who, Inc.). In 1999 he was given the Award for Distinguished College or University Teaching of Mathematics by the Ohio Section of the MAA.

Dr. Edward J. Dudewicz is Professor of Mathematics at Syracuse University, New York. He has been active as instructor, researcher and consultant for over thirty years. He has taught statistics and digital simulation at Syracuse University, The Ohio State University, University of Rochester, University of Leuven (Belgium), and National University of Comahue (Argentina) and served as a staff member of the Instruction and Research Computer Center at The Ohio State University, and as Head Statistician of New Methods Research, Inc. His consulting activities include O.M. Scott and Sons Company, Ohio Bureau of Fiscal Review, Mead Paper Corporation, and Blasland, Bouck, & Lee, Engineers & Geoscientists. Dr. Dudewicz is author, co-author, and editor of over 150 works, including 21 texts and reference works in statistics, simulation, computation, and modeling, as well as handbook chapters on statistical methods (*Quality Control Handbook* and *Magnetic Resonance Imaging*). He serves as editor of two series: *Modern Digital Simulation: Advances in Theory, Application, & Design*, part of the American Series in Mathematical and Management Sciences, and *Modern Mathematical, Management, and Statistical Sciences*.

Dr. Dudewicz holds the bachelor's degree from the Massachusetts Institute of Technology and master's and doctoral degrees from Cornell University. He has been Visiting Scholar and Associate Professor at Stanford University, Visiting Professor at the University of Leuven (Belgium) and at the Science University of Tokyo (Japan), Visiting Distinguished Professor at Clemson

University, and Titular Professor at the National University of Comahue (Argentina) while Fulbright Scholar to Argentina. His editorial posts have included *Technometrics* (Management Committee), *Journal of Quality Technology* (Editorial Review Board), *Statistical Theory and Method Abstracts* (Editor, U.S.A.), *Statistics & Decisions* (Germany) (Editor), and *American Journal of Mathematical and Management Sciences* (Founding Editor and Editor-in-Chief), and he also serves the journal *Information Systems Frontiers* (Executive Editorial Board).

Dr. Dudewicz has served as President, Syracuse Chapter, American Statistical Association; Graduate Committee Chairman, Department of Statistics, The Ohio State University; Chairman, University Statistics Council, Syracuse University; External Director, Advanced Simulation Project, National University of Comahue, Argentina; Awards Chairman, Chemical Division, American Society for Quality; and Founding Editor, *Basic References in Quality Control: Statistical Techniques*, American Society for Quality (the "How To" series).

Recognitions of Dr. Dudewicz include Research Award, Ohio State Chapter, Society of the Sigma Xi; Chancellor's Citation of Recognition, Syracuse University; Jacob Wolfowitz Prize for Theoretical Advances; Thomas L. Saaty Prize for Applied Advances; Co-author, Shewell Award paper; Jack Youden Prize for the best expository paper in *Technometrics*. He has received the Seal of Banares Hindu University (India), where he was Chief Guest and Special Honoree, presenting an Inaugural Address and Keynote Address. He is an elected Fellow of the New York Academy of Sciences, the American Society for Quality, the Institute of Mathematical Statistics, the American Statistical Association, the International Statistical Institute, and the American Association for the Advancement of Science. He is a subject of biographical record in *Who's Who in America* (Marquis Who's Who, Inc.). In 1999 he was awarded the International Francqui Chair in Exact Sciences.

To

Stephen and Maya

and

In loving memory of a loving father, grandfather, and greatgrandfather,
Lt. Edward G. Dudewicz (1912–1999).
May God rest his soul.

Contents

Chapter 1

The Generalized Lambda Family of Distributions

Much of **modern human endeavor**, in wide-ranging fields that include science, technology, medicine, engineering, management, and virtually all of the areas that comprise human knowledge **involves the construction of statistical models** to describe the fundamental variables in those areas.[1] The most basic and widely used model, called the **probability distribution**, relates the values of the fundamental variables to their probability of occurrence. When the variable of interest can take on (subject to the precision of the measuring process) any value in an interval, the probability distribution is called **continuous**. This book deals with how to model (or, **how to fit**) **a continuous probability distribution** to data.

The area of fitting distributions to data has seen explosive growth in recent years.[2] Consequently, few individuals are well versed in the new results that have become available. In many cases these recent developments have solved old problems with the fitting process; they have also provided the practitioner with a confusing array of methods. Moreover, some of these methods, called "asymptotic" methods in the theoretical literature, were developed with very large sample sizes in mind and are ill-suited for many applications. With these facts in mind, this book seeks to

- Give the results, tables, algorithms, and computer programs needed to fit continuous probability distributions to data, using the Generalized Lambda Distribution (GLD) and the Generalized Bootstap (GB) approaches;

---

[1] For example, in actuarial science, see Klugman, Panjer, and Willmot (1998), which while an "applied text" (p. ix), needed to assume "that the reader has a solid background in mathematical statistics" (p. ix).

[2] Recent works include Bowman and Azzalini (1997), Scott (1992), and Simonoff (1996). An overview with 71 references to recent work is given by Müller (1997).

- Bring together in one place the key results on GLD and GB fitting of continuous probability distributions, with emphasis on recent results that make these methods nearly universally applicable.

There are good reasons for using the GLD distribution and GB methods. GLD fits have been used successfully in many fields (e.g., the construction industry, atmospheric data, quality control, medical data, reliability). Moreover, the "bootstrap" is widely used in its non-generalized form, making the GB a method that should appeal to those familiar with the Bootstrap — though perhaps not with its drawbacks that the GB generalization addresses. The inclusion of other methods, as was done in our earlier work (see Part IV of Dudewicz and Karian (1985)), would raise the question: "Which method should be used in practice?" Our answer to this question has always been: Try the GLD and GB first and stop there if the results are acceptable. Combined with the fact that our own recent work has contributed to the refinement and completion of the GLD and GB methods, whose proper exposition alone fills a book, the decision to focus on the GLD and GB methods is a clear one for us.

The plan of the book is as follows:

Chapter 1. Specify the basic GLD family of distributions. The GLD has four parameters that one needs to choose to fit the data at hand.

Chapter 2. Show how to estimate the unknown parameters with the method of moments. This is how the parameters have usually been estimated in applications.

Chapter 3. Develop the extended GLD (EGLD) to cover all possible moments. The limited coverage of moment-space has been a problem in some GLD applications; this problem is now solved by the EGLD.

Chapter 4. Show how to estimate the unknown parameters using percentiles. Since moments do not always exist, and moment-based estimators can have high variability, the recent development of methods using percentiles allows for better fitting distributions in some applications.

Chapter 5. Develop a bivariate version of GLD (and of the EGLD), which is important when two characteristics must be fitted jointly. This method was made available only recently.

Chapter 6. Develop the Generalized Bootstrap method of fitting and contrast it with the Bootstrap method. The latter has important drawbacks that the GB generalization addresses.

Appendices. Give the tables and computer code needed for applications so that although a person may write his or her own code if desired, the code we used will be available for easy implementation. The tables will be sufficient in many applications.

Along the way, we include a wide-ranging variety of applications from a number of areas as illustrations in detail (to aid the practitioner), and proofs of key results (to aid the person interested in the theoretical development, who may wish to teach a course in this area, or to do research in it).

1.1 History and Background

The search for a method of fitting a continuous probability distribution to data is quite old. Pearson (1895) gave a four-parameter system of probability density functions, and fitted the parameters by what he called the "method of moments" (Pearson (1894)). It has been stated (Hald (1998), pp. 649-650) that

> Like Chebyshev, Fechner, Thiele, and Gram he [Pearson] felt the need for a collection of continuous distributions to choose among for describing the ... phenomena he was studying. He wanted a system embracing distributions with finite as well as infinite support and with skewness both to the right and the left.

The system we develop in this book has its origins in the one-parameter lambda distribution proposed by John Tukey[3] (1960). Tukey's lambda distribution was generalized, for the purpose of generating random variates for Monte Carlo simulation studies, to the four-parameter generalized lambda distribution, or GLD, by John Ramberg and Bruce Schmeiser (Ramberg and Schmeiser (1972, 1974)). Subsequently, a system, with the necessary tables, was developed for fitting a wide variety of curve shapes by Ramberg, Tadikamalla, Dudewicz, and Mykytka (1979). Since the early 1970s the GLD has been applied to fitting phenomena in many fields of endeavor with continuous probability density functions. In some cases problems arose, such as a need for tables for a range not yet published or computed, or a need for an extension to the part of moment-space not covered by the GLD; solutions to these problems are given later in this book.

We will cover the **applications** through a number of detailed examples. A few words about some important applications are appropriate in this in-

---

[3]It has been pointed out that the Tukey Lambda was introduced in Hastings, Mosteller, Tukey, and Winsor (1947); however, the usual origin is stated as Tukey (1960).

troduction. In an early application of the GLD (at the time called the RS (or Ramberg-Schmeiser) distribution), Ricer (1980) dealt with **construction industry data**. His concern was to correct for the deviations from normality which occur in construction data, especially in expectancy pricing and in a competitive bidding environment, finding such quantities as the optimum markup.

In another important application area, **meteorology**, it is recognized that many variables have "...non-normality, [so] most climatologists have used empirical distributions..." (Öztürk and Dale (1982), p. 995). As an alternative, fitting of solar radiation data with the GLD was successful due to the "flexibility and generality" (Öztürk and Dale (1982), p. 1003) of the GLD, which "could successfully be used to fit the wide variety of curve shapes observed." In many applications, this means that we can use the GLD to **describe data with a single functional form** by specifying its four parameter values for each case, instead of giving the basic data (which is what the empirical distribution essentially does) for each case. The one functional form allows us to group cases that are similar, as opposed to being overburdened with a mass of numbers or graphs.

Each application area has its own concerns and goals. What they have in common is the need for a flexible, easy-to-work-with, complete system of continuous probability distributions—a need for which the GLD has been found to be quite suitable. We urge those who wish to look up some of the work in areas for which we have not given references to write the first paper applying the GLD to those areas and send us a copy. Some of the more accessible references in several important areas are

- Modeling biological and physical phenomena, Silver (1977)

- Generating random variables for Monte Carlo studies, Hogben (1963), Shapiro and Wilk (1965), Shapiro, Wilk, and Chen (1968), Karian and Dudewicz (1999b)

- Sensitivity studies of statistical methods, Filliben (1969)

- Approximation of percentiles of distributions, Tukey (1960), Van Dyke (1961)

- Testing goodness of fit in logistic regression, Pregibon (1980)

- Modeling complex situations in applied research, such as quantal response in bioassay and economics, Mudholkar and Phatak (1984), Pregibon (1980)

- Quality management, engineering, control, and planning, Dudewicz (1999).

As a prelude, and for some a refresher, before defining the Generalized Lambda Distribution (GLD) family in Section 1.2, we **review some basic notions from statistics**. For fuller details, see Karian and Dudewicz (1999b) or Dudewicz and Mishra (1998).

If the variable X is involved in the model we are constructing, and we have no way of accurately predicting its value from other known variables, then X is usually called a **stochastic variable** or **random variable (r.v.)**.[4] If X can take on only a few discrete values (such as 0 or 1 for failure or success, or $0, 1, 2, 3, \ldots$ as the number of occurrences of some event of interest), then X is called a **discrete random variable**. If the outcome of interest X can take on values in a continuous range (such as all values greater than zero for an engine failure time), then X is called a **continuous random variable**.[5]

For any r.v. of interest, we wish to know: What are the chances of occurrence of the various possible values of the r.v.? For example, what is the probability that X is no larger than 3.2? Or, that X is no larger than 4.5 (or any other value)? Specifying $P(X \leq 3.2), P(X \leq 4.5)$, or in general $P(X \leq x)$, is **one way of specifying the chances** of occurrence of the various values that are possible. This is called giving the **distribution function (d.f.)** of X:

$$F_X(x) = P(X \leq x), \quad -\infty < x < +\infty. \tag{1.1.1}$$

As an example, if X has what is called the **standard normal or $N(0,1)$ distribution function**, then $F_X(x)$ is usually denoted by $\Phi(x)$. There is no simple formula for finding $\Phi(3.2)$, but $\Phi(3.2)$ can be approximated through numerical integration. Values for $\Phi(x)$ are given in Appendix E for various values of x. From the table in Appendix E, we find $\Phi(3.2) = .9993129$, i.e., the chance that X will be 3.2 or less is then 99.93129%. It is possible to plot the function $\Phi(x)$ by graphing a fine grid of points $(x, \Phi(x))$ using x-values included in Appendix E, and joining the points by a smooth curve. This is done in Figure 1.1–1. The plot, while specific to the standard normal d.f.,

---

[4]Here we restrict ourselves to r.v.s that assume a single number, such as $X = 3.2$, called **univariate** r.v.s. The case when X may be a pair, such as $X = (T, H)$ with T=Temperature and H=Humidity, is also important (X is then called a **bivariate** r.v.) and is dealt with in Chapter 5.

[5]There are also **mixed** r.v.s that arise when X takes on some values with positive probability but is otherwise continuous. (E.g., at random choose to flip a fair coin, marked 3 on one side and 5 on the other side, or spin a spinner that points to a number on a circumference marked continuously from 0.0 to 1.0. Then with probability 0.25, $X = 3$; with probability 0.25, $X = 5$; and otherwise X is a number between 0 and 1.) We deal here only with the continuous part of X and its modeling.

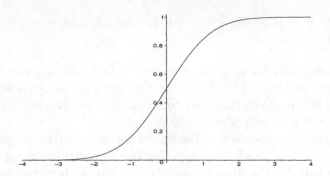

Figure 1.1–1. The d.f. of a $N(0,1)$ r.v.

has some properties in common with all plots of d.f.s of continuous r.v.s: the curve starts near a height of zero at the left (very small values of x), and increases continuously, approaching a height of 1.0 as x gets large.

A **second way of specifying the chances of occurrence of the various values of** X is to give what is called the **probability density function (p.d.f.)** of X. This is a function $f_X(x)$ that is ≥ 0 for all x, integrates to 1 over the range $-\infty < x < +\infty$, and such that for all x,

$$F_X(x) = \int_{-\infty}^{x} f_X(t)dt. \tag{1.1.2}$$

Every d.f. $F_X(\cdot)$ has associated with it a unique[6] p.d.f. $f_X(\cdot)$, so we gain the same information by specifying either one of these functions. As an example, if X has the $N(0,1)$ d.f., then its p.d.f. is usually given the name $\phi(x)$ and has the simple expression

$$\phi(x) = \frac{1}{\sqrt{2\pi}}e^{-x^2/2}, \qquad -\infty < x < +\infty. \tag{1.1.3}$$

It is possible to plot the function $\phi(x)$ by graphing a fine grid of points $(x, \phi(x))$ using any x-value grid we wish and computing $\phi(x)$ from (1.1.3). This is done in Figure 1.1–2. The plot, while specific to the standard normal p.d.f., has some properties in common with all plots of p.d.f.s: the curve is

---

[6]Actually, the p.d.f. is **unique up to sets of measure zero**, i.e. there is what is called an **equivalence class** of such functions. However, this is a notion not needed in our work (and those familiar with it are unlikely to need this review), so we do not dwell on it. One important point that comes from it, however, is that the p.d.f.s $f_1(x) = 1$ if $0 \leq x \leq 1$ (and $= 0$ otherwise), and $f_2(x) = 1$ if $0 < x < 1$ (and $= 0$ otherwise) are not different — they yield the same d.f., and both are called the **uniform p.d.f.** (on $(0,1)$).

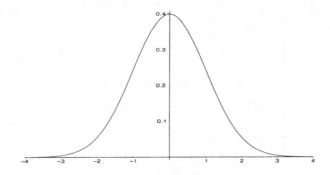

Figure 1.1–2. The p.d.f. of a $N(0,1)$ r.v.

non-negative, the area under the whole curve is 1, and areas under the curve between values give us the probability of that range of values.

A **third way of specifying the chances of occurrence of the various values of** X is to give what is called the **inverse distribution function**, or **percentile function (p.f.)**, of X. This is the function $Q_X(y)$ which, for each y between 0 and 1, tells us the value of x such that $F_X(x) = y$:

$$Q_X(y) = (\text{The value of } x \text{ such that } F_X(x) = y), \quad 0 \le y \le 1. \quad (1.1.4)$$

For example, for the $N(0,1)$ d.f., we know from Appendix E that $\Phi(1.96) = 0.975$. Hence, $Q(0.975) = 1.96$ for the $N(0,1)$ distribution, i.e., the value that is (with probability .975) not exceeded is 1.96. Equivalently, with the $N(0,1)$ distribution one will find values of 1.96 or smaller 97.5% of the time. It is possible to plot the function $Q(y)$ by graphing a fine grid of points $(y, Q(y))$ using any y-value grid over the range 0 to 1. This is done in Figure 1.1–3 for the standard normal case, using numerical methods to converge on the root needed in (1.1.4). The plot, while specific to the standard normal case, has some properties in common with all plots of p.f.s: all "action" takes place for horizontal axis values between 0 and 1 (these represent probabilities of non-exceedance, and probabilities must be between 0 and 1), and the curve is increasing from the smallest value of X (in the limit as y tends to 0), to the largest value of X (in the limit as y tends to 1).

We see that there are three ways to specify the chances of occurrence of a r.v. (the d.f., the p.d.f., and the p.f.). For the $N(0,1)$ example it was not easy to specify the d.f. (we had to numerically calculate it; we may think of this as easy since virtually all statistics books have this table in them, nevertheless there is no simple formula); it was easy to specify the p.d.f.; and it was not easy to specify the p.f. (numerical calculation was necessary;

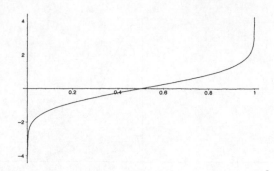

Figure 1.1–3. The p.f. of a $N(0,1)$ r.v.

we may think of this as harder, as $Q(y)$ is given in most statistics books only for a few "holy" values of y, such as 0.90, 0.95, etc., but in fact it is no harder than finding the $N(0,1)$ d.f.). For other examples, it varies which of the d.f., p.d.f., and p.f. is easier to deal with. In particular, for the GLD we will see in Section 1.2 that the p.f. is very easy to obtain, the p.d.f. is also easy to find, and the d.f. needs numerical calculation.

For later reference, we give the p.d.f., d.f., and p.f. for a r.v. with the general normal distribution with mean μ and variance σ^2, $N(\mu, \sigma^2)$. The p.d.f. is

$$f(x) = \frac{1}{\sqrt{2\pi}\,\sigma}\, e^{-(x-\mu)^2/(2\sigma^2)}, \tag{1.1.5}$$

the distribution function (d.f.) is

$$F(x) = P(X \le x) = P\left(\frac{X-\mu}{\sigma} \le \frac{x-\mu}{\sigma}\right) = \Phi\left(\frac{x-\mu}{\sigma}\right), \tag{1.1.6}$$

and the percentile function (p.f.) is

$$Q(y) = (x \text{ such that } F(x) = y)$$

$$= \mu + \sigma\Phi^{-1}(y). \tag{1.1.7}$$

We should note that, in addition to $Q_X(y)$, there are several notations in common use for the p.f.; one usually finds the notation $F_X^{-1}(x)$. This notation, while more common in the literature, is often confused with $1/F_X(y)$, for which the same notation is used. For this reason, we will designate the p.f. by $Q_X(y)$ throughout this book.

1.2 Definition of the Generalized Lambda Distributions

The **generalized lambda distribution family with parameters λ_1, λ_2, λ_3, λ_4, GLD$(\lambda_1, \lambda_2, \lambda_3, \lambda_4)$**, is most easily specified in terms of its percentile function

$$Q(y) = Q(y; \lambda_1, \lambda_2, \lambda_3, \lambda_4) = \lambda_1 + \frac{y^{\lambda_3} - (1-y)^{\lambda_4}}{\lambda_2}, \qquad (1.2.1)$$

where $0 \le y \le 1$. The parameters λ_1 and λ_2 are, respectively, location and scale parameters, while λ_3 and λ_4 determine the skewness and kurtosis of the GLD$(\lambda_1, \lambda_2, \lambda_3, \lambda_4)$. Recall that for the normal distribution there are also restrictions on (μ, σ^2), namely, $\sigma > 0$. The restrictions on $\lambda_1, \ldots, \lambda_4$ that yield a valid GLD$(\lambda_1, \lambda_2, \lambda_3, \lambda_4)$ distribution will be discussed in Section 1.3 and the impact of λ_3 and λ_4 on the shape of the GLD$(\lambda_1, \lambda_2, \lambda_3, \lambda_4)$ p.d.f. will be considered in Section 1.4.

It is relatively easy **to find the probability density function from the percentile function of the GLD**, as we now show.

Theorem 1.2.2. *For the* GLD$(\lambda_1, \lambda_2, \lambda_3, \lambda_4)$, *the probability density function is*

$$f(x) = \frac{\lambda_2}{\lambda_3 y^{\lambda_3 - 1} + \lambda_4 (1-y)^{\lambda_4 - 1}}, \qquad at \ x = Q(y). \qquad (1.2.3)$$

(Note that $Q(y)$ can be calculated from (1.2.1).)

Proof. Since $x = Q(y)$, we have $y = F(x)$. Differentiating with respect to x, we find

$$\frac{dy}{dx} = f(x)$$

or

$$f(x) = \frac{dy}{d(Q(y))} = \frac{1}{\dfrac{d(Q(y))}{dy}}. \qquad (1.2.4)$$

Since we know the form of $Q(y)$ from (1.2.1), we find directly that

$$\frac{dQ(y)}{dy} = \frac{d}{dy}\left(\lambda_1 + \frac{y^{\lambda_3} - (1-y)^{\lambda_4}}{\lambda_2}\right) = \frac{\lambda_3 y^{\lambda_3 - 1} + \lambda_4 (1-y)^{\lambda_4 - 1}}{\lambda_2}, \qquad (1.2.5)$$

from which the theorem follows using (1.2.5) in (1.2.4). ∎

In plotting the function $f(x)$ for a density such as the normal, where $f(x)$ is given as a specific function of x, we proceed by calculating $f(x)$ at a grid of x values, then plotting the pairs $(x, f(x))$ and connecting them with a smooth curve. **For the GLD family, plotting** $f(x)$ proceeds differently since (1.2.3) tells us the value of $f(x)$ at $x = Q(y)$. Thus, we take a grid of y values (such as .01, .02, .03, ..., .99, that give us the 1%, 2%, 3%, ..., 99% points), find x at each of those points from (1.2.1), and find $f(x)$ at that x from (1.2.3). Then, we plot the pairs $(x, f(x))$ and link them with a smooth curve.

As an example of plotting $f(x)$ for a GLD, consider the GLD($\lambda_1, \lambda_2, \lambda_3, \lambda_4$) with parameters

$$\lambda_1 = 0.0305, \quad \lambda_2 = 1.3673, \quad \lambda_3 = 0.004581, \quad \lambda_4 = 0.01020,$$

i.e., the GLD($\lambda_1, \lambda_2, \lambda_3, \lambda_4$) with (see (1.2.1))

$$Q(y) = 0.0305 + \left(y^{0.004581} - (1 - y)^{0.01020}\right)/1.3673. \tag{1.2.6}$$

This GLD arose (see Ramberg, Tadikamalla, Dudewicz, and Mykytka (1979)) as the fit to measurements of the coefficient of friction for a metal. For example, in the process noted in the above paragraph, we find that at $y = 0.25$ the Q(0.25)=0.028013029, from (1.2.6). Next, at $x = 0.028$, using (1.2.3) with the specified values of $\lambda_1, \lambda_2, \lambda_3, \lambda_4$, $f(0.028) = 43.0399612$. Hence, $(0.028, 43.04)$ will be one of the points on the graph of $f(x)$. Proceeding in this way for $y = 0.01, 0.02, ..., 0.99$, we obtain the graph of $f(x)$ given in Figure 1.2–1.

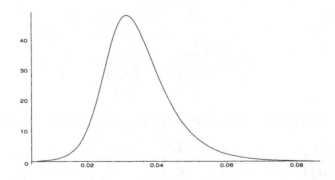

Figure 1.2–1. The p.d.f. of GLD($0.0305, 1.3673, 0.004581, 0.01020$).

1.3 The Parameter Space of the GLD

We noted, following formula (1.2.1), that it does not always specify a valid distribution. The reason is that one cannot just write down any formula and be assured it will specify a distribution without checking the conditions needed for that fact to hold. In particular, **a function $f(x)$ is a probability density function if and only if it satisfies the conditions**

$$f(x) \geq 0 \quad \text{and} \quad \int_{-\infty}^{\infty} f(x)\, dx = 1. \tag{1.3.1}$$

From (1.2.4) we see that for the GLD$(\lambda_1, \lambda_2, \lambda_3, \lambda_4)$, conditions (1.3.1) are satisfied if and only if

$$\frac{\lambda_2}{\lambda_3 y^{\lambda_3-1} + \lambda_4(1-y)^{\lambda_4-1}} \geq 0 \quad \text{and} \quad \int_{-\infty}^{\infty} f(Q(y))\, dQ(y) = 1. \tag{1.3.2}$$

Since from (1.2.4) we know that

$$f(Q(y))\, dQ(y) = dy,$$

and y is on the range $[0, 1]$, the second condition in (1.3.2) follows. Thus, for any $\lambda_1, \lambda_2, \lambda_3, \lambda_4$ the function $f(x)$ will integrate to 1. It remains to show that the first condition in (1.3.2) holds.

Since λ_1 does not enter into the first condition in (1.3.2), this parameter will be unrestricted, leading us to the following theorem.

Theorem 1.3.3. *The GLD$(\lambda_1, \lambda_2, \lambda_3, \lambda_4)$ specifies a valid distribution if and only if*

$$\frac{\lambda_2}{\lambda_3 y^{\lambda_3-1} + \lambda_4(1-y)^{\lambda_4-1}} \geq 0 \tag{1.3.4}$$

for all $y \in [0, 1]$.

The next theorem establishes the role of λ_1 **as a location parameter.**

Theorem 1.3.5. *If the random variable X is GLD$(0, \lambda_2, \lambda_3, \lambda_4)$, then the random variable $X + \lambda_1$ is GLD$(\lambda_1, \lambda_2, \lambda_3, \lambda_4)$.*

Proof. If X is GLD$(0, \lambda_2, \lambda_3, \lambda_4)$, by (1.1.1) we have

$$Q(y) = \frac{y^{\lambda_3} - (1-y)^{\lambda_4}}{\lambda_2}.$$

Now

$$F_{X+\lambda_1}(x) = P[X + \lambda_1 \le x] = P[X \le x - \lambda_1] = F_X(x - \lambda_1), \quad (1.3.6)$$

hence $F_X(x - \lambda_1) = y$ also implies $F_{X+\lambda_1}(x) = y$, yielding

$$x - \lambda_1 = Q_X(y) = \frac{y^{\lambda_3} - (1-y)^{\lambda_4}}{\lambda_2}, \quad x = Q_{X+\lambda_1}(y), \quad (1.3.7)$$

whence

$$Q_{X+\lambda_1}(y) = x = \lambda_1 + Q_X(y) = \lambda_1 + \frac{y^{\lambda_3} - (1-y)^{\lambda_4}}{\lambda_2}. \quad (1.3.8)$$

This proves that $X + \lambda_1$ is a GLD$(\lambda_1, \lambda_2, \lambda_3, \lambda_4)$ random variable. ∎

Since $0 \le y \le 1$ in (1.3.4), we immediately have the following.

Corollary 1.3.9. *The GLD$(\lambda_1, \lambda_2, \lambda_3, \lambda_4)$ of (1.2.1) specifies a valid distribution if and only if*

$$g(y, \lambda_3, \lambda_4) \equiv \lambda_3 y^{\lambda_3 - 1} + \lambda_4 (1-y)^{\lambda_4 - 1} \quad (1.3.10)$$

has the same sign (positive or negative) for all y in $[0, 1]$, as long as λ_2 takes that sign also. In particular, the GLD$(\lambda_1, \lambda_2, \lambda_3, \lambda_4)$ specifies a valid distribution if $\lambda_2, \lambda_3, \lambda_4$ all have the same sign.

To determine the (λ_3, λ_4) pairs that lead to a valid GLD, we consider (λ_3, λ_4)-space in the following regions:

$$\text{Region 1} = \{(\lambda_3, \lambda_4) \mid \lambda_3 \le -1, \lambda_4 \ge 1\} \quad (1.3.11)$$

$$\text{Region 2} = \{(\lambda_3, \lambda_4) \mid \lambda_3 \ge 1, \lambda_4 \le -1\} \quad (1.3.12)$$

$$\text{Region 3} = \{(\lambda_3, \lambda_4) \mid \lambda_3 \ge 0, \lambda_4 \ge 0\} \quad (1.3.13)$$

$$\text{Region 4} = \{(\lambda_3, \lambda_4) \mid \lambda_3 \le 0, \lambda_4 \le 0\} \quad (1.3.14)$$

$$V_1 = \{(\lambda_3, \lambda_4) \mid \lambda_3 < 0, 0 < \lambda_4 < 1\} \quad (1.3.15)$$

$$V_2 = \{(\lambda_3, \lambda_4) \mid 0 < \lambda_3 < 1, \lambda_4 < 0\} \quad (1.3.16)$$

$$V_3 = \{(\lambda_3, \lambda_4) \mid -1 < \lambda_3 < 0, \lambda_4 > 1\} \quad (1.3.17)$$

$$V_4 = \{(\lambda_3, \lambda_4) \mid \lambda_3 > 1, -1 < \lambda_4 < 0\} \quad (1.3.18)$$

The following lemma is a direct consequence of Corollary 1.3.9.

Lemma 1.3.19. *The GLD$(\lambda_1, \lambda_2, \lambda_3, \lambda_4)$ is valid in Regions 3 and 4 specified in (1.3.13) and (1.3.14).*

Next, we consider the other Regions, starting with Regions V_1 and V_2.

Lemma 1.3.20. *The* $GLD(\lambda_1, \lambda_2, \lambda_3, \lambda_4)$ *is not valid in Regions* V_1 *and* V_2.

Proof. By Corollary 1.3.9, the $GLD(\lambda_1, \lambda_2, \lambda_3, \lambda_4)$ is valid at $(\lambda_1, \lambda_2, \lambda_3, \lambda_4)$ if and only if $g(y, \lambda_3, \lambda_4)$, as defined in (1.3.10), has the same sign for all y in $[0, 1]$, and λ_2 takes that same sign. In Region V_1 we have $\lambda_3 < 0$ and $0 < \lambda_4 < 1$. It is easy to see that

$$\lim_{y \to 0+} g(y, \lambda_3, \lambda_4) = -\infty \quad \text{and} \quad \lim_{y \to 1^-} g(y, \lambda_3, \lambda_4) = +\infty,$$

so that $g(y, \lambda_3, \lambda_4)$ cannot keep the same sign over the interval $[0, 1]$, hence the $GLD(\lambda_1, \lambda_2, \lambda_3, \lambda_4)$ is not valid for (λ_3, λ_4) in V_1. The analysis for V_2 is similar (with λ_3 and λ_4 interchanged). ∎

Lemma 1.3.21. *The* $GLD(\lambda_1, \lambda_2, \lambda_3, \lambda_4)$ *is valid in Regions 1 and 2 specified in (1.3.11) and (1.3.12).*

Proof. We will show that for (λ_3, λ_4) in Region 1, $g(y, \lambda_3, \lambda_4)$, (defined in (1.3.10)), is negative for all y in $[0, 1]$. We start by considering

$$\frac{\partial \, g(y, \lambda_3, \lambda_4)}{\partial \, \lambda_3} = y^{\lambda_3 - 1} + \lambda_3 \ln(y) y^{\lambda_3 - 1}.$$

Since this is positive ($\lambda_3 < 0$ and $\ln(y) < 0$), $g(y, \lambda_3, \lambda_4)$ increases as λ_3 increases and

$$g(y, \lambda_3, \lambda_4) \leq g(y, -1, \lambda_4)$$

$$= \frac{-1}{y^2} + \lambda_4 (1 - y)^{\lambda_4 - 1} = h(y, \lambda_4). \tag{1.3.22}$$

Now,

$$\frac{\partial \, h(y, \lambda_4)}{\partial \, \lambda_4} = (1 - y)^{\lambda_4 - 1} + \lambda_4 (1 - y)^{\lambda_4 - 1} \ln(1 - y)$$

$$= (1 - y)^{\lambda_4 - 1} [1 + \lambda_4 \ln(1 - y)]$$

and $\dfrac{\partial \, h(y, \lambda_4)}{\partial \, \lambda_4} \geq 0$ if and only if

$$\lambda_4 \leq \frac{-1}{\ln(1 - y)}. \tag{1.3.23}$$

Case 1: $-1/\ln(1-y) \leq 1$, equivalently, $1 - e^{-1} \leq y \leq 1$.
By (1.3.23) $h(y, \lambda_4)$ increases with λ_4 and

$$g(y, \lambda_3, \lambda_4) \leq h(y, \lambda_4) \leq h(y, 1) = \frac{-1}{y^2} + 1 \leq 0.$$

Case 2: $-1/\ln(1-y) > 1$, equivalently, $0 \leq y < 1 - e^{-1}$.
In this case, $h(y, \lambda_4)$ will be largest when $\lambda_4 = -1/\ln(1-y)$. Therefore,

$$h(y, \lambda_4) \leq h(y, \frac{-1}{\ln(1-y)}) \tag{1.3.24}$$

$$= -\frac{1}{y^2} - \left(\frac{1}{\ln(1-y)}\right)(1-y)^{\left(-1 - \frac{1}{\ln(1-y)}\right)} = f(y).$$

The derivative of $f(y)$, after some simplification, is

$$f'(y) = \frac{2(1-y)^2 \ln^2(1-y) - y^3 e^{-1}(1 + \ln(1-y))}{y^3 \ln^2(1-y)(1-y)^2}.$$

Since the denominator is non-negative, $f'(y)$ has the same sign as

$$\frac{2e(y-1)^2}{y}\left(\frac{\ln(1-y)}{y}\right)^2 - (1 + \ln(1-y)).$$

To show that $f'(y) \geq 0$, it suffices to establish that

$$k(y) = \frac{2e(y-1)^2}{y}\left(\frac{\ln(1-y)}{y}\right)^2 \geq 1$$

because we would then have $\ln(1-y) < 0$, $-(1+\ln(1-y)) > -1$, and $f'(y) = k(y) - (1 + \ln(1-y)) > 0$. The first factor of $k(y)$, $2(y-1)^2$, decreases as y increases on $[0, 1 - e^{-1}]$, the interval to which y is constrained. Therefore,

$$\frac{2e(y-1)^2}{y} \geq \frac{2e^{-1}}{y} \geq \frac{2e^{-1}}{1 - e^{-1}} = \frac{2}{e-1} > 1. \tag{1.3.25}$$

The other factor of $k(y)$, $\ln^2(1-y)/y^2$, increases as y increases on $[0, 1-e^{-1}]$. Hence, its value must exceed

$$\lim_{y \to 0^+} \left(\frac{\ln(1-y)}{y}\right)^2 = 1.$$

This, together with (1.3.25), makes $k(y) > 1$, hence, $f'(y) > 0$. We now have

$$f(y) \leq f(1 - e^{-1}) = \frac{1 - 2e}{(1 - e)^2} < 0,$$

yielding,

$$g(y, \lambda_3, \lambda_4) \leq h(y, \lambda_4) \leq f(y) < 0.$$

A similar argument, with λ_3 and λ_4 interchanged, gives $g(y, \lambda_3, \lambda_4) < 0$ for (λ_3, λ_4) from Region 2 and Corollary 1.3.9 is used to conclude the proof. ∎

The situation is quite different in Regions V_3 and V_4. We start by observing that the GLD($\lambda_1, \lambda_2, \lambda_3, \lambda_4$) is valid at **only some points of V_3**. For example, at $\lambda_3 = -1/2$, $\lambda_4 = 2$, the $g(y, \lambda_3, \lambda_4)$ of (1.3.10) is

$$g(y) = -\frac{1}{2}y^{-3/2} + 2(1 - y),$$

for which $g'(y) = \frac{3}{4}y^{-5/2} - 2$, which is ≤ 0 if and only if $(3/4)y^{-5/2} \leq 2$, i.e., if and only if $3/8 \leq y^{5/2}$, or $y \geq (3/8)^{2/5} = .67548$. Noting that

$$\lim_{y \to 0^+} g(y) = -\infty \text{ and } g(1) = -\frac{1}{2},$$

we see that $g(y)$ increases as y increases from 0 to $(3/8)^{2/5}$, then decreases. Its maximum occurs at $y = (3/8)^{2/5}$, in which case

$$g\left(\left(\frac{3}{8}\right)^{2/5}\right) = -\frac{1}{2}((3/8)^{2/5})^{-3/2} + 2(1 - (3/8)^{2/5})$$
$$= -0.90064 + 0.64904 = -0.25160.$$

Thus, $g(y)$ is negative for all y in $[0, 1]$, and the GLD($\lambda_1, \lambda_2, \lambda_3, \lambda_4$) (with $\lambda_2 < 0$) is valid at $(\lambda_3, \lambda_4) = (-1/2, 2)$.

This can be **contrasted with the point** $(\lambda_3, \lambda_4) = (-1/2, 1)$ where

$$g(y) = 1 - \frac{1}{2y^{3/2}}.$$

In this case

$$\lim_{y \to 0^+} g(y) = -\infty \text{ and } g(1) = \frac{1}{2},$$

which establishes $(\lambda_3, \lambda_4) = (-1/2, 1)$ as **a point of V_3 where the GLD is not valid**.

The following result, due to Karian, Dudewicz and McDonald (1996), gives the **complete characterization of the valid points of Regions V_3 and V_4**.

Lemma 1.3.26. *A point in Region V_3 is valid if and only if*

$$\frac{(1 - \lambda_3)^{1-\lambda_3}}{(\lambda_4 - \lambda_3)^{\lambda_4 - \lambda_3}}(\lambda_4 - 1)^{\lambda_4 - 1} < \frac{-\lambda_3}{\lambda_4}. \tag{1.3.27}$$

Proof. Let $0 < y < 1$ and $-1 < \lambda < 0$ (think of λ as λ_3), and $f(\lambda) = \alpha > 1$ (think of α as λ_4, so we are considering Region V in the second quadrant). Let

$$G(y) = \lambda y^{\lambda - 1} + \alpha(1 - y)^{\alpha - 1}. \tag{1.3.28}$$

Since the GLD$(\lambda_1, \lambda_2, \lambda_3, \lambda_4)$ is valid if $G(y)$ has constant sign for all y in $[0, 1]$, we examine the zeros of G. $G(y) = 0$ is equivalent to

$$\frac{-\lambda}{\alpha} = \frac{(1 - y)^{\alpha - 1}}{y^{\lambda - 1}}. \tag{1.3.29}$$

Through the substitutions

$$\beta = \left(\frac{-\lambda}{\alpha}\right)^{1/(\alpha - 1)} \quad \text{and} \quad \gamma = \frac{1 - \lambda}{\alpha - 1},$$

(1.3.29) can be simplified to

$$\beta = y^{\gamma}(1 - y), \tag{1.3.30}$$

where $\beta, \ \gamma > 0$.

Differentiating with respect to y we obtain a relation for the critical points of $h(\lambda, y) = y^{\gamma}(1 - y)$:

$$\frac{\partial\, h(\lambda, y)}{\partial\, y} = \gamma y^{\gamma - 1}(1 - y) - y^{\gamma}$$

which is zero if and only if $\gamma y^{\gamma - 1} = y^{\gamma}(1 + \gamma)$, i.e., if and only if y has the value $y_c = \gamma/(1 + \gamma)$. At y_c, $h(\lambda, y)$ has a maximum since $h(\lambda, 0) = h(\lambda, 1) = 0$ and $h(\lambda, y) \geq 0$. This maximum is given by

$$h(\lambda, y_c) = y_c^{\gamma}(1 - y_c) = \frac{y_c^{\gamma}}{(1 + \gamma)} = \frac{\gamma^{\gamma}}{(1 + \gamma)^{1 + \gamma}}.$$

Since the difference of the two sides in (1.3.30) will go from positive at $y = 0$ to negative at $y = y_c$, G changes sign on $[0, 1]$ if and only if

$$\frac{\gamma^\gamma}{(1 + \gamma)^{1+\gamma}} \geq \beta. \tag{1.3.31}$$

By restating this in terms of the λ_3 and λ_4 parameters of the GLD, we see that (λ_3, λ_4) fails to yield a valid GLD if and only if

$$\frac{(1 - \lambda_3)^{1-\lambda_3}}{(\lambda_4 - \lambda_3)^{\lambda_4-\lambda_3}}(\lambda_4 - 1)^{\lambda_4-1} \geq \frac{-\lambda_3}{\lambda_4}. \quad \blacksquare \tag{1.3.32}$$

The following theorem, by summarizing the results of Lemmas 1.3.19, 1.3.20, 1.3.21 and 1.3.26, **completely characterizes the (λ_3, λ_4) pairs for which the GLD is valid.**

Theorem 1.3.33. *With a suitable λ_2, the $\mathrm{GLD}(\lambda_1, \lambda_2, \lambda_3, \lambda_4)$ is valid at points (λ_3, λ_4) if and only if (λ_3, λ_4) is in one of the unshaded regions depicted in Figure 1.3–1. The curved boundaries between the valid and nonvalid regions are given by.*

$$\frac{(1 - \lambda_3)^{1-\lambda_3}}{(\lambda_4 - \lambda_3)^{\lambda_4-\lambda_3}}(\lambda_4 - 1)^{\lambda_4-1} = \frac{-\lambda_3}{\lambda_4} \quad \text{(Region 5 in the second quadrant)}$$

and

$$\frac{(1 - \lambda_4)^{1-\lambda_4}}{(\lambda_3 - \lambda_4)^{\lambda_3-\lambda_4}}(\lambda_3 - 1)^{\lambda_3-1} = \frac{-\lambda_4}{\lambda_3} \quad \text{(Region 6 in the fourth quadrant)}.$$

Figure 1.3–1 shows all the (λ_3, λ_4) points (the points in the unshaded region) for which a valid $\mathrm{GLD}(\lambda_1, \lambda_2, \lambda_3, \lambda_4)$ exists. The shaded region consists of the points excluded by Lemma 1.3.20 and Theorem 1.3.33. Therefore, for (λ_3, λ_4) in the shaded region there will not exist a valid $\mathrm{GLD}(\lambda_1, \lambda_2, \lambda_3, \lambda_4)$ distribution.

Theorem 1.3.33 gives an algebraic characterization of the boundary between the (λ_3, λ_4) points of V_3 and V_4 that lead to a valid $\mathrm{GLD}(\lambda_1, \lambda_2, \lambda_3, \lambda_4)$ and those that do not. However, it is not clear from this algebraic formulation that the shape of the shaded region is as depicted in Figure 1.3–1. The next two theorems clarify this point.

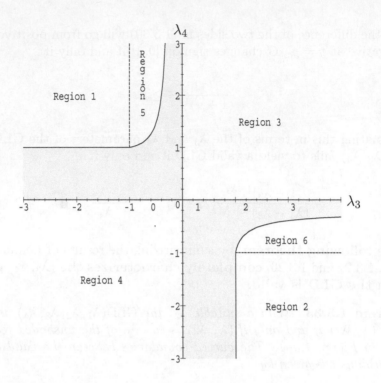

Figure 1.3–1. Regions 1, 2, 3, 4, 5, and 6 where the GLD is valid.

Theorem 1.3.34. *If* $\text{GLD}(\lambda_1, \lambda_2, \lambda_3^*, \lambda_4^*)$ *is valid for a point* $(\lambda_3^*, \lambda_4^*)$ *in* V_3 *and* (λ_3, λ_4^*)*, is a point with* $-1 \leq \lambda_3 \leq \lambda_3^*$*, then* $\text{GLD}(\lambda_1, \lambda_2, \lambda_3, \lambda_4^*)$ *is also valid for* (λ_3, λ_4^*)*.*

Proof. We know from Corollary 1.3.9 that for $\text{GLD}(\lambda_1, \lambda_2, \lambda_3^*, \lambda_4^*)$ to be valid,

$$g(y, \lambda_3^*, \lambda_4^*) = \lambda_3^* y^{\lambda_3^* - 1} + \lambda_4^* (1 - y)^{\lambda_4^* - 1} \qquad (1.3.35)$$

must have the same sign for all y in $[0, 1]$. Since

$$\lim_{y \to 0^+} \lambda_3^* y^{\lambda_3^* - 1} = -\infty \qquad \text{and} \qquad \lim_{y \to 0^+} \lambda_4^* (1 - y)^{\lambda_4^* - 1} = \lambda_4,$$

$g(y, \lambda_3^*, \lambda_4^*) < 0$ for some sufficiently small y. Hence, $g(y, \lambda_3^*, \lambda_4^*) \leq 0$ for all y in $[0, 1]$. We next observe that

$$\frac{\partial \, g(y, \lambda_3, \lambda_4)}{\partial \, \lambda_3} = y^{\lambda_3 - 1}(1 + \lambda_3 \ln y) \qquad (1.3.36)$$

and the right-hand side of (1.3.36) is positive because both λ_3 and $\ln y$ are negative. Hence, for any y in $[0, 1]$, $g(y, \lambda_3, \lambda_4)$ is an increasing function of λ_3 and

$$g(y, \lambda_3, \lambda_4^*) \le g(y, \lambda_3^*, \lambda_4^*) \le 0$$

for all y in $[0, 1]$. By Corollary 1.3.9, $\text{GLD}(\lambda_1, \lambda_2, \lambda_3, \lambda_4^*)$ must be valid. ∎

Theorem 1.3.37. *Given* $-1 < \lambda_3^* < 0$, *there exists a* $\lambda_4^* > 1$ *such that the* $\text{GLD}(\lambda_1, \lambda_2, \lambda_3, \lambda_4)$ *is not valid for points* (λ_3, λ_4) *with* $\lambda_4 \le \lambda_4^*$ *and it is valid for points* (λ_3, λ_4) *with* $\lambda_4 > \lambda_4^*$.

Proof. From (1.3.32), $\text{GLD}(\lambda_1, \lambda_2, \lambda_3^*, \lambda_4)$ is valid if and only if

$$\frac{(1 - \lambda_3^*)^{1-\lambda_3^*}}{(\lambda_4 - \lambda_3^*)^{\lambda_4-\lambda_3^*}} (\lambda_4 - 1)^{\lambda_4-1} < \frac{-\lambda_3^*}{\lambda_4}$$

which is equivalent to

$$-\frac{(1 - \lambda_3^*)^{1-\lambda_3^*}}{\lambda_3^*} < \frac{(\lambda_4 - \lambda_3^*)^{\lambda_4-\lambda_3^*}}{\lambda_4(\lambda_4 - 1)^{\lambda_4-1}} = h(\lambda_4). \tag{1.3.38}$$

Differentiating $h(\lambda_4)$ we have

$$h'(\lambda_4) = \frac{(\lambda_4 - \lambda_3^*)^{\lambda_4-\lambda_3^*}}{\lambda_4(\lambda_4 - 1)^{\lambda_4-1}} [\lambda_4 \ln(\lambda_4 - \lambda_3^*) - \lambda_4 \ln(\lambda_4 - 1)]. \tag{1.3.39}$$

The terms outside of the brackets in (1.3.39) are positive. The expression inside the brackets can be rewritten as

$$\lambda_4 \ln \left(\frac{\lambda_4 - \lambda_3^*}{\lambda_4 - 1} \right)$$

and since $\lambda_4 - \lambda_3^* > \lambda_4 - 1$, the bracketed part of (1.3.39) is also positive. Therefore, $h(\lambda_4)$ is a continuous and increasing function of λ_4 and it must attain all values between

$$h(1) = (1 - \lambda_3^*)^{\lambda_3^*-1} \tag{1.3.40}$$

and

$$\lim_{\lambda_4 \to \infty} h(\lambda_4) = \lim_{\lambda_4 \to \infty} \frac{(\lambda_4 - \lambda_3^*)^{\lambda_4-\lambda_3^*}}{\lambda_4(\lambda_4 - 1)^{\lambda_4-1}} = \infty. \tag{1.3.41}$$

The limit in (1.3.41) is infinite because the degree (in λ_4) of the numerator is $\lambda_4 - \lambda_3^*$ which is larger than λ_4, the degree in the denominator. In particular, since

$$(1 - \lambda_3^*)^{\lambda_3^* - 1} < -\frac{(1 - \lambda_3^*)^{1 - \lambda_3^*}}{\lambda_3^*} < \infty,$$

$h(\lambda_4)$ must attain the value

$$-\frac{(1 - \lambda_3^*)^{1 - \lambda_3^*}}{\lambda_3^*}$$

for some $\lambda_4 = \lambda_4^*$. Since $h(\lambda_4)$ increases with λ_4, when $\lambda_4 > \lambda_4^*$, the inequality of (1.3.38) holds and the GLD($\lambda_1, \lambda_2, \lambda_3, \lambda_4$) is valid at such points. Similarly when $\lambda_4 < \lambda_4^*$, the inequality of (1.3.38) cannot hold and the GLD($\lambda_1, \lambda_2, \lambda_3, \lambda_4$) is not valid at these points. ∎

Theorems 1.3.34 and 1.3.37 **justify the shape of the invalid shaded region of V_3 given in Figure 1.3–1, and of the valid Region 5.** Similar arguments, with λ_3 and λ_4 interchanged, establish analogous results for V_4 and Region 6. One of the consequences of Theorem 1.3.37 is that for a given λ_3 in $(-1, 0)$, there is one and only one λ_4 for which (λ_3, λ_4) is on the boundary curve specified in Theorem 1.3.33. Therefore, the boundary curve can be viewed as the graph of a function (say $\lambda_4 = B(\lambda_3)$) that specifies the points $(\lambda_3, B(\lambda_4))$ on the boundary.

As λ_3 approaches 0 from the left, $B(\lambda_3)$ grows at a surprisingly high rate — this is not apparent in Figure 1.3–1 because $\lambda_3 < -0.3$ in Figure 1.3–1. However, some direct (and difficult) computations yield the following values of $B(\lambda_3)$ for $\lambda_3 = -0.3, \; -0.2, \; -0.1, \; -0.08,$ and -0.04.

$$B(-0.3) = 3.6196669$$
$$B(-0.2) = 25.450660$$
$$B(-0.1) = 476523.97$$
$$B(-0.08) = 1.9935190 \times 10^8$$
$$B(-0.04) = 1.2580865 \times 10^{24}.$$

It can be seen in Figure 1.3–2 (a) that even $\ln(B(\lambda_3))$ rises quite rapidly when λ_3 gets sufficiently close to 0. A significant, although somewhat more moderated, growth can be seen in Figure 1.3–2 (b) which gives the graph of $\ln(\ln(B(\lambda_3)))$.

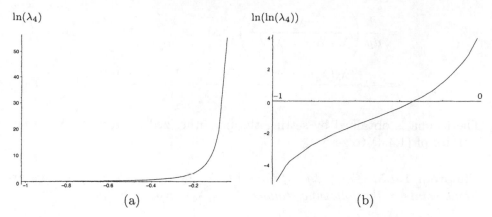

Figure 1.3–2. Growth of $\ln(\lambda_4)$ (a) and $\ln(\ln(\lambda_4))$ (b) as λ_3 increases to 0 on the boundary of valid and non-valid regions in V_3.

1.4 Shapes of the GLD Density Functions

In this section we undertake **an investigation of the possible shapes of the GLD($\lambda_1, \lambda_2, \lambda_3, \lambda_4$) p.d.f.s** by charting, over (λ_3, λ_4)-space, the regions where the p.d.f. has zero, one, or two relative extreme points (points where the GLD($\lambda_1, \lambda_2, \lambda_3, \lambda_4$) p.d.f. has a relative maximum or relative minimum). This is followed by graphs of GLD($\lambda_1, \lambda_2, \lambda_3, \lambda_4$) p.d.f.s that illustrate shapes associated with various regions of (λ_3, λ_4)-space.

Lemma 1.4.1. *The relative extreme points of the* GLD($\lambda_1, \lambda_2, \lambda_3, \lambda_4$) *p.d.f. occur at values of y where*

$$g(y) = \frac{y^{\lambda_3 - 2}}{(1 - y)^{\lambda_4 - 2}} = \frac{\lambda_4(\lambda_4 - 1)}{\lambda_3(\lambda_3 - 1)}. \tag{1.4.2}$$

Proof. It was shown in Section 1.2 (see equation (1.2.3)) that the p.d.f. of a GLD($\lambda_1, \lambda_2, \lambda_3, \lambda_4$) random variable is given by

$$f(x) = \frac{\lambda_2}{\lambda_3 y^{\lambda_3 - 1} + \lambda_4(1 - y)^{\lambda_4 - 1}} \tag{1.4.3}$$

where $x = Q(y)$, or equivalently, $y = F(x)$. Differentiating with respect to x,

$$f'(x) = \frac{d}{dy}\left(\frac{\lambda_2}{\lambda_3 y^{\lambda_3 - 1} + \lambda_4^{\lambda_4 - 1}}\right)\frac{dy}{dx}$$

$$= \frac{d}{dy}\left(\frac{\lambda_2}{\lambda_3 y^{\lambda_3-1} + \lambda_4^{\lambda_4-1}}\right)f(x)$$

$$= -\lambda_2 \frac{(\lambda_3(\lambda_3-1)y^{\lambda_3-2} - \lambda_4(\lambda_4-1)(1-y)^{\lambda_4-2})}{(\lambda_3 y^{\lambda_3-1} + \lambda_4^{\lambda_4-1})^2}f(x). \quad (1.4.4)$$

The lemma is obtained by setting the parenthesized expression in the numerator of (1.4.4) to zero. ∎

Theorem 1.4.5. *The p.d.f. of the* GLD$(\lambda_1,\lambda_2,\lambda_3,\lambda_4)$ *has no relative extreme points in the following regions of* (λ_3,λ_4)-*space.*

$$E_{0,1} = \{(\lambda_3,\lambda_4) \mid 0 < \lambda_3 < 1, \ \lambda_4 > 1\}$$
$$E_{0,2} = \{(\lambda_3,\lambda_4) \mid \lambda_3 > 1, \quad\ 0 < \lambda_4 < 1\}$$
$$E_{0,3} = \{(\lambda_3,\lambda_4) \mid \lambda_3 < 0, \quad\ 0 < \lambda_4 < 1\}$$
$$E_{0,4} = \{(\lambda_3,\lambda_4) \mid 0 < \lambda_3 < 1, \ \lambda_4 < 0\}.$$

Proof. Before starting the proof we note that the GLD is valid for (λ_3,λ_4) in $E_{0,1}$ and $E_{0,2}$ but not valid in $E_{0,3}$ and $E_{0,4}$ (see Figure 1.3–1). The latter two regions are included for the sake of completeness.

The left-hand side of (1.4.2) is always positive. However, in all of the cases listed, the right-hand side is negative. ∎

Theorem 1.4.6. *The* GLD$(\lambda_1,\lambda_2,\lambda_3,\lambda_4)$ *p.d.f. has a unique relative extremum when* $\lambda_3 > 2$ *and* $\lambda_4 > 2$.

Proof. Differentiating the $g(y)$ that was defined at (1.4.2) in Lemma 1.4.1,

$$g'(y) = \frac{(\lambda_3-2)(1-y)^{\lambda_4-2}y^{\lambda_3-3} + (\lambda_4-2)y^{\lambda_3-2}(1-y)^{\lambda_4-3}}{(1-y)^{2(\lambda_4-2)}}$$

$$= \left(\frac{y^{\lambda_3-3}(1-y)^{\lambda_4-3}}{(1-y)^{2(\lambda_4-2)}}\right)((\lambda_3-2)(1-y) + (\lambda_4-2)y). \quad (1.4.7)$$

Hence, $g'(y)$ has the same sign as

$$h(y) = (\lambda_3-2)(1-y) + (\lambda_4-2)y. \quad (1.4.8)$$

When $\lambda_3 > 2$ and $\lambda_4 > 2$, $h(y)$, and consequently $g'(y)$, is positive. Moreover,

$$\lim_{y\to 0^+} g(y) = 0 \qquad \text{and} \qquad \lim_{y\to 1^-} g(y) = \infty,$$

making $g(y)$ a function that increases from 0 to ∞. Therefore, (1.4.2) must hold at exactly one value of y which is the only critical point of the p.d.f. of the GLD($\lambda_1, \lambda_2, \lambda_3, \lambda_4$). Moreover, since the $f'(x)$ of (1.4.4) changes sign from positive to negative as y moves from 0 to 1, this unique extreme point is a relative maximum of the p.d.f. of the GLD($\lambda_1, \lambda_2, \lambda_3, \lambda_4$). ■

Theorem 1.4.9. *The* GLD($\lambda_1, \lambda_2, \lambda_3, \lambda_4$) *p.d.f. has a unique relative extremum when $\lambda_3 < 2$, $\lambda_4 < 2$, and (λ_3, λ_4) is not in one of the regions excluded by Theorem 1.4.5.*

Proof. From (1.4.8), the constraints $\lambda_3 < 2$ and $\lambda_4 < 2$ make $h(y)$, and hence $g'(y)$, negative. Also,

$$\lim_{y \to 0^+} g(y) = \infty \qquad \text{and} \qquad \lim_{y \to 1^-} g(y) = 0.$$

Thus, $g(y)$ decreases from ∞ to 0, yielding a unique solution to (1.4.2). ■

Lemma 1.4.10. *Let*

$$U(\lambda_3, \lambda_4) = \frac{(2 - \lambda_3)^{\lambda_3 - 2}(\lambda_4 - \lambda_3)^{\lambda_4 - \lambda_3}}{(\lambda_4 - 2)^{\lambda_4 - 2}}, \qquad (1.4.11)$$

$$V(\lambda_3, \lambda_4) = \frac{\lambda_4(\lambda_4 - 1)}{\lambda_3(\lambda_3 - 1)}. \qquad (1.4.12)$$

If $\lambda_4 > 2$, and $1 < \lambda_3 < 2$ or $\lambda_3 < 0$, then the GLD($\lambda_1, \lambda_2, \lambda_3, \lambda_4$) *p.d.f. has*

two relative extrema if $U(\lambda_3, \lambda_4) < V(\lambda_3, \lambda_4)$
one relative extremum if $U(\lambda_3, \lambda_4) = V(\lambda_3, \lambda_4)$
no relative extrema if $U(\lambda_3, \lambda_4) > V(\lambda_3, \lambda_4)$.

Proof. We have already established that $g'(y)$ has the same sign as $h(y)$ (as defined in (1.4.8)). $h(y)$ will be zero if and only if

$$y = y_0 = \frac{2 - \lambda_3}{\lambda_4 - \lambda_3}.$$

For the constraints specified in this lemma we see that $h(y)$ is negative, zero, or positive depending on whether y is less than, equal to, or greater than y_0, giving $g(y)$ a "parabolic shape" that opens up. The existence and number of solutions to (1.4.2) will depend on how this parabolic-shaped curve intersects the horizontal line

$$y = \frac{\lambda_4(\lambda_4 - 1)}{\lambda_3(\lambda_3 - 1)} = V(\lambda_3, \lambda_4).$$

There are three possibilities:

$$g(y_0) \begin{cases} < V(\lambda_3, \lambda_4) \text{ and (1.4.2) has two solutions} \\ = V(\lambda_3, \lambda_4) \text{ and (1.4.2) has a unique solution} \\ > V(\lambda_3, \lambda_4) \text{ and (1.4.2) has no solutions.} \end{cases}$$

The following direct computation of $g(y_0)$ completes the proof.

$$g(y_0) = \left(\frac{2 - \lambda_3}{\lambda_4 - \lambda_3}\right)^{\lambda_3 - 2} \Big/ \left(1 - \frac{2 - \lambda_3}{\lambda_4 - \lambda_3}\right)^{\lambda_4 - 2}$$

$$= \frac{(2 - \lambda_3)^{\lambda_3 - 2}}{(\lambda_4 - \lambda_4)^{\lambda_4 - \lambda_3}} \times \frac{(\lambda_4 - \lambda_3)^{\lambda_4 - 2}}{(\lambda_4 - 2)^{\lambda_4 - 2}}$$

$$= \frac{(2 - \lambda_3)^{\lambda_3 - 2}(\lambda_4 - \lambda_3)^{\lambda_4 - \lambda_3}}{(\lambda_4 - 2)^{\lambda_4 - 2}} = U(\lambda_3, \lambda_4). \blacksquare$$

Lemma 1.4.13. *If $\lambda_4 > 2$ and $\lambda_3 < 0$, then $U(\lambda_3, \lambda_4)/V(\lambda_3, \lambda_4)$ is a decreasing function of λ_3.*

Proof. From

$$\frac{U(\lambda_3, \lambda_4)}{V(\lambda_3, \lambda_4)} = \frac{\lambda_3(\lambda_3 - 1)(2 - \lambda_3)^{\lambda_3 - 2}(\lambda_4 - \lambda_3)^{\lambda_4 - \lambda_3}}{\lambda_4(\lambda_4 - 1)(\lambda_4 - 2)^{\lambda_4 - 2}}$$

we have

$$\frac{\partial}{\partial \lambda_3} \frac{U(\lambda_3, \lambda_4)}{V(\lambda_3, \lambda_4)} = -\frac{(\lambda_4 - 2)^{2 - \lambda_4}(\lambda_4 - \lambda_3)^{\lambda_4 - \lambda_3}(2 - \lambda_3)^{\lambda_3 - 2}}{\lambda_4(\lambda_4 - 1)}$$

$$\times \left(\lambda_3^2 \ln\left(\frac{\lambda_4 - \lambda_3}{2 - \lambda_3}\right) - \lambda_3 \ln\left(\frac{\lambda_4 - \lambda_3}{2 - \lambda_3}\right) - 2\lambda_3 + 1\right). \quad (1.4.14)$$

Except for the initial negative sign, all portions of this expression are positive. $\blacksquare$

Theorem 1.4.15. *For each $\lambda_4^* > 2$, there is one and only one λ_3^* between -1 and 0 for which*

$$\frac{(2 - \lambda_3^*)^{\lambda_3^* - 2}(\lambda_4^* - \lambda_3^*)^{\lambda_4^* - \lambda_3^*}}{(\lambda_4^* - 2)^{\lambda_4^* - 2}} = \frac{\lambda_4^*(\lambda_4^* - 1)}{\lambda_3^*(\lambda_3^* - 1)}$$

and the p.d.f. of the $\text{GLD}(\lambda_1, \lambda_2, \lambda_3^*, \lambda_4^*)$ *has a unique relative extremum. At points* (λ_3, λ_4) *with* $\lambda_3 < \lambda_3^*$ *the p.d.f. has no relative extrema and at points with* $\lambda_3 > \lambda_3^*$ *the p.d.f. has exactly two relative extrema.*

Proof. We know from Lemma 1.4.13 that $U(\lambda_3, \lambda_4)/V(\lambda_3, \lambda_4)$ decreases with λ_3. We observe that

$$\lim_{\lambda_3 \to 0^-} \frac{U(\lambda_3, \lambda_4)}{V(\lambda_3, \lambda_4)} = \frac{U(0, \lambda_4)}{\lim_{\lambda_3 \to 0^-} V(\lambda_3, \lambda_4)}$$

$$= \left(\frac{2^{-2} \lambda_4^{\lambda_4}}{(\lambda_4 - 2)^{\lambda_4 - 2}} \right) \Big/ \infty = 0$$

and

$$\lim_{\lambda_3 \to -1^+} \frac{U(\lambda_3, \lambda_4)}{V(\lambda_3, \lambda_4)} = \frac{U(-1, \lambda_4)}{V(-1, \lambda_4)}$$

$$= \frac{2(\lambda_4 + 1)^{\lambda_4 + 1}}{27 \lambda_4 (\lambda_4 - 1)(\lambda_4 - 2)^{\lambda_4 - 2}}. \tag{1.4.16}$$

The derivative of (1.4.16), with respect to λ_4, is

$$\frac{2(\lambda_4 + 1)(\lambda_4 - 2)^{2 - \lambda_4}}{27 \lambda_4 (\lambda_4 - 1)(\lambda_4 + 1)^{\lambda_4}}.$$

Since this derivative is positive, $U(-1, \lambda_4)/V(-1, \lambda_4)$ is an increasing function of λ_4 and $U(-1, \lambda_4)/V(-1, \lambda_4) \geq U(-1, 2)/V(-1, 2) = 1$, making $U(\lambda_3, \lambda_4) > V(\lambda_3, \lambda_4)$ at $\lambda_3 = -1$.

We now have $U(\lambda_3, \lambda_4)/V(\lambda_3, \lambda_4)$ decreasing from a number larger than 1 at $\lambda_3 = -1$ to 0 as $\lambda_3 \to 0^-$. This means that for a fixed $\lambda_4^* > 2$, the surfaces $U(\lambda_3, \lambda_4)$ and $V(\lambda_3, \lambda_4)$ cross exactly once for some λ_3^* between -1 and 0. Moreover, when $\lambda_3 < \lambda_3^*$, $U(\lambda_3, \lambda_4) > V(\lambda_3, \lambda_4)$ indicating, by Lemma 5, the absence of extreme points for the $\text{GLD}(\lambda_1, \lambda_2, \lambda_3, \lambda_4)$ p.d.f.; and, when $\lambda_3 > \lambda_3^*$, $U(\lambda_3, \lambda_4) < V(\lambda_3, \lambda_4)$, indicating the presence of two relative extrema. ∎

Lemma 1.4.17. *If* $\lambda_4 > 2$ *and* $1 < \lambda_3 < 2$, *then* $U(\lambda_3, \lambda_4)/V(\lambda_3, \lambda_4)$ *is a decreasing function of* λ_4

Proof. From (1.4.11) and (1.4.12),

$$\frac{U(\lambda_3, \lambda_4)}{V(\lambda_3, \lambda_4)} = \frac{\lambda_3 (\lambda_3 - 1)(2 - \lambda_3)^{\lambda_3 - 2}(\lambda_4 - \lambda_3)^{\lambda_4 - \lambda_3}}{\lambda_4 (\lambda_4 - 1)(\lambda_4 - 2)^{\lambda_4 - 2}}$$

and we have

$$\frac{\partial}{\partial \lambda_4} \frac{U(\lambda_3, \lambda_4)}{V(\lambda_3, \lambda_4)} = \frac{\lambda_3(\lambda_3 - 1)(2 - \lambda_3)^{\lambda_3 - 2}(\lambda_4 - 2)^{2 - \lambda_4}(\lambda_4 - \lambda_3)^{\lambda_4 - \lambda_3}}{\lambda_4^2(\lambda_4 - 1)^2} \times$$

$$\left[\lambda_4^2 \ln \left(\frac{\lambda_4 - \lambda_3}{\lambda_4 - 2} \right) - \lambda_4 \ln \left(\frac{\lambda_4 - \lambda_3}{\lambda_4 - 2} \right) - 2\lambda_4 + 1 \right]. \quad (1.4.18)$$

It is clear that, with the exception of the expression in brackets, all terms in (1.4.18) are positive. Therefore, $\dfrac{\partial}{\partial \lambda_4} \dfrac{U(\lambda_3, \lambda_4)}{V(\lambda_3, \lambda_4)} < 0$ if and only if

$$\lambda_4^2 \ln \left(\frac{\lambda_4 - \lambda_3}{\lambda_4 - 2} \right) - \lambda_4 \ln \left(\frac{\lambda_4 - \lambda_3}{\lambda_4 - 2} \right) - 2\lambda_4 + 1 < 0. \quad (1.4.19)$$

The inequality in (1.4.19) can be written as

$$\ln \left(\frac{\lambda_4 - \lambda_3}{\lambda_4 - 2} \right) < \frac{2\lambda_4 - 1}{\lambda_4^2 - \lambda_4}$$

and is equivalent to

$$\frac{\lambda_4 - \lambda_3}{\lambda_4 - 2} < e^{(2\lambda_4 - 1)/(\lambda_4^2 - \lambda_4)}.$$

Since $(\lambda_4 - \lambda_3)/(\lambda_4 - 2) < 1$ when $1 < \lambda_3 < 2$, it is sufficient for the proof of this lemma to show that

$$1 < e^{(2\lambda_4 - 1)/(\lambda_4^2 - \lambda_4)} = a(\lambda_4) \quad (1.4.20)$$

when $\lambda_4 > 2$. We establish this by observing that

$$a'(\lambda_4) = -\frac{(2\lambda_4^2 - 2\lambda_4 + 1)e^{(2\lambda_4 - 1)/(\lambda_4^2 - \lambda_4)}}{\lambda_4^2(\lambda_4 - 1)^2} < 0.$$

Hence

$$a(\lambda_4) > \lim_{\lambda_4 \to \infty} a(\lambda_4)$$

$$= \lim_{\lambda_4 \to \infty} e^{(2\lambda_4 - 1)/(\lambda_4^2 - \lambda_4)}$$

$$= 1.$$

Having verified the inequality in (1.4.20), we have the proof of the lemma.

■

Theorem 1.4.21. *If* $\lambda_4 > 2$ *and* $1 < \lambda_3 < 2$, *then the p.d.f. of the* $GLD(\lambda_1, \lambda_2, \lambda_3, \lambda_4)$ *has exactly two relative extrema.*

Proof. From Lemma 1.4.17, we know that for the specified values of λ_3 and λ_4, $U(\lambda_3, \lambda_4)/V(\lambda_3, \lambda_4)$ is a decreasing function of λ_4. Hence,

$$\frac{U(\lambda_3, \lambda_4)}{V(\lambda_3, \lambda_4)} < \lim_{\lambda_4 \to 2^+} \frac{U(\lambda_3, \lambda_4)}{V(\lambda_3, \lambda_4)}$$

$$= \frac{\lim_{\lambda_4 \to 2^+} U(\lambda_3, \lambda_4)}{V(\lambda_3, 2)} = \frac{1}{V(\lambda_3, 2)}$$

$$= \frac{\lambda_3(\lambda_3 - 1)}{2} \leq 1.$$

This makes $U(\lambda_3, \lambda_4) < V(\lambda_3, \lambda_4)$ and the theorem follows from Lemma 1.4.10. ∎

Symmetry and the presence (or absence) of tails are additional shape details of interest for $GLD(\lambda_1, \lambda_2, \lambda_3, \lambda_4)$ **distributions.** We can see from (1.2.3) that if $f(x) = f(Q(y))$, the p.d.f. of a $GLD(\lambda_1, \lambda_2, \lambda_3, \lambda_4)$ distribution can be viewed as a function k of y, λ_1, λ_2, λ_3 and λ_4, and that then

$$k(y, \lambda_1, \lambda_2, \lambda_3, \lambda_4) = k(1 - y, \lambda_1, \lambda_2, \lambda_4, \lambda_3).$$

If x in the domain of f corresponds to y via $x = Q(y)$, we must have, from (1.2.1),

$$x = Q(y, \lambda_1, \lambda_2, \lambda_3, \lambda_4) = \lambda_1 + \frac{y^{\lambda_3} - (1 - y)^{\lambda_4}}{\lambda_2} = \lambda_1 + A.$$

The x-value that corresponds to $1 - y$ when λ_3 and λ_4 are interchanged is

$$Q(1 - y, \lambda_1, \lambda_2, \lambda_4, \lambda_3) = \lambda_1 + \frac{(1 - y)^{\lambda_4} - (y)^{\lambda_3}}{\lambda_2} = \lambda_1 - A.$$

Therefore, **the** $GLD(\lambda_1, \lambda_2, \lambda_4, \lambda_3)$ **is the symmetric image, about the line** $x = \lambda_1$, **of the** $GLD(\lambda_1, \lambda_2, \lambda_3, \lambda_4)$ **p.d.f.** This is illustrated in Figure 1.4–1. These comments are summarized in the following theorem.

Theorem 1.4.22. *The p.d.f.s of* $GLD(\lambda_1, \lambda_2, \lambda_3, \lambda_4)$ *and* $GLD(\lambda_1, \lambda_2, \lambda_4, \lambda_3)$ *are symmetric images of each other and the axis of symmetry is the line* $x = \lambda_1$.

Depending on the choice of λ_3 and λ_4, a $GLD(\lambda_1, \lambda_2, \lambda_3, \lambda_4)$ p.d.f. may have all possible combinations of finite or infinite right and left tails. For a

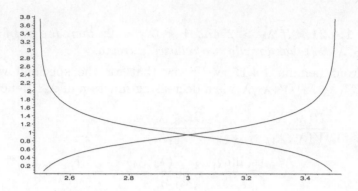

Figure 1.4–1. The GLD$(3, 2, 1.5, .5)$ and GLD$(3, 2, .5, 1.5)$ p.d.f.s; one is the symmetric image of the other about the line $x = 3$.

p.d.f. *not* to have an infinite right tail, its c.d.f. $F(x)$ must attain the value 1 at some finite point $x = x_0$. This means that $Q(1) = x_0$. Since $Q(y) = \lambda_1 + [y^{\lambda_3} - (1-y)^{\lambda_4}]/\lambda_2$ we see that Q is defined at $y = 1$ if and only if $\lambda_4 \geq 0$. Thus, a GLD p.d.f. has an infinite right tail if and only if $\lambda_4 < 0$. A similar argument shows that a GLD$(\lambda_1, \lambda_2, \lambda_3, \lambda_4)$ p.d.f. will have an infinite left tail if and only if $\lambda_3 < 0$. A similar analysis can be applied for other choices of λ_3 and λ_4. For example, if $\lambda_3 = 0$ and $\lambda_4 > 0$,

$$\lim_{y \to 0^+} Q(y) = \lambda_1 \quad \text{and} \quad \lim_{y \to 1^-} Q(y) = Q(1) = \lambda_1 + 1/\lambda_2.$$

We formalize these results on the support of the GLD$(\lambda_1, \lambda_2, \lambda_3, \lambda_4)$ in the next theorem. We emphasize these results since, while their derivation is simple, **the results of Theorem 1.4.23 are easy to give incorrectly**. The results come from an early work, a 1971 M.S. thesis at the University of Iowa, for Regions 1, 2, 3, and 4, and are new for Regions 5 and 6.

Theorem 1.4.23. *The support of the* GLD$(\lambda_1, \lambda_2, \lambda_3, \lambda_4)$ *p.d.f. (i.e., points where the p.d.f. is positive) is as given in the following tables; the first for Regions 1, 2, 5, and 6 and the second for Regions 3 and 4 (the regions are shown in Figure 1.3–1).*

Region	λ_3	λ_4	Support
Region 1	$\lambda_3 < -1$	$\lambda_4 > 1$	$(-\infty, \lambda_1 + 1/\lambda_2]$
Region 2	$\lambda_3 > 1$	$\lambda_4 < -1$	$[\lambda_1 - 1/\lambda_2, \infty)$
Region 5	$-1 < \lambda_3 < 0$	$\lambda_4 > 1$	$(-\infty, \lambda_1 + 1/\lambda_2]$
Region 6	$\lambda_3 > 1$	$-1 < \lambda_4 < 0$	$[\lambda_1 - 1/\lambda_2, \infty)$

Region	λ_3	λ_4	Support
	$\lambda_3 > 0$	$\lambda_4 > 0$	$[\lambda_1 - 1/\lambda_2, \lambda_1 + 1/\lambda_2]$
Region 3	$\lambda_3 > 0$	$\lambda_4 = 0$	$[\lambda_1 - 1/\lambda_2, \lambda_1]$
	$\lambda_3 = 0$	$\lambda_4 > 0$	$[\lambda_1, \lambda_1 + 1/\lambda_2]$
	$\lambda_3 < 0$	$\lambda_4 < 0$	$(-\infty, \infty)$
Region 4	$\lambda_3 < 0$	$\lambda_4 = 0$	$(-\infty, \lambda_1]$
	$\lambda_3 = 0$	$\lambda_4 < 0$	$[\lambda_1, \infty)$

Theorems 1.4.5, 1.4.6, 1.4.9, 1.4.15, and 1.4.21, together with the symmetry properties of the $\text{GLD}(\lambda_1, \lambda_2, \lambda_3, \lambda_4)$ about the line $\lambda_3 = \lambda_4$, completely characterize the regions of (λ_3, λ_4)-space that give rise to $\text{GLD}(\lambda_1, \lambda_2, \lambda_3, \lambda_4)$ p.d.f.s with 0, 1, or 2 relative extreme points. Figure 1.4–2 summarizes these results. The region designated by an "X" in Figure 1.4–2 does not produce valid $\text{GLD}(\lambda_1, \lambda_2, \lambda_3, \lambda_4)$ distributions. All other regions are labeled with the number of relative extreme points associated with the (λ_3, λ_4) points in that region.

Figures 1.4–3 through 1.4–9 show $\text{GLD}(\lambda_1, \lambda_2, \lambda_3, \lambda_4)$ p.d.f.s with (λ_3, λ_4) taken from various regions of (λ_3, λ_4)-space. **Figure 1.4–3a** shows p.d.f.s with $\lambda_1 = 0$, $\lambda_2 = 1$, $\lambda_3 = 2.5$ and $\lambda_4 = 0.5, 0.75, 1.0, \ldots, 2.5$. The graph that rises to the highest point on the left (at $x = -1$) is the one corresponding to $\lambda_4 = 0.5$, the next highest corresponds to $\lambda_4 = 0.75$, and so on. From Theorem 1.4.23, we know that p.d.f.s with these values of λ_3 and λ_4 will not have infinite right or left tails. Moreover, we can observe from Figure 1.4–2 that there will be a transition from zero to one and eventually to two critical points as λ_4 moves from 0.5 to 2.25. It is not apparent in Figure 1.4–3a which of the graphs, if any, have critical points. As we look at the p.d.f.s with $\lambda_4 = 0.5, 0.75, 1.0$ in **Figure 1.4–3b** we see that, consistent with the foregoing analysis, these p.d.f.s do not have critical points whereas the p.d.f.s in **Figure 1.4–3c**, corresponding to $\lambda_4 = 1.25, 1.5, 1.75$, exhibit two critical points (this is perhaps most apparent for $\lambda_4 = 1.5$ and 1.75). The last set of the p.d.f.s from Figure 1.4–3a, those for $\lambda_4 = 2.0, 2.25, 2.5$, are shown, with considerable rescaling, in **Figure 1.4–3d** where the presence of a critical point is clearly visible.

Figure 1.4–4a depicts $\text{GLD}(\lambda_1, \lambda_2, \lambda_3, \lambda_4)$ p.d.f.s for $\lambda_1 = 0$, $\lambda_2 = 1$, $\lambda_3 = 1.5$ and $\lambda_4 = 0.5, 0.75, 1.0, \ldots, 2.5$. As before, the graph that rises to the highest point on the left (at $x = -1$) is the one corresponding to $\lambda_4 = 0.5$, the next highest corresponds to $\lambda_4 = 0.75$, and so on.

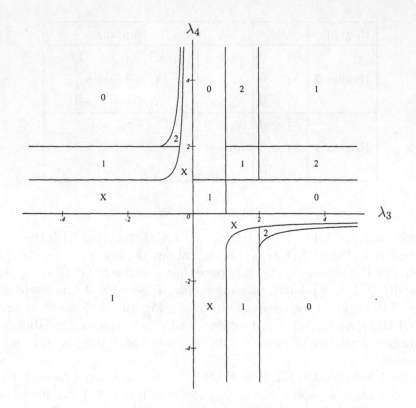

Figure 1.4–2. The number of critical points of the p.d.f.s
of GLD($\lambda_1, \lambda_2, \lambda_3, \lambda_4$) distributions.

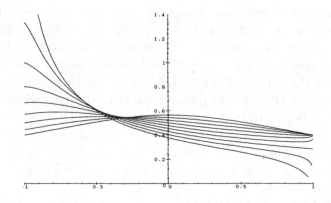

Figure 1.4–3a. GLD($0, 1, \lambda_3, \lambda_4$) p.d.f.s with $\lambda_3 = 2.5$ and
$\lambda_4 = 0.5, 0.75, 1.0, 1.25, 1.5, 1.75, 2.0, 2.25, 2.5$.

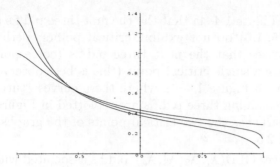

Figure 1.4–3b. GLD$(0, 1, \lambda_3, \lambda_4)$ p.d.f.s with $\lambda_3 = 2.5$ and
$\lambda_4 = 0.5, 0.75, 1.0.$

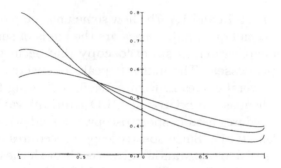

Figure 1.4–3c. GLD$(0, 1, \lambda_3, \lambda_4)$ p.d.f.s with $\lambda_3 = 2.5$ and
$\lambda_4 = 1.25, 1.5, 1.75.$

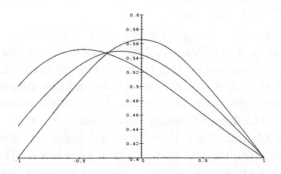

Figure 1.4–3d. GLD$(0, 1, \lambda_3, \lambda_4)$ p.d.f.s with $\lambda_3 = 2.5$ and
$\lambda_4 = 2.0, 2.25, 2.5.$

It is clear from Figure 1.4–4a that the the first three p.d.f.s (corresponding to $\lambda_4 = 0.5, 0.75, 1.0$) do not exhibit critical points; perhaps with some difficulty, we can see that the next three p.d.f.s (corresponding to $\lambda_4 = 1.0, 1.25, 1.5$) have a single critical point (this is best observed on the right side of the graphs in Figure 1.4–4a where these curves "turn up" as x gets close to 1. The remaining three p.d.f.s are replotted in Figure 1.4–4b where one can more clearly see the two critical points of the graphs corresponding to $\lambda_4 = 2.25$ and 2.5.

Figure 1.4–5 gives GLD($\lambda_1, \lambda_2, \lambda_3, \lambda_4$) p.d.f.s associated with $\lambda_1 = 0$, $\lambda_2 = 1$, $\lambda_3 = 0.5$ and $\lambda_4 = 0.25, 0.5, 0.75, 1.0, 1.5, 2.0, 2.5$. In this case the p.d.f. corresponding to $\lambda_4 = 0.25$ is the one that rises to the highest point in the center of the graph, the next highest corresponding to $\lambda_4 = 0.5$, and so on. It is easy to see the transition from one to no critical points as λ_4 goes through the $\lambda_4 = 1$ barrier.

Figures 1.4–6 and 1.4–7a and 1.4–7b show some unusual p.d.f. shapes with (λ_3, λ_4) from the second quadrant. These are the kinds of sharp peaks that are often found in applications in **spectroscopy** that occur with spectra of **autoregressive processes**. The multiple peaks of spectra often look like the overlay of the several curves in Figure 1.4–6, indicating that it may be possible to model them as a **mixture of GLD random variables**. For an overview of the procedures used in spectroscopy, see Dudewicz, Mommaerts, and van der Meulen (1991). Since spectroscopy procedures are not simple, there may be a potential for substantial advances through the use of mixtures of GLDs in this area. Recall that Theorem 1.4.23 indicates the presence of a left tail for these choices of (λ_3, λ_4). In Figure 1.4–6, $\lambda_1 = 0$, $\lambda_2 = -1$, $\lambda_3 = -0.2$ and $\lambda_4 = 27, 30, 35, 50$. The highest curve (in the middle) corresponds to $\lambda_4 = 27$, the next highest corresponds to $\lambda_4 = 30$, and so on. The (λ_3, λ_4) points for all four p.d.f.s fall in the second quadrant region marked with a "2" in Figure 1.4–2.

The (λ_3, λ_4) points for the p.d.f.s of Figure 1.4–7a ($\lambda_1 = 0$, $\lambda_2 = -1$, $\lambda_3 = -0.5$ and $\lambda_4 = 2.91, 2.92, \ldots, 2.99$) cross the boundary between the two regions that give rise to two and then to zero critical points. The graphs are so packed together that it is difficult to distinguish among them. Figure 1.4–7b gives a magnification of the portion of the graphs where critical points seem to appear. The highest graph corresponds to $\lambda_4 = 2.91$, the next highest to $\lambda_4 = 2.92$, and so on. It is clear that there are two critical points when $\lambda_4 = 2.91$ and no critical points when $\lambda_4 = 2.99$. The point on the boundary where the transition occurs is (approximately) $(-0.5, 2.996)$.

Figures 1.4–8a and 1.4–8b show GLD($\lambda_1, \lambda_2, \lambda_3, \lambda_4$) p.d.f.s with $\lambda_1 = 0$, $\lambda_2 = -1$, $\lambda_3 = -3$, and $\lambda_4 = 1.0, 1.25, 1.5, 1.75, 2.0, 6.0, 20$. Figure 1.4–8a

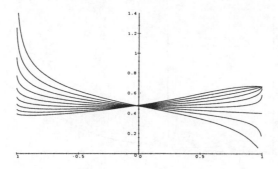

Figure 1.4–4a. GLD$(0, 1, \lambda_3, \lambda_4)$ p.d.f.s with $\lambda_3 = 1.5$ and $\lambda_4 = 0.5, 0.75, 1.0, 1.25, 1.5, 1.75, 2.0, 2.25, 2.5$.

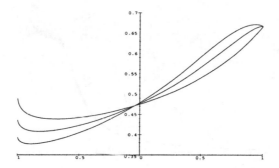

Figure 1.4–4b. GLD$(0, 1, \lambda_3, \lambda_4)$ p.d.f.s with $\lambda_3 = 1.5$ and $\lambda_4 = 2.0, 2.25, 2.5$.

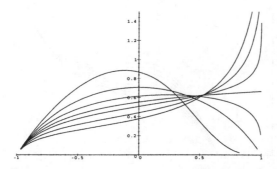

Figure 1.4–5. GLD$(0, 1, \lambda_3, \lambda_4)$ p.d.f.s with $\lambda_3 = 0.5$ and $\lambda_4 = 0.25, 0.5, 0.75, 1.0, 1.5, 2.0, 2.5$.

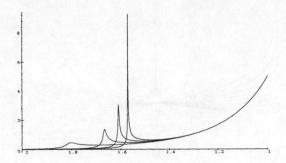

Figure 1.4–6. GLD$(0, -1, \lambda_3, \lambda_4)$ p.d.f.s with $\lambda_3 = -0.2$ and
$\lambda_4 = 27, 30, 35, 50$.

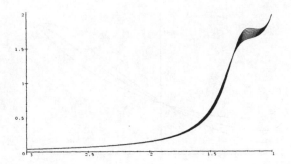

Figure 1.4–7a. GLD$(0, -1, \lambda_3, \lambda_4)$ p.d.f.s with $\lambda_3 = -0.5$ and
$\lambda_4 = 2.91, 2.92, 2.93, \ldots, 2.99$.

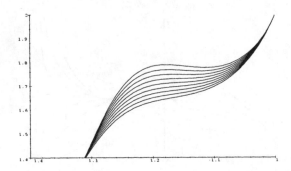

Figure 1.4–7b. GLD$(0, -1, \lambda_3, \lambda_4)$ p.d.f.s with $\lambda_3 = -0.5$ and
$\lambda_4 = 2.91, 2.92, 2.93, \ldots, 2.99$.

shows the first four of these with the p.d.f. with the highest point on the right corresponding to $\lambda_4 = 1.0$, the next highest to $\lambda_4 = 1.25$, etc. The critical point for the p.d.f. with $\lambda_4 = 1.25$ is clearly visible; the ones for $\lambda_4 = 1.5$ and 1.75 are more difficult to spot. Although the p.d.f.s in Figure 1.4–8b correspond to three distinct λ_4 values ($\lambda_4 = 2.0, 6.0, 20.0$), only two graphs are discernible; the graphs for $\lambda_4 = 6.0$ and 20 are so close that they cannot be distinguished.

The p.d.f.s depicted in Figure 1.4–9 have shapes that are frequently encountered in applications. For these p.d.f.s $\lambda_1 = 0$, $\lambda_2 = 1$, $\lambda_3 = -0.25$ and $\lambda_4 = -0.1, -0.2, -0.35, -0.5$. The curve that rises highest corresponds to $\lambda_4 = -0.1$, the next highest to $\lambda_4 = -0.2$, etc. Since this places (λ_3, λ_4) in the third quadrant, by Theorem 1.4.23, they all have infinite left and right tails and by earlier results, each has a unique critical point.

We close this section with some notes on shape and related results from the literature. Freimer, Kollia, Mudholkar, and Lin (1988) used a slightly different parametrization of the GLD than the traditional one used at (1.2.1), but, in the main, their results are similar to those of this section. In their Section 3 they note

> The family ... is very rich in the variety of density and tail shapes. It contains unimodal, U-shaped, J-shaped and monotone p.d.f.s. These can be symmetric or asymmetric and their tails can be smooth, abrupt, or truncated, and long, medium or short ... properties ... relevant in ... modelling data and determining the methods of subsequent analysis.

They then classify the GLD with respect to tail shape and density shape.

Freimer, Kollia, Mudholkar, and Lin (1988) also (in their Section 4) **relate the GLD system to that of Karl Pearson**, adding to what was shown by Ramberg, Tadikamalla, Dudewicz, and Mykytka (1979). As **similarities**, they note

- Both contain the uniform and exponential distributions;

- Both accommodate densities of varied shapes.

As **differences**, they note

- Only the GLD includes the logistic distribution;

- Only the Pearson includes the normal distribution (however, the GLD can come very close to the normal distribution);

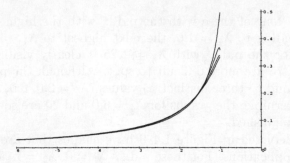

Figure 1.4–8a. GLD$(0, -1, \lambda_3, \lambda_4)$ p.d.f.s with $\lambda_3 = -3$ and
$\lambda_4 = 1.0, 1.25, 1.5, 1.75.$

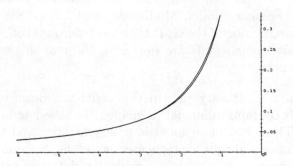

Figure 1.4–8b. GLD$(0, -1, \lambda_3, \lambda_4)$ p.d.f.s with $\lambda_3 = -3$ and
$\lambda_4 = 2.0, 6.0, 20.$

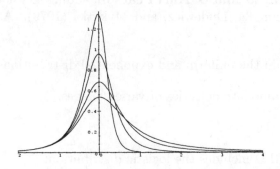

Figure 1.4–9. GLD$(0, 1, \lambda_3, \lambda_4)$ p.d.f.s with $\lambda_3 = -0.25$ and
$\lambda_4 = -0.1, -0.2, -0.35, -0.5.$

- The Pearson covers all skewness and all kurtosis values, the GLD does not. This is a problem in some applications and it is solved by the Extended GLD developed in our Chapter 3.

1.5 GLD Random Variate Generation

As we noted in the Introduction to this Chapter, one of the important applications of the GLD has been the **generation of r.v.s for Monte Carlo studies**. This important application arises due to the confluence of several key results, which we will now state.

Theorem 1.5.1. *If $Q_X(\cdot)$ is the percentile function of a random variable X, and U is a uniform random variable on $(0, 1)$, then $Q_X(U)$ has the same d.f. as does X.*

This result is key in simulation and Monte Carlo studies (for a proof, see p. 156 of Karian and Dudewicz (1999a)), as it allows easy generation of a stream of random variables from any distribution for which the percentile function is readily available. This follows from

Corollary 1.5.2. *If $U_1, U_2, \ldots$ are independent uniform random variables on $(0, 1)$, then*

$$X_1 = Q_X(U_1), \ X_2 = Q_X(U_2), \ldots \tag{1.5.3}$$

are independent random variables each with the same d.f. as X.

For a proof, also see p. 156 of Karian and Dudewicz (1999a). The percentile function is not available in a closed (or easy-to-work-with) form for many of the most important distributions, such as the normal distribution. However the GLD is (see (1.2.1)) **defined** by its p.f., which is a simple-to-calculate expression. Thus, **r.v.s for a simulation study can easily be generated from any distribution that can be modeled by a GLD.**

Example. Suppose we have modeled an important r.v. by an approximate standard normal distribution X. We will show in Section 2.4.1 that a close fit to the standard normal is available via the GLD with

$$(\lambda_1, \lambda_2, \lambda_3, \lambda_4) = (0, 0.1975, 0.1349, 0.1349) \tag{1.5.4}$$

and this GLD has p.f. (see (1.1.1))

$$Q(y) = (y^{0.1349} - (1 - y)^{0.1349})/0.1975$$

$$= 5.06329(y^{0.1349} - (1 - y)^{0.1349}). \tag{1.5.5}$$

Thus, if $U_1, U_2, \ldots$ are independent uniform r.v.s on $(0, 1)$, then

$$Q(U_1), \ Q(U_2), \ \ldots \tag{1.5.6}$$

are independent and (approximately) $N(0, 1)$ r.v.s for the simulation study at hand.

There are a number of good sources of independent uniform r.v.s on $(0, 1)$. For example (see p. 137 of Karian and Dudewicz (1999a)), the generator called URN41 has a period of approximately 5×10^{18} (see Karian and Dudewicz (1999a), pp. 132-133 for the period, Figure 3.5–8 for its FOR-TRAN code, Figure 3.5–9 for its C code, and Appendix G (p. 493) for its first 100 numbers), has passed extensive testing, and yields

$$U_1 = 0.67460162, \ U_2 = 0.15152637, \ \ldots . \tag{1.5.7}$$

Thus, using these in (1.5.5) (see (1.5.6)), we find the approximate normal r.v.s

$$X_1 = 5.06329((0.67460162)^{0.1349} - (1 - 0.67460162)^{0.1349}) = 0.44975078$$
$$X_2 = 5.06329((0.15152637)^{0.1349} - (1 - 0.15126372)^{0.1349}) = -1.026919958.$$

We can continue the stream by generating additional uniform r.v.s on $(0, 1)$, $U_3, U_4, \ldots$ and evaluate $Q(U_3), Q(U_4), \ldots$.

Since we will see in Section 2.2 (especially Figure 2.2–5) that **the GLD covers a broad space of distributions** and can model many different shapes well, it follows that **the GLD is very useful for modeling input to simulation and Monte Carlo studies: one can change the distribution being used by simply altering the lambda vector.**

Problems for Chapter 1

1.1. Suppose that the d.f. of X is $F_X(x) = 0$ if $x \leq 1, = (x^4 - 1)/255$ if $1 < x < 4$, and $= 1$ if $x \geq 4$.

    a. Find the p.d.f. of X, say $f_X(x)$, and graph it. Also graph the d.f. of X.

    b. Find the p.f. of X, say $Q_X(y)$, and graph it. Find $Q_X(.95)$.

1.2. Suppose that the p.d.f. of X is $f_X(x) = 5e^{-5x}$ when x is positive (and $= 0$ otherwise).

   a. Graph the p.d.f. of X. Find and graph the d.f. of X. Find $F_X(5.3)$.

   b. Find the p.f. of X and graph it. Find $Q_X(.75)$.

1.3. Suppose the p.f. of X is $Q(y) = y^2$ (for y between 0 and 1).

   a. Find $f(x)$, the p.d.f. of X, and graph it. (Note that your answer should not have any y variables in it — they need to be eliminated.)

   b. Find $F(x)$, the d.f. of X, and graph it.

   c. Using $Q(y)$, at what x is the probability below x equal to 0.9?

   d. Using $F(x)$, at what x is the probability below x equal to 0.9?

1.4. Identify the p.f. $Q(y) = y^2$ as a member of GLD family, i.e., what are the values of $\lambda_1, \lambda_2, \lambda_3, \lambda_4$ (see (1.2.1))? Then, use Theorem 1.2.2 to find $f(x)$. Do your results agree with those in Theorem 1.4.23 as to when the p.d.f. is positive? Does Corollary 1.3.9 say this is a valid distribution case of the GLD?

1.5. Suppose that X has a valid GLD distribution of the form (1.2.1), i.e., with p.f.
$$Q(y) = \lambda_1 + (y^{\lambda_3} - (1-y)^{\lambda_4})/\lambda_2.$$

In their work on GLDs, Freimer, Kollia, Mudholkar, and Lin (1988) took the p.f. to be

$$Q^*(y) = a_1 + (y^{a_3} - 1)/(a_2 a_3) - ((1-y)^{a_4} - 1)/(a_2 a_4).$$

Suppose that $Q(y) = Q^*(y)$ for all y between 0 and 1. Then what is the relationship between the vectors $(\lambda_1, \lambda_2, \lambda_3, \lambda_4)$ and (a_1, a_2, a_3, a_4)? If you cannot find a relationship that makes $Q(y) = Q^*(y)$ for all y, then in what sense is $Q^*(y)$ a "GLD", or is it simply another similar, but different, family of p.f.s?

1.6. Joiner and Rosenblatt (1971) used the p.f.

$$Q_1(y) = b_1 + y^{b_3}/b_2 - (1-y)^{b_4}/b_5.$$

Can you find relationships that make $Q(y) = Q_1(y)$? That make $Q_1(y) = Q^*(y)$ (for Q, see (1.2.1); for Q^*, see Problem 1.5)?

1.7. Hogben (1963) and Shapiro and Wilk (1965) did sampling studies (Monte Carlo experiments) using the p.f.

$$Q_2(y) = y^{c_3}/c_2 - (1-y)^{c_4}.$$

Relate this to the p.f.s Q, Q^*, and Q_1 (of (1.2.1), Problem 1.5, and Problem 1.6, respectively).

1.8. Gilchrist (1997) states the result: If $R(y)$ and $S(y)$ are each individually valid percentile functions (that may involve various parameters), then

$$T(y) = A + B(R(y) + S(y))$$

is also a valid percentile function. Prove this result.

Chapter 2

Fitting Distributions and Data with the GLD via the Method of Moments

In most practical applications, when constructing a statistical model we do not know the appropriate probability distribution (or do not know it fully). **If the appropriate probability distribution is** fully **known** (e.g., if it is known that X follows the normal distribution with mean 5 feet and standard deviation 6 inches), **then this distribution should be used** in the model. However, if a variable such as the height of females in a certain population is stated in the literature to be normal in distribution with population mean 5 feet and standard deviation 6 inches, **very often these are only estimates obtained from a sample.** Note that the normal distribution for height would be **inappropriate formally** since it gives $P(X < 0) > 0$, when we know that X cannot be negative. This need not be a reason to reject the normal model since for this model

$$P(X < 0) = \Phi(-10) = 0.7619855 \times 10^{-23}$$

(National Bureau of Standards (1953)), and a model that comes close to the true value of zero for the probability is **acceptable as long as it is used intelligently** by realizing that we have a close approximation to the true model, and do not make claims such as "persons of negative height will occur once each $1/(0.7619855 \times 10^{-23}) = 1.3 \times 10^{23}$ people." While such a claim is absurd, and no reasonable person would think a model bad because an unreasonable person could use it to make such a statement, the adage "What is so uncommon as common sense?" is one we have found to have much truth to it.

Even if we know, by theoretical derivation from reasonable assumptions, for example, that X is normal in distribution, we will often not know its parameters. These will need to be estimated from whatever data is available. Suppose that we have a set of data $X_1, X_2, \ldots, X_n$ that are independent and identically distributed random variables. If the data is normally distributed, i.e., if X_i is $N(\mu, \sigma^2)$ for $i = 1, 2, \ldots, n$, then the mean μ and the variance σ^2 are usually estimated, respectively, by

$$\hat{\mu} \equiv \bar{X} = \frac{X_1 + \ldots + X_n}{n}$$

and

$$\hat{\sigma}^2 \equiv s^2 = \sum_{i=1}^{n} (X_i - \bar{X})^2 / (n - 1).$$

Note that a random variable X that is $N(\mu, \sigma^2)$ has

$$\begin{cases} \text{(measure of center)} = E(X) = \mu, \\ \text{(measure of variability)} = E(X - \mu)^2 = \sigma^2, \\ \text{(measure of skewness)} = E(X - \mu)^3 / \sigma^3 = 0, \\ \text{(measure of kurtosis)} = E(X - \mu)^4 / \sigma^4 = 3. \end{cases}$$

If a random variable Y has a distribution other than the normal, we might attempt to approximate it by a random variable X that is $N(\mu, \sigma^2)$ for some μ and σ^2. We can do this successfully for center $E(Y)$ and variability $\text{Var}(Y) = E(Y - E(Y))^2$ by choosing μ and σ^2 in such a way as to match the center and variability of Y with the same center and variability for X. However, after that is done there are no free parameters in the distribution of X, and (unless the skewness and kurtosis of Y are, respectively, 0 and 3) we will not be able to match them in X. Hence, the normal family of distributions cannot be used to match data successfully unless the data is symmetric (so its skewness is 0) and has tail weight similar to that of the normal (so that its kurtosis is near 3). For this reason, families of distributions with additional parameters are often used, allowing us to match more than the center and the variability of Y.

In order to find a moment-based $\text{GLD}(\lambda_1, \lambda_2, \lambda_3, \lambda_4)$ fit to a given dataset $X_1, X_2, \ldots, X_n$, we determine the first four moments ($\bar{X}$, and the second, third, and fourth central moments) of $X_1, X_2, \ldots, X_n$, set these equal to their $\text{GLD}(\lambda_1, \lambda_2, \lambda_3, \lambda_4)$ counterparts, and solve the resulting equations for $\lambda_1, \lambda_2, \lambda_3, \lambda_4$.

In Section 2.1 we consider the first four moments of $\text{GLD}(\lambda_1, \lambda_2, \lambda_3, \lambda_4)$ distributions and in Section 2.2 we determine the possible values that these moments can attain. Fitting a $\text{GLD}(\lambda_1, \lambda_2, \lambda_3, \lambda_4)$ through the method of moments is developed in Section 2.3. Applications of these results for approximating some well-known distributions, and for fitting a $\text{GLD}(\lambda_1, \lambda_2, \lambda_3, \lambda_4)$ to a dataset, are developed in Sections 2.4 and 2.5, respectively.

2.1 The Moments of the GLD Distribution

In this section we develop expressions for the moments of $\text{GLD}(\lambda_1, \lambda_2, \lambda_3, \lambda_4)$ random variables. We start by setting $\lambda_1 = 0$ to simplify this task; next, we obtain the non-central moments of the $\text{GLD}(\lambda_1, \lambda_2, \lambda_3, \lambda_4)$; and finally, we derive the central $\text{GLD}(\lambda_1, \lambda_2, \lambda_3, \lambda_4)$ moments.

Theorem 2.1.1. *If X is a $\text{GLD}(\lambda_1, \lambda_2, \lambda_3, \lambda_4)$ random variable, then $Z = X - \lambda_1$ is $\text{GLD}(0, \lambda_2, \lambda_3, \lambda_4)$.*

Proof. Since X is $\text{GLD}(\lambda_1, \lambda_2, \lambda_3, \lambda_4)$,

$$Q_X(y) = \lambda_1 + \frac{y^{\lambda_3} - (1-y)^{\lambda_4}}{\lambda_2},$$

and

$$F_{X-\lambda_1}(x) = P[X - \lambda_1 \le x] = P[X \le x + \lambda_1] = F_X(x + \lambda_1). \quad (2.1.2)$$

If we set $F_X(x + \lambda_1) = y$, we obtain

$$x + \lambda_1 = Q_X(y) = \lambda_1 + \frac{y^{\lambda_3} - (1-y)^{\lambda_4}}{\lambda_2}, \quad x = Q_{X-\lambda_1}(y). \quad (2.1.3)$$

From (2.1.2) we also have $F_{X-\lambda_1}(x) = y$ which with (2.1.3) yields

$$Q_{X-\lambda_1}(y) = x = \frac{y^{\lambda_3} - (1-y)^{\lambda_4}}{\lambda_2},$$

proving that $X - \lambda_1$ is $\text{GLD}(0, \lambda_2, \lambda_3, \lambda_4)$. ∎

Having established λ_1 as a location parameter, we now determine the non-central moments (when they exist) of the $\text{GLD}(\lambda_1, \lambda_2, \lambda_3, \lambda_4)$.

Theorem 2.1.4. *If Z is $\text{GLD}(0, \lambda_2, \lambda_3, \lambda_4)$, then $E(Z^k)$, the expected value of Z^k, is given by*

$$E(Z^k) = \frac{1}{\lambda_2^k} \sum_{i=0}^{k} \left[\binom{k}{i} (-1)^i \beta(\lambda_3(k-i) + 1, \lambda_4 i + 1) \right] \quad (2.1.5)$$

where $\beta(a, b)$ is the beta function defined by

$$\beta(a, b) = \int_0^1 x^{a-1}(1-x)^{b-1} \, dx. \quad (2.1.6)$$

Proof.

$$E(Z^k) = \int_{-\infty}^{\infty} z^k f(z)\, dz = \int_0^1 (Q(y))^k\, dy \tag{2.1.7}$$

$$= \int_0^1 \left(\frac{y^{\lambda_3} - (1-y)^{\lambda_4}}{\lambda_2}\right)^k dy = \frac{1}{\lambda_2^k} \int_0^1 \left(y^{\lambda_3} - (1-y)^{\lambda_4}\right)^k dy.$$

By the binomial theorem,

$$\left(y^{\lambda_3} - (1-y)^{\lambda_4}\right)^k = \sum_{i=0}^k \left[\binom{k}{i}(y^{\lambda_3})^{k-i}(-(1-y)^{\lambda_4})^i\right]. \tag{2.1.8}$$

Using (2.1.8) in the last expression of (2.1.7), we get

$$E(Z^k) = \frac{1}{\lambda_2^k} \sum_{i=0}^k \left[\binom{k}{i}(-1)^i \int_0^1 y^{\lambda_3(k-i)}(1-y)^{\lambda_4 i}\, dy\right]$$

$$= \frac{1}{\lambda_2^k} \sum_{i=0}^k \left[\binom{k}{i}(-1)^i \beta(\lambda_3(k-i)+1, \lambda_4 i + 1)\right],$$

completing the proof of the theorem. ∎

Before continuing with our investigation of the GLD($\lambda_1, \lambda_2, \lambda_3, \lambda_4$) moments, we note three properties of the beta function that will be useful in our subsequent work.

Properties of the beta function

1. The integral in (2.1.6) that defines the beta function will converge if and only if a and b are positive (this can be verified by choosing c from the $(0,1)$ interval and considering the integral over the subintervals $(0,c)$ and $(c,1)$).

2. When a and b are positive, $\beta(a,b) = \beta(b,a)$. Using the substitution $y = 1 - x$ in the integral for $\beta(a,b)$ will transform it to the integral for $\beta(b,a)$.

3. By direct evaluation of the integral in (2.1.6), it can be determined that for $u > -1$,

$$\beta(u+1, 1) = \beta(1, u+1) = \frac{1}{u+1}. \tag{2.1.9}$$

The first of these observations, along with (2.1.5) of Theorem 2.1.4, helps us determine the conditions under which the $GLD(\lambda_1, \lambda_2, \lambda_3, \lambda_4)$ moments exist.

Corollary 2.1.10. *The k-th $GLD(\lambda_1, \lambda_2, \lambda_3, \lambda_4)$ moment exists if and only if $\lambda_3 > -1/k$ and $\lambda_4 > -1/k$.*

Proof. From Theorem 2.1.1, $E(X^k)$ will exist if and only if $E(Z^k) = E((X - \lambda_1)^k)$ exists, which, by Theorem 2.1.4, will exist if and only if

$$\lambda_3(k - i) + 1 > 0 \text{ and } \lambda_4 i + 1 > 0, \quad \text{for } i = 0, 1, \ldots, k.$$

This condition will prevail if and only if $\lambda_3 > -1/k$ and $\lambda_4 > -1/k$. ∎

Since, ultimately, we are going to be interested in the first four moments of the $GLD(\lambda_1, \lambda_2, \lambda_3, \lambda_4)$, we will **need to impose the condition $\lambda_3 > -1/4$ and $\lambda_4 > -1/4$ throughout the remainder of this chapter.** The next theorem gives an explicit formulation of the first four centralized $GLD(\lambda_1, \lambda_2, \lambda_3, \lambda_4)$ moments.

Theorem 2.1.11. *If X is $GLD(\lambda_1, \lambda_2, \lambda_3, \lambda_4)$ with $\lambda_3 > -1/4$ and $\lambda_4 > -1/4$, then its first four moments, $\alpha_1, \alpha_2, \alpha_3, \alpha_4$ (mean, variance, skewness, and kurtosis, respectively), are given by*

$$\alpha_1 = \mu = E(X) = \lambda_1 + \frac{A}{\lambda_2}, \tag{2.1.12}$$

$$\alpha_2 = \sigma^2 = E\left[(X - \mu)^2\right] = \frac{B - A^2}{\lambda_2^2}, \tag{2.1.13}$$

$$\alpha_3 = E(X - E(X))^3/\sigma^3 = \frac{C - 3AB + 2A^3}{\lambda_2^3 \sigma^3}, \tag{2.1.14}$$

$$\alpha_4 = E(X - E(X))^4/\sigma^4 = \frac{D - 4AC + 6A^2B - 3A^4}{\lambda_2^4 \sigma^4}, \tag{2.1.15}$$

where

$$A = \frac{1}{1 + \lambda_3} - \frac{1}{1 + \lambda_4}, \tag{2.1.16}$$

$$B = \frac{1}{1 + 2\lambda_3} + \frac{1}{1 + 2\lambda_4} - 2\beta(1 + \lambda_3, 1 + \lambda_4), \tag{2.1.17}$$

$$C = \frac{1}{1 + 3\lambda_3} - \frac{1}{1 + 3\lambda_4} - 3\beta(1 + 2\lambda_3, 1 + \lambda_4)$$
$$+ 3\beta(1 + \lambda_3, 1 + 2\lambda_4), \tag{2.1.18}$$

$$D = \frac{1}{1 + 4\lambda_3} + \frac{1}{1 + 4\lambda_4} - 4\beta(1 + 3\lambda_3, 1 + \lambda_4)$$
$$+ 6\beta(1 + 2\lambda_3, 1 + 2\lambda_4) - 4\beta(1 + \lambda_3, 1 + 3\lambda_4). \tag{2.1.19}$$

Proof. Let Z be a GLD$(0, \lambda_2, \lambda_3, \lambda_4)$ random variable. By Theorem 2.1.1,

$$E(X^k) = E((Z + \lambda_1)^k).$$

We first express $E(Z^i)$, for $i = 1, 2, 3$, and 4, in terms of A, B, C, and D. To do this for $E(Z)$, we use Theorem 2.1.4 to obtain

$$E(Z) = \frac{1}{\lambda_2} \left(\beta(\lambda_3 + 1, 1) - \beta(1, \lambda_4 + 1) \right),$$

and from (2.1.9) we get

$$E(Z) = \frac{1}{\lambda_2} \left(\frac{1}{\lambda_3 + 1} - \frac{1}{\lambda_4 + 1} \right) = \frac{A}{\lambda_2}. \tag{2.1.20}$$

For $E(Z^2)$ we again use Theorem 2.1.4 and the simplification allowed by (2.1.9) to get

$$E(Z^2) = \frac{1}{\lambda_2^2} \left(\beta(2\lambda_3 + 1, 1) - \beta(\lambda_3 + 1, \lambda_4 + 1) + \beta(1, 2\lambda_4 + 1) \right)$$
$$= \frac{1}{\lambda_2^2} \left(\frac{1}{2\lambda_3 + 1} - \frac{1}{2\lambda_4 + 1} - 2\beta(\lambda_3 + 1, \lambda_4 + 1) \right) = \frac{B}{\lambda_2^2}. \tag{2.1.21}$$

Similar arguments, with somewhat more complicated algebraic manipulations, for $E(Z^3)$ and $E(Z^4)$ produce

$$E(Z^3) = \frac{C}{\lambda_2^3} \tag{2.1.22}$$

$$E(Z^4) = \frac{D}{\lambda_2^4}. \tag{2.1.23}$$

We now use (2.1.20) to derive (2.1.12):

$$\alpha_1 = E(X) = E(Z + \lambda_1) = \lambda_1 + E(Z) = \lambda_1 + \frac{A}{\lambda_2}.$$

Next, we consider (2.1.13):

$$\alpha_2 = E(X^2) - \alpha_1^2 = E((Z + \lambda_1)^2) - \alpha_1^2$$
$$= E(Z^2) + 2\lambda_1 E(Z) + \lambda_1^2 - \alpha_1^2. \tag{2.1.24}$$

Substituting A/λ_2 for $E(Z)$ and $\lambda_1 + A/\lambda_2$ for α_1 in 2.1.24 and using (2.1.21), we get

$$\alpha_2 = E(Z^2) - \frac{A^2}{\lambda_2^2} = \frac{B - A^2}{\lambda_2^2}.$$

The derivations of (2.1.14) and (2.1.15) are similar but algebraically more involved. ∎

Corollary 2.1.25. *If $\alpha_1, \alpha_2, \alpha_3, \alpha_4$ are the first four moments of* $\mathrm{GLD}(\lambda_1, \lambda_2, \lambda_3, \lambda_4)$, *then the first four moments of* $\mathrm{GLD}(\lambda_1, \lambda_2, \lambda_4, \lambda_3)$ *will be*

$$\alpha_1 - \frac{2A}{\lambda_2}, \qquad \alpha_2, \qquad -\alpha_3, \qquad \alpha_4. \tag{2.1.26}$$

Proof. The exchange of λ_3 and λ_4 in the expressions for A, B, C, and D in (2.1.16) through (2.1.20) changes the signs of A and C and leaves B and D intact. Thus from (2.1.12), the first moment of $\mathrm{GLD}(\lambda_1, \lambda_2, \lambda_4, \lambda_3)$ will be $\lambda_1 - A/\lambda_2 = \alpha_1 - 2A/\lambda_2$.

Since B and A^2 are not affected by the exchange of λ_3 and λ_4, from (2.1.13), α_2 will be the second moment of $\mathrm{GLD}(\lambda_1, \lambda_2, \lambda_4, \lambda_3)$. C, A, and A^3 of (2.1.14) all change signs when λ_3 and λ_4 are switched, making $-\alpha_3$ the third moment of $\mathrm{GLD}(\lambda_1, \lambda_2, \lambda_4, \lambda_3)$. Since D, AC, A^2B, and A^4 of (2.1.15) are not affected by the exchange of λ_3 and λ_4, $\mathrm{GLD}(\lambda_1, \lambda_2, \lambda_4, \lambda_3)$ will have α_4 for its fourth moment. ∎

2.2   The (α_3^2, α_4)-**Space Covered by the GLD Family**

If a random variable Y has a distribution other than the GLD, we might try to approximate it by a random variable X that is $\mathrm{GLD}(\lambda_1, \lambda_2, \lambda_3, \lambda_4)$ for some $\lambda_1, \lambda_2, \lambda_3, \lambda_4$. Suppose that the first four moments of Y are $\alpha_1 = \mu$, $\alpha_2 = \sigma^2$, α_3, and α_4. If we can choose λ_3, λ_4 so that a $\mathrm{GLD}(0, 1, \lambda_3, \lambda_4)$

has third and fourth moments α_3 and α_4, then we can let λ_1 and λ_2 be solutions of the equations

$$\mu = \lambda_1 + \frac{A}{\lambda_2}, \quad \sigma^2 = \frac{B - A^2}{\lambda_2^2}. \tag{2.2.1}$$

It follows from Theorem 2.1.11 that the resulting λ_1, λ_2, λ_3, λ_4 specify a GLD with the desired first four moments. Here, we note that A, B, C, D are functions only of λ_3, λ_4, and that (2.2.1) can be solved for any μ and any $\sigma^2 > 0$. We have, therefore, established the following consequence of Theorem 2.1.11.

Corollary 2.2.2. *The* GLD$(\lambda_1, \lambda_2, \lambda_3, \lambda_4)$ *can match any first two moments* μ *and* σ^2, *and some third and fourth moments* α_3 *and* α_4.

The larger the set of (α_3, α_4) that the GLD$(\lambda_1, \lambda_2, \lambda_3, \lambda_4)$ can generate, the more useful the GLD$(\lambda_1, \lambda_2, \lambda_3, \lambda_4)$ family will be in fitting a broad range of datasets and approximating a variety of other random variables. So we next consider the spectrum of values that α_3 and α_4 can attain. From Corollary 2.1.25, we know that if (α_3, α_4) can be attained, then so can $(-\alpha_3, \alpha_4)$ (by switching λ_3 and λ_4), allowing us to consider the (α_3^2, α_4)-space associated with the GLD$(\lambda_1, \lambda_2, \lambda_3, \lambda_4)$.

Figures 2.2–1, 2.2–2, and 2.2–3 show the (α_3^2, α_4) contour plots for (λ_3, λ_4) from Regions 3, 4, and 5 and 6, respectively (recall that these regions were defined in Section 1.3 and illustrated in Figure 1.3–1). The curves in Figure 2.2–1 are associated with a sequence of values of λ_4, with λ_3 ranging on the interval $(0, 15)$ for each of these choices of λ_4. For example, the curve labeled "0.02" is obtained by plotting the (α_3^2, α_4) pairs when λ_4 is set to 0.02 and λ_3 is taken from the interval $(0, 15)$. All the curves of Figure 2.2–1 are obtained in a similar manner with $\lambda_4 = 0.02, 0.07, 0.12, \ldots, 0.52, 0.6, 0.7, \ldots, 1.0$ and λ_3 from the interval $(0, 15)$.

The construction of Figure 2.2–2 is similar to that of Figure 2.2–1, with (λ_3, λ_4) taken from Region 4. Note that in this case we must have

$$-1/4 < \lambda_3 < 0 \quad \text{and} \quad -1/4 < \lambda_4 < 0$$

since otherwise (see Corollary 2.1.10) α_3 or α_4 or both may not exist. Some of the λ_4 values associated with these curves are given in Figure 2.2–2; the other values of λ_4 are $-0.0125, -0.0250, \ldots, -0.075$. In Figure 2.2–3 the roles of λ_3 and λ_4 are reversed in the sense that λ_3 is fixed (to the values shown in the figure) and λ_4 is allowed to range upward (in the case of Region 5) from the boundary that defines the region (see Theorem 1.3.33).

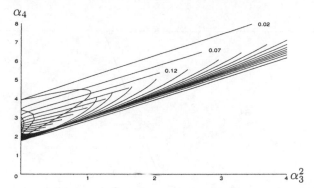

Figure 2.2–1. (α_3^2, α_4)-space generated by (λ_3, λ_4) from Region 3 (see (1.3.13)).

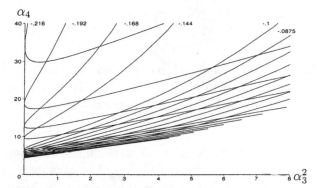

Figure 2.2–2. (α_3^2, α_4)-space generated by (λ_3, λ_4) from Region 4 (see (1.3.14)).

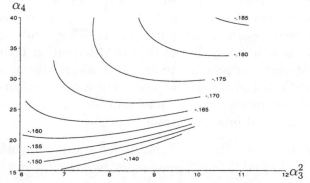

Figure 2.2–3. (α_3^2, α_4)-space generated by (λ_3, λ_4) from Regions 5 and 6 (defined in Theorem 1.3.33).

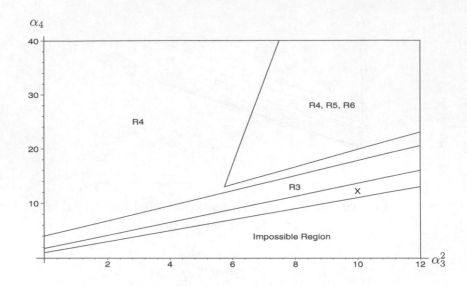

Figure 2.2–4. Regions of (α_3^2, α_4) moment-space that the GLD
can attain (R3, R4, R5, R6); cannot attain (X);
impossible for any distribution.

Figure 2.2–4 gives a comprehensive view of the connection between the
GLD (α_3^2, α_4)-space and the regions of (λ_3, λ_4). The area marked "Impossible
Region" is where $\alpha_4 \leq 1 + \alpha_3^2$, an impossibility since the inequality

$$\alpha_4 > 1 + \alpha_3^2 \tag{2.2.3}$$

holds for all distributions (it is less well-known than the inequality $E(X^2) \geq \mu^2$, which follows from $\mathrm{Var}(X) = E(X^2) - \mu^2 \geq 0$). Moreover, since $\alpha_4 > 1 + \alpha_3^2$ always holds, α_4 necessarily exceeds 1. This result, given in some
classical books, is well-known in the field of distribution fitting but is not
covered in most texts on probability and statistics. Moreover, we are not
aware of any texts that give a simple proof. For a brief indication of how an
advanced proof may proceed, see Kendall and Stuart (1969), p. 92, Exercise
3.19. We now state and prove this result.

Theorem 2.2.4.   *For any r.v. X for which these moments exist, $\alpha_4 > 1 + \alpha_3^2$. (Note that equality is not possible for any continuous distribution.)*

Proof. Since

$$\alpha_i = \frac{E(X - E(X))^i}{\sigma_X^i} = E\left(\frac{X - E(X)}{\sigma_X}\right)^i,$$

we can assume without loss of generality that $E(X) = 0$ and $\sigma_X = 1$, so that $E(X^2) = 1$ (because α_i involves only $X^* = (X - E(X))/\sigma_X$, which has $E(X^*) = 0$ and $E(X^{*2}) = 1$). Now the Schwarz inequality (see, for example, Dudewicz and Mishra (1988), p. 240, Theorem 5.3.23) says that for any r.v.s, X and Y, for which the expectation exists, $(E(XY))^2 \leq E(X^2)E(Y^2)$. If we take the two r.v.s in the Schwarz inequality to be X and $X^2 - 1$, then

$$\left(E(X(X^2 - 1))\right)^2 \leq E(X^2)E((X^2 - 1)^2)$$

$$\left(E(X^3 - X)\right)^2 \leq 1 \cdot E(X^4 - 2X^2 + 1)$$

$$\left(E(X^3) - 0\right)^2 \leq E(X^4) - 2 + 1$$

$$\alpha_3^2 \leq \alpha_4 - 1.$$

The proof will be complete when we show that equality is not possible for any continuous distribution.

The Schwarz inequality also asserts that equality occurs if and only if for some constant, a, we have $Y = aX$. In our case, this implies that equality holds if and only if $X^2 - 1 = aX$ or

$$X = \frac{a \pm \sqrt{a^2 + 4}}{2}.$$

Thus, we can have equality if and only if X is a r.v. that takes on only the two values

$$\frac{a - \sqrt{a^2 + 4}}{2} \quad \text{and} \quad \frac{a + \sqrt{a^2 + 4}}{2}.$$

Suppose these values are attained with probabilities p and $1 - p$, respectively $(0 \leq p \leq 1)$. Then $E(X) = 0$ implies

$$p = +\frac{1}{2}\left(1 + \frac{a}{\sqrt{a^2 + 4}}\right). \tag{2.2.5}$$

Since $0 \leq p \leq 1$, $-1 \leq a/\sqrt{a^2 + 4} \leq 1$. And $E(X^2) = 1$ also implies (2.2.5). So $\alpha_3^2 = \alpha_4 - 1$ iff for some a

$$P\left(X = \frac{a - \sqrt{a^2 + 4}}{2}\right) = \frac{1}{2}\left(1 + \frac{a}{\sqrt{a^2 + 4}}\right)$$

and

$$P\left(X = \frac{a + \sqrt{a^2 + 4}}{2}\right) = \frac{1}{2}\left(1 - \frac{a}{\sqrt{a^2 + 4}}\right).$$

For general a, $(\alpha_3, \alpha_4) = (a, a^2 + 1)$. When $a = 0$, $P(X = -1) = 0.5 = P(X = +1)$ and $(\alpha_3, \alpha_4) = (0, 1)$. ∎

Immediately above the Impossible Region in Figure 2.2–4 there is a narrow "sliver" marked by "X." The GLD$(\lambda_1, \lambda_2, \lambda_3, \lambda_4)$ does not produce (α_3^2, α_4) in this region. We saw in Figures 2.2–1, 2.2–2, and 2.2–3 that Regions 3, 4, 5, and 6 cannot yield points in area X. That Regions 1 and 2 also cannot follows from the fact that $\lambda_3 \leq -1$ (Region 1) and $\lambda_4 \leq -1$ (Region 2) violate the conditions $\lambda_3 > -0.25$ and $\lambda_4 > -0.25$ needed for the third and fourth moments to exist (see Corollary 2.1.10). Other distributions (e.g., the beta distribution) do have their (α_3^2, α_4) in this area. In Chapter 3 we give an extension of the GLD that covers this portion of (α_3^2, α_4)-space. (While area X of Figure 2.2–4 may look "small," it is important in a variety of applications.)

The remaining portions of Figure 2.2–4 are marked with "R3," designating that (λ_3, λ_4) has to be chosen from Region 3 for this portion of (α_3^2, α_4)-space, and with "R4 R5 R6," designating that (λ_3, λ_4) is to be chosen from one of Regions 4, 5 or 6 to generate (α_3^2, α_4) in this area. The boundaries between the various portions of Figure 2.2–4 are drawn reasonably accurately, except for the boundaries that enclose the "R4, R5, R6" area; these are rough approximations obtained through numeric computations from the curves in Figure 2.2–3.

The (α_3^2, α_4)-space covered by the GLD already includes the moments of such distributions as the uniform, Student's t, normal, Weibull, gamma, lognormal, exponential, and some beta distributions, among others. Thus, it is a rich class in terms of moment coverage. To put this in context, we show in **Figure 2.2–5** the (α_3^2, α_4) pairs associated with a number of well-known distributions. **The shaded region is the region covered by the (α_3^2, α_4) pairs of the** GLD$(\lambda_1, \lambda_2, \lambda_3, \lambda_4)$.

The lines that are designated by "W," "L-N," "G," and "S" (the latter refers to the line defined by $\alpha_3^2 = 0$) show the (α_3^2, α_4) pairs for the Weibull, lognormal, gamma, and Student's t distributions, respectively. The area designated by **"B E T A   R E G I O N"** shows the (α_3^2, α_4) **points that can be produced by the beta distribution. This region extends from the Impossible Region to slightly beyond the line marked for the lognormal distribution**. The point designated by a small square with label "u" and located at $(0, 1.8)$ represents the uniform distribution; the point at $(0, 3)$ labeled with "n" represents the normal distribution, $N(\mu, \sigma^2)$; and the point located at $(4, 9)$ and labeled with "e" gives the (α_3^2, α_4) point associated with the exponential distribution.

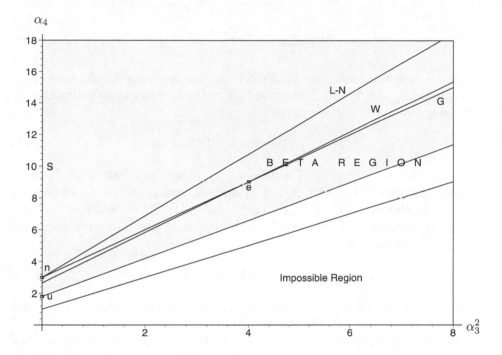

Figure 2.2–5. (α_3^2, α_4) points of some distributions (the **shaded region** consists of GLD (α_3^2, α_4) points).

2.3 Fitting the GLD Through the Method of Moments

As stated at the beginning of the chapter, our intention is to fit a GLD to a dataset by equating $\alpha_1, \alpha_2, \alpha_3, \alpha_4$ to $\hat{\alpha}_1, \hat{\alpha}_2, \hat{\alpha}_3, \hat{\alpha}_4$, the sample statistics corresponding to $\alpha_1, \alpha_2, \alpha_3, \alpha_4$, and solving the equations for $\lambda_1, \lambda_2, \lambda_3, \lambda_4$. For a dataset $X_1, X_2, \ldots, X_n$, the **sample moments** corresponding to $\alpha_1, \alpha_2, \alpha_3, \alpha_4$ are denoted $\hat{\alpha}_1, \hat{\alpha}_2, \hat{\alpha}_3, \hat{\alpha}_4$ and are defined by

$$\hat{\alpha}_1 = \overline{X} = \sum_{i=1}^{n} x_i/n, \tag{2.3.1}$$

$$\hat{\alpha}_2 = \hat{\sigma}^2 = \sum_{i=1}^{n} (X_i - \overline{X})^2/n, \tag{2.3.2}$$

$$\hat{\alpha}_3 = \sum_{i=1}^{n} (X_i - \overline{X})^3/(n\hat{\sigma}^3), \tag{2.3.3}$$

$$\hat{\alpha}_4 = \sum_{i=1}^{n} (X_i - \overline{X})^4 / (n\hat{\sigma}^4). \tag{2.3.4}$$

These are not the maximum likelihood estimators (those would have some ns replaced by $n-1$), but correspond to method-of-moments estimators.

Solving the system of equations

$$\alpha_i = \hat{\alpha}_i \qquad \text{for } i = 1, 2, 3, 4 \tag{2.3.5}$$

for $\lambda_1, \lambda_2, \lambda_3, \lambda_4$ is simplified somewhat by observing that A, B, C, D of (2.1.16) through (2.1.19) are free of λ_1 and λ_2, and λ_2 drops out of (2.1.14) and (2.1.15) because (see (2.1.13)) $\lambda_2^i \sigma^i = (B - A^2)^{i/2}$ for $i = 3$ and 4. Thus, α_3 and α_4 depend only on λ_3 and λ_4. Hence, if λ_3 and λ_4 can be obtained by solving the subsystem

$$\alpha_3 = \hat{\alpha}_3 \quad \text{and} \quad \alpha_4 = \hat{\alpha}_4 \tag{2.3.6}$$

of two equations in the two variables λ_3 and λ_4, then using (2.1.13) and (2.1.12) successively will yield λ_2 and λ_1.

Unfortunately, (2.3.6) is complex enough to prevent exact solutions, forcing us to appeal to numerical methods to obtain approximate solutions. Algorithms for finding numerical solutions to systems of equations such as (2.3.6) are generally designed to "search" for a solution by checking if an initial set of values ($\lambda_3 = \lambda_3^*$, $\lambda_4 = \lambda_4^*$ in the case of (2.3.6)) can be considered an approximate solution. This determination is made by checking if

$$\max(|\alpha_3 - \hat{\alpha}_3|, |\alpha_4 - \hat{\alpha}_4|) < \epsilon, \tag{2.3.7}$$

when $\lambda_3 = \lambda_3^*$ and $\lambda_4 = \lambda_4^*$. The positive number ϵ represents the accuracy associated with the approximation; if it is determined that the initial set of values $\lambda_3 = \lambda_3^*$, $\lambda_4 = \lambda_4^*$ does not provide a sufficiently accurate solution, the algorithm searches for a better choice of λ_3 and λ_4 and iterates this process until a suitable solution is discovered (i.e., one that satisfies (2.3.7)). In algorithms of this type there is no assurance that the algorithm will terminate successfully nor that greater accuracy will be attained in successive iterations. Therefore, such searching algorithms are designed to terminate (unsuccessfully) if (2.3.7) is not satisfied after a fixed number of iterations.

2.3.1   Fitting through Direct Computation

The outcome of searching algorithms (success or failure, and in the latter case the particular solution) depends on the equations themselves, the ϵ of

(2.3.7), the maximum number of iterations allowed, and the initial starting point for the search. Such algorithms usually have built-in specifications for ϵ and the maximal number of iterations, leaving the **choice of the starting point as the only real option for the user**. To get some insight into where to look for solutions (i.e., how to choose a starting point), consider a specific case where

$$\hat{\alpha}_1 = 0, \quad \hat{\alpha}_2 = 1, \quad \hat{\alpha}_3^2 = 0.025, \quad \hat{\alpha}_4 = 2. \tag{2.3.8}$$

It is clear from Figures 2.2–1, 2.2–2, and 2.2–3 that the only region of (λ_3, λ_4)-space that can produce a solution is Region 3 (this can also be observed from Figure 2.2–4). The equation $\alpha_4 = 2$, represented by a curve in (λ_3, λ_4), is shown in Figure 2.3–1. The (λ_3, λ_4) points on this curve satisfy the second equation; hence, they represent *potential* solutions to (2.3.6). The actual solutions will also have to be on the curve specified by $\alpha_3^2 = 0.025$. We see from the intersection of these two curves, shown in Figure 2.3–2, that there seem to be four solutions with (λ_3, λ_4) roughly

$$(0.02, 0.75), \ (0.1, 0.85), \ (0.5, 0.8), \ (2.25, 3.25) \tag{2.3.9}$$

and an additional four solutions when λ_3 and λ_4 are exchanged. The symmetry of solutions about the $\lambda_3 = \lambda_4$ line is due to the presence of two values of $\hat{\alpha}_3 = \pm\sqrt{0.025}$. In practice, of course, we have either $\hat{\alpha}_3 = +\sqrt{0.025}$ or $\hat{\alpha}_3 = -\sqrt{0.025}$ and then find 4 solutions for each case.

The procedure FindLambdasM, given in Appendix A, was devised specifically to produce solutions to (2.3.5). It requires $\hat{\alpha}_1, \hat{\alpha}_2, \hat{\alpha}_3, \hat{\alpha}_4$ and an initial (λ_3, λ_4) as input parameters. By default, a maximum of 10 iterations is used by FindLambdasM but this default can be reset if an additional parameter is passed to the program indicating the number of iterations that need to be used. The program uses $\epsilon = 10^{-5}$; to change the value of ϵ, the two occurrences of 0.00001 will have to be suitably adjusted in the program itself. When FindLambdasM terminates successfully, it returns the $\lambda_1, \lambda_2, \lambda_3, \lambda_4$ of the fitted distribution.

FindLambdasM, along with the other programs listed in Appendix A, is available on the CRC Press website through

http://www.crcpress.com.

This program is written in *Maple* and the command

> FindLambdasM([0, 1, sqrt(0.025), 2], 0.02, 0.75);

produces

[-1.332725170, .2919002955, .02981696805, .7181437677]

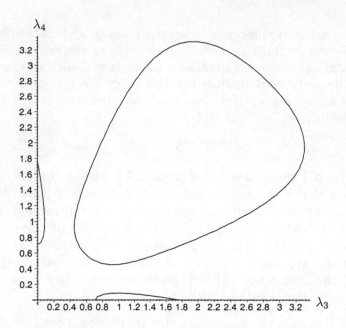

Figure 2.3–1. The contour curve for $\alpha_4 = 2$.

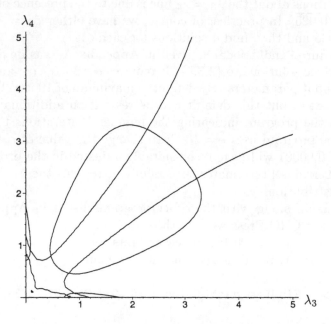

Figure 2.3–2. The contour curves for $\alpha_3^2 = 0.025$ and $\alpha_4 = 2$.

indicating that the fitted distribution has

$$(\lambda_1, \lambda_2, \lambda_3, \lambda_4) = (-1.332725170, .2919002955, .02981696805, .7181437677)$$

and $|\hat{\alpha}_i - \alpha_i| = 0$ (to 5 decimal places).

The search for this solution corresponds to $\hat{\alpha}_3 = +\sqrt{0.025}$ and it was initiated at $(0.02, 0.75)$. Using `FindLambdasM` with $\hat{\alpha}_3 = +\sqrt{0.025}$ and all four of the possibilities in (2.3.9) gives the following GLD fits.

$$\text{GLD}_1 = \text{GLD}(-1.33272663, .2919002520, .0298168180, .7181444379)$$
$$\text{GLD}_2 = \text{GLD}(.2562450422, .4992390619, .8014867557, .4640761247)$$
$$\text{GLD}_3 = \text{GLD}(1.189623323, .3463943247, .9383793840, .0776168109)$$
$$\text{GLD}_4 = \text{GLD}(.1267486891, .5357244180, 3.225675940, 2.283523402)$$

The p.d.f.s of these GLD fits are shown in Figure 2.3–3 where GLD_i is labeled with (i). In GLD_2, GLD_3 and GLD_4, $\lambda_3 > \lambda_4$ but in GLD_1, $\lambda_3 < \lambda_4$. It can be seen, perhaps with some difficulty, from Figure 2.3–3 that GLD_1 is skewed to the left while GLD_2, GLD_3, and GLD_4 are skewed to the right.

Note that **had the original $\hat{\alpha}_3^2$ and $\hat{\alpha}_4$ been 0.5 and 2**, respectively, we would easily determine from Figures 2.2–1, 2.2–2, and 2.2–3 (or, with some difficulty from Figure 2.2–4) that (2.3.6) would not have any solutions. This can be seen even more convincingly in Figure 2.3–4 which shows that the contour curves of the two equations do not intersect when (λ_3, λ_4) is in Region 3.

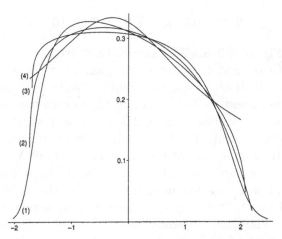

Figure 2.3–3. Four GLD fits for the $\hat{\alpha}_1, \hat{\alpha}_2, \hat{\alpha}_3, \hat{\alpha}_4$ specified by (2.3.8).

Figure 2.3–5 shows a family of **contour curves for Region 3 for** α_4 with values

$$1.825, \ 1.85, \ 1.9, \ 2, \ 2.1, \ 2.25, \ 2.5, \ 2.75, \ 3, \ 3.5, \ 4.$$

The curve associated with $\alpha_4 = 1.825$ consists of the innermost oval and the lowest branches along the λ_3 and λ_4 axes, the curve for $\alpha_4 = 1.85$ consists of the next larger oval and the next higher branches along the axes, and so on. Figure 2.3–6 gives **contour curves for Region 3 for** α_3^2 with values

$$0.005, \ 0.01, \ 0.015, \ 0.025, \ 0.05, \ 0.1, \ 0.2, \ 0.3, \ 0.5, \ 0.75, \ 1, \ 1.25, \ 1.5, \ 2.$$

The curve closest to the line $\lambda_3 = \lambda_4$ (not shown) and on either side of this line is associated with $\alpha_3^2 = 0.005$ and subsequent curves moving away from $\lambda_3 = \lambda_4$ represent increasing values of α_3^2. The "dense" set of curves in the lower left corner are branches of the curves in the larger portion of Figure 2.3–6. The one farthest from the origin is associated with $\alpha_3^2 = 0.005$ and is actually connected with the rest of the curve for $\alpha_3^2 = 0.005$. As α_3^2 increases the curves become disconnected and move closer to the origin. These two families of curves (the α_4 curves of Figure 2.3–5 and the α_3^2 curves of Figure 2.3–6) are shown in Figure 2.3–7 and provide a rough guide for determining initial searching points when Region 3 solutions are sought.

The **contour plots for Region 4** with α_4 taking values

$$6, \ 6.5, \ 7, \ 8, \ 9, \ 10, \ 12, \ 14, \ 17, \ 20, \ 25, \ 30, \ 45, \ 65$$

and α_3^2 taking values

$$0.2, \ 0.4, \ 0.7, \ 1, \ 1.5, \ 2, \ 3, \ 5, \ 7, \ 10, \ 13, \ 17$$

are given in Figure 2.3–8. The curves that are open near the origin are associated with α_4 and those that are open away from the origin are the curves for α_3^2. In both cases, the smallest values (of α_4 or α_3^2) produce curves that are closest to $\lambda_3 = \lambda_4$ with the curves moving away from this line with increasing values. Figure 2.3–8 not only provides starting points for the search of solutions to (2.3.6), but it also indicates that, if a solution exists, it will be unique, except for the interchange of λ_3 and λ_4.

It seems from Figure 2.3–8 that there should be a unique (up to symmetry) GLD fit with (λ_3, λ_4) from Region 4 for the $\hat{\alpha}_1, \hat{\alpha}_2, \hat{\alpha}_3, \hat{\alpha}_4$ specified in (2.3.8). Moreover, the (λ_3, λ_4) associated with this fit should be close to the origin. The many attempts made through FindLambdasM to find a solution in Region 4 failed. When search algorithms such as the one implemented in FindLambdasM fail, they fail because the surfaces associated with

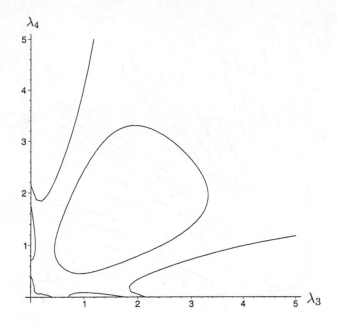

Figure 2.3–4. The contour curves for and $\alpha_3^2 = 0.5$ and $\alpha_4 = 2$.

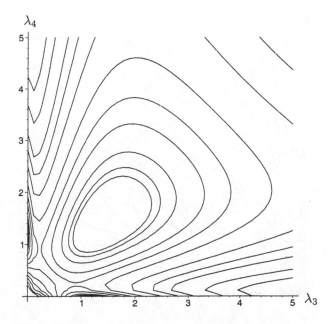

Figure 2.3–5. Contour curves of α_4 with (λ_3, λ_4) from Region 3.

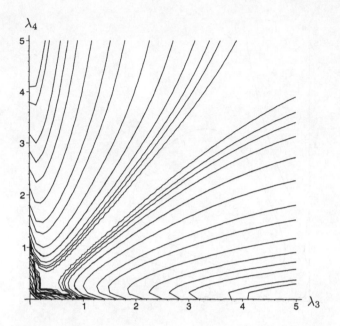

Figure 2.3–6. Contour curves of α_3^2 with (λ_3, λ_4) from Region 3.

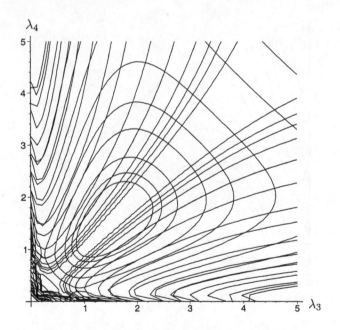

Figure 2.3–7. Contour curves of α_3^2 and α_4 with (λ_3, λ_4) from Region 3.

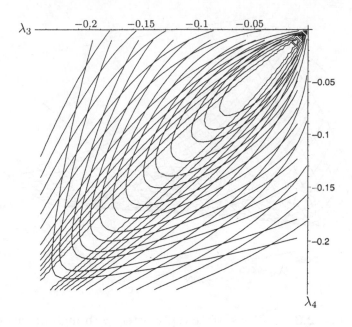

Figure 2.3–8. Contour curves of α_3^2 and α_4 with (λ_3, λ_4) from Region 4.

the equations have sharp corners or points where differentiability fails. This is not the case in our situation. FindLambdasM cannot find a solution in Region 4 because of the proximity of the solution to the origin. When α_3 and α_4, particularly α_4, are calculated with (λ_3, λ_4) near the origin, unless *very* high levels of computational precision are used, the computational errors at intermediate levels could get magnified throughout the search path of the algorithm. This phenomenon is illustrated in Figures 2.3–9 and 2.3–10. In Figure 2.3–9 the surface α_4 is plotted using 25 digits of precision for all computations (this is well beyond the precision allowed by most hardware-based floating point operations). We can see that the surface is smooth and well-behaved. In Figure 2.3–10 the same surface is plotted with only 10 digits of precision, a rather common level of precision in most computing environments. It is clear that substantial errors are produced when (λ_3, λ_4) is near the origin. If FindLambdasM is unable to obtain a solution within the specified number of iterations and error tolerance, ϵ, it gives the approximate values of $\lambda_1, \lambda_2, \lambda_3, \lambda_4$ that it has computed, along with an appropriate warning.

A final word of caution: when FindLambdasM returns $\lambda_1, \lambda_2, \lambda_3, \lambda_4$, there is no assurance that the GLD associated with $\lambda_1, \lambda_2, \lambda_3, \lambda_4$ is valid. Thus,

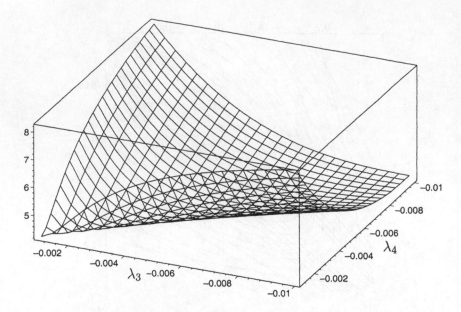

Figure 2.3–9. The surface α_4, plotted with high precision
computation, near the origin in Region 4.

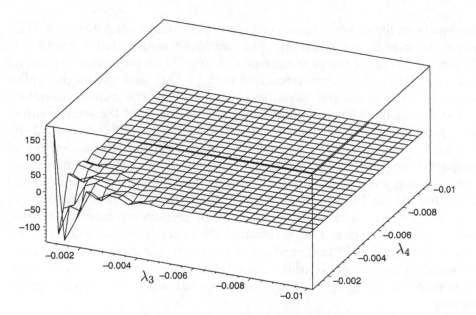

Figure 2.3–10. The surface α_4, plotted with ordinary precision
computation, near the origin in Region 4.

one needs to check that (λ_3, λ_4) is in one of the valid regions of Figure 1.3–1.

2.3.2 Fitting by the Use of Tables

Some readers may not have sufficient expertise in programming or adequate programming support to use the type of analysis that was illustrated in Section 2.3.1. For this reason, a number of investigators have provided tables for the estimation of $\lambda_1, \lambda_2, \lambda_3, \lambda_4$. The first of these was given by Ramberg and Schmeiser (1974); more comprehensive tables (in the sense of coverage of (α_3^2, α_4)-space) were provided subsequently by Ramberg, Tadikamalla, Dudewicz and Mykytka (1979); Cooley (1991) used greater computational precision to improve previous tables; and Dudewicz and Karian (1996) provide the most accurate and comprehensive tables to date. The latter are given in Appendix B. To capture as much precision as possible within the table of Appendix B, the notation a^b is used for the entries of the table to mean $a \times 10^{-b}$. For example, an entry of 0.1417^2 represents 0.001417.

Unless some simplifications are used, tabulated results for determining $\lambda_1, \lambda_2, \lambda_3, \lambda_4$ from $\hat{\alpha}_1, \hat{\alpha}_2, \hat{\alpha}_3, \hat{\alpha}_4$ would require a "four-dimensional" display, a decidedly impractical undertaking. To make the tabulation manageable, we first use not $(\hat{\alpha}_1, \hat{\alpha}_2, \hat{\alpha}_3, \hat{\alpha}_4)$ but $(0, 1, |\hat{\alpha}_3|, \hat{\alpha}_4)$ and obtain a solution $(\lambda_1(0,1), \lambda_2(0,1), \lambda_3, \lambda_4)$ to (2.3.5). Note that interchanging λ_3 and λ_4 would change the signs of A and C in (2.1.12) through (2.1.15), changing the sign of α_3 and necessitating a sign change for $\lambda_1(0,1)$. Therefore, when $\hat{\alpha}_3 < 0$, we interchange λ_3 and λ_4 and change the sign of $\lambda_1(0,1)$. Next, we obtain the solution to (2.3.5) associated with $\hat{\alpha}_1, \hat{\alpha}_2, \hat{\alpha}_3, \hat{\alpha}_4$ by setting

$$\lambda_1 = \lambda_1(0,1)\sqrt{\hat{\alpha}_2} + \hat{\alpha}_1 \quad \text{and} \quad \lambda_2 = \lambda_2(0,1)/\sqrt{\hat{\alpha}_2}.$$

We summarize this process in the GLD–M algorithm below.

Algorithm GLD–M: Fitting a GLD distribution to data by the method of moments.

GLD–M–1. Use (2.3.1) through (2.3.4) to compute $\hat{\alpha}_1, \hat{\alpha}_2, \hat{\alpha}_3, \hat{\alpha}_4$;

GLD–M–2. Find the entry point in a table of Appendix B closest to $(|\hat{\alpha}_3|, \hat{\alpha}_4)$;

GLD–M–3. Using $(|\hat{\alpha}_3|, \hat{\alpha}_4)$ from Step GLD–M–2 extract $\lambda_1(0,1), \lambda_2(0,1), \lambda_3,$ and λ_4 from the table;

GLD–M–4. If $\hat{\alpha}_3 < 0$, interchange λ_3 and λ_4 and change the sign of $\lambda_1(0,1)$;

GLD–M–5. Compute $\lambda_1 = \lambda_1(0,1)\sqrt{\hat{\alpha}_2} + \hat{\alpha}_1 \quad \text{and} \quad \lambda_2 = \lambda_2(0,1)/\sqrt{\hat{\alpha}_2}.$

To illustrate the use of Algorithm GLD–M and the table of Appendix B suppose that $\hat{\alpha}_1, \hat{\alpha}_2, \hat{\alpha}_3, \hat{\alpha}_4$ have been computed to have values

$$\hat{\alpha}_1 = 2, \quad \hat{\alpha}_2 = 3, \quad \hat{\alpha}_3 = -\sqrt{0.025}, \quad \hat{\alpha}_4 = 2. \qquad (2.3.10)$$

Note that $\hat{\alpha}_4$ has been taken to be the same as, and $\hat{\alpha}_3$ has been taken to be the negative of, the previous values from (2.3.8) used in the FindLambdasM procedure. Step GLD–M–1 is taken care of since $\hat{\alpha}_1, \hat{\alpha}_2, \hat{\alpha}_3, \hat{\alpha}_4$ is given. For Step GLD–M–2, we observe that $\hat{\alpha}_3 = -0.15811$; hence, the closest point to $(|\hat{\alpha}_3|, \hat{\alpha}_4)$ in the Table of Appendix B is $(0.15, 2.0)$, giving us

$$\lambda_1(0,1) = -1.3231, \quad \lambda_2(0,1) = 0.2934, \quad \lambda_3 = 0.03145, \quad \lambda_4 = 0.7203.$$

The instructions on the use of the table in Appendix B indicate that a superscript of b in a table entry designates a factor of 10^{-b}. In this case, an entry of 0.3145^1 for λ_3 indicates a value of $0.3145 \times 10^{-1} = 0.03145$. Since $\alpha_3 < 0$, Step GLD–M–4 readjusts these to

$$\lambda_1(0,1) = 1.3231, \quad \lambda_2(0,1) = 0.2934, \quad \lambda_3 = 0.7203, \quad \lambda_4 = 0.03145.$$

With the computations in Step GLD–M–5 we get

$$\lambda_1 = 4.2917, \quad \lambda_2 = 0.1694, \quad \lambda_3 = 0.7203, \quad \lambda_4 = 0.03145.$$

2.3.3   Limitations of the Method of Moments

The wide applicability of the methods developed in Sections 2.3.1 and 2.3.2 will be apparent when we use GLD distributions to approximate a number of commonly encountered distributions (Section 2.4) and when we fit GLD distributions to several datasets (Section 2.5). However, it is worth keeping in mind that most methods have limitations and we discuss the limitations associated with fitting a GLD through the method of moments here.

Algorithm GLD–M, through the table of Appendix B, enables us to fit a $\text{GLD}(\lambda_1, \lambda_2, \lambda_3, \lambda_4)$ when (α_3^2, α_4) is confined by

$$1.8(\hat{\alpha}_3^2 + 1) \leq \hat{\alpha}_4 \leq 1.8\hat{\alpha}_3^2 + 15. \qquad (2.3.11)$$

The upper restriction $\hat{\alpha}_4 \leq 1.8\hat{\alpha}_3^2 + 15$ is forced on us by limitations of table space and difficulties associated with computations when this restriction is removed. We can see from Figures 2.2–1 through 2.2–3 that the GLD is capable of generating distributions with (α_3^2, α_4) beyond this constraint (see also the shaded region of possible (α_3^2, α_4) pairs in Figure 2.2–5). If needed, it is quite likely that, perhaps with some difficulty, we would be able to find a suitable fit in this region.

The lower restriction of (2.3.11), $1.8(\hat{\alpha}_3^2+1) \leq \hat{\alpha}_4$, cannot be removed since it is a limitation of the $GLD(\lambda_1, \lambda_2, \lambda_3, \lambda_4)$ family of distributions. Recall (Theorem 2.2.4) that for all distributions we must have $\alpha_4 > 1 + \alpha_3^2$. Thus, while the upper restriction of (2.3.11) may be overcome through greater computational effort, the lower restriction eliminates the possibility of fitting a $GLD(\lambda_1, \lambda_2, \lambda_3, \lambda_4)$ when

$$1 + \hat{\alpha}_3^2 < \hat{\alpha}_4 < 1.8(\hat{\alpha}_3^2 + 1).$$

While analyses of actual data by Wilcox (1990), Pearson and Please (1975), and Micceri (1989), indicate that values of $|\hat{\alpha}_3|$ up to 4 and values of $\hat{\alpha}_4$ up to 50 are realistic, it is most common for data to produce (α_3^2, α_4) with

$$1 + \hat{\alpha}_3^2 < \hat{\alpha}_4 \leq 1.8\hat{\alpha}_3^2 + 15,$$

making the lower constraint of (2.3.11) a more serious limitation. In Chapter 3 we develop the EGLD system, the Extended GLD, to address this problem.

A different problem arises when we try to find a $GLD(\lambda_1, \lambda_2, \lambda_3, \lambda_4)$ approximation to a distribution when (some of) the first four moments of the distribution do not exist. This type of fitting problem will also arise in a less obvious form if we encounter data that is a random sample from such a distribution. We address this difficulty in Chapter 4 by devising a $GLD(\lambda_1, \lambda_2, \lambda_3, \lambda_4)$ fitting method that depends on percentiles rather than moments.

In terms of a preference between the two approaches discussed in Sections 2.3.1 (direct computation) and 2.3.2 (use of tables), we note that the unavailability of the proper computing environment and, to a lesser extent, simplicity are the principle advantages of using tables. There are, however, two disadvantages:

1. Because of length limitations, existing tables provide at most one solution even when multiple solutions may exist.

2. Results obtained through Algorithm GLD–M, because of their dependence on tables, are less accurate than estimations of $\lambda_1, \lambda_2, \lambda_3, \lambda_4$ by direct computation. We know, for example, from the GLD_1 fit of Section 2.3.1 that 0.7181 and 0.02982 would be more precise values for λ_3 and λ_4, respectively.

If one has access to a computational system that can provide solutions to equations like (2.3.5), it is possible to use the tables of Appendix B to determine a good starting point for the search so that an accurate solution may be obtained with considerable efficiency.

Of course, the ultimate criterion is **goodness of fit** of the fitted distribution to the true (unknown) distribution. Methods of assessing this are discussed in Section 2.5.1.

2.4 GLD Approximations of Some Well-Known Distributions

In Section 1.4 we saw the large variety of shapes that the GLD($\lambda_1, \lambda_2, \lambda_3, \lambda_4$) p.d.f. can attain. For the GLD($\lambda_1, \lambda_2, \lambda_3, \lambda_4$) to be useful for fitting distributions to data, it should be able to provide good fits to many of the distributions the data may come from. In this section we see that the GLD($\lambda_1, \lambda_2, \lambda_3, \lambda_4$) fits well many of the most important distributions.

We apply three checks on each occasion where we fit a GLD to a distribution. The **first check** considers the closeness of $\hat{f}(x)$, the approximating GLD p.d.f., to $f(x)$, the p.d.f. of the distribution being approximated. The proximity of $\hat{f}(x)$ to $f(x)$ is determined by approximating $\sup |\hat{f}(x) - f(x)|$. The p.d.f. of the distribution we will be approximating, $f(x)$, is available to us; therefore, there is no difficulty with computing $f(x)$. To compute $\hat{f}(x)$,

- we take 249 equispaced points $y_i = i/250$, for $i = 1, 2, \ldots, 249$ from the interval $(0, 1)$;

- using (1.2.1) we compute $x_i = Q(y_i)$ for $i = 1, 2, \ldots, 249$;

- using (1.2.3) we compute $\hat{f}(x_i)$ for $i = 1, 2, \ldots, 249$.

We now use the approximation

$$\sup |\hat{f}(x) - f(x)| \approx \max_{1 \leq i \leq 249} |\hat{f}(x) - f(x)|.$$

In actual practice we found that using 249 points does not limit the accuracy of the $\sup |\hat{f}(x) - f(x)|$ that we compute. We have obtained essentially the same (i.e., within 10^{-3}) values for $\sup |\hat{f}(x) - f(x)|$ when the 249 points have been increased to 4999 points.

For a **second check**, we look at the proximity of the approximating and approximated d.f.s, $\hat{F}(x)$ and $F(x)$, respectively. While p.d.f. differences are less easy to interpret, differences in d.f.s have an immediate meaning in the probability assigned to easy-to-interpret events. To check the closeness of

$\hat{F}(x)$ to $F(x)$, we follow the same idea (and use the same points x_i) as in the computation of $\hat{f}(x)$ and use the approximation

$$\sup |\hat{F}(x) - F(x)| \approx \max_{1 \leq i \leq 249} |\hat{F}(x) - F(x)|.$$

Again, in practice, $\sup |\hat{F}(x) - F(x)|$ does not change much (less than 10^{-4}) when a much larger number of points is used.

For the **third check** we consider the closeness of the $\hat{\alpha}_1, \hat{\alpha}_2, \hat{\alpha}_3, \hat{\alpha}_4$ of the fitted GLD to the $\alpha_1, \alpha_2, \alpha_3, \alpha_4$ of the distribution that is approximated. Since in all cases FindLambdasM will be used to obtain the approximating GLD, we are assured, subject to a warning message from FindLambdasM, that both $|\hat{\alpha}_3 - \alpha_3|$ and $|\hat{\alpha}_4 - \alpha_4|$ are less than 10^{-5}. From (2.1.12) and (2.1.13) we know that there are no difficulties associated with the computations of $\hat{\alpha}_1$ and $\hat{\alpha}_2$; therefore, we expect $|\hat{\alpha}_1 - \alpha_1|$ and $|\hat{\alpha}_2 - \alpha_2|$ to be no larger than 10^{-4}. Since the use of FindLambdasM takes care of this check, we do not explicitly mention it as we look for fits to distributions in the following sections.

2.4.1 The Normal Distribution

The normal distribution, with mean μ and variance σ^2 ($\sigma > 0$), $N(\mu, \sigma^2)$, has p.d.f.

$$f(x) = \frac{1}{\sigma \sqrt{2\pi}} \exp\left[-\frac{(x-\mu)^2}{2\sigma^2}\right], \quad -\infty < x < \infty.$$

Since all normal distributions can be obtained by a location and scale adjustment to $N(0,1)$, we consider a GLD($\lambda_1, \lambda_2, \lambda_3, \lambda_4$) fit to $N(0,1)$ for which

$$\alpha_1 = 0, \quad \alpha_2 = 1, \quad \alpha_3 = 0, \quad \alpha_4 = 3.$$

Appendix B suggests $(\lambda_3, \lambda_4) = (0.13, 0.13)$ as a starting point for our invocation of FindLambdasM which yields

$$\text{GLD}(0, 0.1975, 0.1349, 0.1349).$$

For our **first check** of the fit, we observe that the graphs of the $N(0,1)$ and the fitted GLD p.d.f.s, given in Figure 2.4–1 (a), show the two p.d.f.s to be "nearly identical" (the $N(0,1)$ p.d.f. is slightly higher at the center). Specifically, we compute

$$\sup |\hat{f}(x) - f(x)| = 0.002812,$$

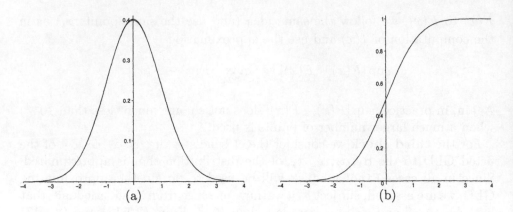

Figure 2.4–1. The $N(0,1)$ with its fitted GLD; the GLD p.d.f. rises
higher at the center (a). The two d.f.s are so close
that they appear as a single curve (b).

where $f(x)$ and $\hat{f}(x)$ are the p.d.f.s of the $N(0,1)$ and the fitted distributions,
respectively. As a **second check** of the fit, we observe that the graphs
of the $N(0,1)$ and the fitted d.f.s, given in Figure 2.4–1 (b), cannot be
distinguished. Specifically,

$$\sup |\hat{F}(x) - F(x)| = 0.001085,$$

where $F(x)$ and $\hat{F}(x)$ are the d.f.s of the $N(0,1)$ and the fitted distributions,
respectively. This means that the probability that X is at most x differs from
its approximation by no more than 0.001085 at any x.

Since (λ_3, λ_4) is from Region 3, GLD$(0, 0.1975, 0.1349, 0.1349)$ has finite
support (see Theorem 1.4.23) in the form of the interval $[-5.06, 5.06]$. This
may or may not be desirable — see the discussion at the beginning of this
chapter.

2.4.2   The Uniform Distribution

The continuous uniform distribution on the interval (a, b) with $a < b$ has
p.d.f.

$$f(x) = \begin{cases} \dfrac{1}{b-a}, & \text{if } a < x < b \\ 0, & \text{otherwise.} \end{cases}$$

For simplicity we consider the uniform distribution on the interval $(0, 1)$ for which

$$\alpha_1 = \frac{1}{2}, \quad \alpha_2 = \frac{1}{12}, \quad \alpha_3 = 0, \quad \alpha_4 = \frac{9}{5}.$$

The indication from Appendix B is to use $(\lambda_3, \lambda_4) = (1, 1)$ as a starting point for FindLambdasM. This, in turn, yields

$$\text{GLD}(0.5, 2.0000, 1.0000, 1.0000).$$

In this case, the fit is perfect because using these values of $\lambda_1, \lambda_2, \lambda_3, \lambda_4$, in (1.2.1) gives $Q(y) = y$ where $Q(y)$ is the inverse distribution function of the fitted GLD. Therefore, we must also have $F(y) = y$, matching the distribution function of the uniform distribution on the $(0, 1)$ interval. Hence, our **checks**: $(\alpha_1, \alpha_2, \alpha_3, \alpha_4) = (1/2, 1/2, 0, 9/5)$ exactly; the p.d.f. graphs are identical (i.e., $\sup |\hat{f}(x) - f(x)| = 0$); and the d.f.s are identical (i.e., $\sup |\hat{F}(x) - F(x)| = 0$).

For the general uniform distribution on (a, b) we have

$$\alpha_1 = \frac{a+b}{2}, \quad \alpha_2 = \frac{(b-a)^2}{12}, \quad \alpha_3 = 0, \quad \alpha_4 = \frac{9}{5}.$$

Since α_3 and α_4 do not change, to fit the uniform distribution on (a, b) we need only adjust λ_1 and λ_2.

If $(\lambda_1, \lambda_2, \lambda_3, \lambda_4) = (1/2, 2, 1, 1)$ yields $(\alpha_1, \alpha_2, \alpha_3, \alpha_4) = (1/2, 1/12, 0, 9/5)$, then by Theorem 2.1.11, changing to $(\lambda_1, \lambda_2, \lambda_3, \lambda_4) = (1/2 + \delta_1, 2\delta_2, 1, 1)$ will yield (since in this case $A = 0$)

$$(\alpha_1, \alpha_2, \alpha_3, \alpha_4) = \left(\frac{1}{2} + \delta_1, \frac{1}{12\delta_2^2}, 0, \frac{9}{5} \right).$$

Solving

$$\frac{1}{2} + \delta_1 = \frac{a+b}{2}, \quad \frac{1}{12\delta_2^2} = \frac{(b-a)^2}{12},$$

we find

$$\delta_1 = \frac{a+b-1}{2}, \quad \delta_2 = \frac{1}{b-a}.$$

This establishes that

$$(\lambda_1, \lambda_2, \lambda_3, \lambda_4) = \left(\frac{a+b}{2}, \frac{(b-a)^2}{12}, 1, 1 \right)$$

will produce

$$(\alpha_1, \alpha_2, \alpha_3, \alpha_4) = \left(\frac{a+b}{2}, \frac{(b-a)^2}{12}, 0, \frac{9}{5} \right),$$

the exact moments of the uniform distribution on (a, b). As was the case when $a = 0$ and $b = 1$, the fit will be exact.

2.4.3   The Student's t Distribution

The Student's t distribution with ν degrees of freedom, $t(\nu)$, has p.d.f.

$$f(x) = \frac{\Gamma\left(\frac{\nu+1}{2}\right)}{\sqrt{\pi\nu}\,\Gamma\left(\frac{\nu}{2}\right)\left(1 + \frac{x^2}{\nu}\right)^{(\nu+1)/2}}, \quad -\infty < x < \infty.$$

The specification of the $t(\nu)$ p.d.f. uses the gamma function, $\Gamma(t)$, which for $t > 0$ is defined by

$$\Gamma(t) = \int_0^\infty y^{t-1} e^{-y}\, dy.$$

Some properties of $\Gamma(t)$ will be developed in Section 3.1 (for a more detailed discussion see Artin (1964)).

The existence of the i-th moment of $t(\nu)$ depends on the relative sizes of i and ν. For the i-th moment to exist, the integral

$$\int_{-\infty}^\infty \frac{x^i \Gamma\left(\frac{\nu+1}{2}\right)}{\sqrt{\pi\nu}\,\Gamma\left(\frac{\nu}{2}\right)\left(1 + \frac{x^2}{\nu}\right)^{(\nu+1)/2}}\, dx$$

must converge. The power of x in the integrand is $i - \nu - 1$. Therefore, the integral will converge if and only if $i - \nu - 1 < -1$ or $i < \nu$.

Since we need the first four moments of a distribution to apply the method of moments, we can only consider t distributions with $\nu \geq 5$. (In Chapter 4 we develop other methods that can be used for approximating t distributions with small ν.) We expect that the limiting case $\nu = 5$ may be the one where a $\text{GLD}(\lambda_1, \lambda_2, \lambda_3, \lambda_4)$ fit will be most difficult.

The first four moments of $t(\nu)$ are

$$\alpha_1 = 0, \quad \alpha_2 = \frac{\nu}{\nu - 2}, \quad \alpha_3 = 0, \quad \alpha_4 = 3\frac{\nu - 2}{\nu - 4} \qquad (2.4.1)$$

which, when $\nu = 5$, become

$$\alpha_1 = 0, \quad \alpha_2 = \frac{5}{3}, \quad \alpha_3 = 0, \quad \alpha_4 = 9.$$

To fit $t(5)$, we use $(\lambda_3, \lambda_4) = (-0.13, -0.13)$ (suggested in Appendix B) as a starting point for `FindLambdasM`. The resulting fit is

$$\text{GLD}(0.1613 \times 10^{-17}, -0.2481, -0.1359, -0.1359).$$

For our **first check** we observe the p.d.f.s of this GLD and $t(5)$, shown in Figure 2.4-2 (a) (the one that rises higher at the center is the GLD p.d.f.), and compute

$$\sup |\hat{f}(x) - f(x)| = 0.03581.$$

Our **second check** leads us to consider the graphs of the d.f.s of $t(5)$ and its fitted distribution. These are shown in Figure 2.4-2 (b) (the $t(5)$ d.f. rises slightly higher on the left and is slightly lower on the right) and yield

$$\sup |\hat{F}(x) - F(x)| = 0.01488.$$

While this seems to be a reasonably good fit, it does not look as good as the $N(0, 1)$ fit. Also, unlike the $N(0, 1)$ fit, the support of this GLD fit is $(-\infty, \infty)$ — as it is for $t(5)$.

As ν gets large, the $\text{GLD}(\lambda_1, \lambda_2, \lambda_3, \lambda_4)$ fits for $t(\nu)$ get better. For $\nu = 6, 10$, and 30 we get, respectively, the following fits.

$$(\lambda_1, \lambda_2, \lambda_3, \lambda_4) = (3 \times 10^{-18}, -0.1376, -0.08020, -0.08020) \text{ with}$$
$$\sup |\hat{f}(x) - f(x)| = 0.02311, \text{ and } \sup |\hat{F}(x) - F(x)| = 0.009513;$$

$$(\lambda_1, \lambda_2, \lambda_3, \lambda_4) = (6 \times 10^{-15}, 0.02335, 0.01476, 0.01476), \text{ with}$$
$$\sup |\hat{f}(x) - f(x)| = 0.01041 \text{ and } \sup |\hat{F}(x) - F(x)| = 0.004160;$$

$$(\lambda_1, \lambda_2, \lambda_3, \lambda_4) = (0, 0.1452, 0.09701, 0.09701), \text{ with}$$
$$\sup |\hat{f}(x) - f(x)| = 0.004544 \text{ and } \sup |\hat{F}(x) - F(x)| = 0.001766.$$

2.4.4 The Exponential Distribution

The exponential distribution with parameter $\theta > 0$ has p.d.f.

$$f(x) = \begin{cases} \dfrac{1}{\theta} e^{-\frac{x}{\theta}}, & \text{if } x > 0 \\ 0, & \text{otherwise.} \end{cases}$$

For this distribution

$$\alpha_1 = \theta, \quad \alpha_2 = \theta^2, \quad \alpha_3 = 2, \quad \alpha_4 = 9.$$

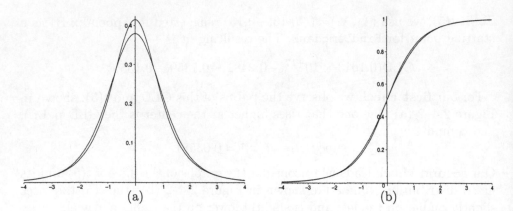

Figure 2.4–2. The p.d.f. and d.f. of $t(5)$ with the fitted GLD. The GLD p.d.f. rises higher at the center (a). The $t(5)$ d.f. rises higher on the left and is lower on the right (b)).

We can see that α_3 and α_4 do not change because θ is a scale parameter. Therefore, if a specific exponential distribution, say with $\theta = 1$, can be fitted, then other exponential distributions can be fitted using the λ_3 and λ_4 from the fit obtained for $\theta = 1$.

For $\theta = 1$,

$$\alpha_1 = 1, \quad \alpha_2 = 1, \quad \alpha_3 = 2, \quad \alpha_4 = 9.$$

Since $(\alpha_3, \alpha_4) = (2, 9)$ is an actual entry point in the table of Appendix B, we can use the table values without the concern about interpolation errors. This produces, after the adjustments mandated in Step GLD–M–5 of Algorithm GLD–M,

$$\text{GLD}(0.006862, -0.0010805, -0.4072 \times 10^{-5}, -0.001076).$$

For our **first check**, we consider the exponential p.d.f., with $\theta = 1$, and the fitted GLD p.d.f., plotted together in Figure 2.4–3 (the curve that turns down near the vertical axis is the GLD p.d.f.). The explanation for the surprising result

$$\sup |\hat{f}(x) - f(x)| = 0.6771$$

is that, although not evident from Figure 2.4–3, the GLD p.d.f. has infinite tails in both directions and for negative values close to 0, it assumes values near 0.7 where the exponential p.d.f. is zero. This yields a large difference over a small range for the p.d.f.s (though, as our second check below shows, the d.f.s are very close).

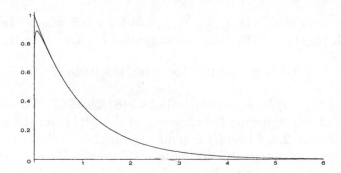

Figure 2.4–3. The exponential p.d.f. with $\theta = 1$ and its fitted GLD (the GLD p.d.f. turns down on the left).

For our **second check** we observe that the d.f.s of the GLD fit and the exponential distribution with $\theta = 1$ cannot be visually distinguished and a graphic comparison of these d.f.s is not given. We do note that for $\theta = 1$

$$\sup |\hat{F}(x) - F(x)| = 0.009801$$

and when $\theta = 3$,

$$(\lambda_1, \lambda_2, \lambda_3, \lambda_4) = (0.02100, -0.0003603, -0.4072 \times 10^{-5}, -0.001076),$$

with

$$\sup |\hat{f}(x) - f(x)| = 0.2257, \qquad \sup |\hat{F}(x) - F(x)| = 0.009801.$$

2.4.5 The Chi-Square Distribution

The p.d.f. of the $\chi^2(\nu)$ distribution with ν degrees of freedom is given by

$$f(x) = \begin{cases} \dfrac{x^{\nu/2-1}e^{-x/2}}{\Gamma(\nu/2)2^{\nu/2}}, & \text{if } x \geq 0 \\ 0, & \text{otherwise.} \end{cases}$$

The first four moments of $\chi^2(\nu)$ are

$$\alpha_1 = \nu, \quad \alpha_2 = 2\nu, \quad \alpha_3 = 2\frac{\sqrt{2}}{\sqrt{\nu}}, \quad \alpha_4 = 3 + \frac{12}{\nu}.$$

We first illustrate the GLD($\lambda_1, \lambda_2, \lambda_3, \lambda_4$) fit for $\nu = 5$ where `FindLambdasM` is used with (λ_3, λ_4)= (0.01, 0.05), as suggested by Appendix B, to provide

$$\text{GLD}(2.6040, 0.01756, 0.009469, 0.05422)$$

with support $[\lambda_1 - 1/\lambda_2, \lambda_1 + 1/\lambda_2] = [-54.3482, 59.5582]$ (Theorem 1.4.23).

The **first check** regarding the closeness of the $\chi^2(5)$ and the fitted p.d.f.s is shown in Figure 2.4–4 (the GLD p.d.f. rises higher at the center). We can see from Figure 2.4–4 that this GLD fits $\chi^2(5)$ reasonably well, and has long but finite left and right tails. To complete this check, we determine

$$\sup |\hat{f}(x) - f(x)| = 0.02115.$$

For our **second check**, we obtain

$$\sup |\hat{F}(x) - F(x)| = 0.01357$$

but we do not illustrate the $\chi^2(5)$ and the fitted d.f.s because the two graphs cannot be visually distinguished.

When we try to fit $\chi^2(1)$ by the methods of this chapter, we run into difficulties. For $\nu = 1$,

$$\alpha_1 = 1, \quad \alpha_2 = 2, \quad \alpha_3 = 2.8284, \quad \alpha_4 = 15,$$

placing (α_3^2, α_4) well outside the range of the table of Appendix B and our computational capability. When $\nu = 2$, $\chi^2(2)$ is the same as the exponential distribution with $\theta = 2$ and fitting this distribution was covered in Section 2.4.4. There are no difficulties associated with fitting $\chi^2(\nu)$ distributions with $\nu \geq 3$. For $\nu = 3$, $\nu = 10$, and $\nu = 30$, we find, respectively,

$$(\lambda_1, \lambda_2, \lambda_3, \lambda_4) = (0.8596, 0.009543, 0.002058, 0.02300), \text{ with}$$
$$\sup |\hat{f}(x) - f(x)| = 0.05312 \text{ and } \sup |\hat{F}(x) - F(x)| = 0.0156;$$

$$(\lambda_1, \lambda_2, \lambda_3, \lambda_4) = (7.1747, 0.02168, 0.02520, 0.09388), \text{ with}$$
$$\sup |\hat{f}(x) - f(x)| = 0.007783 \text{ and } \sup |\hat{F}(x) - F(x)| = 0.01182;$$

$$(\lambda_1, \lambda_2, \lambda_3, \lambda_4) = (26.4479, 0.01896, 0.05578, 0.1366), \text{ with}$$
$$\sup |\hat{f}(x) - f(x)| = 0.002693 \text{ and } \sup |\hat{F}(x) - F(x)| = 0.007649.$$

We know that for large ν, the closeness of $\chi^2(\nu)$ to $N(\nu, 2\nu)$ would produce a reasonably good fit.

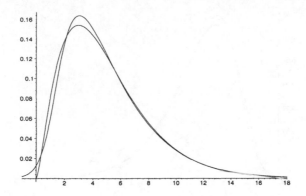

Figure 2.4–4. The $\chi^2(5)$ p.d.f. with its fitted GLD (the GLD p.d.f. rises higher at the center).

2.4.6 The Gamma Distribution

The p.d.f. of the Gamma distribution, $\Gamma(\alpha, \theta)$, with parameters $\alpha > 0$ and $\theta > 0$ is given by

$$f(x) = \begin{cases} \dfrac{x^{\alpha-1}e^{-x/\theta}}{\Gamma(\alpha)\theta^\alpha}, & \text{if } x \geq 0 \\ 0, & \text{otherwise.} \end{cases}$$

If $\alpha = 1$, this is the exponential distribution; if $\alpha = \nu/2$ and $\theta = 2$, it is the $\chi^2(\nu)$. The $\alpha_1, \alpha_2, \alpha_3, \alpha_4$ for $\Gamma(\alpha, \theta)$ are given by

$$\alpha_1 = \alpha\theta, \quad \alpha_2 = \alpha\theta^2, \quad \alpha_3 = \frac{2}{\sqrt{\alpha}}, \quad \alpha_4 = 3 + \frac{6}{\alpha}.$$

For the purpose of this illustration we take $\alpha = 5$ and $\theta = 3$ for which

$$\alpha_1 = 15, \quad \alpha_2 = 45, \quad \alpha_3 = 2\frac{\sqrt{5}}{5}, \quad \alpha_4 = \frac{21}{5}.$$

Here $\alpha_3 \approx 0.89$ and $\alpha_4 = 4.2$ and Appendix B for this entry point suggests an initial search point of $(0.03, 0.1)$. The GLD fit that results from FindLambdasM is

$$\text{GLD}(10.7620, 0.01445, 0.02520, 0.09388).$$

For the **first check**, we consider the two p.d.f.s shown in Figure 2.4–5 (the curve that rises higher in the middle is the GLD p.d.f.). This seems to be a

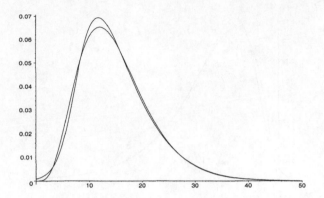

Figure 2.4–5. The gamma p.d.f. with $\alpha = 5$ and $\theta = 3$ and its fitted GLD (the GLD p.d.f. rises higher at the center).

reasonable fit for which

$$\sup |\hat{f}(x) - f(x)| = 0.005188.$$

The fit is somewhat deceptive in that its support is, by Theorem 1.4.23, the interval $[-58.43, 79.96]$. For the **second check**, we cannot distinguish the graphs of the d.f.s of $\Gamma(5,3)$ and its fitted GLD and

$$\sup |\hat{F}(x) - F(x)| = 0.01182.$$

2.4.7 The Weibull Distribution

A Weibull random variable with parameters $\alpha > 0$ and $\beta > 0$ has p.d.f.

$$f(x) = \begin{cases} \alpha\beta x^{\beta-1} e^{-\alpha x^{\beta}}, & \text{if } x \geq 0 \\ \\ 0, & \text{otherwise.} \end{cases}$$

An excellent reference on the Weibull distribution, its multivariate generalizations, as well as reliability-related distributions is Harter (1993), which contains several hundred references on the Weibull distribution alone.

The moments of the Weibull distribution are

$$\alpha_1 = \alpha^{(\beta-1)^{-1}} \Gamma(1 + \beta^{-1}),$$

$$\alpha_2 = \alpha^{2(\beta-1)^{-1}} \Gamma(\frac{\beta+2}{\beta}) - \alpha^{2(\beta-1)^{-1}} \left(\Gamma(\frac{\beta+1}{\beta})\right)^2,$$

$$\alpha_3 = \left(-3\,\Gamma(3\,\beta^{-1})\beta^2 + 6\,\Gamma(\beta^{-1})\Gamma(2\,\beta^{-1})\beta - 2\left(\Gamma(\beta^{-1})\right)^3\right) \Big/$$
$$\left(2\Gamma(2\,\beta^{-1})\beta - \left(\Gamma(\beta^{-1})\right)^2\right)\beta\sqrt{\left(2\Gamma(2\beta^{-1})\beta - (\Gamma(\beta^{-1}))^2\right)\beta^{-2}},$$

$$\alpha_4 = -\left\{-\Gamma(\frac{\beta+4}{\beta}) + 4\,\Gamma(\frac{3+\beta}{\beta})\Gamma(\frac{\beta+1}{\beta}) - 6\,\Gamma(\frac{\beta+2}{\beta})\left(\Gamma(\frac{\beta+1}{\beta})\right)^2\right.$$
$$\left. + 3\left(\Gamma(\frac{\beta+1}{\beta})\right)^4\right\} \Big/ \left(\Gamma(\frac{\beta+2}{\beta}) - \left(\Gamma(\frac{\beta+1}{\beta})\right)^2\right)^2.$$

Note that α_3 and α_4 do not depend on the parameter α. If we take $\alpha = 1$ and $\beta = 5$, the moments of the Weibull distribution will be

$$\alpha_1 = 0.91816, \quad \alpha_2 = 0.04423, \quad \alpha_3 = -0.2541, \quad \alpha_4 = 2.8802.$$

From a search with `FindLambdasM` with the starting point $(0.1, 0.2)$ that is obtained from Appendix B, we get the GLD fit

$$GLD(0.9935, 1.0491, 0.2121, 0.1061).$$

The support of the fitted GLD is the interval $[0.0403, 1.947]$. For the **first check** we see in Figure 2.4–6 that the Weibull and the fitted p.d.f.s are "nearly identical" and calculate

$$\sup |\hat{f}(x) - f(x)| = 0.03538.$$

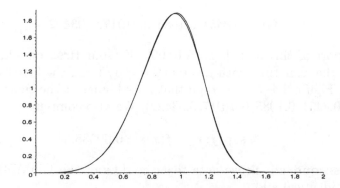

Figure 2.4–6. The Weibull p.d.f. with $\alpha = 5$ and $\beta = 2$ and its fitted GLD (the GLD p.d.f. is to the right near its peak).

As our **second check** we note that the graphs of the Weibull and the fitted distributions appear to be identical and

$$\sup |\hat{F}(x) - F(x)| = 0.002726.$$

2.4.8   The Lognormal Distribution

The p.d.f. of the lognormal distribution with parameters μ and $\sigma > 0$ is

$$f(x) = \begin{cases} \dfrac{1}{x\sigma\sqrt{2\pi}} \exp\left[-\dfrac{(\ln(x) - \mu)^2}{2\sigma^2} \right], & \text{if } x \geq 0, \\ 0, & \text{otherwise.} \end{cases}$$

The moments of the lognormal distributions are

$$\alpha_1 = e^{\mu + \sigma^2/2},$$
$$\alpha_2 = (e^{\sigma^2} - 1)e^{2\mu + \sigma^2},$$
$$\alpha_3 = \sqrt{e^{\sigma^2} - 1}\left(e^{\sigma^2} + 2 \right),$$
$$\alpha_4 = e^{4\sigma^2} + 2e^{3\sigma^2} + 3e^{2\sigma^2} - 3.$$

In the special case when $\mu = 0$ and $\sigma = 1/3$, the moments, obtained by direct computation, are

$$\alpha_1 = 1.0571, \quad \alpha_2 = 0.1313, \quad \alpha_3 = 1.0687, \quad \alpha_4 = 5.0974.$$

With the starting point of $(0.01, 0.03)$ indicated by Appendix B, we get, using `FindLambdasM`, the GLD fit

$$\text{GLD}(0.8451, 0.1085, 0.01017, 0.03422).$$

The support of this fit is $[-8.37, 10.06]$. For our **first check**, we plot the p.d.f.s of the lognormal with $\mu = 0$ and $\sigma = 1/3$ and the fitted GLD. This is shown in Figure 2.4-7 (the graph that rises higher at the center is the p.d.f. of GLD$(0.8451, 0.1085, 0.01017, 0.03422)$. We also compute

$$\sup |\hat{f}(x) - f(x)| = 0.09535.$$

For our **second check** we note that the d.f.s of the two distributions are virtually identical and

$$\sup |\hat{F}(x) - F(x)| = 0.01235.$$

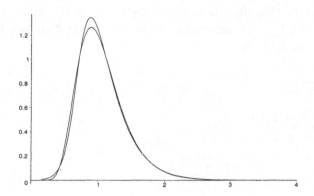

Figure 2.4–7. The lognormal p.d.f. with $\mu = 0$ and $\sigma = 1/3$ and its fitted GLD (the GLD p.d.f. rises higher at the center).

The choice of parameters in this, as well as previous distributions, has been quite arbitrary. Generally, GLD fits can be obtained with most, but not all, choices of parameters. In the case of the lognormal distribution, had we chosen $\mu = 0$ and $\sigma = 1$, the resulting moments would have been

$$\alpha_1 = 1.6487, \quad \alpha_2 = 4.6708, \quad \alpha_3 = 6.1849, \quad \alpha_4 = 113.9364,$$

making the search for a solution, if indeed one exists, quite difficult. In general, for (α_3^2, α_4) to be within the range of computation, we must have $0 < \sigma \leq 0.55$. When σ is small (e.g., $\sigma = 0.1$) the hazard rate of the lognormal is increasing; when σ is moderate (e.g., $\sigma = 0.5$) it increases and then slowly decreases; when σ is large (e.g., $\sigma = 1.0$) it is decreasing. The latter case arises when $X = \ln(Y)$ and Y is $N(0,1)$, making X lognormal with $\mu = 0$ and $\sigma = 1$. Thus, it is possible to fit the GLD in the ranges of most use in reliability applications (see, for example, Nelson (1982), p. 35).

2.4.9 The Beta Distribution

A random variable has the beta distribution if for parameters $\beta_3, \beta_4 > -1$, it has p.d.f.

$$f(x) = \begin{cases} \dfrac{x^{\beta_3}(1-x)^{\beta_4}}{\beta(\beta_3+1, \beta_4+1)}, & \text{if } 0 \leq x \leq 1, \\ 0, & \text{otherwise.} \end{cases}$$

The notation of β_3 and β_4 for the parameters of the beta distribution is used for reasons that will become clear in Chapter 3, when we consider a generalization of this distribution. If $\beta_3 = \beta_4 = 0$, this is the uniform distribution on $(0, 1)$. The moments of the beta distribution (these will be derived in Section 3.1) are

$$\alpha_1 = \frac{\beta_3 + 1}{\beta_3 + \beta_4 + 2},$$

$$\alpha_2 = \frac{(\beta_3 + 1)(\beta_4 + 1)}{(\beta_3 + \beta_4 + 2)^2(\beta_3 + \beta_4 + 3)},$$

$$\alpha_3 = \frac{2(\beta_4 - \beta_3)\sqrt{\beta_3 + \beta_4 + 3}}{(\beta_3 + \beta_4 + 4)\sqrt{(\beta_3 + 1)(\beta_4 + 1)}},$$

$$\alpha_4 = \frac{3(\beta_3 + \beta_4 + 3)\left(\beta_3\beta_4(\beta_3 + \beta_4 + 2) + 3\beta_3^2 + 5\beta_3 + 3\beta_4^2 + 5\beta_4 + 4\right)}{(\beta_3 + \beta_4 + 5)(\beta_3 + \beta_4 + 4)(\beta_3 + 1)(\beta_4 + 1)}.$$

If we take the specific beta distribution with $\beta_3 = \beta_4 = 1$, we get

$$\alpha_1 = \frac{1}{2}, \qquad \alpha_2 = \frac{1}{20}, \qquad \alpha_3 = 0, \qquad \alpha_4 = \frac{15}{7}$$

and using the starting point of $(0.4, 0.4)$ (obtained from Appendix B) in FindLambdasM, we obtain the fit

$$\text{GLD}(0.5000, 1.9693, 0.4495, 0.4495).$$

Our **first check** indicates that this is an excellent fit (with support the interval $[-.008, 1.008]$) to the chosen beta distribution, as can be seen from Figure 2.4–8 where the *two* p.d.f.s are indistinguishable. Moreover,

$$\sup |\hat{f}(x) - f(x)| = 0.009298.$$

The quality of this fit is confirmed by our **second check** where the d.f.s of the beta distribution with parameters $\beta_3 = \beta_4 = 1$ and its fitted GLD are virtually identical and

$$\sup |\hat{F}(x) - F(x)| = 0.0003734.$$

In spite of this good fit, in some cases the GLD is unable to provide a good approximation to beta distributions because the (α_3^2, α_4) points of these distributions are outside the range of the GLD (they lie in the region marked X in Figure 2.2–4). Consider, for example, the beta distribution with $\beta_3 = -1/2$, $\beta_4 = 1$, for which

$$\alpha_1 = \frac{1}{5}, \qquad \alpha_2 = \frac{8}{175}, \qquad \alpha_3 = -\frac{\sqrt{14}}{3}, \qquad \alpha_4 = \frac{42}{11}.$$

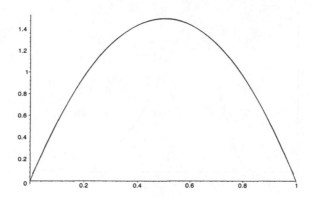

Figure 2.4–8. The beta p.d.f. with $\beta_3 = \beta_4 = 1$ and its fitted
GLD (the *two* p.d.f.s are nearly indistinguishable).

In this case (α_3^2, α_4) lies below the region that the GLD moments cover (see (2.3.11) and Figure 2.2–5). Only some of the (α_3^2, α_4) points in the BETA REGION of Figure 2.2–5 are in the GLD (α_3^2, α_4) region that was given in (2.3.11). This motivates (in Chapter 3) extending the GLD to cover the portion of (α_3^2, α_4)-space below the region described in (2.3.11).

2.4.10   The Inverse Gaussian Distribution

The p.d.f. of the inverse Gaussian distribution with parameters $\mu > 0$ and $\lambda > 0$ is given by

$$f(x) = \begin{cases} \sqrt{\dfrac{\lambda}{2\pi x^3}} \exp\left[-\dfrac{\lambda(x-\mu)^2}{2\mu^2 x}\right], & \text{if } x > 0, \\ 0, & \text{otherwise.} \end{cases}$$

This distribution has applications to problems associated with the times required to cover fixed distances in linear Brownian motion. The reader may wish to consult Govindarajulu (1987), p. 611. The moments of this distribution are

$$\alpha_1 = \mu, \qquad \alpha_2 = \frac{\mu^3}{\lambda}, \qquad \alpha_3 = 3\sqrt{\frac{\mu}{\lambda}}, \qquad \alpha_4 = 3 + \frac{15\mu}{\lambda}.$$

We first consider the special case of $\mu = 0.5$ and $\lambda = 6$ with

$$\alpha_1 = \frac{1}{2}, \qquad \alpha_2 = \frac{1}{48}, \qquad \alpha_3 = \frac{\sqrt{3}}{2}, \qquad \alpha_4 = \frac{17}{4}.$$

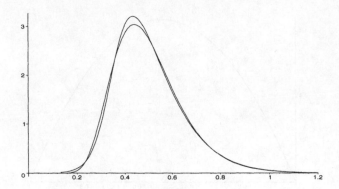

Figure 2.4–9. The inverse Gaussian p.d.f. ($\mu = 0.5$ and $\lambda = 6$) and its fitted GLD (the GLD p.d.f. rises higher at the center).

Use of `FindLambdasM` with the starting point $(0.01, 0.03)$ gives the fit

$$\text{GLD}(0.4164, 0.6002, 0.02454, 0.08009).$$

We note that the support of this GLD is $[-1.2497, 2.0825]$ and proceed to our **first check** by considering the graphs of the p.d.f.s of this inverse Gaussian distribution and its fitted GLD (given in Figure 2.4–9) and determine

$$\sup |\hat{f}(x) - f(x)| = 0.1899.$$

For our **second check**, we note that it is impossible to graphically distinguish the d.f. of this inverse Gaussian distribution from that of its fitted GLD and

$$\sup |\hat{F}(x) - F(x)| = 0.01052.$$

2.4.11   The Logistic Distribution

The p.d.f. of the logistic distribution, with parameters μ and $\sigma > 0$, is given by

$$f(x) = \frac{e^{-(x-\mu)/\sigma}}{\sigma \left(1 + e^{-(x-\mu)/\sigma}\right)^2} \quad \text{for} \quad -\infty < x < \infty.$$

For this distribution,

$$\alpha_1 = \mu, \quad \alpha_2 = \frac{\pi^2 \sigma^2}{3}, \quad \alpha_3 = 0, \quad \alpha_4 = \frac{21}{5}.$$

The α_3 and α_4 do not depend on the parameters of the logistic distribution and $(\alpha_3, \alpha_4) = (0, 4.2)$ is an entry in Table B–1 of Appendix B.

In the special case of $\mu = 0$ and $\sigma = 1$, we use Table B–1 of Appendix B to obtain the fit

$$\text{GLD}(0, -0.0003637, -0.0003630, -0.0003630)$$

and note that the support of the fitted distribution is $(-\infty, \infty)$. Next, we observe that the p.d.f.s of this logistic distribution and its fitted GLD are graphically indistinguishable, as shown in Figure 2.4–10. Moreover,

$$\sup |\hat{f}(x) - f(x)| = 0.0004191.$$

This takes care of our **first check**. For our **second check**, we consider the d.f.s of the two distributions and observe that they look identical and

$$\sup |\hat{F}(x) - F(x)| = 0.0003541.$$

Since for the logistic distribution $(\alpha_3, \alpha_4) = (0, 4.2)$ for all admissible μ and σ, the entry of Table B–1 of Appendix B that we used in this specific case can be used in all situations, guaranteeing a GLD fit for all members of this family of distributions. The logistic is used as a life distribution (see Nelson (1982)), and is not a member of the Pearson family (see Freimer, Kollia, Mudholkar, and Lin (1988), p. 3560).

Since the **loglogistic** distribution deals with $Y = e^X$ where X is logistic, by taking logarithms, we can reduce the loglogistic to the logistic.

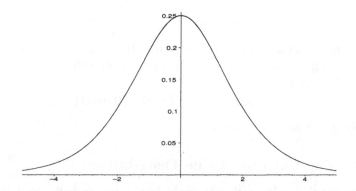

Figure 2.4–10. The logistic p.d.f. ($\mu = 0$ and $\sigma = 1$) and its fitted GLD. The two p.d.f.s cannot be distinguished.

2.4.12   The Largest Extreme Value Distribution

The largest extreme value distribution, with parameters μ and $\sigma > 0$, has
p.d.f.

$$f(x) = \frac{1}{\sigma} e^{-(x-\mu)/\sigma} \exp\left[-e^{-(x-\mu)/\sigma}\right], \quad \text{for} \quad -\infty < x < \infty.$$

This distribution can be used to describe human life (see Nelson (1982),
p. 40) and has moments

$$\alpha_1 = \mu + \gamma\sigma, \quad \alpha_2 = \frac{1}{6}\pi^2\sigma^2, \quad \alpha_3 = \sqrt{1.29857}, \quad \alpha_4 = 5.4,$$

where $\gamma \approx 0.57722$ is Euler's constant.

Since α_3 and α_4 are independent of the parameters of the distribution,
we can consider fitting a specific distribution, knowing that fits for other
distributions of this family can be obtained in a similar manner. If we set
$\mu = 0$ and $\sigma = 1$,

$$\alpha_1 = .5772, \quad \alpha_2 = 1.6449, \quad \alpha_3 = 1.1395, \quad \alpha_4 = 5.4.$$

Using the initial search point of $(0.007, 0.001)$, we obtain the fit

$$\text{GLD}(-0.1857, 0.02107, 0.006696, 0.02326)$$

and note that the support of this GLD is $[-47.647, 47.275]$. We compare
the p.d.f. of this distribution and that of its fitted GLD. This comparison,
illustrated in Figure 2.4–11, yields

$$\sup|\hat{f}(x) - f(x)| = 0.02215$$

for our **first check**. Next, we compare the d.f.s of the extreme value distribution with $\mu = 0$ and $\sigma = 1$ to observe that the d.f.s cannot be visually
distinguished and

$$\sup|\hat{F}(x) - F(x)| = 0.01004,$$

completing our **second check**.

2.4.13   The Extreme Value Distribution

This distribution, also called the smallest extreme value distribution, is important in some applications. For example, Weibull (1951) reported that
the strength of a certain material follows an extreme value distribution with
$\mu = 108\text{kg/cm}^2$ and $\sigma = 9.27\text{kg/cm}^2$ (also see Nelson (1982), p. 41).

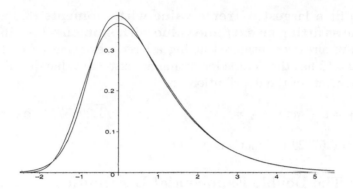

Figure 2.4–11. The largest extreme value p.d.f. ($\mu = 0$ and $\sigma = 1$) and its fitted GLD (the GLD p.d.f. rises higher at the center).

Also, if X is Weibull, then $\ln(X)$ has an extreme value distribution, making this relationship similar to the one between the lognormal and the normal distributions. It is also known (see Kendall and Stuart (1969), pp. 85, 335, 344) that if X has the extreme value distribution then the p.d.f. of $(X-\mu)/\sigma$ is $f(x) = e^{-x-e^{-x}}$. The extreme value distribution is also used in life and failure data analysis, "weakest link" situations, temperature minima, rainfall in droughts, human mortality of the aged, etc., and often represents the first failure (which fails the unit).

The extreme value distribution has parameters μ and $\sigma > 0$ and p.d.f.

$$f(x) = \frac{1}{\sigma}e^{(x-\mu)/\sigma} \exp\left[-e^{(x-\mu)/\sigma}\right], \quad \text{for} \quad -\infty < x < \infty.$$

If Y is extreme value with parameters μ_Y and σ_Y, **then $X = -Y$ is largest extreme value** with parameters $\mu_X = -\mu_Y$ and $\sigma_X = \sigma_Y$. This follows since

$$P(Y \leq y) = 1 - e^{-e^{(y-\mu_Y)/\sigma_Y}}.$$

Hence,

$$P(X \leq x) = P(-Y \leq x) = P(Y \geq -x) = 1 - P(Y \leq -x)$$

$$= 1 - \left(1 - e^{-e^{(-x-\mu_Y)/\sigma_Y}}\right) = e^{-e^{-(x+\mu_Y)/\sigma_Y}} = e^{-e^{-(x-\mu_X)/\sigma_X}}.$$

The moments of the two distributions have the relation

$$\alpha_1(X) = -\alpha_1(Y), \quad \alpha_2(X) = \alpha_2(Y), \quad \alpha_3(X) = -\alpha_3(Y), \quad \alpha_4(X) = \alpha_4(Y).$$

Thus, **to fit a largest extreme value with moments** $(5, 16, 0.5, 9.2)$ **is the same as fitting an extreme value with moments** $(-5, 16, -0.5, 9.2)$. As we have already considered fitting a largest extreme value distribution, Section 2.4.12 has the details for fitting an extreme value distribution.

The moments of the distribution are

$$\alpha_1 = \mu - \gamma\sigma, \quad \alpha_2 = \frac{1}{6}\pi^2\sigma^2, \quad \alpha_3 = -\sqrt{1.29857}, \quad \alpha_4 = 5.4,$$

where $\gamma \approx 0.57722$ is Euler's constant.

2.4.14   The Double Exponential Distribution

The double exponential distribution with parameter $\lambda > 0$ has p.d.f.

$$f(x) = \frac{e^{-|x|/\lambda}}{2\lambda} \quad \text{for} \quad -\infty < x < \infty.$$

The moments of this distribution are

$$\alpha_1 = 0, \quad \alpha_2 = 2\lambda^2, \quad \alpha_3 = 0, \quad \alpha_4 = 6.$$

Since α_3 and α_4 are constants, we will be able to find $\text{GLD}(\lambda_1, \lambda_2, \lambda_3, \lambda_4)$ fits to this distribution if we can find a fit when $\lambda = 1$ and the moments are

$$\alpha_1 = 0, \quad \alpha_2 = 2, \quad \alpha_3 = 0, \quad \alpha_4 = 6.$$

Initiating a search at $(-0.1, -0.1)$ `FindLambdasM` provides the fit

$$\text{GLD}(7.5505 \times 10^{-17}, -0.1192, -0.08020, -0.08020).$$

Note that the support of this GLD is $(-\infty, \infty)$. We now compare the p.d.f.s of the double exponential distribution with $\lambda = 1$ to that of the fitted GLD. The graphs of these p.d.f.s are given in Figure 2.4–12 (a) where the graph of the double exponential p.d.f. rises higher in the center. Our observation that

$$\sup |\hat{f}(x) - f(x)| = 0.1457$$

completes the **first check**. For the **second check**, we compute

$$\sup |\hat{F}(x) - F(x)| = 0.02871$$

and obtain a plot of the d.f.s of the two distributions, shown in Figure 2.4–12 (b). Note that in contrast to most of the fits provided so far, there is an observable difference between the two d.f.s (the double exponential d.f. is lower immediately to the left of the center and higher immediately to the right of the center).

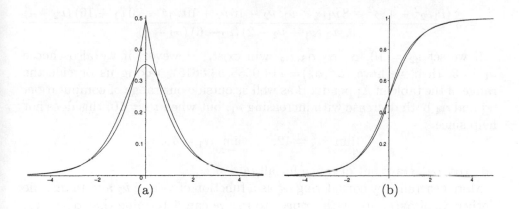

Figure 2.4–12. The p.d.f.s (a) and d.f.s (b) of the double exponential ($\lambda = 1$) and its fitted GLD.

2.4.15   The F-Distribution

The p.d.f. of the F distribution, with $\nu_1 > 0$ and $\nu_2 > 0$ degrees of freedom, $F(\nu_1, \nu_2)$, is given by

$$f(x) = \frac{\Gamma\left((\nu_1 + \nu_2)/2\right)(\nu_1/\nu_2)^{\nu_1/2}}{\Gamma(\nu_1/2)\Gamma(\nu_2/2)} \times \frac{x^{(\nu_1-2)/2}}{\left(1 + x\frac{\nu_1}{\nu_2}\right)^{(\nu_1+\nu_2)/2}}$$

when $x > 0$ and $f(x) = 0$ when $x \le 0$.

The power of x in the p.d.f. of the F distribution is $-1 - \nu_2/2$. Therefore for the i-th moment to exist, we must have $i - 1 - \nu_2/2 < -1$ or $\nu_2 > 2i$. This immediately restricts the moment-based GLD($\lambda_1, \lambda_2, \lambda_3, \lambda_4$) fits to those F distributions with $\nu_2 > 8$.

The moments of the F distribution (when $\nu_2 > 8$), are

$$\alpha_1 = \frac{\nu_2}{\nu_2 - 2},$$

$$\alpha_2 = 2\frac{\nu_2{}^2 (\nu_1 + \nu_2 - 2)}{\nu_1 (\nu_2 - 4)(\nu_2 - 2)^2},$$

$$\alpha_3 = \frac{2\sqrt{2}(\nu_2 - 2 + 2\nu_1)}{\nu_1(\nu_2 - 6)\sqrt{\frac{\nu_1+\nu_2-2}{\nu_1(\nu_2-4)}}},$$

$$\alpha_4 = 3 \frac{(\nu_1\nu_2{}^2 + 4\nu_2{}^2 + 8\nu_1\nu_2 + \nu_1{}^2\nu_2 - 16\nu_2 + 10\nu_1{}^2 - 20\nu_1 + 16)\,(\nu_2 - 4)}{\nu_1\,(\nu_1 + \nu_2 - 2)\,(\nu_2 - 6)\,(\nu_2 - 8)}$$

If we set $\nu_2 = 10$, $\alpha_1, \alpha_2, \alpha_3, \alpha_4$ will exist. However, if we also choose $\nu_1 = 5$, then we have $(\alpha_3^2, \alpha_4) = (14.9538, 53.8615)$, taking us outside the range of the table in Appendix B as well as outside our range of computation. α_3^2 and α_4 both decrease with increasing ν_1, but when $\nu_2 = 10$, this does not help since

$$\lim_{\nu_1 \to \infty} \alpha_3^2 = 12, \qquad \lim_{\nu_1 \to \infty} \alpha_4 = 45,$$

and (α_3^2, α_4) stays out of range for all $\nu_1 > 0$.

More generally, by considering α_3^2 as a function of ν_1 and ν_2 and taking the derivative of $\alpha_3^2(\nu_1, \nu_2)$ with respect to ν_2, we can determine that $\alpha_3^2(\nu_1, \nu_2)$ decreases with increasing ν_2 when $\nu_2 > 8$. This implies that for $8 \leq \nu_2 \leq 15$,

$$\alpha_3^2(\nu_1, \nu_2) \geq \alpha_3^2(\nu 1, 15) = \frac{88(2\nu_1 + 13)^2}{81(\nu_1(\nu_1 + 13)}.$$

This last expression is a decreasing function of ν_1. Hence,

$$\alpha_3^2(\nu_1, \nu_2) \geq \lim_{\nu_1 \to \infty} \frac{88(2\nu_1 + 13)^2}{81(\nu_1(\nu_1 + 13)} \approx 4.35,$$

placing (α_3^2, α_4) outside of our computational range whenever $\nu_2 \leq 15$.

With similar analyses, we can determine that in order to have $\alpha_3^2 \leq 4$ when $\nu_2 = 16$, we must have $\nu_1 \geq 28$; when $\nu_2 = 20$, we must have $\nu_1 \geq 6.26$; and when $\nu_2 = 30$, we must have $\nu_1 \geq 3.52$.

It is, therefore, possible to obtain fits for large ν_2. For example, if we let $\nu_2 = 25$ and $\nu_1 = 6$, we get

$$\alpha_1 = 1.0869, \qquad \alpha_2 = 0.5439, \qquad \alpha_3 = 1.8101, \qquad \alpha_4 = 9.1986$$

and the GLD$(\lambda_1, \lambda_2, \lambda_3, \lambda_4)$ fit

$$\text{GLD}(0.5898, -0.09063, -0.01095, -0.05314)$$

by using the entries of Appendix B. We observe that the support of this fit is $(-\infty, \infty)$. For the **first check**, we compare the p.d.f. of $F(6, 25)$ with that of its GLD fit (see Figure 2.4–13) and compute

$$\sup |\hat{f}(x) - f(x)| = 0.1155.$$

For the **second check**, we note that d.f.s of $F(6, 25)$ and the fitted GLD cannot be visually distinguished and

$$\sup |\hat{F}(x) - F(x)| = 0.01633.$$

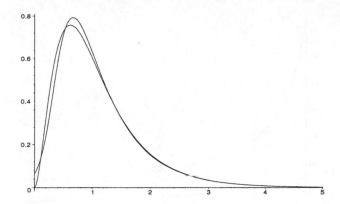

Figure 2.4–13. The p.d.f.s of $F(6, 25)$ and its fitted GLD (the one that rises higher at the center is the p.d.f. of the fitted GLD).

2.4.16 The Pareto Distribution

The Pareto distribution, with parameters $\beta > 0$ and $\lambda > 0$, has p.d.f.

$$f(x) = \begin{cases} \dfrac{\beta \lambda^\beta}{x^{\beta+1}}, & \text{if } x > \lambda, \\ \\ 0, & \text{otherwise.} \end{cases}$$

The moments of the Pareto distributions are

$$\alpha_1 = \frac{\beta \lambda}{\beta - 1},$$

$$\alpha_2 = \frac{\beta \lambda^2}{(\beta - 1)^2 (\beta - 2)},$$

$$\alpha_3 = \frac{2 (\beta + 1)}{(\beta - 3) \sqrt{\frac{\beta}{\beta-2}}},$$

$$\alpha_4 = 3 \frac{(3 \beta^2 + \beta + 2) (\beta - 2)}{\beta (\beta - 3) (\beta - 4)}.$$

We see that α_3 and α_4 depend only on the single parameter β and for $\alpha_1, \alpha_2, \alpha_3, \alpha_4$ to exist, we must have $\beta > 4$. For almost all reasonable choices of β, α_3 and α_4 are out of range for us to be able to compute $\lambda_1, \lambda_2, \lambda_3, \lambda_4$. For example, when $\beta = 5$, $\alpha_3 = 4.6476$ and $\alpha_4 = 73.8000$. As β increases

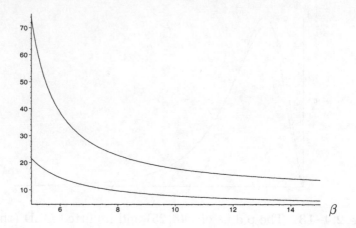

Figure 2.4–14. The α_3^2 and α_4 of the Pareto distribution as
functions of β (α_4 is the higher curve).

both α_3 and α_4 decrease. However,

$$\lim_{\beta \to \infty} \alpha_3 = 2$$

and

$$\lim_{\beta \to \infty} \alpha_4 = 9$$

so β has to get quite large for α_3 and α_4 to come near our tabulated range.
This can be seen from the graphs of α_3^2 and α_4 given in Figure 2.4–14 where
α_4 is the higher of the two curves.

2.4.17 Summary of Distributions and their GLD fits

In Table 2.4–15 we summarize the distributions considered in this section,
giving the section reference, the (λ_3, λ_4) of the fitted $GLD(\lambda_1, \lambda_2, \lambda_3, \lambda_4)$, the
$\sup |\hat{f}(x) - f(x)|$, and the $\sup |\hat{F}(x) - F(x)|$ for each distribution. The range
of (α_3^2, α_4) of these "standard" distributions represents a rather small portion
of the space that the GLD makes available. Therefore, the GLD provides
a much richer variety of distributions than the "standard" distributions. In
light of the results about (α_3^2, α_4) found in real datasets by Micceri (1989),
Pearson and Pease (1975), and Wilcox (1990), use of the GLD is indicated
in many application areas.

Table 2.4–15. Distributions considered in this section, with section reference, $GLD(\lambda_1, \lambda_2, \lambda_3, \lambda_4)$ fit, $\sup|\hat{f}(x) - f(x)|$ and $\sup|\hat{F}(x) - F(x)|$.

Distribution	Section	$\lambda_1, \lambda_2, \lambda_3, \lambda_4$ for GLD Fit	$\sup|\hat{f}(x) - f(x)|$	$\sup|\hat{F}(x) - F(x)|$
$N(0,1)$	2.4.1	$(0, 0.1975, 0.1349, 0.1349)$	0.002812	0.001085
Uniform on $(0,1)$	2.4.2	$(0.5, 2, 1, 1)$	0	0
Student's t, $\nu = 5$	2.4.3	$(0, -0.2481, -0.1359, -0.1359)$	0.03581	0.01488
$\nu = 6$	2.4.3	$(0, -0.1376, -0.08020, -0.08020)$	0.02311	0.009513
$\nu = 10$	2.4.3	$(0, 0.02335, 0.01476, 0.01476)$	0.01041	0.004160
$\nu = 30$	2.4.3	$(0, 0.1452, 0.09701, 0.09701)$	0.004544	0.001766
Exponential, $\theta = 1$	2.4.4	$(0.006862, -0.0010805, -0.4072 \times 10^{-5}, -0.001076)$	0.6771	0.009801
$\theta = 3$	2.4.4	$(0.02100, -0.0003603, -0.4072 \times 10^{-5}, -0.001076)$	0.2257	0.009801
Chi-square, $\nu = 3$	2.4.5	$(0.8596, 0.0095443, 0.002058, 0.02300)$	0.05312	0.0156
$\nu = 5$	2.4.5	$(2.6040, 0.01756, 0.009469, 0.05422)$	0.02115	0.01357
$\nu = 10$	2.4.5	$(7.1747, 0.02168, 0.02520, 0.09388)$	0.007783	0.01182
$\nu = 30$	2.4.5	$(26.4479, 0.01896, 0.05578, 0.1366)$	0.002693	0.007649
Gamma, $\alpha = 5$, $\theta = 3$	2.4.6	$(10.7620, 0.01445, 0.02520, 0.09388)$	0.005188	0.01182
Weibull, $\alpha = 1$, $\beta = 5$	2.4.7	$(0.9935, 1.0491, 0.2121, 0.1061)$	0.03538	0.002726
Lognormal, $\mu = 0$, $\sigma = 1/3$	2.4.8	$(0.8451, 0.1085, 0.01017, 0.03422)$	0.09535	0.01235
Beta, $\beta_3 = \beta_4 = 1$	2.4.9	$(0.5000, 1.9693, 0.4495, 0.4495)$	0.009298	0.0003734
Inverse Gaussian, $\mu = 1/2$, $\lambda = 6$	2.4.10	$(0.4164, 0.6002, 0.02454, 0.08009)$	0.1899	0.01052
Logistic, $\mu = 0$, $\sigma = 1$	2.4.11	$(0, -0.0003637, -0.0003630, -0.0003630)$	0.0004217	0.0003613
Largest Ext. Val., $\mu = 0$, $\sigma = 1$	2.4.12	$(-0.1857, 0.02107, 0.006696, 0.02326)$	0.02215	0.01004
Extreme Value	2.4.13	Fits are similar to those of the Largest Extreme Value Distribution		
Double Exponential, $\lambda = 1$	2.4.14	$(7.5505 \times 10^{-17}, -0.01192, -0.080102, -0.08020)$	0.1457	0.02871
F, $\nu_1 = 6$, $\nu_2 = 25$	2.4.15	$(0.5898, -0.09063, -0.01095, -0.05314)$	0.1155	0.01633
Pareto	2.4.16	(α_3, α_4) out of table and computation range		

2.5   Examples: GLD Fits of Data, Method of Moments

We have already seen details of the shapes of GLD distributions (Section 1.4), a GLD fitted to an actual dataset (the fit to measurements of the coefficient of friction of a metal at (1.2.6)), and fits to many of the most important distributions encountered in applications in various areas (summarized in Table 2.4–9).

In this section we consider **fits of the GLD to actual datasets**. A number of examples will be given, **from a variety of fields**, in order to illustrate a spectrum of nuances that arise in the fitting process. In this section, we will fit datasets for which the estimated (α_3^2, α_4) pair (denoted by $(\hat{\alpha}_3^2, \hat{\alpha}_4)$) is in the region covered by the GLD that was given in Figure 2.2–4. The reasons for this restriction are twofold: first, as we saw in Section 2.4.9, when we attempt to fit distributions that have (skewness^2, kurtosis) that is not in the area covered by the GLD (wide though this is) the fit will usually not be excellent; second, in Chapter 3 we will extend the GLD to an EGLD that covers the (α_3^2, α_4) points not covered by the GLD so it makes sense to consider such examples in Chapter 3.

It is true that there is variability of sampling in the estimates of (skewness^2, kurtosis): even if the true (α_3^2, α_4) point is in the GLD region, the estimate from the data might not be or if the true point is not in the GLD region, the estimate from the data might be. Thus, one may, in applications, end up fitting by a model that cannot cover the true distribution very well. For this reason, when we take up these additional examples in Chapter 3, after we extend the GLD to the EGLD, we will attempt to fit both a GLD and an EGLD and compare the two fits.

We should also note, as we will see in detail in Chapter 3, that the method of extension of the GLD is such that there is a zone of overlap of the (α_3^2, α_4) points of the two models. This means that we in fact have a zone where both model types will fit the data well, so "wrong zone" (skewness^2, kurtosis) should be less of a problem in applications when $(\hat{\alpha}_3^2, \hat{\alpha}_4)$ is "near" the boundary.

In the examples we discuss we have both real datasets from the literature, and datasets with simulated (Monte Carlo) data. While with a real dataset we can assess how well the model fits the data, we cannot assess how well it fits the underlying true distribution (as that distribution is not known). With simulated datasets we can do both, which is why we have included them.

2.5.1 Assessment of Goodness-of-Fit

There are a number of aspects of "goodness-of-fit," a topic on which many papers and books have been written that are relevant to the subject area of this book. We will cover them mainly in this section. **The situation** we have **is as follows:**

S–1. We are seeking to model a phenomenon X of interest in some area of research.

S–2. There are certain distributions $F_1, F_2, \ldots, F_k$ that we wish to include, in some form, among those we wish to consider to describe X. These may be ones that it is believed X truly follows (e.g., they are derived from assumptions X is believed to obey), or they may simply be ones that yielded reasonable approximations in the previous studies.

S–3. We have available data on the phenomenon X in some form, for example,

    1. Independent observations $X_1, X_2, \ldots, X_n$ on the phenomenon.

    2. A histogram of the distribution of X based on data (but the data itself is not available to us).

    3. The sample moments $\hat{\alpha}_1, \hat{\alpha}_2, \hat{\alpha}_3, \hat{\alpha}_4$ based on data (but we do not have the data).

    4. Some other form of information based on data is available.

S–4. We are considering using a family of distributions $G(\omega)$ to fit the phenomenon's p.d.f./d.f./p.f., where ω is the vector of parameters which (when chosen) picks a particular member of the family $G(\omega)$. For example, if G is the GLD family then $\omega = (\lambda_1, \lambda_2, \lambda_3, \lambda_4)$.

We **make the following points** regarding the circumstances described in S–1 through S–4.

Point 1. In most cases, the F_is of **S–2 will merely be approximations to a true (and unknown) distribution** of X. If they are truly the distributions that X follows, and have one or more unknown parameters, then one should act as though X is in the family $F_i(\omega)$ and use techniques which optimally choose ω for that family.

Point 2. In light of Point 1 **it is not necessarily required that the fitted G distribution have all of the characteristics of F_i of S–2.** For example, in the first paragraph of this chapter we noted that often the support (range of possible values) of F_i may be ones we know cannot occur; in such a setting we certainly do not want to require that G have the same support as F_i.

Point 3. In light of S–1, we will **want the family $G(\omega)$ to be one that can come reasonably close to including $F_1, F_2, \ldots, F_k$ when ω** is chosen appropriately. For this reason, in Section 2.4 we described how close the GLD family can come to various widely used distributions, using a fit based on matching the first four moments of the GLD to the first four moments of the particular F_i. We might also fit with the EGLD of Chapter 3, or with a Method of Percentiles as in Chapter 4, especially if we did not find the fit to F_i satisfactory. **If F_i is fitted with G_0, then the question arises: Is G_0 a good approximation to F_i?** Some ways to approach this question include

- Assess the adequacy and appropriateness of the support of G_0.

- Compare the p.d.f.s, d.f.s, and p.f.s of G_0 with those of F_i.

- Compare G_0 with F_i on other measures that "matter" for the application under consideration; for example, moments, probabilities of key events, hazard functions, and so on.

These are not simple questions for which there is a simple formula that can provide a correct answer — they are some of the most difficult questions that arise in modeling, and usually require serious interaction between the statistician and the subject matter specialists — but it is essential that they be carefully and fully considered if we are to have confidence in the model.

Point 4. With data of any of the forms described in S–3, we can use the Method of Moments to fit a GLD. In case of S–3–2, the moments are approximated using the midpoint assumption; with S–3–1 and S–3–2, the Method of Percentiles (presented in Chapter 4) can be used to fit a GLD; in case of S–3–2, the percentiles needed are approximated (an analog of case S–3–3 is to have available certain sample percentiles). What can be done with situation S–3–4 depends on the specifics of the information available.

Point 5. After we fit a distribution G to the data the question will arise: **Is this a good fit to the true underlying distribution?** Since the true distribution is unknown, this is a difficult question. In case S–3–3 we can reasonably ask that G have values close to the specified sample moments (or, if sample percentiles were given, to the sample percentiles). In case S–3–2 we should **overplot the histogram of the data and p.d.f. g of the d.f. G**

and examine key shape elements. In case S–3–1 we can **overplot the d.f.** G **and the empiric d.f. of the data** and examine their closeness.

Point 6. There are many **statistical tests of the hypothesis** that the data $X_1, X_2, \ldots, X_n$ come from the distribution G. These are in addition to **the eyeball test**, which should always be used (and in which many experimenters place the most faith). Based on the theorem that follows we assert: The hypothesis that the data come from G is **equivalent to the hypothesis that** $G(X_1), G(X_2), \ldots, G(X_n)$ **are independent uniform r.v.s on** $(0, 1)$.

Theorem 2.5.1. *If a r.v. Y has continuous d.f. $H(y)$, then the r.v. $Z = H(Y)$ has the uniform distribution on* $(0, 1)$.

There are many statistical tests for this hypothesis such as the Kolmogorov–Smirnov D, Cramér–von Mises W^2, Kuiper V, Watson U^2, Anderson–Darling A^2, log–statistic Q, χ^2, entropy, as well as other tests of uniformity. Some references are Dudewicz and van der Meulen (1981) (where the entropy test is developed, and which has references to the literature) and Shapiro (1980) (which, while oriented to testing normality, has excellent comments on goodness-of-fit testing in general).

Of the goodness-of-fit tests, the **oldest is the chi-square test** proposed by Pearson (1900). The idea of the test is to divide the range of the distribution into k **cells** and compare the **observed number** in each cell to the number that would be **expected** if the assumed distribution is true. Under certain construction of the cells, the resulting **test statistic has (approximately) a chi-square distribution with degrees of freedom** $k - 1 - t$ where t is the number of parameters estimated (e.g., $t = 4$ for the GLD). Drawbacks of the test are

- low power (ability to reject an incorrect fit) vs. other tests;

- loss of information when the data are grouped into cells;

- arbitrariness of the choice of the cells.

For reference on the chi-square test, see Moore (1977). We illustrate the chi-square test in the examples that we consider below.

We should note **why some experimenters place the most faith in the eyeball test**, though they may also do a formal test of the hypothesis:

- the chi-square test itself is approximate (and assumes use of the method of maximum likelihood, rather than the method of moments);

- for any test, failure to reject does not mean the hypothesis is true; it could, for example, be that the sample size is too small to detect differences that exist between the true and hypothesized models;

- for any test, rejection does not mean that the hypothesized model is inappropriate for the purpose intended.

However, the eyeball test requires experience. Therefore, **we recommend starting with a hypothesis test and** if we then wish to go against the conclusion of the test based on examination of the overplot of hypothesis and data, **make the case for reversal**. This is analogous to a lawyer making a case at an appellate level. As Gibbons (1997, p. 81) states, "the investigator hopes to be able to accept the null hypothesis, even when it appears to be only nearly true." The eyeball test is not a license to follow a personal whim, and it is important to guard against any lack of rigor in so important a decision as choice of distributional model for the phenomenon under study. The eyeball test is sometimes called a **graphical test** (see Ricer (1980), p. 18). If the plot is of the e.d.f. and the fitted G, we might look for criss-crossing of the two rather than one staying consistently on one side of the other. We should always be aware that there is subjectivity to the analysis.

The **Kolmogorov–Smirnov** or KS test is based on the largest difference (in absolute value) between the e.d.f. and its hypothesized counterpart. This can be interpreted as a quantification of the eyeball test. Only the biggest deviation is used; if one curve is uniformly on one side of the other but does not reach very far away at any one point, the test will not reject; for this reason, the test is sometimes viewed as lacking in power. A good comparison of the KS and chi-square tests is given by Gibbons (1997, pp. 80–82). The use of the KS statistic to give a confidence interval for the true d.f. (see Gibbons (1997), Section 3.2) is also of some interest.

There is a large literature on quantification of eyeball tests; one article, with references, worth considering is Gan, Koehler, and Thompson (1991).

2.5.2   Example: Cadmium in Horse Kidneys

Elinder, Jönsson, Piscator and Rahnster (1981) investigated histopathological changes due different levels of metal concentrations in horse kidneys. We list below part of their data dealing with the average cadmium concentrations in the kidney cortex of horses.

11.9	16.7	23.4	25.8	25.9	27.5	28.5	31.1
32.5	35.4	38.3	38.5	41.8	42.9	50.7	52.3

52.5	52.6	54.5	54.7	56.6	56.7	58.0	60.8
61.8	62.3	62.5	62.6	63.0	67.7	68.5	69.7
73.1	76.0	76.9	77.7	78.2	80.3	93.7	101.0
104.5	105.4	107.0					

For this data we find

$$\hat{\alpha}_1 = 57.2442, \quad \hat{\alpha}_2 = 576.0741, \quad \hat{\alpha}_3 = 0.2546, \quad \hat{\alpha}_4 = 2.5257$$

and obtain the GLD fit

$$GLD(41.7897, 0.01134, 0.09853, 0.3606)$$

whose support is $[-46.355, 129.935]$.

Figure 2.5–1 (a) shows the p.d.f. of the fitted GLD with a histogram of the data; Figure 2.5–1 (b) shows the e.d.f. of the data with the d.f. of the fitted GLD. These figures indicate a reasonably good fit which is substantiated with a chi-square test. The test uses the classes

$$(-\infty, 30), \quad [30, 50), \quad [50, 60), \quad [60, 70), \quad [70, 85), \quad [85, \infty)$$

whose respective frequencies are

$$7, \quad 7, \quad 9, \quad 9, \quad 6, \quad 5.$$

Calculation of expected frequencies yields

$$5.5633, \quad 12.3032, \quad 6.5387, \quad 5.8075, \quad 6.6775, \quad 6.1097$$

and the chi-square statistic and the p-value that result are

$$5.6087 \quad \text{and} \quad 0.01787.$$

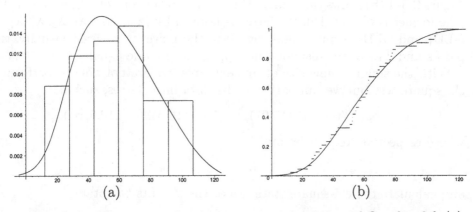

Figure 2.5–1. Histogram of cadmium concentrations and fitted p.d.f. (a); the e.d.f. of the data with the d.f. of its fitted GLD (b).

2.5.3   Example: Brain (Left Thalamus) MRI Scan Data

Dudewicz, Levy, Lienhart, and Wehrli (1989) give data on the brain tissue MRI scan parameter, AD. It should be noted that the term "parameter" is used differently in brain scan studies — it is used to designate what we would term random variables. In the cited study the authors show that AD^{-2} has a normal distribution while AD does not, and report the following 23 observations associated with scans of the left thalamus.

108.7	107.0	110.3	110.0	113.6	99.2	109.8	104.5
108.1	107.2	112.0	115.5	108.4	107.4	113.4	101.2
98.4	100.9	100.0	107.1	108.7	102.5	103.3	

We compute the $\hat{\alpha}_1, \hat{\alpha}_2, \hat{\alpha}_3, \hat{\alpha}_4$ for AD to obtain

$$\hat{\alpha}_1 = 106.8349, \quad \hat{\alpha}_2 = 22.2988, \quad \hat{\alpha}_3 = -0.1615, \quad \hat{\alpha}_4 = 2.1061.$$

With these $\hat{\alpha}_1, \hat{\alpha}_2, \hat{\alpha}_3, \hat{\alpha}_4$ Karian, Dudewicz and McDonald (1996) fitted this data by following Algorithm GLD–M of Section 2.3, using the entry at $(|\hat{\alpha}_3|, \hat{\alpha}_4) = (0.15, 2.1)$ to obtain the fit

$$GLD_1(112.1335, 0.06374, 0.6387, 0.05478)$$

with support [96.445, 127.822].

Here we appeal to FindLambdasM to determine the more precise fit

$$GLD_2(102.8998, 0.08060, 0.1475, 0.8041)$$

whose support is [90.493, 115.307]. Figure 2.5–2 (a) shows the p.d.f.s of GLD_1 (labeled (1)), and GLD_2 (labeled (2)), and a histogram of the data; Figure 2.5–1 (b) shows the e.d.f. of the data with the d.f.s of GLD_1 and GLD_2 (the former is not labeled, the latter is labeled by (2)). The $\lambda_1, \lambda_2, \lambda_3, \lambda_4$ of GLD_1 and GLD_2 seem to differ significantly; however, at least visually, the p.d.f.s and d.f.s of the distributions appear to provide equally valid fits.

With the small sample size of this example, we are not able to perform a chi-square test but we can partition the data into classes, such as

$$(-\infty, 103), \quad [103, 107), \quad [107, 110.5), \quad [110.5, \infty),$$

whose respective frequencies are

$$7, \quad 6, \quad 5, \quad 6,$$

and calculate the chi-square statistics of the two fits to obtain

$$1.3122 \quad \text{and} \quad 1.6405$$

for GLD_1 and GLD_2, respectively.

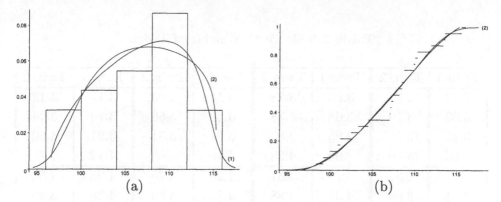

Figure 2.5–2. Histogram of AD and GLD_1 and GLD_2 p.d.f.s, designated by (1) and (2) in (a); e.d.f.s of AD with the d.f.s of GLD_1 and GLD_2 in (b).

2.5.4 Example: Human Twin Data for Quantifying Genetic (vs. Environmental) Variance

There is variability in most human characteristics: people differ in height, weight, and other variables. One question of interest in fields such as health, clothing manufacture, etc. (see Dudewicz, Chen, and Taneja (1989), pp. 223–228) is how much of the variability is due to genetic makeup, and how much is due to environmental influences. There is little dispute that some of the variability is from each source; there is a great controversy about how much of the variability should be attributed to each source. One approach to quantifying this variability is through so-called "twin studies," where human twins are studied in various scenarios. In such studies, normality of the variables is often assumed when this assumption may not always be valid (see Williams and Zhou (1998)). Some references to this interesting field include Christian, Carmelli, Castelli, Fabsitz, Grim, Meaney, Norton, Reed, Williams, and Wood (1990), and Christian, Kang, and Norton (1974). Interesting datasets in this area come from the Indiana Twin Study. We focus on one dataset[1] which is given in sorted form in Table 2.5–3.

The sample moments of X are

$$\hat{\alpha}_1 = 5.48589, \quad \hat{\alpha}_2 = 1.3082, \quad \hat{\alpha}_3 = -0.04608, \quad \hat{\alpha}_4 = 2.7332.$$

---

[1]This data comes from the P.h.D. thesis of Dr. Cynthia Moore, under the supervision of Dr. Joseph C. Christian, Department of Medical and Molecular Genetics, Indiana University School of Medicine, kindly shared with us by Dr. Moore. The data collection was supported by the National Institutes of Health Individual Research Fellowship Grant–"Twin Studies in Human Development." PHS-5-F32-HD06869, 1987–1990.

Table 2.5–3. Birth Weights of Twins

Twin 1	Twin 2	Twin 1	Twin 2	Twin 1	Twin 2	Twin 1	Twin 2
2.44	2.81	3.00	3.78	3.15	2.93	3.17	4.13
3.63	3.19	3.68	5.38	3.69	3.56	3.74	3.24
3.75	3.16	3.83	3.83	3.91	3.81	3.91	4.60
4.00	3.66	4.00	4.28	4.10	5.00	4.12	4.75
4.12	6.31	4.19	4.31	4.20	4.75	4.28	4.50
4.31	3.66	4.31	3.88	4.31	4.69	4.38	3.40
4.38	5.00	4.44	5.13	4.49	4.15	4.53	3.83
4.56	4.31	4.56	5.38	4.63	4.12	4.69	4.63
4.75	4.56	4.75	4.63	4.81	4.38	4.81	4.44
4.91	5.13	4.94	4.78	4.95	4.22	5.00	5.38
5.00	6.16	5.03	4.94	5.13	3.81	5.15	5.00
5.16	5.75	5.16	6.69	5.19	4.94	5.22	4.75
5.25	5.38	5.25	5.63	5.25	5.81	5.25	6.10
5.25	6.25	5.31	4.69	5.38	5.69	5.38	6.81
5.41	5.69	5.44	5.75	5.47	5.13	5.47	6.75
5.50	5.38	5.56	4.44	5.56	5.48	5.56	6.31
5.59	6.22	5.61	6.18	5.63	4.69	5.63	4.88
5.63	4.97	5.66	5.38	5.72	5.06	5.75	5.63
5.81	5.19	5.81	5.94	5.84	5.56	5.88	4.88
5.88	5.69	5.88	5.88	5.88	5.88	5.91	5.50
5.94	4.81	5.94	5.41	5.94	5.75	5.94	6.31
5.97	5.63	6.06	5.31	6.10	5.19	6.10	6.13
6.16	5.85	6.19	4.44	6.19	5.50	6.31	5.81
6.31	6.10	6.33	8.14	6.38	5.19	6.38	6.05
6.56	6.38	6.59	6.16	6.60	6.53	6.63	6.19
6.63	6.19	6.63	6.38	6.66	7.10	6.69	5.81
6.69	6.13	6.75	6.56	6.78	6.22	6.81	6.19
6.81	6.60	6.81	7.41	6.88	6.06	6.88	6.63
6.94	5.50	6.95	5.72	7.06	7.31	7.25	8.00
7.31	4.58	7.31	7.31	7.46	7.22	7.69	7.25
7.72	6.44	8.13	7.75	8.44	6.31		

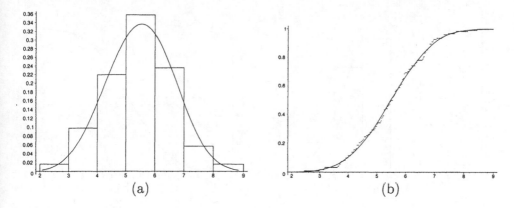

Figure 2.5–4. The histogram of X and its fitted p.d.f. (a); the empirical d.f. of X and the d.f. of its fitted GLD (b).

Using a search with initial point$(0.2, 0.2)$, FindLambdasM determines the fit

$$\text{GLD}(5.5872, 0.2266, 0.2089, 0.1762)$$

and indicates that $|\hat{\alpha}_i - \alpha_i| < 10^{-5}$ for $i = 1, 2, 3$, and 4. To get a sense of the quality of this fit we look at the histogram of X with the superimposed graph of the fitted p.d.f. (Figure 2.5–4 (a)) as well as the e.d.f. of X with the superimposed d.f. of the GLD fit (Figure 2.5–4 (b)). Next, we apply the chi-square test with the 8 classes

$$(-\infty, 4), \quad [4, 4.5), \quad [4.5, 5), \quad [5, 5.5), \quad [5.5, 6), \quad [6, 6.5), \quad [6.5, 7), \quad [7, \infty)$$

for which the observed frequencies, respectively, are

$$12, \quad 15, \quad 12, \quad 21, \quad 25, \quad 11, \quad 18, \quad 9.$$

The chi-square test yields expected frequencies of

$$12.6118, \; 12.0611, \; 16.8280, \; 20.0078, \; 20.2769, \; 17.3427, \; 12.2938, \; 11.5779$$

for these 8 classes, giving us

$$8.8225 \quad \text{and} \quad 0.03175$$

for the chi-square statistic and p-value for this test.

With a similar analysis of Y, using the same classes, we find

$$\hat{\alpha}_1 = 5.3666, \quad \hat{\alpha}_2 = 1.2033, \quad \hat{\alpha}_3 = -0.01219, \quad \hat{\alpha}_4 = 2.7665,$$

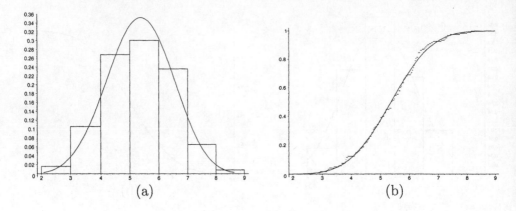

Figure 2.5–5. The histogram of Y and its fitted p.d.f. (a); the
empirical d.f. of Y and the d.f. of its fitted GLD (b).

$$GLD(5.3904, 0.2293, 0.1884, 0.1807).$$

The comparative p.d.f. and d.f. plots for Y are given in Figures 2.5–5 (a)
and (b).

When we apply the chi-square test we obtain observed frequencies of

$$15, \quad 11, \quad 19, \quad 19, \quad 21, \quad 22, \quad 7, \quad 9,$$

expected frequencies of

$$13.5767, \quad 13.4924, \quad 18.6506, \quad 21.4160, \quad 20.4944, \quad 16.2641, \quad 10.5725, \quad 8.5333,$$

and chi-square statistic and p-value of

$$4.1568 \quad \text{and} \quad 0.2450.$$

2.5.5   Example: Rainfall Distributions

Statistical modeling plays an essential role in the study of rainfall and the
relationships between rainfall at multiple sites (Shimizu (1993)). Lognormal distributions have been used extensively in this work, and in univariate
cases have worked well; however, with multiple sites the many rejections of
lognormality (e.g., see the Rs in "Table 2. Test for lognormality" on p. 168
of Shimizu (1993)) indicate a need for more modeling flexibility. While the
data for Shimizu's studies was from sites in Japan and is no longer available,
similar data for U.S. sites is readily available from the U.S. National Oceanic

and Atmospheric Administration. From that data, shown in Table 2.5–6 is data from the period May 1998 to October 1998 in Rochester, New York and Syracuse, New York. This data is for the 47 days on which both cities had positive rainfall measured. (The study of Shimizu (1993) looks at all 4 combinations of "rain, no rain" possibilities; however, for this example's purpose we concentrate on only the case of rain at both sites.)

The moments of X are

$$\hat{\alpha}_1 = 0.4913, \quad \hat{\alpha}_2 = 0.4074, \quad \hat{\alpha}_3 = 1.8321, \quad \hat{\alpha}_4 = 5.7347,$$

and those of Y are

$$\hat{\alpha}_1 = .3906, \quad \hat{\alpha}_2 = 0.1533, \quad \hat{\alpha}_3 = 1.6163, \quad \hat{\alpha}_4 = 5.2245.$$

In both cases, $(\hat{\alpha}_3^2, \hat{\alpha}_4)$ is located outside of the region of the table of Appendix B and outside of the region for reliable computations. We will return to this example in Chapters 3 and 4 and fit $\text{GLD}(\lambda_1, \lambda_2, \lambda_3, \lambda_4)$ distributions to both X and Y via the EGLD and through a percentile-based method.

Table 2.5–6. Rainfall (in inches) at Rochester, N.Y. (X) and Syracuse, N.Y. (Y) from May to October of 1998, on days when both sites had positive rainfall.

X	Y	X	Y	X	Y
.03	.05	.11	.07	.07	.04
.09	.14	.01	.14	.69	.72
.62	.60	.08	.21	.61	.36
.08	.09	.50	.54	.23	.18
1.27	.22	.03	1.26	.09	.10
1.20	.21	1.65	.94	1.87	.17
.15	.05	2.61	.94	.85	.37
1.35	.48	.15	.28	.31	.38
.16	.48	.27	.31	.06	.04
.23	1.09	1.38	.42	.74	1.79
2.51	1.25	.13	.11	.10	.16
.24	1.05	.76	.36	.03	.42
.29	.08	.02	.17	.07	.02
.17	.19	.47	.04	.41	.79
.03	.37	.15	.18	.01	.09
.02	.19	.19	.22		

2.6   Moment-Based GLD Fit to Data from a Histogram

In the examples of Section 2.5 the actual data on the phenomenon of interest, $X_1, X_2, \ldots, X_n$, was available to us. We acted assuming that these were independent and identically distributed observations — an assumption that should be verified. In many cases the data are given in the form of a **histogram**. While it would in general be preferable to have the actual data (going to a histogram involves a certain loss of information, such as the distribution of the observations within the classes), it nevertheless is possible to proceed with fitting of a distribution using data in the form of a histogram.

One could **seek the original data** from the authors of the histogram, but this is often difficult: the author may have moved, or may not be able to release the data without a time-consuming approval process, or may not have retained the basic data. As an example of use of histograms, one may look in virtually any scientific journal or newspaper. For example, Dahl-Jensen, Mosegaard, Gundestrup, Clow, Johnsen, Hansen, and Balling (1998) in their Figure 2 give six histograms, each based on 2000 Monte Carlo observations. As they use between 5 and 15 classes in each histogram, they are easily able to meet the rules for an informative histogram: there should be **no more than two classes with a frequency less than 5, and classes should be of equal width.** While histograms **with classes of unequal width** can be used, they are often misused. In order not to be misleading, **the heights of the bars must be adjusted so the area of the bar is proportional to the frequency.** Thus, if class 1 has a width of 2.5 and 150 observations, and its height is 150 in the frequency histogram, and if class 2 has width 5 and also 150 observations, then its height must be 75. Below we give two examples of the use of a histogram to fit a distribution.

Sometimes the histogram data is presented to us in the form of a table such as Table 2.6–1, which comes from Ramberg, Tadikamalla, Dudewicz, and Mykytka (1979, p. 207). The classes have equal widths (assuming the lowest class is 0.010 to 0.015 and the highest class is 0.060 to 0.065), but there are more than 2 classes with fewer than 5 observations in them, so we combine the two lowest and the two highest classes. Now there are no classes with fewer than 5 observations in them. Classes with low frequencies have highly variable estimates of true p.d.f. heights, due to inadequate amount of observation, and it is preferable to avoid them in order to avoid a bias in the fitting.

We use the mid-point assumption that every data point is located at the

Table 2.6–1. Observed Coefficient of Friction Frequencies.

Coefficient of Friction	Observed Frequency
less than 0.015	1
0.015 – 0.020	9
0.020 – 0.025	30
0.025 – 0.030	44
0.030 – 0.035	58
0.035 – 0.040	45
0.040 – 0.045	29
0.045 – 0.050	17
0.050 – 0.055	9
0.055 – 0.060	4
0.060 or more	4
Total	250

center of its class interval; for the first class, we assume that its entries are located at 0.0125 and for the last class we assume that the entries have value 0.0625. With these assumptions, we can compute approximate moments for this data and obtain

$$\hat{\alpha}_1 = 0.03448, \quad \hat{\alpha}_2 = 9.2380 \times 10^{-5}, \quad \hat{\alpha}_3 = 0.5374, \quad \hat{\alpha}_4 = 3.2130,$$

and subsequently, the GLD fit

$$GLD_1(0.02889, 18.1935, 0.05744, 0.1850).$$

We note that the original data are in fact given in Hahn and Shapiro (1967), and from the original data (not a histogram) Hahn and Shapiro computed the sample moments as

$$\hat{\alpha}_1 = .0345, \quad \hat{\alpha}_2 = 0.00009604, \quad \hat{\alpha}_3 = .87, \quad \hat{\alpha}_4 = 4.92. \qquad (2.6.1)$$

We now also fit a GLD using (2.6.1) to illustrate the change that results from having the original data vs. only the histogram of the data. The GLD_2 associated with the moments given in (2.6.1) is given by

$$GLD_2(0.03031, 1.5638, 0.005133, 0.01179).$$

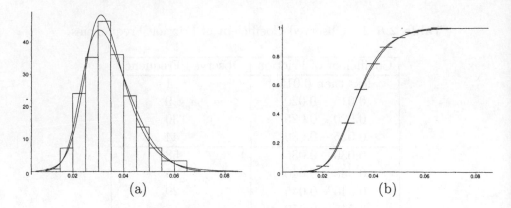

(a)   (b)

Figure 2.6–2. The histogram of coefficients of friction data and its
fitted p.d.f. (a). The empirical d.f. of coefficients of
friction and the d.f. of its fitted GLD (b).

The histogram of the data with the fitted GLD_1 and GLD_2 p.d.f.s is given
in Figure 2.6–2 (a) (the GLD_2 p.d.f. rises higher at the center). The e.d.f of
the data together with the two fitted d.f.s is given in Figure 2.6–2 (b) (the
GLD_2 d.f. extends farther to the left).

We do a chi-square test using the classes

$$(-\infty, 0.020), \quad [0.020, 0.025), \quad [0.025, 0.030), \quad [0.030, .035),$$
$$[0.035, 0.040), \quad [0.040, 0.045), \quad [0.045, 0.050), \quad [0.050, 0.055), \quad [0.055, \infty)$$

for which the observed frequencies are, respectively,

$$10, \quad 30, \quad 44, \quad 58, \quad 45, \quad 29, \quad 17, \quad 9, \quad 8.$$

We get the following expected frequencies:

$$GLD_1 : \ 9.9332, \quad 28.9050, \quad 49.9837, \quad 52.8583,$$
$$42.8412, \quad 29.6866, \quad 18.2559, \quad 10.0063, \quad 7.5298$$
$$GLD_2 : \ 9.6062, \quad 25.1726, \quad 51.3265, \quad 58.5142,$$
$$44.7824, \quad 27.8634, \quad 15.6573, \quad 8.34979, \quad 8.7277.$$

These give the following chi-square statistics and p-values for the two fits:

$$GLD_1 : \chi^2 \text{ statistic } = 1.6000, \quad p\text{–value } = 0.8088$$
$$GLD_2 : \chi^2 \text{ statistic } = 2.2661, \quad p\text{–value } = 0.6869.$$

We conclude that at least in this example (with a relatively large sample size, and a well-constructed histogram), the effect of the lack of the original data has been negligible: both GLD_1 and GLD_2 models fit the data quite well.

2.7 The GLD and Design of Experiments

In many studies including, but not limited to, simulation studies and theoretical studies of sensitivity of results to distributions, there is a distribution one thinks of as the "best fit" to the true underlying distribution. For example, in the coefficient of friction data of the example in Section 2.6, the **underlying "true" state of nature is estimated to have moments**

$$(\alpha_1, \alpha_2, \alpha_3, \alpha_4) = (0.03448, 9.238 \times 10^{-5}, 0.5374, 3.2130) \tag{2.7.1}$$

and the GLD p.f.

$$Q(y) = 0.02889 + (y^{0.05744} - (1 - y)^{0.1850})/18.1935. \tag{2.7.2}$$

However, as these are only estimates from the data, very often it will be desired to **re-do the analysis (e.g., re-run the simulation program, or refine the theoretical analysis) with a changed distribution** (a distribution in the neighborhood of the best estimate, but varied as much as seems reasonable in the setting using the best available engineering information).

This type of analysis is relatively simple to **perform using the GLD and notions of statistical design of experiments** (e.g., see Section 6.4.2 of Karian and Dudewicz (1999a)). This approach is superior to the "grab bag of distributions" approach sometimes used, wherein one haphazardly adds alternative distributions to the set used. To vary the distribution from the fitted GLD in (2.7.2), we vary the parameters $(\lambda_1, \lambda_2, \lambda_3, \lambda_4)$, but these do not have intrinsic value to us (as we may not easily understand the meaning of a change of 0.5 units in λ_4, for example). So, instead we **vary the $(\alpha_1, \alpha_2, \alpha_3, \alpha_4)$, fit a new GLD for each desired set of α-values, and run our study with the new fitted GLD.**

In the coefficient of friction example, suppose we are comfortable with the fitted mean (0.03448) and variance (9.2380×10^{-5}) and do not believe that perturbations of reasonable size in these values will impact the variables of interest to us. However, we believe that the shape parameters (skewness α_3 and kurtosis α_4) may have an impact when varied within reasonable bounds. Further, suppose that we have reason to believe that α_3 may vary from 0.5374 by up to ± 0.2, and α_4 may vary from 3.2130 by up to ± 0.5. That reason can either be based on **previous studies** in the area, or on **theoretical considerations**, or on **statistical reasons for which we might take these bounds to be ± 2 standard deviations (estimated) of the estimates** in (2.7.1). Note here that one needs to estimate (and take twice the square root of) the variances

$$\text{Var}(\hat{\alpha}_1) = \frac{\alpha_2}{n}, \tag{2.7.3}$$

$$\text{Var}(\hat{\alpha}_2) = \frac{(n-1)^2}{n^3}\mu_4 - \frac{(n-1)(n-3)}{n^3}\alpha_2^2 \approx \frac{\mu_4 - \alpha_2^2}{n}, \tag{2.7.4}$$

$$\text{Var}(\hat{\alpha}_3) \approx (\alpha_2^2\mu_6 - 3\alpha_2\mu_3\mu_5 - 6\alpha_2^3\mu_4 \\ + 2.25\mu_3^2\mu_4 + 8.75\alpha_2^2\mu_3^2 + 9\alpha_2^5)/(n\alpha_2^5), \tag{2.7.5}$$

$$\text{Var}(\hat{\alpha}_4) \approx (\alpha_2^2\mu_8 - 4\alpha_2\mu_4\mu_6 - 8\alpha_2^2\mu_3\mu_5 + 4\mu_4^3 \\ - \alpha_2^2\mu_4^2 + 16\alpha_2\mu_3^2\mu_4 + 16\alpha_2^3\mu_3^2)/(n\alpha_2^6). \tag{2.7.6}$$

In the above we already know how to estimate α_2 by $\hat{\alpha}_2$. The new expressions $\mu_3, \mu_4, \mu_5, \mu_6, \mu_8$ are related to the $\alpha_1, \alpha_2, \alpha_3, \alpha_4$ that we are familiar with by the equations

$$\mu_3 = \alpha_2^{1.5}\alpha_3, \quad \mu_4 = \alpha_2^2\alpha_4, \quad \mu_5 = \alpha_2^{2.5}\alpha_5, \quad \mu_6 = \alpha_2^3\alpha_6, \quad \mu_8 = \alpha_2^4\alpha_8. \tag{2.7.7}$$

Hence, they can be estimated by replacing right-hand side terms by their estimates. The approximations in (2.7.5) and (2.7.6) are up to order $O(n^{-3/2})$. A good reference is Cramér (1946, pp. 354, 357). In this example, we get

$$\alpha_5 = 4.5553,$$
$$\alpha_6 = 17.0073,$$
$$\alpha_8 = 114.2131,$$
$$\mu_3 = 4.7716 \times 10^{-7},$$
$$\mu_4 = 2.7142 \times 10^{-8},$$
$$\mu_5 = 3.7365 \times 10^{-10},$$
$$\mu_6 = 1.3408 \times 10^{-11},$$
$$\mu_8 = 8.3182 \times 10^{-15},$$
$$\text{Var}(\alpha_3) = 0.01600,$$
$$\text{Var}(\alpha_4) = 0.07150,$$
$$\sqrt{\text{Var}(\alpha_3)} = 0.1265,$$

$$\sqrt{\mathrm{Var}(\alpha_4)} = 0.2674.$$

Hence, $2\sqrt{\mathrm{Var}(\alpha_3)} = 0.253$ and $2\sqrt{\mathrm{Var}(\alpha_4)} = 0.536$. A study that varies α_3 by up to ± 0.25, and α_4 by up to ± 0.54, is reasonable. We use ± 0.2 and ± 0.5 for simplicity below.

We now choose the (α_3, α_4) for our experiments using what is called a **Central Composite Design in two variables** (α_3 and α_4), with a multiplier of 1.5. The sample point $(0.5374, 3.2130)$ is at the center of our experiments, and is called a **Center Point**. We go out from this Center Point to a distance $\pm 1.5d_1 = \pm 0.2$ on α_3, and $\pm 1.5d_2 = \pm 0.5$ on α_4, so we have $d_1 = .1333$, $d_2 = .3333$.

This sets the **Star Points** as

$$(0.5374 - 0.2, 3.2130) = (0.3374, 3.2130)$$
$$(0.5374 + .02, 3.2130) = (0.7374, 3.2130)$$
$$(0.5374, 3.2130 - 0.5) = (0.5374, 2.7130) \qquad (2.7.8)$$
$$(0.5374, 3.2130 + 0.5) = (0.5374, 3.7130).$$

The **Factorial Points** are taken as

$$(0.5374 - d_1, 3.2130 - d_2) = (0.4041, 2.8797)$$
$$(0.5374 - d_1, 3.2130 + d_2) = (0.4041, 3.5463)$$
$$(0.5374 + d_1, 3.2130 - d_2) = (0.6707, 2.8797) \qquad (2.7.9)$$
$$(0.5374 + d_1, 3.2130 + d_2) = (0.6707, 3.5463).$$

This is illustrated on coordinate axes in Figure 2.7–1. Also see Figure 6.4–2 of Karian and Dudewicz (1999b) for the general case and see Dudewicz and Karian (1985, pp. 189, 196, 206, 233) for details for more than two variables; that case would arise if we varied all of $\alpha_1, \alpha_2, \alpha_3, \alpha_4$, and there are ways to cut the number of points used from the full $2^k + 2k + 1$ with k variables without losing much information in a precise sense. This is desirable since when $k = 2$ we have only 9 points, but this would, if not reduced, become 25 with $k = 4$ and grow rapidly with k.

In Figure 2.7–1 note that the factorial points lie on the corners of a box (shown by "x"), and the star points (shown by "*") lie on perpendiculars from the center point and go out a distance ± 0.2 (horizontally), ± 0.5 (vertically) from the center of the design area. One should also be careful that the specified variations continue to fall in the possible region of Figure 2.2–4; if they move outside the GLD area, one may need to utilize the EGLD of Chapter 3.

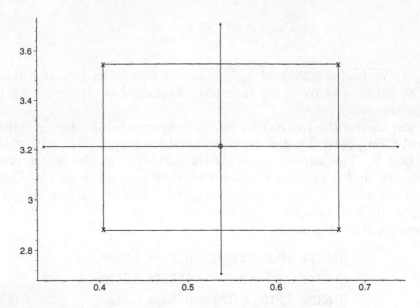

Figure 2.7–1. The design points for the sensitivity study
of the coefficient of friction example.

We now fit a GLD for each of the 9 points (we already have this for the
center point, so there are 8 additional points to be fitted). This process yields
the following (α_3, α_4) and their associated GLDs (the GLDs associated with
star points are designated with the superscript "*," those associated with
factorial points are similarly marked with an "x," and the one associated
with the center point is designated by "o").

$(\alpha_3, \alpha_4) = (0.5374, 3.2130)$, $GLD^o(0.02902, 18.1399, 0.05850, 0.1825)$;

$(\alpha_3, \alpha_4) = (0.3374, 3.2130)$, $GLD^*(0.03136, 17.0249, 0.07594, 0.1412)$;

$(\alpha_3, \alpha_4) = (0.7374, 3.2130)$, $GLD^*(0.02595, 19.0831, 0.03251, 0.2410)$;

$(\alpha_3, \alpha_4) = (0.5374, 2.7130)$, $GLD^*(0.02547, 23.6743, 0.04238, 0.3404)$;

$(\alpha_3, \alpha_4) = (0.5374, 3.7130)$, $GLD^*(0.03080, 10.8659, 0.04175, 0.08697)$;

$(\alpha_3, \alpha_4) = (0.4041, 2.8797)$, $GLD^x(0.02895, 22.5726, 0.07766, 0.2452)$;

$(\alpha_3, \alpha_4) = (0.4041, 3.5463)$, $GLD^x(0.03160, 11.9256, 0.05164, 0.09109)$;

$(\alpha_3, \alpha_4) = (0.6707, 2.8797)$, $GLD^x(0.02450, 21.8367, 0.02642, 0.3223)$;

$(\alpha_3, \alpha_4) = (0.6707, 3.5463)$, $GLD^x(0.02881, 14.7763, 0.04340, 0.1434)$.

The graphs of these p.d.f.s are shown in Figure 2.7–2, where the fit associated
with the center point is shown by diamond-shaped points. From Figure 2.7–

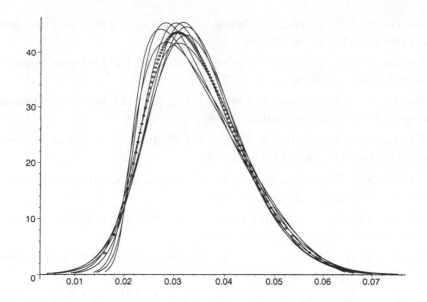

Figure 2.7–2. The p.d.f.s associated with the design points for the sensitivity study of the coefficient of friction.

2 we can observe the spread of the distributions chosen for the sensitivity study about the distribution associated with the center point.

As a final point, note that if we are interested in an output of an experiment (simulation or other experiment), we can now fit a metamodel via regression. That will show the effect of changes in α_3 and α_4 from the center on the output, and is a useful way of summarizing the information in the experiments run at the design points. For a detailed example, including SAS code and contour graphs for interpretation, see Section 8.3 of Karian and Dudewicz (1999a).

Problems for Chapter 2

2.1. For the GLD studied in (1.2.6), find the value of $\alpha = (\alpha_1, \alpha_2, \alpha_3, \alpha_4)$, and identify (α_3^2, α_4) in Figure 2.2–4.

2.2. For the two GLDs of Figure 1.4–1, find the first four moments, and identify the distributions in the plot of Figure 2.2–4.

2.3. For the 9 GLDs plotted in Figure 1.4–3a, find the first four moments, and identify their points in Figure 2.2–4.

2.4. For the 9 GLDs plotted in Figure 1.4–4a, find the first four moments, and identify their points in Figure 2.2–4.

2.5. For the 7 GLDs plotted in Figure 1.4–5, find the first four moments, and identify their points in Figure 2.2–4.

2.6. For the 4 GLDs plotted in Figure 1.4–6, find the first four moments, and identify their points in Figure 2.2–4.

2.7. For the 9 GLDs plotted in Figure 1.4–7a, find the first four moments, and identify their points in Figure 2.2–4.

2.8. For the 4 GLDs plotted in Figure 1.4–8a, find the first four moments, and identify their points in Figure 2.2–4.

2.9. For the 3 GLDs plotted in Figure 1.4–8b, find the first four moments, and identify their points in Figure 2.2–4.

2.10. For the 4 GLDs plotted in Figure 1.4–9, find the first four moments, and identify their points in Figure 2.2–4.

2.11. In Chapter 1 (e.g., see Figures 1.4–6, 1.4–7a, 1.4–8a) plots were given of GLDs with parameters in Regions 1 and 5 of Figure 1.3–1, but no plots for Regions 2 and 6. Construct similar plots for Regions 2 and 6, and find their (α_3^2, α_4) points (if they exist). (Hint: See Theorem 1.4.22.)

2.12. Ramberg (1975) developed the model

$$Q(y) = \lambda_1 + (\lambda_4 y^{\lambda_3} - (1 - y)^{\lambda_3})/\lambda_2,$$

and gave a set of tables. Fit some examples with the GLD and with the above model and compare. Which do you prefer, and why?

2.13. In Figure 2.5–4 (b) we plotted the e.d.f. and the fitted GLD d.f. on the same axes. Plot a 95% confidence band on the same axes to quantify the quality of the fit. (The confidence band uses the e.d.f. data to yield a band in which the true d.f. will, with high confidence, lie. See Gan, Koehler, and Thompson (1991).)

Chapter 3

The Extended GLD System, the EGLD: Fitting by the Method of Moments

In this chapter we address one of the two limitations of the GLD (when fitted via the method of moments) discussed in Section 2.3.3: the inability to provide fits for (α_3^2, α_4) in the region $1 + \alpha_3^2 < \alpha_4 \leq 1.8(\alpha_3^2 + 1)$. We do this by developing and appending a Generalized Beta Distribution (GBD) that covers this region not covered by the $GLD(\lambda_1, \lambda_2, \lambda_3, \lambda_4)$ and using the method of moments to fit the resulting Extended GLD (the EGLD) distributions to data.

The method described in the following sections is due to Karian, Dudewicz and McDonald (1996). In Section 3.1 we define the beta distribution and derive its moments. Next, in Section 3.2, we develop the GBD family of distributions as a generalization of the beta distribution and in Section 3.3, we use the method of moments to estimate the parameters of the GBD/EGLD. Illustrations for approximating known distributions and for fitting the EGLD to a dataset follow in Sections 3.4 and 3.5, respectively.

3.1 The Beta Distribution and its Moments

The beta distribution, discussed in many statistics texts (see, for example, Dudewicz and Mishra (1988), p. 137), with parameters $\beta_3 > -1$ and $\beta_4 > -1$, is defined through its p.d.f. by

$$f(x) = \begin{cases} \dfrac{x^{\beta_3}(1-x)^{\beta_4}}{\beta(\beta_3 + 1, \beta_4 + 1)}, & \text{for } 0 \leq x \leq 1 \\ 0, & \text{otherwise} \end{cases} \tag{3.1.1}$$

where $\beta(a, b)$ is the beta function in (2.1.6).

Before proceeding to determine the moments of a beta random variable, we establish several useful results related to beta and gamma functions. For $a > 0$, the gamma function encountered in Section 2.4 is defined by

$$\Gamma(a) = \int_0^\infty t^{a-1} e^{-t} \, dt. \tag{3.1.2}$$

Two of the properties of the gamma function are given in Lemmas 3.1.3 and 3.1.4 and its most basic connection to the beta function is given in Theorem 3.1.5. The moments of the beta distribution are derived in Theorem 3.1.9, its Corollary 3.1.15, and Theorem 3.1.16.

Lemma 3.1.3. *For $a > 0$,*

$$\Gamma(a+1) = a\Gamma(a).$$

Proof. Integrating $\Gamma(a+1)$ by parts gives

$$\Gamma(a+1) = \int_0^\infty t^a e^{-t} \, dt = \left[-t^a e^{-t} \right]_{t=0}^{t=\infty} + \int_0^\infty e^{-t} \, d(t^a)$$

$$= 0 + \int_0^\infty a e^{-t} t^{a-1} \, dt = a\Gamma(a). \quad \blacksquare$$

Lemma 3.1.4. *For $a > 0$,*

$$\Gamma(a) = 2 \int_0^\infty u^{2a-1} e^{-u^2} \, du.$$

Proof. The proof is a direct consequence of making the substitution $t = u^2$ in (3.1.2). $\quad \blacksquare$

Theorem 3.1.5. *For a and b positive,*

$$\beta(a, b) = \frac{\Gamma(a)\Gamma(b)}{\Gamma(a+b)}.$$

Proof. We start by considering $\Gamma(a)\Gamma(b)$ which, by Lemma 3.1.4, can be written as

$$\Gamma(a)\Gamma(b) = \left(2 \int_0^\infty u^{2a-1} e^{-u^2} \, du \right) \left(2 \int_0^\infty v^{2b-1} e^{-v^2} \, dv \right)$$

$$= 4 \int_0^\infty \int_0^\infty u^{2a-1} v^{2b-1} e^{-(u^2+v^2)} \, du \, dv.$$

Using the substitution $u = r \cos \theta$, $v = r \sin \theta$ in the above integral, we get

$$\Gamma(a)\Gamma(b) = 4 \int_0^\infty \int_0^{\frac{\pi}{2}} r^{2a-1} \cos^{2a-1} \theta r^{2b-1} \sin^{2b-1} \theta e^{-r^2} r \, dr \, d\theta$$

$$= 4 \int_0^\infty r^{2a+2b-1} e^{-r^2} \, dr \int_0^{\frac{\pi}{2}} \cos^{2a-1} \theta \sin^{2b-1} \theta \, d\theta. \qquad (3.1.6)$$

We now evaluate each of the two integrals in (3.1.6) separately. In the first integral we use the substitution $x = r^2$ to get

$$\int_0^\infty r^{2a+2b-1} e^{-r^2} \, dr = \int_0^\infty \frac{x^{a+b-\frac{1}{2}} e^{-x}}{2\sqrt{x}} \, dx$$

$$= \int_0^\infty \frac{x^{a+b-1}}{2} e^{-x} \, dx = \frac{1}{2} \Gamma(a+b). \qquad (3.1.7)$$

The second integral of (3.1.6), through the substitution $y = \sin^2 \theta$ (noting that this makes $dy = 2 \sin \theta \cos \theta d\theta$), yields

$$\int_0^{\frac{\pi}{2}} \cos^{2a-1} \theta \sin^{2b-1} \theta \, d\theta = \frac{1}{2} \int_0^{\frac{\pi}{2}} \cos^{2a-2} \theta \sin^{2b-2} \theta (2 \sin \theta \cos \theta) \, d\theta$$

$$= \frac{1}{2} \int_0^1 (1-y)^{a-1} y^{b-1} \, dy = \frac{1}{2} \beta(b,a)$$

$$= \frac{1}{2} \beta(a,b). \qquad (3.1.8)$$

The proof of the theorem is completed by substituting the results of (3.1.7) and (3.1.8) in (3.1.6). ∎

Theorem 3.1.9. *If X is a beta random variable with parameters β_3 and β_4 and if $B_i = \beta_3 + \beta_4 + i$, $i = 1, 2, \ldots$, then $E(X^k)$, the k-th moment of X, is given by*

$$E(X^k) = \frac{\beta(\beta_3 + k + 1, \beta_4 + 1)}{\beta(\beta_3 + 1, \beta_4 + 1)} \qquad (3.1.10)$$

and also by the recursive relation

$$E(X^k) = \left(\frac{\beta_3 + k}{B_{k+1}} \right) E(X^{k-1}) \text{ for } k = 1, 2, 3, \ldots, \quad E(X^0) = 1. \qquad (3.1.11)$$

Proof. By definition

$$E(X^k) = \int_0^1 \frac{x^k x^{\beta_3}(1-x)^{\beta_4}}{\beta(\beta_3+1,\beta_4+1)}\,dx$$

$$= \frac{\beta(\beta_3+k+1,\beta_4+1)}{\beta(\beta_3+1,\beta_4+1)} \int_0^1 \frac{x^{\beta_3+k}(1-x)^{\beta_4}}{\beta(\beta_3+k+1,\beta_4+1)}\,dx. \quad (3.1.12)$$

Since the integral in (3.1.12) is the integral of the beta p.d.f. with parameters β_3+k and β_4, its value must be 1. This establishes the part of the theorem given in (3.1.10).

For the second part of the theorem, we consider $E(X^k)/E(X^{k-1})$, which, from (3.1.10), can be written as

$$\frac{E(X^k)}{E(X^{k-1})} = \frac{\beta(\beta_3+k+1,\beta_4+1)}{\beta(\beta_3+k,\beta_4+1)}. \quad (3.1.13)$$

Applying Theorem 3.1.5 to the numerator and denominator of the ratio in (3.1.13), we have

$$\frac{E(X^k)}{E(X^{k-1})} = \frac{\Gamma(\beta_3+k+1)\Gamma(\beta_4+1)}{\Gamma(\beta_3+\beta_4+k+2)} \times \frac{\Gamma(\beta_3+\beta_4+k+1)}{\Gamma(\beta_3+k)\Gamma(\beta_4+1)}. \quad (3.1.14)$$

After cancelling $\Gamma(\beta_4+1)$, and using Lemma 3.1.3 on $\Gamma(\beta_3+k+1)$ and $\Gamma(\beta_3+\beta_4+k+2)$ in (3.1.14), we obtain

$$\frac{E(X^k)}{E(X^{k-1})} = \frac{\beta_3+k}{\beta_3+\beta_4+k+1} = \frac{\beta_3+k}{B_{k+1}},$$

which completes the proof of the theorem by establishing (3.1.11). ∎

Corollary 3.1.15. *If X is a beta random variable with parameters β_3 and β_4 and we let $B_i = \beta_3 + \beta_4 + i$, $i = 1, 2, \ldots$, then*

$$E(X^k) = \frac{(\beta_3+1)(\beta_3+2)\cdots(\beta_3+k)}{B_2 B_3 \cdots B_{k+1}}.$$

Proof. In this proof by mathematical induction, we first observe that when $k = 1$ the result follows from the substitution of $k = 1$ in (3.1.11). Now assume that the Corollary holds for $k - 1$, i.e., assume that

$$E(X^{k-1}) = \frac{(\beta_3+1)(\beta_3+2)\cdots(\beta_3+k-1)}{B_2 B_3 \cdots B_k}$$

and use (3.1.12) to express $E(X^k)$ by

$$E(X^k) = \left(\frac{\beta_3 + k}{B_{k+1}}\right) E(X^{k-1})$$

$$= \left(\frac{\beta_3 + k}{B_{k+1}}\right) \frac{(\beta_3 + 1)(\beta_3 + 2) \cdots (\beta_3 + k - 1)}{B_2 B_3 \cdots B_k}$$

$$= \frac{(\beta_3 + 1)(\beta_3 + 2) \cdots (\beta_3 + k)}{B_2 B_3 \cdots B_{k+1}}. \quad \blacksquare$$

Theorem 3.1.16. *If X is a beta random variable with parameters β_3 and β_4 and $B_i = \beta_3 + \beta_4 + i$, $i = 1, 2, \ldots$, then*

$$\alpha_1 = E(X) = \mu = \frac{\beta_3 + 1}{B_2}, \tag{3.1.17}$$

$$\alpha_2 = E\left[(X - \mu)^2\right] = \sigma^2 = \frac{(\beta_3 + 1)(\beta_4 + 1)}{B_2^2 B_3}, \tag{3.1.18}$$

$$\sigma^3 \alpha_3 = E\left[(X - \mu)^3\right] = 2\frac{(\beta_3 + 1)(\beta_4 + 1)(\beta_4 - \beta_3)}{B_2^3 B_3 B_4}, \tag{3.1.19}$$

$$\sigma^4 \alpha_4 = E\left[(X - \mu)^4\right] = 3\left(\frac{(\beta_3 + 1)(\beta_4 + 1)}{B_2^4 B_3 B_4 B_5}\right)$$

$$\times \left(\beta_3 \beta_4 B_2 + 3\beta_3^2 + 3\beta_4^2 + 5\beta_3 + 5\beta_4 + 4\right). \tag{3.1.20}$$

Proof. Setting $k = 1$ in Corollary 3.1.15 gives the first assertion. To obtain (3.1.18), the second assertion, we note that

$$\sigma^2 = E(X^2) - [E(X)]^2$$

and by Corollary 3.1.15,

$$\sigma^2 = \frac{(\beta_3 + 1)(\beta_3 + 2)}{B_2 B_3} - \left(\frac{\beta_3 + 1}{B_2}\right)^2,$$

which can be simplified to (3.1.18).

In (3.1.19),

$$E\left[(X - \mu)^3\right] = E\left[X^3 - 3\mu X^2 + 3\mu^2 X - \mu^3\right]$$

$$= E(X^3) - 3\mu E(X^2) + 2\mu^3.$$

Using Corollary 3.1.15, we get

$$E\left[(X-\mu)^3)\right] = \frac{(\beta_3+1)(\beta_3+2)(\beta_3+3)}{B_2 B_3 B_4} - 3\frac{(\beta_3+1)^2(\beta_3+2)}{B_2^2 B_3}$$

$$+2\left(\frac{\beta_3+1}{B_2}\right)^3 = \left(\frac{\beta_3+1}{B_2^3 B_3 B_4}\right)$$

$$\times \left((\beta_3+2)(\beta_3+3)B_2^2 - 3(\beta_3+1)(\beta_3+2)B_2 B_4 + 2(\beta_3+1)^2 B_3 B_4\right)$$

$$=2\frac{(\beta_3+1)(\beta_4+1)(\beta_4-\beta_3)}{B_2^3 B_3 B_4}.$$

For the proof of (3.1.20), we note that

$$E\left[(X-\mu)^4\right] = E\left[X^4 - 4\mu X^3 + 6\mu^2 X^2 - 4\mu^3 X + \mu^4\right]$$

$$= E(X^4) - 4\mu E(X^3) + 6\mu^2 E(X^2) - 3\mu^4.$$

From Corollary 3.1.15, we have

$$E\left[(X-\mu)^4\right] = \frac{(\beta_3+1)(\beta_3+2)(\beta_3+3)(\beta_3+4)}{B_2 B_3 B_4 B_5}$$

$$-4\left(\frac{\beta_3+1}{B_2}\right)\left(\frac{(\beta_3+1)(\beta_3+2)(\beta_3+3)}{B_2 B_3 B_4}\right)$$

$$+6\left(\frac{\beta_3+1}{B_2}\right)^2\left(\frac{(\beta_3+1)(\beta_3+2)}{B_2 B_3}\right) - 3\left(\frac{\beta_3+1}{B_2}\right)^4$$

$$= \left(\frac{\beta_3+1}{B_2^4 B_3 B_4 B_5}\right)$$

$$\times \Big[(\beta_3+2)(\beta_3+3)(\beta_3+4)B_2^3 - 4(\beta_3+1)(\beta_3+2)(\beta_3+3)B_2^2 B_5$$

$$+ 6(\beta_3+1)^2(\beta_3+2)B_2 B_4 B_5 - (\beta_3+1)^3 B_3 B_4 B_5\Big].$$

Through cumbersome but straightforward algebraic manipulations the last expression in the brackets can be equated to

$$3\left(\beta_3\beta_4 B_2 + 3\beta_3^2 + 3\beta_4^2 + 5\beta_3 + 5\beta_4 + 4\right),$$

completing the proof of (3.1.20).   ∎

3.2 The Generalized Beta Distribution and its Moments

To be able to use the beta distribution for fitting datasets, we need to make the distribution more flexible so that its support is not confined to the interval $(0, 1)$. This can be done through the introduction of location and scale parameters, β_1 and β_2, respectively. The transformation given in the following definition establishes a generalized beta random variable.

Definition 3.2.1. *If X is a beta random variable with parameters β_3 and β_4, for $\beta_2 > 0$, and any β_1, the random variable $Y = \beta_1 + \beta_2 X$ is said to have a Generalized Beta Distribution, GBD($\beta_1, \beta_2, \beta_3, \beta_4$).*

We now proceed to establish the p.d.f. of a GBD($\beta_1, \beta_2, \beta_3, \beta_4$) random variable.

Theorem 3.2.2. *The p.d.f. of the GBD($\beta_1, \beta_2, \beta_3, \beta_4$) random variable is*

$$f(x) = \begin{cases} \dfrac{(x - \beta_1)^{\beta_3} (\beta_1 + \beta_2 - x)^{\beta_4}}{\beta(\beta_3 + 1, \beta_4 + 1)\beta_2^{(\beta_3+\beta_4+1)}}, & \text{for } \beta_1 \leq x \leq \beta_1 + \beta_2 \\ 0, & \text{otherwise.} \end{cases}$$

Proof. Let Y be a GBD($\beta_1, \beta_2, \beta_3, \beta_4$) random variable. From Definition 3.2.1, $Y = \beta_1 + \beta_2 X$ where X is a beta random variable with parameters β_3 and β_4. The distribution function (d.f.) of Y is

$$F_Y(y) = P(Y \leq y) = P(\beta_1 + \beta_2 X \leq y) = P\left(X \leq \frac{y - \beta_1}{\beta_2}\right).$$

Since X is a beta random variable, from (3.1.1) we have

$$F_Y(y) = \int_0^{(y-\beta_1)/\beta_2} \frac{x^{\beta_3}(1 - x)^{\beta_4}}{\beta(\beta_3 + 1, \beta_4 + 1)} \, dx. \qquad (3.2.3)$$

Using the substitution $u = \beta_1 + \beta_2 x$ in (3.2.3), we get

$$F_Y(y) = \int_{\beta_1}^{y} \frac{(u - \beta_1)^{\beta_3}(\beta_1 + \beta_2 - u)^{\beta_4}}{\beta(\beta_3 + 1, \beta_4 + 1)\beta_2^{(\beta_3+\beta_4+1)}} \, du,$$

which, through differentiation gives us the p.d.f. as stated. $\blacksquare$

In order to fit a GBD($\beta_1, \beta_2, \beta_3, \beta_4$) to a dataset using the method of moments, we will need to match $\hat{\alpha}_1, \hat{\alpha}_2, \hat{\alpha}_3, \hat{\alpha}_4$ (as defined in (2.3.1) through

(2.3.4)) to their GBD($\beta_1, \beta_2, \beta_3, \beta_4$) counterparts, α_1, α_2, α_3, α_4. These $\alpha_1, \alpha_2, \alpha_3, \alpha_4$, for GBD($\beta_1, \beta_2, \beta_3, \beta_4$) distributions, are given in the following theorem.

Theorem 3.2.4.   *Let Y be a GBD($\beta_1, \beta_2, \beta_3, \beta_4$) random variable with mean μ_Y and variance σ_Y^2. Then*

$$\alpha_1 = \mu_Y = \beta_1 + \frac{\beta_2(\beta_3 + 1)}{B_2}, \tag{3.2.5}$$

$$\alpha_2 = E\left[(Y - \mu_Y)^2\right] = \sigma_Y^2 = \frac{\beta_2^2(\beta_3 + 1)(\beta_4 + 1)}{B_2^2 B_3}, \tag{3.2.6}$$

$$\alpha_3 = \frac{E\left[(Y - \mu_Y)^3\right]}{\sigma_Y^3} = \frac{2(\beta_4 - \beta_3)\sqrt{B_3}}{B_4\sqrt{(\beta_3 + 1)(\beta_4 + 1)}}, \tag{3.2.7}$$

$$\alpha_4 = \frac{E\left[(Y - \mu_Y)^4\right]}{\sigma_Y^4} = \frac{3B_3\left(\beta_3\beta_4 B_2 + 3\beta_3^2 + 5\beta_3 + 3\beta_4^2 + 5\beta_4 + 4\right)}{B_4 B_5(\beta_3 + 1)(\beta_4 + 1)}, \tag{3.2.8}$$

where $B_i = \beta_3 + \beta_4 + i$ for $i = 1, \ldots$.

Proof. We know from Definition 3.2.1 that $Y = \beta_1 + \beta_2 X$ where X has a beta distribution with parameters β_3 and β_4. From (3.1.17),

$$\alpha_1 = \mu_Y = E(Y) = \beta_1 + \beta_2 E(X) = \beta_1 + \frac{\beta_2(\beta_3 + 1)}{B_2}.$$

To prove the theorem for α_2, α_3, and α_4, we note that for $i = 2, 3, 4$,

$$\alpha_i = E\left[(Y - \mu_Y)^i\right] = E\left[\left(\beta_1 + \beta_2 X - \beta_1 - \frac{\beta_2(\beta_3 + 1)}{B_2}\right)^i\right]$$

$$= \beta_2^i E\left[\left(X - \frac{\beta_3 + 1}{B_2}\right)^i\right] = \beta_2^i E\left[(X - \mu_X)^i\right] \tag{3.2.9}$$

where μ_X denotes the mean of the beta random variable X. We see from (3.2.9) that $\alpha_2 = \beta_2^2 \sigma_X^2$ where σ_X^2 is the variance of X and (3.2.6) follows from (3.1.18).

To prove the results for α_3 and α_4 we divide by the appropriate power of σ_Y and use (3.1.19) or (3.1.20). ∎

We now proceed to show that the (α_3^2, α_4)-space of the GBD($\beta_1, \beta_2, \beta_3, \beta_4$) distributions contains the region

$$1 + \alpha_3^2 < \alpha_4 < 1.8(1 + \alpha_3^2)$$

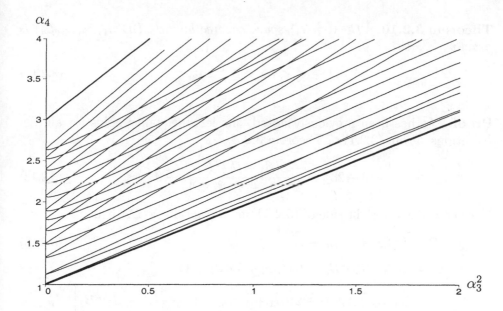

Figure 3.2–1. Contour plots of the (α_3^2, α_4) for $\beta_3 = -.99, -.9, -.7,$
$-.5, -.25, 0, .25, .75, 1.5, 2.5, 4, 6$ and for $-1 < \beta_4 < \infty$.

that the GLD$(\lambda_1, \lambda_2, \lambda_3, \lambda_4)$ did not cover. The contour plots of the (α_3^2, α_4)-points of the GBD$(\beta_1, \beta_2, \beta_3, \beta_4)$ depicted in Figure 3.2–1 give a preliminary indication that the desired region is covered. The curves of Figure 3.2–1 correspond to values of $\beta_3 = -.99, -.9, -.7, -.5, -.25, 0, .25, .75, 1.5, 2.5, 4, 6$ and $-1 < \beta_4 < \infty$. The lines shown in heavy print are the boundary lines $\alpha_4 = 1 + \alpha_3^2$ and $\alpha_4 = 3 + 2\alpha_3^2$. The "turning points" of the curves occur when $\beta_3 = \beta_4$ and we can see from (3.2.7) that at these points $\alpha_3 = 0$ and from (3.2.8) that

$$\alpha_4 = 3\frac{2\beta_3 + 3}{2\beta_3 + 5} \times \frac{2\beta_3^3 + 8\beta_3^2 + 10\beta_3 + 4}{(2\beta_3 + 4))(2\beta_3 + 1)^2} = 3\frac{2\beta_3 + 3}{2\beta_3 + 5}.$$

More specifically, when $\beta_3 = \beta_4 = -1$, the turning point is at $\alpha_4 = 1$ and when $\beta_3 = \beta_4 = 0$, the turning point is at $\alpha_4 = 1.8$. Moreover, when $\beta_3 = \beta_4$,

$$\lim_{\beta_3 = \beta_4 \to \infty} \alpha_4 = 3.$$

The following theorem gives the (α_3^2, α_4)-space of the GBD$(\beta_1, \beta_2, \beta_3, \beta_4)$ distributions.

Theorem 3.2.10. *The (α_3^2, α_4)-space covered by the* GBD$(\beta_1, \beta_2, \beta_3, \beta_4)$ *is exactly*

$$1 + \alpha_3^2 < \alpha_4 < 3 + 2\alpha_3^2. \tag{3.2.11}$$

Proof. As has been previously noted, the left inequality holds for all distributions. From (3.2.7) of Theorem 3.2.4,

$$3 + 2\alpha_3^2 = \frac{8(\beta_4 - \beta_3)^2 B_3}{B_4^2(\beta_3 + 1)(\beta_4 + 1)} + 3. \tag{3.2.12}$$

If we consider the right side of (3.2.11) as a fraction N/D, then

$$D = B_4^2(\beta_3 + 1)(\beta_4 + 1), \tag{3.2.13}$$

$$N = 8(\beta_4 - \beta_3)^2 B_3 + 3B_4^2(\beta_3 + 1)(\beta_4 + 1)$$

$$= 3\left[(B_3 + 1)(B_2 + 2)(\beta_3 + 1)(\beta_4 + 1) + \frac{8}{3}(\beta_4 - \beta_3)^2 B_3\right]$$

$$= 3B_3\left[(B_2 + 2)(\beta_3 + 1)(\beta_4 + 1) + \frac{8}{3}(\beta_4 - \beta_3)^2\right] + R$$

where

$$R = 3B_3(B_2 + 1)(\beta_3 + 1)(\beta_4 + 1) > 0. \tag{3.2.14}$$

Therefore,

$$N > 3B_3\left[B_2(\beta_3 + 1)(\beta_4 + 1) + 2(\beta_3 + 1)(\beta_4 + 1) + \frac{8}{3}(\beta_4 - \beta_3)^2\right]$$

$$= 3B_3\left[B_2\beta_3\beta_4 + 3\beta_3^2 + 3\beta_4^2 + 5(\beta_3 + \beta_4) + 4 + S\right],$$

with

$$S = \frac{2}{3}\beta_3^2 + \frac{2}{3}\beta_4^2 - \frac{4}{3}\beta_3\beta_4 = \frac{2}{3}(\beta_4 - \beta_3)^2 \geq 0.$$

This implies

$$N > 3B_3\left[B_2\beta_3\beta_4 + 3\beta_3^2 + 3\beta_4^2 + 5(\beta_3 + \beta_4) + 4\right]$$

and from (3.2.13) and (3.2.14),

$$3 + 2\alpha_3^2 = \frac{N}{D} > \frac{3B_3\left[B_2\beta_3\beta_4 + 3\beta_3^2 + 3\beta_4^2 + 5(\beta_3 + \beta_4) + 4\right]}{B_4^2(\beta_3 + 1)(\beta_4 + 1)} = \alpha_4.$$

Since α_3^2 and α_4 are continuous functions of β_3 and β_4, the (α_3^2, α_4)-space covered by the GBD$(\beta_1, \beta_2, \beta_3, \beta_4)$ is the entire space specified by the theorem. We have now shown that the (α_3^2, α_4) of every GBD$(\beta_1, \beta_2, \beta_3, \beta_4)$ satisfies (3.2.11). To complete the proof we need to show that for every pair $(\alpha_3^{2*}, \alpha_4^*)$ satisfying (3.2.11), there is a vector $(\beta_1^*, \beta_2^*, \beta_3^*, \beta_4^*)$ such that the GBD$(\beta_1^*, \beta_2^*, \beta_3^*, \beta_4^*)$ has $\alpha_3^2 = \alpha_3^{2*}$ and $\alpha_4 = \alpha_4^*$. This would show that the GBD$(\beta_1, \beta_2, \beta_3, \beta_4)$ "covers" the space described by (3.2.11). This is established numerically in Section 3.3. ∎

3.3 Estimation of GBD$(\beta_1, \beta_2, \beta_3, \beta_4)$ Parameters

To use the method of moments to estimate GBD$(\beta_1, \beta_2, \beta_3, \beta_4)$ parameters, we need to solve the system of equations

$$\alpha_i = \hat{\alpha}_i \quad \text{for } i = 1, 2, 3, 4$$

for $\beta_1, \beta_2, \beta_3, \beta_4$, where $\hat{\alpha}_i$ are specified in (2.3.1) through (2.3.4) and α_i are given in Theorem 3.2.4. This task is simplified somewhat by the fact that the subsystem of equations for $i = 3$ and 4 depends only on β_3 and β_4. Thus, if β_3 and β_4 could be obtained from $\alpha_3 = \hat{\alpha}_3$ and $\alpha_4 = \hat{\alpha}_4$, then $\alpha_2 = \hat{\alpha}_2$ would yield β_2 and subsequently, $\alpha_1 = \hat{\alpha}_1$ would give us β_1.

The complexity of the third and fourth equations makes it impossible to find closed form solutions; therefore, as in the case of the GLD$(\lambda_1, \lambda_2, \lambda_3, \lambda_4)$ distributions, we appeal to numeric methods based on search algorithms.

To develop some insight about where to begin the search for a solution we start by considering a specific case by letting $\hat{\alpha}_3^2 = 0.25$ and $\hat{\alpha}_4 = 2.5$. Figure 3.2–1 shows the contour curves for this choice of $\hat{\alpha}_3^2$ and $\hat{\alpha}_4$ (the "oval-shaped" curve is for $\hat{\alpha}_4 = 2.5$). The intersection of these two curves, roughly at $(2, 0.5)$ and $(0.5, 2)$, indicates the existence of solutions with (β_3, β_4) near these points when $(\hat{\alpha}_3, \hat{\alpha}_4) = (\pm 0.5, 2.5)$. For $\hat{\alpha}_3 = +\sqrt{0.25} = 0.5$ we have the solution near $(0.5, 2)$ and for $\hat{\alpha}_3 = -0.5$ we have the solution near $(2, 0.5)$.

A more precise solution can be obtained through the use of the *Maple* program `FindBetasM`. This program, whose listing is given in Appendix A, is similar to the `FindLambdasM` program discussed in Section 2.3. Suppose that for some $\hat{\alpha}_1$ and $\hat{\alpha}_2$ (say, $\hat{\alpha}_1 = 0$ and $\hat{\alpha}_2 = 1$), $\hat{\alpha}_3 = 0.5$ and $\alpha_4 = 2.5$, we want to obtain a solution for $\beta_1, \beta_2, \beta_3, \beta_4$. Since we know that (β_3, β_4) must be in the proximity of $(0.5, 2)$, we start the search for a solution at this

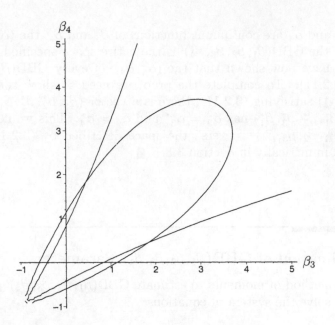

Figure 3.3–1. The contour curves for $\alpha_3^2 = 0.25$ and $\alpha_4 = 2.5$.

point. The command we use is

> $\texttt{FindBetasM}([0, 1, 0.5, 2.5], 0.5, 2)$;

and the result from this command is the $\text{GBD}(\beta_1, \beta_2, \beta_3, \beta_4)$ fit

$$\text{GBD}(-1.6439, 4.8592, 0.4499, 1.8358). \qquad (3.3.1)$$

Figures 3.3–2 and 3.3–3 give α_3^2 and α_4 contour curves, respectively. The curves in Figure 3.3–2 are for $\alpha_3^2 = 0.01, 0.05, 0.1, 0.2, 0.4, 1, 2, 3, 4, 5$, and 6. Each curve has two "branches" arranged symmetrically about the line $\beta_3 = \beta_4$. The innermost pair (those on either side of $\beta_3 = \beta_4$ and closest to it) is associated with $\alpha_3^2 = 0.01$, the next pair away from $\beta_3 = \beta_4$ is associated with $\alpha_3^2 = 0.05$, and so on. The contour curves given in Figure 3.3–3 have $\alpha_4 = 1.3, 1.6, 1.8, 1.9, 2, 2.1, 2.2, 2.3, 2.4, 2.5, 2.75, 3.5$, and 4.5. The innermost loop (hardly visible, near the point $(-1, -1)$) is for $\alpha_4 = 1.3$, the next is for $\alpha_4 = 1.6$, and so on.

To provide a graphic guide for solutions to $\hat{\alpha}_3 = \alpha_3$ and $\hat{\alpha}_4 = \alpha_4$, Figure 3.3–4 superimposes the sets of contours from Figures 3.3–2 and 3.3–3. The curves given in solid lines are the contours for α_4 and those in dotted lines are the contours for α_3^2.

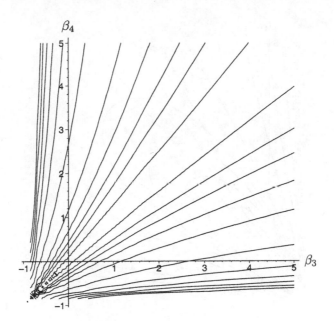

Figure 3.3–2. Contour curves for $\alpha_3^2 = .01, .05, .1, .2, .4, 1, 2, 3, 4, 5, 6.$

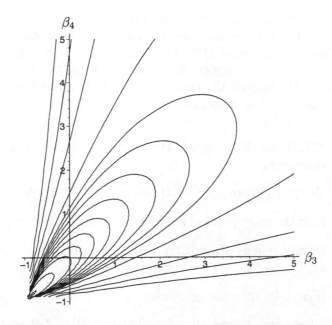

Figure 3.3–3. Contour curves for $\alpha_4 = 1.3, 1.6, 1.8, 1.9,$
$2, 2.1, 2.2, 2.3, 2.4, 2.5, 2.75, 3.5, 4.5, 6.$

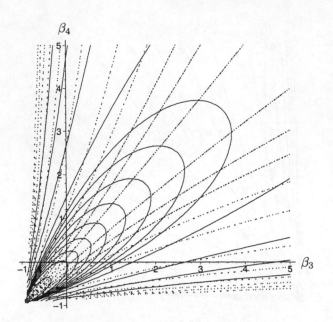

Figure 3.3–4. The contour curves for α_3^2 and α_4.

Those who do not have access to a computing system that (through *Maple* or some other software) can provide solutions to $\alpha_3 = \hat{\alpha}_3$ and $\alpha_4 = \hat{\alpha}_4$ can consult Dudewicz and Karian (1996) for extensive tabled values of (β_3, β_4) for (α_3, α_4) that cover the region of the GBD$(\beta_1, \beta_2, \beta_3, \beta_4)$ stipulated in Theorem 3.2.10. We include the table from Dudewicz and Karian (1996) in Appendix C and give an algorithm for using this table to obtain approximate solutions of β_3 and β_4.

Algorithm GBD–M: Fitting a GBD distribution to data by the method of moments.

GBD–M–1. Use (2.3.1) through (2.3.4) to compute $\hat{\alpha}_1, \hat{\alpha}_2, \hat{\alpha}_3, \hat{\alpha}_4$;

GBD–M–2. Find the entry point in a table of Appendix C closest to $(|\hat{\alpha}_3|, \hat{\alpha}_4)$;

GBD–M–3. Using $(|\hat{\alpha}_3|, \hat{\alpha}_4)$ from Step GBD–M–2 extract β_3 and β_4 from the table;

GBD–M–4. If $\hat{\alpha}_3 < 0$, interchange β_3 and β_4;

GBD–M–5. Substitute β_3 and β_4 (obtained in GBD–M–3 or GBD–M–4) and $\hat{\alpha}_2$, for β_3, β_4, and α_2, respectively, in (3.2.6) and solve it for β_2.

Equivalently, compute β_2 from

$$\beta_2 = (\beta_3 + \beta_4 + 2)\sqrt{\frac{(\beta_3 + \beta_4 + 3)\hat{\alpha}_2}{(\beta_3 + 1)(\beta_4 + 1)}}.$$

GBD–M–6. Substitute β_2, β_3 and β_4 (obtained in GBD–M–3 or GBD–M–4 and GBD–M–5) and $\hat{\alpha}_1$, for $\beta_2, \beta_3, \beta_4$, and α_1, respectively, in (3.2.5) and solve it for β_1. Equivalently, compute β_1 from

$$\beta_1 = \hat{\alpha}_1 - \frac{\beta_2(\beta_3 + 1)}{\beta_3 + \beta_4 + 2}.$$

To illustrate the use of Algorithm GBD–M and the table of Appendix C, consider the $\hat{\alpha}_1, \hat{\alpha}_2, \hat{\alpha}_3, \hat{\alpha}_4$ that were discussed earlier in the second paragraph of this section:

$$\hat{\alpha}_1 = 0, \quad \hat{\alpha}_2 = 1, \quad \hat{\alpha}_3 = 0.5, \quad \hat{\alpha}_4 = 2.5.$$

The entry point to the table of Appendix C is $\hat{\alpha}_3 = 0.5$ and $\hat{\alpha}_4 = 2.49$ and this produces $\beta_3 = 0.4253$ and $\beta_4 = 1.7781$. Next, following Step GBD–M–5, we compute $\beta_2 = 4.8186$. Finally, we follow Step GBD–M–6 and compute $\beta_1 = -1.6339$. As expected, this fit,

$$\text{GBD}(-1.6339, 4.8186, 0.4253, 1.7781), \tag{3.3.2}$$

obtained from Algorithm GBD–M and Appendix C, has $\beta_1, \beta_2, \beta_3, \beta_4$ close to the more accurate fit in (3.3.1) that was obtained through direct computation. For the fit in (3.3.2), we have

$$|\hat{\alpha}_1 - \alpha_1| = 3.0 \times 10^{-6}, \quad |\hat{\alpha}_2 - \alpha_2| = 1.7 \times 10^{-5},$$
$$|\hat{\alpha}_3 - \alpha_3| = 2.2 \times 10^{-5}, \quad |\hat{\alpha}_4 - \alpha_4| = 1.0 \times 10^{-2},$$

and for the fit in (3.3.1), we have

$$|\hat{\alpha}_1 - \alpha_1| = 2.1 \times 10^{-5}, \quad |\hat{\alpha}_2 - \alpha_2| = 6.6 \times 10^{-6},$$
$$|\hat{\alpha}_3 - \alpha_3| = 2.0 \times 10^{-5}, \quad |\hat{\alpha}_4 - \alpha_4| = 2.7 \times 10^{-5}.$$

However, if we use the full accuracy provided by `FindBetasM`, instead of the truncations shown in (3.3.1), we obtain

$$|\hat{\alpha}_1 - \alpha_1| < 10^{-10}, \quad |\hat{\alpha}_2 - \alpha_2| = 2.0 \times 10^{-10},$$
$$|\hat{\alpha}_3 - \alpha_3| = 2.0 \times 10^{-10}, \quad |\hat{\alpha}_4 - \alpha_4| = 1.0 \times 10^{-8}.$$

We now combine the results of this chapter with those of Chapter 2 to form the **system of distributions whose** (α_3^2, α_4)**-space (actually** $(\alpha_1, \alpha_2, \alpha_3, \alpha_4)$**-space since any** α_1 **and** α_2 **are possible) covers the entire range of possibilities.**

Definition 3.3.4. *The Extended Generalized Lambda Distribution (EGLD) system consists of the* $\mathrm{GLD}(\lambda_1, \lambda_2, \lambda_3, \lambda_4)$ *and* $\mathrm{GBD}(\beta_1, \beta_2, \beta_3, \beta_4)$ *families of distributions.*

Appendices B and C provide tabled solutions for some, but not all, of the (α_3^2, α_4) pairs arising from EGLD distributions. Figure 3.3–5 shows the (α_3^2, α_4)-space (restricted to $\alpha_3^2 \leq 4$ and $\alpha_4 \leq 22$) and various components of this space. The $\mathrm{GLD}(\lambda_1, \lambda_2, \lambda_3, \lambda_4)$ covers the shaded region, where $\alpha_3^2 \leq 4$ and $-1.9\alpha_3^2 + \alpha_4 \leq 15$ and solutions are available from the table of Appendix B. It also covers the region above the shaded area where tabled solutions are not available and computations become increasingly difficult as (α_3^2, α_4) moves away from the shaded area.

The portion of (α_3^2, α_4)-space covered by the $\mathrm{GBD}(\beta_1, \beta_2, \beta_3, \beta_4)$ is enclosed between two thick lines and marked "B E T A R E G I O N" in Figure 3.3–5. The GLD and GBD regions overlap and within this intersection both $\mathrm{GLD}(\lambda_1, \lambda_2, \lambda_3, \lambda_4)$ and $\mathrm{GBD}(\beta_1, \beta_2, \beta_3, \beta_4)$ fits are possible. This possibility has benefits as discussed in Section 2.5 and reduces the impact of the variability in (α_3^2, α_4). The table of Appendix C provides approximate solutions for some of the region covered by the $\mathrm{GBD}(\beta_1, \beta_2, \beta_3, \beta_4)$, with $\alpha_3^2 \leq 4$. When the (α_3^2, α_4) within this region gets close to the boundaries, computations become difficult, making it impossible to extend the table of Appendix C to (α_3^2, α_4) that are close to the boundaries. This is especially true for the upper boundary $(\alpha_4 = 3 + 2\alpha_3^2)$ where at least one of β_3 or β_4 gets very large when (α_3^2, α_4) gets close to the boundary. However, when (α_3^2, α_4) is close to the upper boundary we are also in the GLD region and have in Appendix B the table needed to fit a GLD. This, coupled with the fact that there are no computational problems in the non-overlap GBD region, means we have no problem fitting an EGLD anywhere in the combined regions.

The portion of Figure 3.3–5, bounded by the dotted lines, that straddles the GLD and GBD regions represents the (α_3^2, α_4) points of the F-distribution with $\alpha_3^2 < 4$. The lines that show the (α_3^2, α_4) points of a number of distributions considered in Section 2.4 are also shown in Figure 3.3–5 (some of these lines are actually curves when extended beyond $\alpha_3^2 = 4$). The lines of (α_3^2, α_4) points of the gamma, inverse Gaussian, lognormal, Student's t, and Weibull distributions are labeled with "G," "IG," "LN," "T,"

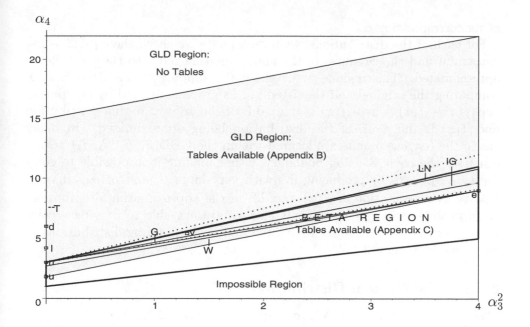

Figure 3.3–5. The (α_3^2, α_4)-space covered by the EGLD system.

and "W," respectively. Of the distributions considered in Section 2.4, the double exponential, exponential, logistic, normal, uniform, largest extreme value (hence, also the extreme value) have single points for their (α_3^2, α_4). These are designated in Figure 3.3–5, respectively, by "d," "e," "l," "n," "u," and "v." Only two of the distributions considered in Section 2.4 are not designated in Figure 3.3–5. The chi-square distribution is a special case of the gamma, hence its (α_3^2, α_4) points are included among those of the gamma distribution, and the Pareto distribution does not have any (α_3^2, α_4) points that are within the α_3^2 and α_4 limitations of Figure 3.3–5.

3.4 GBD Approximations of Some Well-Known Distributions

In this section we attempt to fit a $\text{GBD}(\beta_1, \beta_2, \beta_3, \beta_4)$ to the distributions already considered in Section 2.4, where $\text{GLD}(\lambda_1, \lambda_2, \lambda_3, \lambda_4)$ fits were found. When we use **FindBetasM** we are assured, in the absence of a warning message, that each of the $\hat{\alpha}_1, \hat{\alpha}_2, \hat{\alpha}_3, \hat{\alpha}_4$ produced by **FindBetasM** is within 10^{-6}

of its corresponding α_i.

For each of the distributions, we approximate we check the $\alpha_4 < 3 + 2\alpha_3^2$ constraint and the closeness of the approximating p.d.f. to the p.d.f. being approximated. This is done by checking the support of the fitted EGLD, comparing the original and the fitted p.d.f.s graphically, and by computing $\sup |\hat{f}(x) - f(x)|$ where $\hat{f}(x)$ is the p.d.f. of the approximating distribution and $f(x)$ is the p.d.f. of the distribution being approximated. In many cases, the (α_3^2, α_4) points are located within the GBD$(\beta_1, \beta_2, \beta_3, \beta_4)$ region but near the $\alpha_4 = 3 + 2\alpha_3^2$ boundary, either making it impossible to compute $\beta_1, \beta_2, \beta_3, \beta_4$ or producing fits with very large β_3 and/or β_4. In such situations computation of $\hat{F}(x)$, the d.f. of the approximating distribution, is impossible or extremely difficult and we are not able to provide a comparison of the d.f.s of the approximated and approximating distributions as was done in Section 2.4.

3.4.1   The Normal Distribution

The first four moments of $N(\mu, \sigma^2)$, given in Section 2.4.1, are

$$\alpha_1 = \mu, \qquad \alpha_2 = \sigma^2, \qquad \alpha_3 = 0, \qquad \alpha_4 = 3.$$

Since these satisfy $\alpha_4 = 3 + 2\alpha_3^2$, this (α_3^2, α_4) is located on the boundary of the (α_3^2, α_4) pairs that are attainable by the GBD$(\beta_1, \beta_2, \beta_3, \beta_4)$. We saw earlier that when $\beta_3 = \beta_4$ (a necessary condition for symmetric distributions such as $N(0, 1)$),

$$\alpha_4 = 3\frac{2\beta_3 + 3}{2\beta_3 + 5}.$$

Therefore, for large β_3, and $\beta_4 = \beta_3$, a GBD$(\beta_1, \beta_2, \beta_3, \beta_4)$ will have its (α_3^2, α_4) close to $(0, 3)$, the (α_3^2, α_4) of $N(0, 1)$. Using FindBetasM with $(\alpha_1, \alpha_2, \alpha_3, \alpha_4) = (0, 1, 0, 3)$ and an initial search point of $(10^5, 10^5)$ for (β_3, β_4) gives the fit

GBD$(-1.0119 \times 10^4,\ 2.0239 \times 10^4,\ 5.1201 \times 10^7,\ 5.1201 \times 10^7).$

The graph of the p.d.f. of the approximating GBD$(\beta_1, \beta_2, \beta_3, \beta_4)$ is indistinguishable from that of the $N(0, 1)$ distribution, giving a visual indication that this fit may be even better than the rather good GLD$(\lambda_1, \lambda_2, \lambda_3, \lambda_4)$ fit that was obtained in Section 2.4.1 (see Figure 2.4–1, where the $N(0, 1)$ and the fitted GLD$(\lambda_1, \lambda_2, \lambda_3, \lambda_4)$ p.d.f.s are *almost* indistinguishable). The support of the GBD$(\beta_1, \beta_2, \beta_3, \beta_4)$ approximation is $[-10119,\ 10119]$, contrasted with the support of $[-5.06,\ 5.06]$ for the GLD$(\lambda_1, \lambda_2, \lambda_3, \lambda_4)$ fit;

which one is preferable depends on the application on hand, as discussed at the beginning of Chapter 2. For this fit we have

$$\sup |\hat{f}(x) - f(x)| = 0.002364.$$

It is interesting to note that although the $(\alpha_3^2, \alpha_4) = (0, 3)$ could not be attained by a GBD$(\beta_1, \beta_2, \beta_3, \beta_4)$, the approximation that was obtained has $(\alpha_3^2, \alpha_4) = (1.930030232 \times 10^{-10}, 2.999999941)$.

3.4.2 The Uniform Distribution

The moments of the uniform distribution on the interval (a, b) are, as noted in Section 2.4.2,

$$\alpha_1 = \frac{a+b}{2}, \qquad \alpha_2 = \frac{(b-a)^2}{12}, \qquad \alpha_3 = 0, \qquad \alpha_4 = \frac{9}{5}.$$

Since

$$\alpha_4 = \frac{9}{5} < 3 = 3 + 2\alpha_3^2$$

for all uniform distributions, we should expect GBD$(\beta_1, \beta_2, \beta_3, \beta_4)$ approximations that are reasonably good. In the specific case of the uniform distribution on the $(0, 1)$ interval, we search via `FindBetasM` with the $(0, 0)$ initial point suggested by the table of Appendix C, to obtain the fit

$$\text{GBD}(0, 1, 0, 0).$$

This yields the GBD$(\beta_1, \beta_2, \beta_3, \beta_4)$ p.d.f. of 1 on the $(0, 1)$ interval. As was the case with the GLD$(\lambda_1, \lambda_2, \lambda_3, \lambda_4)$ fit obtained in Section 2.4–2, we have a perfect fit for the uniform distribution on the $(0, 1)$ interval. In general, for the uniform distribution on the interval (a, b), the fit (again, this will be an exact fit) will be

$$\text{GBD}(a, b - a, 0, 0).$$

3.4.3 The Student's t Distribution

We recall from the discussion of $t(\nu)$ (the t distribution with ν degrees of freedom) in Section 2.4.3 that for the i-th moment of $t(\nu)$ to exist, ν must exceed i. Whenever the first 4 moments of $t(\nu)$ exist (i.e., when $\nu \geq 5$), these moments, as given in Section 2.4.3, will be

$$\alpha_1 = 0, \qquad \alpha_2 = \frac{\nu}{\nu - 2}, \qquad \alpha_3 = 0, \qquad \alpha_4 = 3\frac{\nu - 2}{\nu - 4}.$$

For all $t(\nu)$ distributions,

$$\alpha_4 = 3\frac{\nu - 2}{\nu - 4} > 3 = 3 + 2\alpha_3^2,$$

locating (α_3^2, α_4) outside of the region covered by the $\text{GBD}(\beta_1, \beta_2, \beta_3, \beta_4)$. This makes it impossible to find good $\text{GBD}(\beta_1, \beta_2, \beta_3, \beta_4)$ approximation to $t(\nu)$, particularly when ν is small. For large ν, the (α_3^2, α_4) of $t(\nu)$ will come close to the boundary $\alpha_4 = 3 + 2\alpha_3^2$ and we may be able to find a $\text{GBD}(\beta_1, \beta_2, \beta_3, \beta_4)$, such as the one for $N(0, 1)$ in Section 3.4.1, that provides a reasonable approximation to such a $t(\nu)$.

3.4.4   The Exponential Distribution

The exponential distribution with parameters α and θ, defined in Section 2.4.4, has

$$\alpha_1 = \theta, \qquad \alpha_2 = \theta^2, \qquad \alpha_3 = 2, \qquad \alpha_4 = 9$$

and

$$\alpha_4 = 9 < 11 = 3 + 2\alpha_3^2.$$

It is clear from the table of Appendix C that a $\text{GBD}(\beta_1, \beta_2, \beta_3, \beta_4)$ approximation of the exponential distribution must have large β_4 and β_3 near zero. For $\theta = 3$, we set $(\alpha_1, \alpha_2, \alpha_3, \alpha_4) = (3, 9, 2, 9)$ and use the initial point of $(\beta_3, \beta_4) = (0, 10^5)$ (again, it may take a number of trials and a bit of luck to arrive at an initial point that produces an approximation) in `FindBetasM` to find the approximation

$$\text{GBD}(-1.4630 \times 10^{-7}, \ 6.1517 \times 10^7, \ 5.1370 \times 10^{-13}, \ 2.0506 \times 10^7).$$

The p.d.f. of this distribution is indistinguishable from that of the exponential distribution with $\theta = 3$, providing an approximation that seems to be superior to the $\text{GLD}(\lambda_1, \lambda_2, \lambda_3, \lambda_4)$ obtained in Section 2.4.5 (see Figure 2.4–4) with

$$\sup |\hat{f}(x) - f(x)| = 0.001894.$$

The support for the $\text{GBD}(\beta_1, \beta_2, \beta_3, \beta_4)$ approximation is roughly $[-1.4 \times 10^{-7}, 6.2 \times 10^7]$ whereas the support for the $\text{GLD}(\lambda_1, \lambda_2, \lambda_3, \lambda_4)$ approximation of Section 2.4.4 is $(-\infty, \infty)$.

3.4.5   The Chi-Square Distribution

As noted in Section 2.4.5, $\alpha_1, \alpha_2, \alpha_3, \alpha_4$ for $\chi^2(\nu)$, the chi-square distribution with ν degrees of freedom, are

$$\alpha_1 = \nu, \qquad \alpha_2 = 2\nu, \qquad \alpha_3 = \frac{2\sqrt{2}}{\sqrt{\nu}}, \qquad \alpha_4 = 3 + \frac{12}{\nu}.$$

Since

$$\alpha_4 = 3 + \frac{12}{\nu} < 3 + \frac{16}{\nu} = 3 + 2\alpha_3^2,$$

the (α_3^2, α_4) pairs of the $\chi^2(\nu)$ are within the region covered by the GBD with (α_3^2, α_4) approaching the boundary of this region as $\nu \to \infty$. The contrast worthy of note between the $t(\nu)$ and $\chi^2(\nu)$ distributions is that the (α_3^2, α_4) pairs approach the boundary from opposite sides; from inside the region covered by the GBD$(\beta_1, \beta_2, \beta_3, \beta_4)$ in the case of $\chi^2(\nu)$, and from outside this region for $t(\nu)$.

To fit a GBD$(\beta_1, \beta_2, \beta_3, \beta_4)$ to a specific $\chi^2(\nu)$ distribution, say for $\nu = 5$, we use **FindBetasM** with

$$\alpha_1 = 5, \qquad \alpha_2 = 10, \qquad \alpha_3 = \frac{2\sqrt{10}}{5}, \qquad \frac{27}{5}$$

and the initial search point $(\beta_3, \beta_4) = (1.5, 5000)$ to obtain

$$\text{GBD}(-2.1381 \times 10^{-7},\ 8.2257 \times 10^{7},\ 1.5000,\ 4.1128 \times 10^{7}).$$

It is not simple to determine a suitable starting point. From Figure 3.3–4 it seems that we need a large β_4 to accommodate $\alpha_4 = 5.4$ with $0 < \beta_3 < 2$. After a few trials we can find an initial search point (not necessarily $(\beta_3, \beta_4) = (1.5, 5000)$) that provides a GBD$(\beta_1, \beta_2, \beta_3, \beta_4)$ approximation (not necessarily the one that is given here). Another way of finding an initial search point is to consult the table of Appendix C. The closest point to $(\alpha_3, \alpha_4) = (1.265, 5.4)$ in Appendix C is $(\alpha_3, \alpha_4) = (1.275, 4.96)$, for which $(\beta_3, \beta_4) = (0.4516, 12.1859)$. It is also clear from this table that to find a solution for $(\alpha_3, \alpha_4) = (1.265, 5.4)$ we need to take β_4 considerably larger than the tabled 12.1859 and take β_3 somewhat larger than the tabled value of 0.4516. With a few trials and some luck, we may be able to arrive at a suitable initial search point.

When the $\chi^2(5)$ and the fitted GBD$(\beta_1, \beta_2, \beta_3, \beta_4)$ p.d.f.s are plotted, it is not possible to distinguish the two graphs, making the GBD$(\beta_1, \beta_2, \beta_3, \beta_4)$ approximation, at least visually, superior to the GLD$(\lambda_1, \lambda_2, \lambda_3, \lambda_4)$ approximation obtained in Section 2.4.5 (see Figure 2.4–3). The support of the

GBD$(\beta_1, \beta_2, \beta_3, \beta_4)$ approximation is roughly $[-0.1 \times 10^{-6}, \ 8 \times 10^7]$ (that of the GLD$(\lambda_1, \lambda_2, \lambda_3, \lambda_4)$ is $[-54.3, \ 59.6]$) and

$$\sup |\hat{f}(x) - f(x)| = 0.0006473.$$

3.4.6 The Gamma Distribution

The gamma distribution with parameters α and θ, defined in Section 2.4.6, has

$$\alpha_1 = \theta\alpha, \qquad \alpha_2 = \theta^2\alpha, \qquad \alpha_3 = \frac{2}{\sqrt{\alpha}}, \qquad \alpha_4 = 3 + \frac{6}{\alpha}.$$

Since

$$\alpha_4 = 3 + \frac{6}{\alpha} < 3 + \frac{8}{\alpha} = 3 + 2\alpha_3^2,$$

GBD$(\beta_1, \beta_2, \beta_3, \beta_4)$ approximations should be available for all gamma distributions. For the specific case where $\alpha = 5$ and $\theta = 3$ considered in Section 2.4.6, we have

$$\alpha_1 = 15, \qquad \alpha_2 = 45, \qquad \alpha_3 = 2\frac{\sqrt{5}}{5}, \qquad \alpha_4 = \frac{21}{5}.$$

The table of Appendix C indicates a β_3 in the vicinity of 3 and a large β_4. Using the initial point $(\beta_3, \beta_4) = (3, 5000)$ in FindBetasM, we obtain the approximation

$$\text{GBD}(-6.5062 \times 10^{-7}, \ 2.0330 \times 10^8, \ 4.0000, \ 6.7766 \times 10^7).$$

The graphs of the p.d.f.s of the approximation and of the gamma distribution with $\alpha = 5$ and $\theta = 3$ are virtually identical with

$$\sup |\hat{f}(x) - f(x)| = 0.00003560.$$

The support of the GBD$(\beta_1, \beta_2, \beta_3, \beta_4)$ approximation is $[-6.5 \times 10^{-7}, \ 2.0 \times 10^8]$ whereas the support for the GLD$(\lambda_1, \lambda_2, \lambda_3, \lambda_4)$ approximation obtained in Section 2.4.6 is $[-58.4, 80.0]$.

3.4.7 The Weibull Distribution

The $\alpha_1, \alpha_2, \alpha_3, \alpha_4$ for the Weibull distribution with parameters α and β are given in Section 2.4.7. It is difficult to determine analytically the values of α, if there are any, for which $\alpha_4 < 3 + 2\alpha_3^2$. From the graph of $\alpha_4 - 3 - 2\alpha_3^2$ for $0 < \alpha \leq 1$ given in Figure 3.4–1 (a), we see that $\alpha_4 > 3 + 2\alpha_3^2$ for small

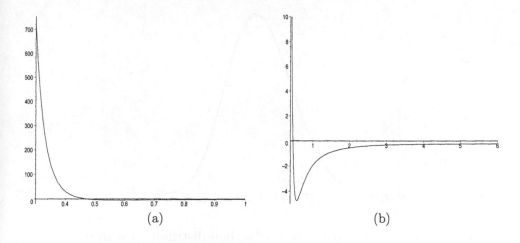

Figure 3.4–1. Graph of $\alpha_4 - 3 - 2\alpha_3^2$ for $0 \leq \alpha \leq 1$ (a)
and for $0.4 \leq \alpha \leq 6$ (b).

values of α and the inequality may be reversed when α is closer to 1 than to 0. The latter assertion is substantiated in Figure 3.4–1 (b) where the same graph is given for $0.4 \leq \alpha \leq 6$. An approximation of the specific value α_0 of α where $\alpha_4 = 3 + 2\alpha_3^2$ can now be found (using Newton's Method or the bisection method), giving us $\alpha_0 = 0.4709331608$. Thus, we conclude that a GBD$(\beta_1, \beta_2, \beta_3, \beta_4)$ approximation to a Weibull distribution is likely to be good if and only if $\alpha > \alpha_0 = 0.47$.

For the Weibull distribution considered in Section 2.4.7, with $\alpha = 1$ and $\beta = 5$, a GBD$(\beta_1, \beta_2, \beta_3, \beta_4)$ should be attainable. The moments of this distribution (from Section 2.4.7) are

$$\alpha_1 = 1.8363, \qquad \alpha_2 = 0.1769, \qquad \alpha_3 = -0.2541, \qquad \alpha_4 = 2.8803$$

and the table of Appendix C indicates a possible solution in the vicinity of $(\beta_3, \beta_4) = (6, 10)$. Note that in this case $\alpha_3 < 0$ and once β_3 and β_4 are determined, they must be interchanged.

The GBD$(\beta_1, \beta_2, \beta_3, \beta_4)$ approximation

$$\text{GBD}(-1.1593, \; 4.5402, \; 15.5955, \; 7.5567)$$

provided by `FindBetasM` seems to be quite good as can be seen in Figure 3.4–2 where the p.d.f.s of the distribution, its GBD$(\beta_1, \beta_2, \beta_3, \beta_4)$ approximation, and its GLD$(\lambda_1, \lambda_2, \lambda_3, \lambda_4)$ approximation (from Section 2.4.7) are plotted. At the point where the three curves peak, the lowest of the three curves is

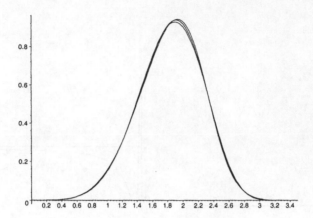

Figure 3.4–2. The p.d.f.s of the Weibull distribution with $\alpha = 1$,
$\beta = 5$, its GBD and GLD approximations.

the GBD$(\beta_1, \beta_2, \beta_3, \beta_4)$ approximation, the middle one is the Weibull distribution, and the highest curve is the GLD$(\lambda_1, \lambda_2, \lambda_3, \lambda_4)$, approximation. For the GBD$(\beta_1, \beta_2, \beta_3, \beta_4)$ fit, we have

$$\sup |\hat{f}(x) - f(x)| = 0.01598.$$

We also note that the supports of the GBD and the GLD approximations are $[-1.1593, 3.3809]$ and $[.0806, 3.8934]$, respectively.

3.4.8 The Lognormal Distribution

Closed form expressions for the first four moments of the lognormal distribution were given in Section 2.4.8. In that section, where the lognormal distribution was defined, we considered $\mu = 0$ and $\sigma = 1/3$, which yielded

$$\alpha_1 = 1.0571, \qquad \alpha_2 = 0.1313, \qquad \alpha_3 = 1.0687, \qquad \alpha_4 = 5.0974.$$

Indeed, $\alpha_4 = 5.0974 < 5.2842 = 3 + 2\alpha_3^2$, placing (α_3^2, α_4) within the region covered by the GBD$(\beta_1, \beta_2, \beta_3, \beta_4)$. However, (α_3^2, α_4) is close to the boundary and computations for $\beta_1, \beta_2, \beta_3, \beta_4$ produce only rough approximations with very large values of β_2 and β_4 (sometimes in excess of 10^{100}) or produce errors. For example, when

```
> FindBetasM([1.0571,.1313, 1.0687, 4.8], 50, 50);
```

is used, computation is shifted to the complex domain and the "fit"

$$\text{GBD}(-5.5155 \times 10^{149} - 4.8615 \times 10^{169}i, 4.1548 \times 10^{306} - 2.9178 \times 10^{306}i,$$
$$-1.8000^{338} + 3.5931 \times 10^{318}i, 1.0803 \times 10^{475} + 1.5384 \times 10^{475}i)$$

is obtained, with a warning that $|\hat{\beta}_3 - \beta_3| = 1.0687$ and $|\hat{\beta}_4 - \beta_4| = 1.8000$. When

$$> \texttt{FindBetasM}([1.0571, .1313, 1.0687, 4.8], 3, 50);$$

is used, computation remains in the real domain and the "fit"

$$\text{GBD}(0.3722, -0.2528 \times 10^{112}, 2.5728, -1.3189 \times 10^{114})$$

is obtained with a warning indicating that $|\hat{\beta}_3 - \beta_3| = 0.01060$ and $|\hat{\beta}_4 - \beta_4| = 0.1206$.

Investigations of other choices of μ and σ reveal that the situation encountered above does not change appreciably as long as σ stays small (i.e., (α_3^2, α_4) remains in the region covered by the $\text{GBD}(\beta_1, \beta_2, \beta_3, \beta_4)$ but close enough to the boundary to make computations unreliable). Moreover, as σ gets large (α_3 and α_4 do not depend on μ), (α_3^2, α_4) moves well into the region covered by the $\text{GLD}(\lambda_1, \lambda_2, \lambda_3, \lambda_4)$. For example, when $\mu = 0$ and $\sigma = 1$,

$$\alpha_1 = 1.6487, \qquad \alpha_2 = 4.6708, \qquad \alpha_3 = 6.1849, \qquad \alpha_4 = 113.9364$$

and when $\mu = 0$ and $\sigma = 2$,

$$\alpha_1 = 7.3890, \qquad \alpha_2 = 2926.3598, \qquad \alpha_3 = 414.3593, \qquad \alpha_4 = 9.2206 \times 10^6,$$

and in both cases, (α_3^2, α_4) is well outside of the region covered by the $\text{GBD}(\beta_1, \beta_2, \beta_3, \beta_4)$.

3.4.9   The Beta Distribution

The moments of the beta distribution derived in Section 3.1 are given in equations (3.1.17) through (3.1.20). It was also established through the definition of the GBD (see Definition 3.2.1) that if X is beta with parameters β_3 and β_4, then X is also GBD with parameters $\beta_1 = 0$, $\beta_2 = 1$, β_3, β_4. Therefore, the $\text{GBD}(0, 1, \beta_3, \beta_4)$ is an exact fit for the beta distribution with parameters β_3 and β_4.

3.4.10   The Inverse Gaussian Distribution

The definition of this distribution and its moments are given in Section 2.4.10. The inverse Gaussian distribution with parameters μ and λ has moments

$$\alpha_1 = \mu, \qquad \alpha_2 = \frac{\mu^3}{\lambda}, \qquad \alpha_3 = 3\sqrt{\frac{\mu}{\lambda}}, \qquad \alpha_4 = 3 + \frac{\mu}{\lambda}.$$

In this case

$$3 + 2\alpha_2^2 = \frac{3\lambda + 18\mu}{\lambda} > \frac{3\lambda + 15\mu}{\lambda} = \alpha_4,$$

placing (α_3^2, α_4) inside the space covered by the GBD for all $\mu > 0$ and $\lambda > 0$. However, as can be seen from the table of Appendix C, (α_3^2, α_4) pairs obtained from an inverse Gaussian distribution are beyond the entries of the table, and generally, beyond computation range.

3.4.11   The Logistic Distribution

We saw in Section 2.4.11, where this distribution is defined, that $(\alpha_3^2, \alpha_4) = (0, 4.2)$ for this distribution. This is outside the (α_3^2, α_4)-space covered by the GBD$(\beta_1, \beta_2, \beta_3, \beta_4)$; therefore, GBD cannot be fitted to a logistic distribution through the method of moments.

3.4.12   The Largest Extreme Value Distribution

Section 2.4.12 gives the definition and moments of this distribution, for which $(\alpha_3^2, \alpha_4) = (1.1395, 5.4)$. Since

$$3 + 2\alpha_3^2 = 5.59714 > 5.4 = \alpha_4,$$

the (α_3^2, α_4) of the largest extreme value distribution is inside the region covered by the GBD$(\beta_1, \beta_2, \beta_3, \beta_4)$. However, as was the case with the inverse Gaussian, this point is in the region that is outside our range of computation.

3.4.13   The Extreme Value Distribution

This distribution, defined in Section 2.4.13, has the same (α_3^2, α_4) as the largest extreme value distribution, producing the same circumstances discussed in Section 3.4.12.

3.4.14 The Double Exponential Distribution

Section 2.4.14 defines this distribution and gives its moments. For all values of the parameter $\lambda > 0$ of this distribution we have $\alpha_3 = 0$ and $\alpha_4 = 6$. Since (α_3^2, α_4) is outside the space covered by the $\text{GBD}(\beta_1, \beta_2, \beta_3, \beta_4)$, it is not possible to fit a GBD to a double exponential distribution through the method of moments.

3.4.15 The F-Distribution

This distribution and its moments are given in Section 2.4.15. It is clear from Figure 3.3–5 that the (α_3^2, α_4)-space of the F-distribution overlaps the space covered by the $\text{GBD}(\beta_1, \beta_2, \beta_3, \beta_4)$. But in almost all cases, even when (α_3^2, α_4) is within the intersection of the F-distribution and $\text{GBD}(\beta_1, \beta_2, \beta_3, \beta_4)$ regions, the (α_3^2, α_4) obtained from an F-distribution, such as the $(\alpha_3^2, \alpha_4) = (1.8101, 9.1986)$ of Section 2.4.15, is beyond the computation range for the $\text{GBD}(\beta_1, \beta_2, \beta_3, \beta_4)$.

3.4.16 The Pareto Distribution

Section 2.4.16 gives the definition and the moments of the Pareto distribution. The analysis of α_3 and α_4 in Section 2.4.16 makes it clear that (α_3^2, α_4) produced from this distribution cannot be in the region covered by the $\text{GBD}(\beta_1, \beta_2, \beta_3, \beta_4)$. Therefore, it is not possible to fit a $\text{GBD}(\beta_1, \beta_2, \beta_3, \beta_4)$ to a Pareto distribution through the method of moments.

3.5 Examples: GBD Fits of Data, Method of Moments

In this section we illustrate the use of the EGLD in fitting datasets. For the first two examples, we use data that is simulated from $\text{GBD}(\beta_1, \beta_2, \beta_3, \beta_4)$ distributions. In one case, we are able to obtain a GBD fit but no GLD fits; in the other case, we obtain GLD and GBD fits. In the next two examples in Sections 3.5.3 and 3.5.4, we reconsider two examples from Chapter 2. For the data on cadmium concentrations in horse kidneys, a $\text{GBD}(\beta_1, \beta_2, \beta_3, \beta_4)$ fit is obtained, giving us two fits for this data. The rainfall data of Section 2.5.4 did not yield any fits through the GLD but we are now able to find a $\text{GBD}(\beta_1, \beta_2, \beta_3, \beta_4)$ fit in the EGLD system for both X and Y components of this data. In Section 3.5.5 we introduce an example from forestry on the stand heights and diameters of trees.

3.5.1 Example: Fitting a GBD to Simulated Data from GBD(3, 5, 0, −0.5)

The following data, considered by Karian, Dudewicz and McDonald (1996), was generated from the GBD(3, 5, 0, −0.5) distribution.

7.88	6.77	7.98	6.73	6.79	6.99	3.25	4.10	7.25	5.74
5.03	5.22	7.99	7.41	7.21	7.81	7.87	7.73	8.00	5.75
6.81	7.79	7.42	6.17	7.96	4.63	7.43	7.77	3.55	4.31
8.00	6.82	7.72	7.90	7.43	6.96	6.71	4.50	4.88	6.03
6.06	4.56	4.06	4.47	6.21	3.87	6.46	3.29	5.99	7.59
7.96	7.59	7.13	5.13	7.89	7.24	7.00	8.00	6.65	4.52
7.98	7.84	7.98	8.00	7.95	5.20	7.85	7.95	6.79	6.84
5.19	5.72	7.99	7.33	7.34	3.09	7.41	5.60	4.76	5.03

The $\hat{\alpha}_1, \hat{\alpha}_2, \hat{\alpha}_3, \hat{\alpha}_4$ for this data is

$$\hat{\alpha}_1 = 6.4975, \quad \hat{\alpha}_2 = 2.0426, \quad \hat{\alpha}_3 = -0.7560, \quad \hat{\alpha}_4 = 2.3536.$$

The point $(\hat{\alpha}_3^2, \hat{\alpha}_4) = (0.5715, 2.3536)$ is in the region covered by both the $GLD(\lambda_1, \lambda_2, \lambda_3, \lambda_4)$ and $GBD(\beta_1, \beta_2, \beta_3, \beta_4)$. However, as can be seen from the table of Appendix B, this $(\hat{\alpha}_3^2, \hat{\alpha}_4)$ is outside the computation range for the GLD. To obtain a $GBD(\beta_1, \beta_2, \beta_3, \beta_4)$ fit, we use **FindBetasM** with initial search point $(-0.5, 0.1)$ to obtain

$$GBD(3.0551, 4.9615, 0.08247, -0.5223).$$

A histogram of the data with the p.d.f.s of GBD(3, 5, 0, −0.5), the distribution from which the data was generated, and the fitted GBD are shown in Figure 3.5–1 (a) (the fitted p.d.f. is slightly lower near the point 3). Figure 3.5–1 (b) gives the empirical d.f. together with the d.f.s of the two GBDs, which cannot be visually distinguished.

If we partition this data into the 7 classes

$$(-\infty, 4.5), \ [4.5, 5.5), \ [5.5, 6.5), \ [6.5, 7.0), \ [7.0, 7.5), \ [7.5, 7.9), \ [7.9, \infty),$$

we observe frequencies of

$$9, \quad 12, \quad 10, \quad 11, \quad 12, \quad 12, \quad 14.$$

The expected frequencies from the fitted distribution,

$$10.6415, \quad 9.7209, \quad 12.4782, \quad 8.0789, \quad 10.7142, \quad 14.4046, \quad 13.8844,$$

give a chi-square statistic and p-value of

$$2.8925 \quad \text{and} \quad 0.2354,$$

respectively.

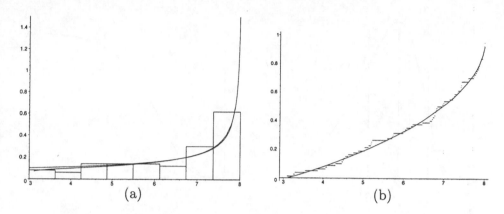

Figure 3.5–1. Histogram of data generated from GBD(3, 5, 0, −0.5) and
the p.d.f.s of the original and fitted GBDs (a); the
e.d.f. of the data with the d.f. of its fitted GBD (b).

3.5.2 Example: Fitting a GBD to Data Simulated from GBD(2, 7, 1, 4)

The following data, from Karian, Dudewicz and McDonald (1996), is generated from the GBD(2,7,1,4) distribution.

3.88	5.31	3.26	3.65	6.78	3.31	6.09	3.42	3.45	4.83
4.01	4.37	3.34	4.20	2.23	3.17	3.20	3.12	5.50	5.45
4.28	4.77	4.49	3.74	4.59	3.64	3.16	4.66	6.44	2.04
3.87	3.37	2.35	4.29	6.25	2.77	3.73	7.28	4.22	4.48
3.54	3.79	3.78	2.28	4.09	5.02	3.37	3.00	2.75	4.12
4.86	4.17	5.94	5.14	3.29	3.79	3.59	4.79	3.18	4.84

For this data

$$\hat{\alpha}_1 = 4.1053, \quad \hat{\alpha}_2 = 1.2495, \quad \hat{\alpha}_3 = 0.6828, \quad \hat{\alpha}_4 = 3.2998.$$

The point $(\hat{\alpha}_3^2, \hat{\alpha}_4) = (0.4663, 3.2998)$ is in the region where the GBD and
GLD overlap. When we use FindLambdasM with the initial search point
$(0.04, 0.2)$, we get the fit

$$GLD(3.2841, 0.1539, 0.04245, 0.2007),$$

and when we use FindBetasM with the initial search point $(1.5, 7)$, we get

$$GBD(1.8696, 10.4885, 1.9344, 9.8320).$$

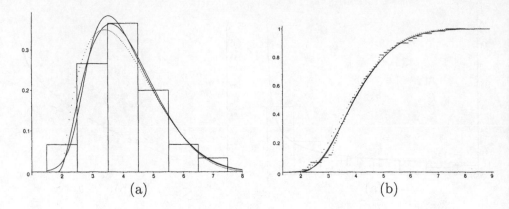

(a)   (b)

Figure 3.5–2. Histogram of data generated from GBD(2, 7, 1, 4) and the
p.d.f.s of the original (dotted line) and fitted GLD and
GBD (a); the e.d.f. of the data with the d.f.s of the
original (dotted line) and its fitted GLD and GBD (b).

Figure 3.5–2 (a) shows a histogram of the data with three p.d.f.s: one for
the distribution from which the data was generated (shown as a dotted line),
one for the GLD fit (the one that rises higher near the center), and one for
the GBD fit. Figure 5.3–2b shows the empirical d.f. of the data with the d.f.s
of the three p.d.f.s depicted in Figure 3.5–2a. The d.f. of the distribution
from which the data was generated is shown as a dotted curve and the other
two d.f.s cannot be distinguished.

To apply a chi-square test, the data is partitioned into the 6 intervals

$$(-\infty, 3), \quad [3, 3.5), \quad [3.5, 4.0), \quad [4.0, 4.5), \quad [4.5, 5.0), \quad [5.0, \infty),$$

producing the observed frequencies

$$6, \quad 14, \quad 11, \quad 11, \quad 7, \quad 11.$$

The expected frequencies for the GLD fit,

$$9.1915, \quad 10.9071, \quad 11.1862, \quad 9.4086, \quad 7.1169, \quad 12.1896,$$

yield a chi-square statistic and p-value of

$$2.3754 \quad \text{and} \quad 0.1233,$$

respectively. The GBD fit gives expected frequencies of

$$9.8047, \quad 10.2981, \quad 10.7894, \quad 9.4861, \quad 7.3604, \quad 12.2612,$$

producing a chi-square statistic and p-value of

$$3.2002 \quad \text{and} \quad 0.07363,$$

respectively.

3.5.3 Example: Cadmium in Horse Kidneys

In this section we reconsider the data of Section 2.5.2 and fit it with a $\text{GBD}(\beta_1, \beta_2, \beta_3, \beta_4)$. Recall that in Section 2.5.2 we gave the moments,

$$\hat{\alpha}_1 = 57.2419, \quad \hat{\alpha}_2 = 576.1894, \quad \hat{\alpha}_3 = 0.2545, \quad \hat{\alpha}_4 = 2.5253,$$

of the data and the GLD fit

$$\text{GLD}(41.7767, 0.01135, 0.09851, 2.3608).$$

To fit a $\text{GBD}(\beta_1, \beta_2, \beta_3, \beta_4)$ to this data we use `FindBetasM` with the initial search point $(2, 4)$ to obtain

$$\text{GBD}(-0.1827, 144.3347, 2.0502, 3.6193).$$

The p.d.f.s of the two fits together with a histogram of the data are shown in Figure 3.5–3 (a) (the GLD p.d.f. rises higher near the center). The e.d.f. of the data and the d.f.s of the fitted GLD and GBD are shown in Figure 3.5–3 (b).

Next, we partition the data into the 6 classes

$$(-\infty, 30), \quad [30, 50), \quad [50, 60), \quad [60, 70), \quad [70, 85), \quad [85, \infty)$$

and obtain the following frequencies in these classes

$$7, \quad 7, \quad 9, \quad 9, \quad 6, \quad 5.$$

From the expected frequencies for the GBD fit (those for the GLD fit are given in Section 2.5.2)

$$5.8577, \quad 11.7590, \quad 6.5332, \quad 5.9843, \quad 6.9084, \quad 5.9574,$$

we compute the chi-square statistic and corresponding p-value to obtain

$$4.8731 \quad \text{and} \quad 0.02728,$$

respectively. For comparison, we note that a chi-square statistic of 5.6095 and a p-value of 0.01786 were obtained for the GLD fit of Section 2.5.2.

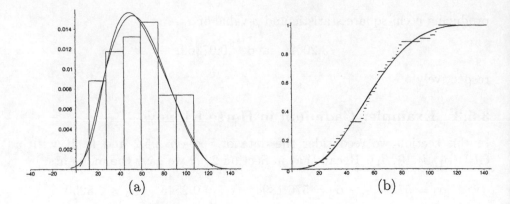

Figure 3.5–3. Histogram of the cadmium data and the p.d.f.s of
the fitted GLD and GBD (a); the e.d.f. of the data
with the d.f.s of the fitted GLD and GBD (b).

3.5.4 Example: Rainfall Data of Section 2.5.5

Table 2.5–6 of Section 2.5.5 gives the Rainfall (in inches) at Rochester (X)
and Syracuse (Y), New York, from May to October of 1998. In Section
2.5.5, after computing the sample moments for X and Y, we were not able
to find a GLD($\lambda_1, \lambda_2, \lambda_3, \lambda_4$) fit to either X or Y. We now attempt to fit
a GBD($\beta_1, \beta_2, \beta_3, \beta_4$) to X and Y. We start with the moments of X (from
Section 2.5.5),

$$\hat{\alpha}_1 = 0.4913, \quad \hat{\alpha}_2 = 0.4074, \quad \hat{\alpha}_3 = 1.8321, \quad \hat{\alpha}_4 = 5.7347,$$

and observe that $(\hat{\alpha}_3^2, \hat{\alpha}_4)$ lies in the region covered by the GBD($\beta_1, \beta_2, \beta_3, \beta_4$).
We use the initial point $(-0.7, 0.6)$ to obtain the fit

$$\text{GBD}(0.06129, 3.0802, -0.7491, 0.5463)$$

from `FindBetasM`. Figure 3.5–4 (a) shows a histogram of X and its fitted
GBD and Figure 3.5–4 (b) shows the e.d.f. of X with the d.f. of its fitted
GBD.

We partition the data into the classes

$$(-\infty, 0.07), \quad [0.07, 0.1), \quad [0.1, 0.2), \quad [0.2, 0.45), \quad [0.45, 1.0), \quad [1.0, \infty)$$

and get the observed expected frequencies

$$9, \quad 6, \quad 9, \quad 7, \quad 8, \quad 8.$$

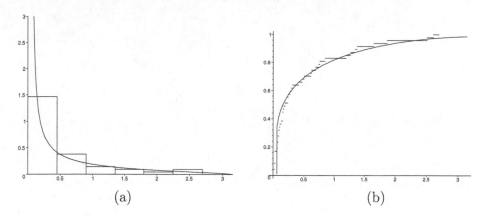

(a) (b)

Figure 3.5–4. Histogram of X and the p.d.f. of the fitted GBD (a); the e.d.f. of X with the d.f. of its fitted GBD (b).

From the fitted GBD, the expected frequencies

$$12.4465, \quad 5.6345, \quad 6.7382, \quad 7.0311, \quad 7.0502, \quad 8.0888$$

are obtained; these lead, respectively, to the following chi-square statistic and p-value

$$1.8663 \quad \text{and} \quad 0.1719.$$

The moments of Y are

$$\hat{\alpha}_1 = 0.3906, \quad \hat{\alpha}_2 = 0.1533, \quad \hat{\alpha}_3 = 1.6163, \quad \hat{\alpha}_4 = 5.2245.$$

With $(-0.5, 1.7)$ as an initial search point, FindBetasM produces the fit

$$\text{GBD}(0.07209, 2.1729, -0.5816, 1.4355).$$

A histogram of Y and the fitted p.d.f. are given in Figure 3.5–5 (a) and the e.d.f. of Y and the d.f. of the fitted GBD are given in Figure 3.5.–5 (b).

When Y is partitioned into the classes

$$(-\infty, 0.08), \quad [0.08, 0.15), \quad [0.15, 0.22), \quad [0.22, 0.38), \quad [0.38, 0.73), \quad [0.73, \infty),$$

we have observed frequencies of

$$7, \quad 7, \quad 9, \quad 8, \quad 8, \quad 8.$$

The expected frequencies from the fitted $\text{GBD}(\beta_1, \beta_2, \beta_3, \beta_4)$

$$6.9494, \quad 10.9353, \quad 5.1887, \quad 7.3122, \quad 8.5232, \quad 8.0677,$$

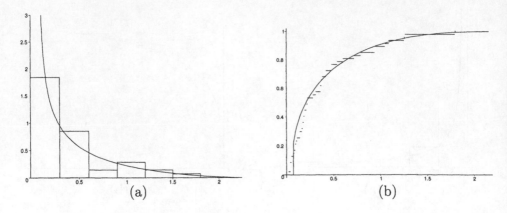

Figure 3.5–5. Histogram of Y and the p.d.f. of the fitted GBD (a);
the e.d.f. of Y with the d.f. of its fitted GBD (b).

produce the chi-square statistic and p-value of

$$3.0371 \quad \text{and} \quad 0.0814,$$

respectively. We thus see that the EGLD, through its GBD part, has succeeded in completing the GLD coverage.

3.5.5 Example: Tree Stand Heights and Diameters in Forestry

The estimation of tree stand volume is an important problem in forestry. Such estimations are based on measurements of tree diameters and heights (see Schreuder and Hafley (1977)). Unpublished studies of the $(\hat{\alpha}_3^2, \hat{\alpha}_4)$ of the two variables DBH (diameter, in inches, at breast height) and H (tree height, in feet) have shown that for the DBH variable, about 2% of stands have $(\hat{\alpha}_3^2, \hat{\alpha}_4)$ in the GLD but not the GBD region, about 70% of stands have $(\hat{\alpha}_3^2, \hat{\alpha}_4)$ in the overlapping region, and the other 28% have $(\hat{\alpha}_3^2, \hat{\alpha}_4)$ pairs in the GBD region only.

For the H variable, these three percentages are 0%, 20%, and 80%, respectively.

Table 3.5–6 gives DBH and H values for tree stands of 0.1 acres of Douglas fir in Idaho. (We are grateful for the communication of this data by Professor Lianjun Zhang of the Faculty of Forestry, SUNY College of Environmental Science and Forestry, Syracuse, New York, and Professor James A. Moore, Director of the Inland Tree Nutrition Cooperative (IFTNC), of the Department of Forest Resources, University of Idaho, and for permission to quote this example and data from their studies of tree growth in the Northwest, USA.)

For the DBH data in Table 3.5–6, we have

$$\hat{\alpha}_1 = 6.7404, \quad \hat{\alpha}_2 = 6.6721, \quad \hat{\alpha}_3 = 0.4544, \quad \hat{\alpha}_4 = 2.7450.$$

The $(\hat{\alpha}_3^2, \hat{\alpha}_4)$ lies in the overlap region of the GLD and GBD and we can invoke FindLambdasM with initial search point $(0.05, 0.5)$, and FindBetasM with initial search point $(2, 5)$, to obtain the fits

$$GLD(4.7744, 0.08911, 0.06257, 0.3056)$$

and

$$GBD(1.3522, 16.7462, 1.6296, 4.5429),$$

respectively. The support of the GLD fit, the interval $[-6.45, 16.00]$, covers the spread of the DBH data $[2.2, 14.8]$ but allows the possibility of values, including negative ones, well below the minimum of the data. The support of the GBD fit, $[1.3, 18.1]$, provides a better coverage of the spread of the data.

A histogram of DBH together with the p.d.f.s of the GLD (rising higher at the center) and GBD fits is shown in Figure 3.5–7 (a) and the e.d.f. of DBH with the d.f.s of the two fitted distributions is shown in Figure 3.5–7 (b).

To perform a chi-square test for these two fits, we partition the data for DBH into the classes

$$(-\infty, 3.75), \quad [3.75, 4.5), \quad [4.5, 5.0), \quad [5.0, 6.0), \quad [6.0, 7.0),$$
$$[7.0, 8.0), \quad [8.0, 9.0), \quad [9.0, 10.25), \quad [10.25, \infty)$$

and obtain the frequencies

$$10, \quad 9, \quad 11, \quad 11, \quad 7, \quad 10, \quad 12, \quad 10, \quad 9,$$

and expected frequencies of

10.0864, 8.4738, 6.5326, 13.6000, 12.7848, 11.0069, 8.8806, 8.1155, 9.5195

and

10.9161, 8.1165, 6.1359, 13.0145, 12.7265, 11.2992, 9.2219, 8.3249, 9.2445

Table 3.5–6. Tree diameters DBH (in inches), and heights H (in feet).

DBH	H	DBH	H	DBH	H	DBH	H	DBH	H
3.30	28.0	5.90	43.0	3.80	30.0	7.00	64.0	5.80	50.0
4.40	48.0	4.60	50.0	8.20	70.0	3.30	30.0	10.20	75.0
11.30	86.0	4.80	34.0	9.10	72.0	4.40	44.0	5.60	30.0
8.50	73.0	8.20	63.0	4.70	52.0	8.00	64.0	8.20	66.0
7.70	64.0	10.00	78.0	4.80	40.0	7.10	60.0	5.50	44.0
12.80	90.0	8.90	78.0	9.80	80.0	7.10	60.0	6.10	64.0
10.00	76.0	8.10	71.0	2.20	14.0	7.30	47.0	10.80	74.0
5.60	60.0	6.40	47.0	4.30	49.0	9.50	80.0	7.70	72.0
7.40	67.0	3.60	48.0	5.50	52.0	6.50	59.0	10.30	82.0
4.90	62.0	5.80	62.0	14.80	82.0	6.40	66.0	4.90	49.0
10.00	80.0	3.40	35.0	6.30	42.0	8.90	85.0	8.10	78.0
10.30	78.0	3.00	28.0	2.40	19.0	4.70	30.0	4.10	34.0
4.00	31.0	4.20	26.0	10.70	75.0	4.00	35.0	2.50	18.0
10.20	80.0	3.50	30.0	9.30	78.0	8.60	68.0	9.10	80.0
8.10	68.0	6.90	67.0	5.50	50.0	5.80	58.0	5.50	39.0
10.40	94.0	4.40	45.0	4.70	55.0	4.70	55.0	4.50	55.0
4.90	55.0	3.00	24.0	10.30	80.0	5.30	40.0	8.80	80.0
7.80	66.0	6.50	40.0	7.20	68.0	7.20	67.0		

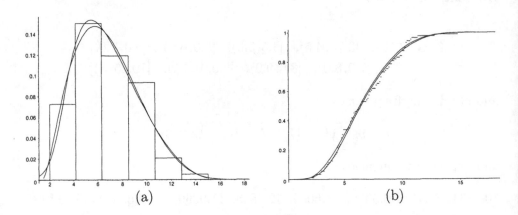

Figure 3.5–7. Histogram of DBH and the p.d.f.s of the fitted
GLD and GBD (a); the e.d.f. of DBH with the
d.f.s of its fitted GLD and GBD (b).

for the GLD and GBD fits, respectively. The chi-square statistic and p-value of the test for the GLD fit are

$$7.8569 \quad \text{and} \quad 0.09696,$$

and for the GBD fit these are

$$8.2471 \quad \text{and} \quad 0.08293.$$

For the H data of Table 3.5–6, we have

$$\hat\alpha_1 = 57.1348, \quad \hat\alpha_2 = 365.3751, \quad \hat\alpha_3 = -0.2825, \quad \hat\alpha_4 = 2.1046,$$

and again, $(\hat\alpha_3^2, \hat\alpha_4)$ is in the GLD and GBD overlap region. With the initial search point $(0.02, 0.6)$ FindLambdasM yields the fit

$$\text{GLD}(82.0495, 0.01442, 0.6212, 0.02459)$$

and with the initial search point of $(0.05, 0.5)$ FindBetasM yields the fit

$$\text{GBD}(11.3851, 77.9226, 0.7781, 0.2504).$$

The support of the GLD fit, $[12.70, 151.40]$, covers the spread of the data, which is $[14.0, 94.0]$. However, the GBD fit, with support $[11.4, 89.3]$ does not cover the spread of the data (see Problem 3.1).

The p.d.f.s of the fitted GLD and GBD and a histogram of H are shown in Figure 3.5–8 (a) (the GLD p.d.f. rises higher near the center) and the fitted d.f.s and the e.d.f. of H are shown in Figure 3.5–8 (b).

To perform a chi-square test the data is partitioned into the classes

$$(-\infty, 30), \quad [30, 37.5), \quad [37.5, 47.5), \quad [47.5, 55), \quad [55, 62.5),$$
$$[62.5, 67.5), \quad [67.5, 75), \quad [75, 80), \quad [80, \infty),$$

with observed frequencies of

$$7, \quad 10, \quad 11, \quad 9, \quad 11, \quad 11, \quad 9, \quad 8, \quad 13.$$

The expected frequencies of the GLD fit are

$$9.3662, 7.2608, 11.8326, 10.2410, 11.2074, 7.8919, 12.1651, 7.9312, 11.1037,$$

leading to the chi-square statistic and p-value of

$$4.2160 \quad \text{and} \quad 0.3776.$$

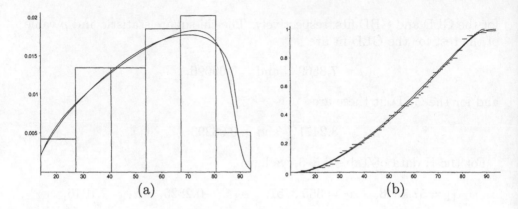

(a)   (b)

Figure 3.5–8.   Histogram of H and the p.d.f.s of the fitted
GLD and GBD (a); the e.d.f. of H with the
d.f.s of its fitted GLD and GBD (b).

For the GBD fit we get expected frequencies of

8.9201, 7.2814, 12.0188, 10.3900, 11.2445, 7.8001, 11.8147, 7.6531, 11.5860

and a chi-square statistic and p-value of

$$3.8776 \quad \text{and} \quad 0.4228,$$

respectively.

This example is continued in Section 5.5.4, where we consider the most
important aspect of this application: the fitting of a bivariate GLD–2 distribution to the pair (DBH, H).

3.6   EGLD Random Variate Generation

We saw in Section 1.4 that the GLD is very useful in simulation and Monte
Carlo studies as it can model a wide range of distribution shapes with a
simple model from which it is easy to generate data.   It is also easy to
change the distribution (e.g., for sensitivity studies) by simply changing the
$(\lambda_1, \lambda_2, \lambda_3, \lambda_4)$.   The EGLD completes the coverage of moment-space, as
we saw in Section 3.3, especially Figure 3.3–5.   To complete this gain for
simulation and Monte Carlo studies, we now specify **how to generate the
GBD part of the EGLD**.

Since (see the proof of Theorem 3.2.2) a GBD r.v., Y, can be represented as

$$Y = \beta_1 + \beta_2 X, \tag{3.6.1}$$

where X has the beta distribution of (3.1.1), **to generate values of Y we can generate values of X, say X_1, X_2, ... and then calculate**

$$Y_1 = \beta_1 + \beta_2 X_1, \; Y_2 = \beta_1 + \beta_2 X_2, \; \ldots . \tag{3.6.2}$$

Three methods of generating X_1, X_2, ... are given in Karian and Dudewicz (1999b), Section 4.6.8. Their Method 1 deals with the case where β_3 and β_4 are both from the set of values -0.5, 0, 0.5, 1, 1.5, 2, 2.5, ...; Method 2 uses the p.f. of a beta r.v., with the incomplete beta function and Newton-Raphson iteration; Method 3 works when β_3 and β_4 are positive. For other methods, some of them exact and fast, see Tadikamalla (1984, pp. 207–208).

Problems for Chapter 3

3.1. In the forestry example of Section 3.5.5, we found a GBD fit to the H (height) data that had support $[11.4, 89.3]$. However, the data ranged from a minimum of 18.0 to a maximum of 94.0. Thus, as noted in the example, the fit had the undesirable property that it could not "explain" all of the observed data. One way to handle such problems is to **constrain the search** for a fit to GBDs that satisfy the condition:

$$\beta_1 \leq 14 \quad \text{and} \quad \beta_1 + \beta_2 \geq 94 \tag{$*$}$$

since then **the support of the fitted GBD will be at least that of the data**.

Perform such a constrained search to obtain a GBD fit to the height H data that has support that covers the range of the data. [Hint: One way to accomplish this is to modify `FindBetasM` by adding a "penalty" to the quantity to be minimized when $(*)$ is not satisfied. This allows you to start the process with the usual initial point, but direct the process to seek solutions that have the desired support.]

3.2. For the constrained fit obtained in Problem 3.1, perform a chi-square test of fit and compare with the results from the unconstrained fit given in Example 3.5.5.

3.3. Modify the routines `FindLambdasM` and `FindBetasM` to obtain procedures `FindLambdasMR` and `FindBetasMR` ("R" for restricted) that produce
fits whose supports cover the support of the data. Find or construct a
dataset for which the original routines yield a support that is "too small"
and obtain fits through `FindLambdasMR` and `FindBetasMR`. (See the hint of
Problem 3.1.)

3.4. For the MRI data of Section 2.5.3, investigate fitting a GBD to the
data and compare your result with the GLD fits of Section 2.5.3.

3.5. For the twin birth-weight data of Section 2.5.4, investigate fitting a
GBD to each of X and Y and compare your results with the GLD fits of
Section 2.5.4.

3.6. For the coefficient of friction data of Section 2.6 given in the form of a
histogram, investigate fitting a GBD to the data and compare your results
with the GLD fits of Section 2.6.

3.7. For a dataset of your choice (not given in this book), investigate fitting
a GBD and compare your result with GLD fits and other fits given in the
literature.

Chapter 4

A Percentile-Based Approach to Fitting Distributions and Data with the GLD

In this chapter we consider a $GLD(\lambda_1, \lambda_2, \lambda_3, \lambda_4)$ fitting process that is based exclusively on percentiles. The concept and name "percentile" (also, "quartile" and "decile") are due to Galton (1875) (see Hald (1998, pp. 602, 604)), who in his 1875 paper proposed to characterize a distribution by its location (median), and its dispersion (half the interquartile range). These are basically (4.1.2) and half of a special case of (4.1.3) below. However, these do not include the two shape measures (4.1.4) and (4.1.5). Mykytka (1979) initially used percentiles for the estimation of $GLD(\lambda_1, \lambda_2, \lambda_3, \lambda_4)$ parameters through methods that relied on a mixture of moments and percentiles. The percentile-based approach that we will describe was proposed in Karian and Dudewicz (1999a). It fits a $GLD(\lambda_1, \lambda_2, \lambda_3, \lambda_4)$ distribution to a given dataset by specifying four percentile-based sample statistics and equating them to their corresponding $GLD(\lambda_1, \lambda_2, \lambda_3, \lambda_4)$ statistics. The resulting equations are then solved for $\lambda_1, \lambda_2, \lambda_3, \lambda_4$, with the constraint that the resulting GLD be a valid distribution.

To make the percentile approach an acceptable alternative to the method of moments and to provide the necessary computational support for the use of percentile-based fits, Dudewicz and Karian (1999a) give extensive tables for estimating the parameters of the fitted $GLD(\lambda_1, \lambda_2, \lambda_3, \lambda_4)$ distribution. These tables are reproduced in Appendix D.

There are three principal **advantages to the use of percentiles:**

1. There is a large class of $GLD(\lambda_1, \lambda_2, \lambda_3, \lambda_4)$ distributions that have fewer than four moments and these distributions are excluded from consideration when one uses parameter estimation methods that require moments. On the occasions when moments do not exist or may be out of table range, percentiles can still be used to estimate parameters and obtain $GLD(\lambda_1, \lambda_2, \lambda_3, \lambda_4)$ fits.

2. The equations associated with the percentile method that we will consider are simpler and the computational techniques required for solving them provide greater accuracy.

3. The relatively large variability of sample moments of orders 3 and 4 can make it difficult to obtain accurate $\text{GLD}(\lambda_1, \lambda_2, \lambda_3, \lambda_4)$ fits through the method of moments.

4.1   The Use of Percentiles

For a given dataset, $X_1, X_2, \ldots, X_n$, let $\hat{\pi}_p$ denote the $(100p)$th percentile of the data. $\hat{\pi}_p$ is computed by first writing $(n+1)p$ as $r + (a/b)$, where r is a positive integer and a/b is a proper fraction, possibly zero. If $Y_1, Y_2, \ldots, Y_n$ are the order statistics of $X_1, X_2, \ldots, X_n$, then $\hat{\pi}_p$ can be obtained from

$$\hat{\pi}_p = Y_r + \frac{a}{b}\left(Y_{r+1} - Y_r\right). \tag{4.1.1}$$

This definition of the **$(100p)$th data percentile** differs from the usual definition. Consider, for example, $p = 0.5$ where the sample median is usually defined as $M_n = Y_k$ if $n = 2k + 1$ for some integer k and $M_n = (Y_k + Y_{k+1})/2$ if $n = 2k$ for some integer k. By contrast, the sample quantile of order 0.5 is usually defined as $Z_{0.5} = Y_{[0.5n]+1}$ where $[0.5n]$ denotes the largest integer less than $0.5n$. Since the sample quantile is defined as a function of a single order statistic, it is mathematically somewhat simpler. However, the sample median is a better estimate of the population median. The $\hat{\pi}_p$ that we have defined is the generalization of the definition of the sample median to the case $p \neq 0.5$ as described in Hogg and Tanis (1997), p. 25.

The sample statistics that we will use are defined by

$$\hat{\rho}_1 = \hat{\pi}_{0.5}, \tag{4.1.2}$$

$$\hat{\rho}_2 = \hat{\pi}_{1-u} - \hat{\pi}_u, \tag{4.1.3}$$

$$\hat{\rho}_3 = \frac{\hat{\pi}_{0.5} - \hat{\pi}_u}{\hat{\pi}_{1-u} - \hat{\pi}_{0.5}}, \tag{4.1.4}$$

$$\hat{\rho}_4 = \frac{\hat{\pi}_{0.75} - \hat{\pi}_{0.25}}{\hat{\rho}_2}, \tag{4.1.5}$$

where u is an arbitrary number between 0 and 1/4. These statistics have the following interpretations (where for ease of discussion we momentarily assume $u = 0.1$).

1. $\hat{\rho}_1$ is the **sample median**;

2. $\hat{\rho}_2$ is the **inter-decile range**, i.e., the range between the 10th percentile and 90th percentile;

3. $\hat{\rho}_3$ is the **left-right tail-weight ratio**, a measure of relative tail weights of the left tail to the right tail (distance from median to the 10th percentile in the numerator and distance from 90th percentile to the median in the denominator);

4. $\hat{\rho}_4$ is the **tail-weight factor** or the ratio of the inter-quartile range to the inter-decile range, which cannot exceed 1 and measures the tail weight (values that are close to 1 indicate the distribution is not greatly spread out in its tails, while values close to 0 indicate the distribution has long tails).

In the case of $N(\mu, \sigma^2)$, **the normal distribution** with mean μ and variance σ^2, we have

$$\hat{\rho}_1 = \mu, \qquad \hat{\rho}_2 = 2.56\sigma, \qquad \hat{\rho}_3 = 1, \qquad \hat{\rho}_4 = 1.36/2.56 = 0.53.$$

This indicates, respectively, that the median of $N(\mu, \sigma^2)$ is μ, the middle 80% of the probability is in the range of about two-and-a-half standard deviations from the median, left and right tail weights are equal, and the inter-quartile range is 53% of the inter-decile range.

From the definition of the GLD($\lambda_1, \lambda_2, \lambda_3, \lambda_4$) inverse distribution function (1.1.1), we now define $\rho_1, \rho_2, \rho_3, \rho_4$, the GLD counterparts of $\hat{\rho}_1, \hat{\rho}_2, \hat{\rho}_3, \hat{\rho}_4$, as

$$\rho_1 = Q\left(\frac{1}{2}\right) = \lambda_1 + \frac{(\frac{1}{2})^{\lambda_3} - (\frac{1}{2})^{\lambda_4}}{\lambda_2}, \tag{4.1.6}$$

$$\rho_2 = Q(1-u) - Q(u) = \frac{(1-u)^{\lambda_3} - u^{\lambda_4} + (1-u)^{\lambda_4} - u^{\lambda_3}}{\lambda_2}, \tag{4.1.7}$$

$$\rho_3 = \frac{Q(\frac{1}{2}) - Q(u)}{Q(1-u) - Q(\frac{1}{2})} = \frac{(1-u)^{\lambda_4} - u^{\lambda_3} + (\frac{1}{2})^{\lambda_3} - (\frac{1}{2})^{\lambda_4}}{(1-u)^{\lambda_3} - u^{\lambda_4} + (\frac{1}{2})^{\lambda_4} - (\frac{1}{2})^{\lambda_3}}, \tag{4.1.8}$$

$$\rho_4 = \frac{Q(\frac{3}{4}) - Q(\frac{1}{4})}{\rho_2} = \frac{(\frac{3}{4})^{\lambda_3} - (\frac{1}{4})^{\lambda_4} + (\frac{3}{4})^{\lambda_4} - (\frac{1}{4})^{\lambda_3}}{(1-u)^{\lambda_3} - u^{\lambda_4} + (1-u)^{\lambda_4} - u^{\lambda_3}}. \tag{4.1.9}$$

The following are direct consequences of the definitions of ρ_1, ρ_2, ρ_3, ρ_4:

1. Since λ_1 may assume any real value, we can see from (4.1.6) that this is also true for ρ_1.

2. Since $0 < u < 1/4$, we have $u < 1 - u$ and from (4.1.7) we see that $\rho_2 \geq 0$.

3. The numerator and denominator of ρ_3 in (4.1.8) are both positive; therefore, $\rho_3 \geq 0$.

4. In (4.1.9), because of the restriction on u, the denominator of ρ_4 must be greater than or equal to its numerator, confining ρ_4 to the unit interval.

In summary, the definitions of ρ_1, ρ_2, ρ_3, ρ_4 lead to the restrictions:

$$-\infty < \rho_1 < \infty, \quad \rho_2 \geq 0, \quad \rho_3 \geq 0, \quad 0 \leq \rho_4 \leq 1. \qquad (4.1.10)$$

If we consider $\rho_3 = \rho_3(u, \lambda_3, \lambda_4)$ and $\rho_4 = \rho_4(u, \lambda_3, \lambda_4)$ as functions of u, λ_3 and λ_4, we see that

$$\rho_3(u, \lambda_3, \lambda_4) = \frac{1}{\rho_3(u, \lambda_4, \lambda_3)} \qquad (4.1.11)$$

and

$$\rho_4(u, \lambda_3, \lambda_4) = \rho_4(u, \lambda_4, \lambda_3). \qquad (4.1.12)$$

A given set of values for ρ_1, ρ_2, ρ_3, ρ_4, subject to the restrictions given in (4.1.10), can be attained from *some* inverse distribution function but it is not clear that it can be attained from a $\text{GLD}(\lambda_1, \lambda_2, \lambda_3, \lambda_4)$ inverse distribution. To develop some insight into **the possible ρ_1, ρ_2, ρ_3, ρ_4 that are attainable from $\text{GLD}(\lambda_1, \lambda_2, \lambda_3, \lambda_4)$ distributions**, we let $f(u, \lambda_3, \lambda_4)$ be the denominator of ρ_4 given in (4.1.9), i.e.,

$$f(u, \lambda_3, \lambda_4) = (1 - u)^{\lambda_3} - u^{\lambda_4} + (1 - u)^{\lambda_4} - u^{\lambda_3}.$$

Differentiating $f(u, \lambda_3, \lambda_4)$ with respect to u and simplifying we get

$$\frac{\partial}{\partial u} f(u, \lambda_3, \lambda_4) =$$
$$-\lambda_3 \left((1 - u)^{\lambda_3 - 1} + u^{\lambda_3 - 1} \right) - \lambda_4 \left((1 - u)^{\lambda_4 - 1} + u^{\lambda_4 - 1} \right). \qquad (4.1.13)$$

This derivative will be negative if both λ_3 and λ_4 are positive, making $f(u, \lambda_3, \lambda_4)$ a decreasing function of u when $\lambda_3 \geq 0$ and $\lambda_4 \geq 0$. This makes ρ_4 an increasing function of u. Moreover, ρ_4 will attain a value of 1 if $u = 1/4$, regardless of the values of λ_3 and λ_4. Figure 4.1–1 gives a more precise view of the impact of u on the surfaces $\rho(u, \lambda_3, \lambda_4)$. Three such surfaces (from the lowest to the highest) for $u = 0.01$, 0.15, and 0.23, respectively, are shown in Figure 4.1–1.

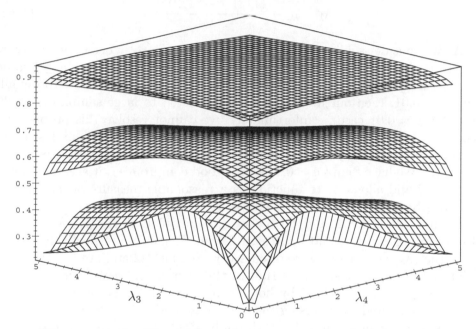

Figure 4.1–1. The surfaces (in rising order) $\rho_4(0.01, \lambda_3, \lambda_4)$, $\rho_4(0.15, \lambda_3, \lambda_4)$, and $\rho_4(0.23, \lambda_3, \lambda_4)$.

4.2 The (ρ_3, ρ_4)-space of $\mathbf{GLD}(\lambda_1, \lambda_2, \lambda_3, \lambda_4)$

The fitting of a $GLD(\lambda_1, \lambda_2, \lambda_3, \lambda_4)$ to a given dataset $X_1, X_2, \ldots, X_n$ is done by solving the system of equations $\hat{\rho}_i = \rho_i (i = 1, 2, 3, 4)$ for $\lambda_1, \lambda_2, \lambda_3, \lambda_4$. The definitions of $\hat{\rho}_1$, $\hat{\rho}_2$, $\hat{\rho}_3$, $\hat{\rho}_4$ in (4.1.2) through (4.1.5) may have seemed strange or arbitrary to this point. However, we now observe the main advantage of these definitions: the subsystem $\hat{\rho}_3 = \rho_3$ and $\hat{\rho}_4 = \rho_4$ involves only λ_3 and λ_4, allowing us to first solve this subsystem for λ_3 and λ_4 and use these values of λ_3 and λ_4 in $\hat{\rho}_2 = \rho_2$ to obtain λ_2 from

$$\lambda_2 = \frac{(1 - u)^{\lambda_3} - u^{\lambda_4} + (1 - u)^{\lambda_4} - u^{\lambda_3}}{\hat{\rho}_2}, \tag{4.2.1}$$

and finally, using the values of λ_2, λ_3 and λ_4 in $\hat{\rho}_1 = \rho_1$ to obtain

$$\lambda_1 = \hat{\rho}_1 - \frac{(1/2)^{\lambda_3} - (1/2)^{\lambda_4}}{\lambda_2}. \tag{4.2.2}$$

As we consider solving the system $\hat{\rho}_3 = \rho_3$, $\hat{\rho}_4 = \rho_4$, it becomes necessary to give u a specific value. For a particular u we must have $(n+1)u \geq 1$ to be able to compute $\hat{\pi}_u$ and $\hat{\pi}_{1-u}$, and eventually $\hat{\rho}_2$, $\hat{\rho}_3$, and $\hat{\rho}_4$. If u is too small, say $u = 0.01$, then our method will be restricted to large samples ($n \geq 99$ for the $u = 0.01$ case). Unfortunately, we cannot resolve this problem by choosing u close to $1/4$ because then the denominator of ρ_4 will be close to its numerator, making ρ_4 close to 1 and rendering it useless as a measure of "tail weight." We have found $u = 0.1$ to be a good compromise: it accommodates all $n \geq 9$ and allows ρ_4 to function as a reasonable measure of tail weight. **Throughout the rest of this chapter we take** $u = 0.1$.

Figures 4.2–1, 4.2–2, and 4.2–3 show contour plots of (ρ_3, ρ_4) for (λ_3, λ_4) from Regions 1, 2, 5 and 6, Region 3, and Region 4, respectively (recall that the (λ_3, λ_4) regions were discussed in Section 1.2 and are illustrated in Figure 1.2–1 of that section). Half of the contour curves of Figures 4.2–1 through 4.2–3 are obtained by holding λ_3 fixed and varying λ_4; the other half are generated by reversing the roles of λ_3 and λ_4.

It is clear from Figures 4.2–1 through 4.2–3 that there are constraints on (ρ_3, ρ_4). In some cases (e.g., when $(\rho_3, \rho_4) = (0.1, 0.7)$), the equations $\hat{\rho}_3 = \rho_3$ and $\hat{\rho}_4 = \rho_4$ have no solutions; in other cases (e.g., when $(\rho_3, \rho_4) = (0.3, 0.45)$), there seem to be multiple solutions. We can also see that the bulk of (ρ_3, ρ_4)-space is covered by (λ_3, λ_4) chosen from Regions 3 and 4 and very little of (ρ_3, ρ_4)-space is covered by (λ_3, λ_4) from the other regions. This prompts us to give preference to numerical searches for solutions from Regions 3 and 4.

The reasons for the existence of multiple solutions can be understood by considering the solutions associated with a fixed $(\hat{\rho}_3^*, \hat{\rho}_4^*)$. Any solution to $\rho_3 = \hat{\rho}_3^*$ and $\rho_4 = \hat{\rho}_4^*$ must be located simultaneously on these two surfaces. For purposes of illustration let us suppose that $(\hat{\rho}_3^*, \hat{\rho}_4^*) = (0.4, 0.5)$. Since a solution must be located on the intersection of the surface $\rho_4 = 0.5$ and the plane at 0.5 (Figure 4.2–4 illustrates this for (λ_3, λ_4) from Region 3), all

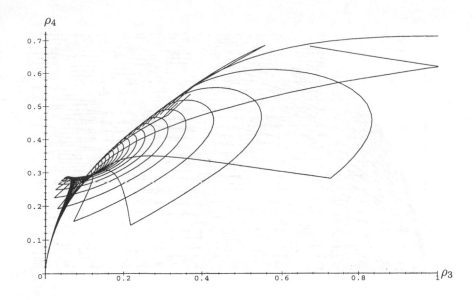

Figure 4.2–1. (ρ_3, ρ_4) generated by (λ_3, λ_4) from Regions 1, 2, 5, and 6 (see Figure 1.2–1).

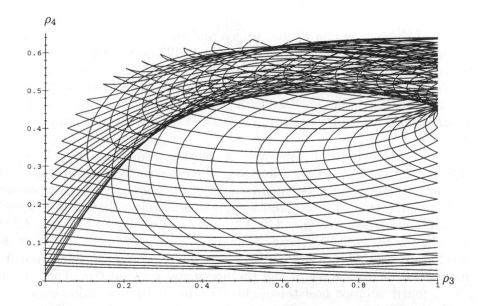

Figure 4.2–2. (ρ_3, ρ_4) generated by (λ_3, λ_4) from Region 3 (see Figure 1.2–1).

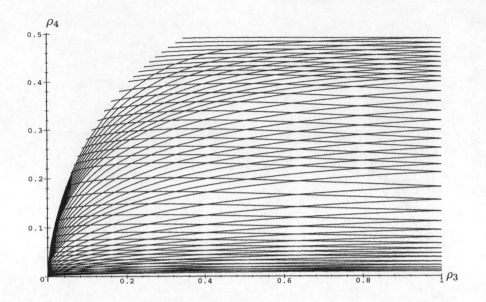

Figure 4.2–3. (ρ_3, ρ_4) generated by (λ_3, λ_4) from Region 4 (see Figure 1.2–1).

points of the intersection of the surface and the plane in Figure 4.2–4 are *possible* solutions. Similar intersection curves will also arise from considering the surface $\rho_3 = 0.4$ and the horizontal plane at 0.4. The actual solutions will be located at the points where these curves themselves intersect. Figure 4.2–5 shows the curves associated with $(\hat{\rho}_3^*, \hat{\rho}_4^*) = (0.4, 0.5)$ (the curves drawn in heavy lines are associated with $\rho_4 = 0.5$ and the more lightly drawn curves are for $\rho_3 = 0.4$).

Figure 4.2–5 indicates the presence of solutions near the points $(\lambda_3, \lambda_4) = (3, 21)$ and $(\lambda_3, \lambda_4) = (6, 1)$. Moreover, there seems to be a possibility of another solution in the lower right portion of the graph. When that portion of the graph is magnified (Figure 4.2–6), it becomes clear that there is indeed a third solution near $(\lambda_3, \lambda_4) = (40, 0.2)$. We can also observe in Figure 4.2–4 that the surface ρ_4 has a "crown" when (λ_3, λ_4) is near $(0, 0)$. It is possible that because of the limited resolution of the figure, we are not able to see the intersection of the surface ρ_4 and the horizontal plane at 0.5. There is, in fact, a fourth solution near the origin. Another possible solution suggested by Figure 4.2–5 is one near $(0.1, 40)$; the λ_4 of such a solution, if it exists, will have to be larger than 60. It is possible that there may be additional solutions that were not part of this analysis (Regions 1, 2, 4, 5, and 6).

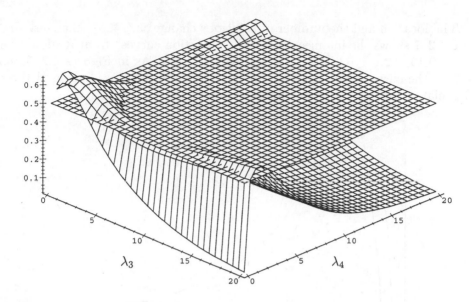

Figure 4.2–4. The surface ρ_4 with horizontal plane at .5 (Region 3).

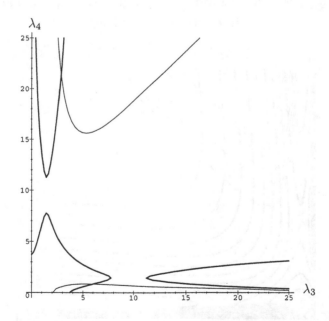

Figure 4.2–5. Solutions to $\rho_3 = .4$, $\rho_4 = .5$ at intersection points (Region 3)

The location and the number of solutions change as $(\hat{\rho}_3^*, \hat{\rho}_4^*)$ changes. Figure 4.2–7 shows the intersections of the "solution curves" from Region 3 for $\rho_3 = 0.15, 0.2, \ldots, 0.6$ and $\rho_4 = 0.1, 0.15, \ldots, 0.6$. As in Figures 4.2–5 and 4.2–6, the curves associated with ρ_4 are shown with heavy lines and those associated with ρ_3 with more lightly drawn lines.

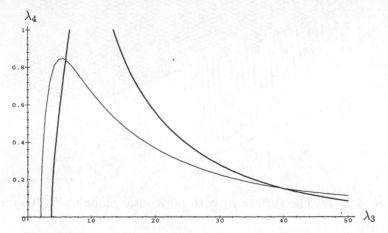

Figure 4.2–6. Solutions to $\rho_3 = 0.4$, and $\rho_4 = 0.5$ at intersection points (Region 3).

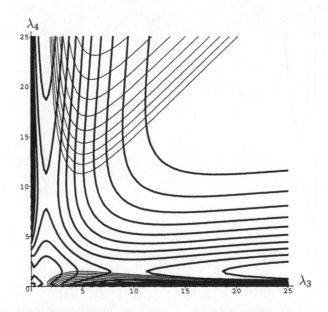

Figure 4.2–7. Intersections of ρ_3 and ρ_4 curves (Region 3).

4.3 Estimation of GLD Parameters Through a Method of Percentiles

As was the case with the equations of Chapters 2 and 3, $\hat{\rho}_3 = \rho_3$ and $\hat{\rho}_4 = \rho_4$ cannot be solved in closed form. We use a program written in *Maple* and similar to the searching programs of Chapters 2 and 3 to obtain approximate solutions to these equations. The listing of this program, called `FindLambdasP`, is given in Appendix A. In this case the equations are simpler, and good approximations with

$$\max(|\rho_3 - \hat{\rho}_3|, |\rho_4 - \hat{\rho}_4|) < 10^{-6}$$

are generally obtained within 3 or 4 iterations.

Solutions for a given $(\hat{\rho}_3, \hat{\rho}_4)$ can be found in various regions of (λ_3, λ_4)-space (depicted in Figure 1.2–1). Depending on the precise values of $\hat{\rho}_3$ and $\hat{\rho}_4$, as many as four solutions may exist in just one region (this occurs in Region 3). The following are more detailed observations that can be made regarding the presence of solutions in various regions of (λ_3, λ_4)-space.

1. It is clear from (4.1.12) that the ρ_4 curves are symmetric with respect to the line $\lambda_3 = \lambda_4$. We can also see from (4.1.11) that symmetric images of the ρ_3 curves (about the line $\lambda_3 = \lambda_4$) would be added in Figure 4.2–7 if $\hat{\rho}_3$ were allowed to exceed 1. When $\hat{\rho}_3 > 1$, we can obtain solutions by replacing $\hat{\rho}_3$ with $1/\hat{\rho}_3$ and exchanging the λ_3 and λ_4 that result from a solution associated with $(1/\hat{\rho}_3, \hat{\rho}_4)$.

2. For a substantial portion of $(\hat{\rho}_3, \hat{\rho}_4)$ with $0 < \hat{\rho}_3 < 1$ and $0 < \hat{\rho}_4 < 0.625$, there are solutions from Region 3.

   (a) Table D–1 of Appendix D gives solutions from Region 3 for a large subset of $(\hat{\rho}_3, \hat{\rho}_4)$ with this constraint.

   (b) For a subportion of the (ρ_3, ρ_4)-space with $0.1 < \rho_4 < 0.625$ and ρ_3 relatively close to ρ_4, there is another solution in Region 3. Table D–2 of Appendix D gives such solutions.

   (c) In the region $3.8 \leq \rho_3 \leq 1$ and $0.54 \leq \rho_4 < 0.625$, depending on the value of ρ_4, there may also be a solution (λ_3, λ_4) near $(0, 0)$; as ρ_3 and ρ_4 get large (i.e., as ρ_3 gets closer to 1 and ρ_4 gets closer to 0.625), these solutions gradually move away from the origin. Such solutions are tabulated in Table D–3 of Appendix D.

(d) In yet a smaller region within $0.66 \leq \rho_3 \leq 1$ and $0.586 \leq \rho_4 \leq 0.625$, it may be possible to obtain an additional solution. Such solutions are given in Table D–4 of Appendix D.

(e) It is also possible that for certain $(\hat{\rho}_3, \hat{\rho}_4)$ there may be yet another solution in Region 3. Such a solution, on the rare occasions that it exists, will have a small λ_3 (generally $\lambda_3 < 0.2$) and a considerably larger λ_4 (generally $\lambda_4 > 60$). These solutions are not tabulated.

3. When $0 < \hat{\rho}_3 < 1$ and $0 < \hat{\rho}_4 < 0.48$, there is a unique solution from Region 4. In contrast to the Region 3 solutions, these will produce $\mathrm{GLD}(\lambda_1, \lambda_2, \lambda_3, \lambda_4)$ distributions with infinite left and right tails (see Section 1.3 for a comprehensive discussion of $\mathrm{GLD}(\lambda_1, \lambda_2, \lambda_3, \lambda_4)$ shapes). Solutions from Region 4 are given in Table D–5 of Appendix D.

4. Solutions from Regions 1, 2, 5, and 6 are also possible. But, as can be observed from Figure 4.2–1, these are associated with a very limited set of $(\hat{\rho}_3, \hat{\rho}_4)$. Such solutions are not tabulated.

These observations are summarized in Figure 4.3–1 which shows the (ρ_3, ρ_4) pairs for which solutions are tabulated in Tables D–1, D–2, D–3, D–4, and D–5 of Appendix D. These regions are designated in Figure 4.3–1 by T1, T2, T3, T4, and T5, respectively. Boundaries are designated for T1, T2, T3, and T5, but not for T4.

1. The region for Table D–4 is not designated by its own boundary since it represents a rather small neighborhood of the point $(0.8, 0.6)$.

2. **The boundary of the region for Table D–2**, except for the small portion on the right where $\rho_3 = 1$, **is depicted with two thick curves**.

3. We can see from Figure 4.3–1 that most of the time when solutions exist there will be at least two solutions. Moreover, on some occasions (e.g., when $\rho_3 = 0.9$ and $\rho_4 = 0.61$), Tables 1, 2, 3, and 4 will provide four distinct solutions. On somewhat rare occasions (e.g., near the point that has $\rho_3 = 0.4$ and $\rho_4 = 0.6$), only a single solution will be given in Table 1.

4. As can be observed from Figures 4.2–1, 4.2–2, and 4.2–3, solutions will not exist in any GLD parameter Region (1, 2, 3, 4, 5, or 6) in the area marked "No Solutions" in Figure 4.3–1.

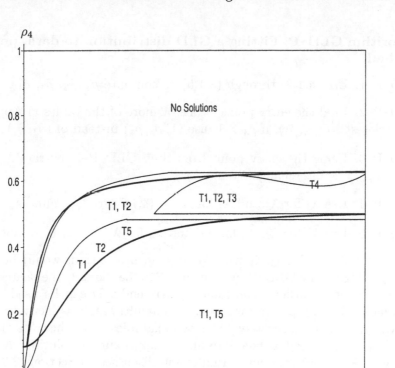

Figure 4.3–1. (ρ_3, ρ_4)-space covered by Tables 1, 2, and 3 (denoted (by T1, T2, and T3, respectively). The $N(0,1)$ and Cauchy points are $(1, .526)$ and $(1, .325)$, respectively.

5. The curves in Figure 4.3–1 are obtained by smoothing the "edge points" of the tables; consequently, the points on the curves represent approximations of these edge points. **The "upper" boundaries of the regions for Tables 1 and 5 are shown by lightly drawn curves** (the higher curve is the boundary for the region of Table 1). On the right, these two regions are bounded by $\rho_3 = 1$ and below they are bounded by $\rho_4 = 0$.

6. The boundary of Table 3 consists of the line $\rho_3 = 1$, a lightly drawn horizontal line at about $\rho_4 = 0.5$, and a lightly drawn curved upper portion.

With the availability of the tables of Appendix D, the algorithm below shows how to obtain numerical values of $\lambda_1, \lambda_2, \lambda_3, \lambda_4$ for a GLD fit.

Algorithm GLD–P: Fitting a GLD distribution to data, percentile method.

GLD–P–1. Use (4.1.2) through (4.1.5) to compute $\hat{\rho}_1$, $\hat{\rho}_2$, $\hat{\rho}_3$, $\hat{\rho}_4$.

GLD–P–2. Find the entry point in one or more of the tables of Appendix D closest to $(\hat{\rho}_3, \hat{\rho}_4)$; if $\hat{\rho}_3 > 1$, use $(1/\hat{\rho}_3, \hat{\rho}_4)$ instead of $(\hat{\rho}_3, \hat{\rho}_4)$.

GLD–P–3. Using the entry point from Step GLD–P–2, extract $\hat{\lambda}_3$ and $\hat{\lambda}_4$; if $\hat{\rho}_3 > 1$, interchange $\hat{\lambda}_3$ and $\hat{\lambda}_4$.

GLD–P–4. Use $\hat{\lambda}_3$ for λ_3 and $\hat{\lambda}_4$ for λ_4 in (4.1.7) to determine $\hat{\lambda}_2$.

GLD–P–5. Use $\hat{\lambda}_2$ for λ_2, $\hat{\lambda}_3$ for λ_3, and $\hat{\lambda}_4$ for λ_4 in (4.1.6) to obtain $\hat{\lambda}_1$.

To estimate $\lambda_1, \lambda_2, \lambda_3, \lambda_4$ with greater accuracy than would be possible through Algorithm GLD–P, we can use `FindLambdasP`, the searching program mentioned earlier. The tables of Appendix D can be used to obtain reasonable starting points for a search that ultimately is likely to estimate $\lambda_1, \lambda_2, \lambda_3, \lambda_4$ more accurately. The two examples that illustrate this below use the program `FindLambdasP` to find an approximate solution. As was the case with `FindLambdasM` and `FindBetasM` discussed in Sections 2.3 and 3.3, respectively, this program is written in *Maple* and requires three arguments: a list consisting of $\hat{\rho}_1$, $\hat{\rho}_2$, $\hat{\rho}_3$, $\hat{\rho}_4$, an initial estimate of λ_3 and an initial estimate of λ_4. By default, `FindLambdasP` uses $u = 0.1$; however, through an optional 4th argument, the user can specify another value for u.

Example 1. Suppose we want to find percentile-based fits when

$$\hat{\rho}_1 = 0, \quad \hat{\rho}_2 = 1, \quad \hat{\rho}_3 = 0.3, \quad \hat{\rho}_4 = 0.45.$$

From the location of $(0.3, 0.45)$ in Figure 4.3–1, we should expect three solutions to arise from entries of Tables D–1, D–2, and D–5 of Appendix D. These solutions, extracted from the tables, have, respectively,

$$(\lambda_3, \lambda_4) = (3.5362, 22.3488), \ (6.97905, 0.528742), \ (-0.0089, -0.2455).$$

The more precise computed estimators of $\lambda_1, \lambda_2, \lambda_3, \lambda_4$ can now be obtained through the Maple interaction:

```
> FindLambdasP([0, 1, 0.3, 0.45], 3.5, 22);

> FindLambdasP([0, 1, 0.3, 0.45], 7, 0.5);

> FindLambdasP([0, 1, 0.3, 0.45], -0.01, -0.2);
```

The three fits that result are, in order,

$$\text{GLD}_1(-0.1072482198, .7829248492, 3.574024953, 22.14595157),$$

$$\text{GLD}_2(0.6190898659, 1.113639228, 7.217498806, 0.5225051496),$$

and

$$\text{GLD}_5(-0.2281927712, -0.8542746993, -0.01319056636, -0.2679842550).$$

The fact that the supports of GLD_1, GLD_2, and GLD_3 are, respectively,

$$[-1.38, 1.17], \quad [-0.28, 1.52], \quad \text{and} \quad (-\infty, \infty),$$

gives us an indication that we are likely to have three rather diverse fits to choose from for our application. This is even more apparent from Figure 4.3–2 which shows the GLD_1, GLD_2, and GLD_5 p.d.f.s, designated by "(1)," "(2)," and "(5)" respectively. Because the computations of FindLambdasP guarantee that $|\hat{\rho}_3 - \rho_3|$ and $|\hat{\rho}_4 - \rho_4|$ are less than 10^{-8} and $\hat{\rho}_1$ and $\hat{\rho}_2$ can be easily obtained from (4.1.6) and (4.1.7), we can be certain that for all three fits $(\hat{\rho}_1, \hat{\rho}_2, \hat{\rho}_3, \hat{\rho}_4) = (0, 1, 0.3, 0.45)$.

Example 2. Let us consider

$$\hat{\rho}_1 = 0, \quad \hat{\rho}_2 = 1, \quad \hat{\rho}_3 = 0.9, \quad \hat{\rho}_4 = 0.61.$$

We can see from Figure 4.3–1 that four solutions may be attained from entries of Tables D–1, D–2, D–3, and D–4 with

$$(\lambda_3, \lambda_4) = (1.2099, 32.4923), \quad (2.48046, 2.05430), \quad (0.01419, 0.84334),$$
$$(0.91871, 0.67829).$$

Since there are actual entries for $(\rho_3, \rho_4) = (0.9, 0.61)$, the exact values of our $(\hat{\rho}_3, \hat{\rho}_4)$, we can obtain accurate estimators of (λ_3, λ_4) either through FindLambdasP or through Algorithm GLD–P. In either case, we obtain the following four fits.

$$\text{GLD}_1(-0.5078313823, 0.8512441110, 1.209933725, 32.49226836),$$

$$\text{GLD}_2(0.03939060205, 1.563264059, 2.480456830, 2.054298999),$$

$$\text{GLD}_3(-0.5395883345, 0.8022004003, 0.01419333590, 0.8433388506),$$

and

$$\text{GLD}_4(.06359226582, 1.508433272, .9187136135, .6782908239).$$

As in Example 1, the supports of these fits

$$[-1.68, 0.67], \quad [-0.60, 0.68], \quad [-1.79, 0.71], \quad \text{and} \quad [-0.60, 0.73]$$

for GLD_1, GLD_2, GLD_3 and GLD_4, respectively, indicate the possibility of a variety of p.d.f. shapes. This is substantiated in Figure 4.3–3 where the GLD_1, GLD_2, GLD_3, and GLD_4 p.d.f.s are shown (designated, respectively, by "(1)," "(2)," "(3)," "(4)").

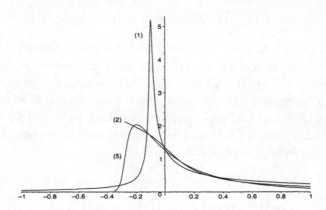

Figure 4.3–2. The GLD_1, GLD_2, and GLD_5 p.d.f.s of Example 1, marked by "(1)," "(2)," and "(5)," respectively.

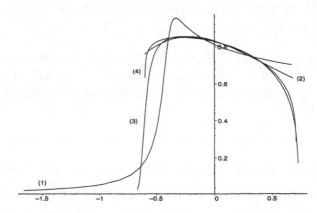

Figure 4.3–3. The GLD_1, GLD_2, GLD_3, and GLD_4 p.d.f.s of Example 2, marked by "(1)," "(2)," "(3)," and "(4)," respectively.

4.4 GLD Approximations of Some Well-Known Distributions

In this section we use the percentile-based method described in Sections 4.1 through 4.3 to fit GLD distributions to the important distributions encountered in applications that were considered in Sections 2.4 and 3.4. It follows from the discussion of Section 4.3, and particularly from Figure 4.3–1, that in most cases we will be able to fit several GLDs to a given distribution. Of course, this does not mean that all the fits will be good ones. In the following sections as we consider the distributions of Sections 2.4 and 3.4, we will discover that, generally, the percentile method produces three fits, of which one is clearly superior to the others. In our first example we fit the $N(0, 1)$ distribution, obtain three fits, and give all the details associated with each fit. In subsequent sections we concentrate, on the best fit.

4.4.1 The Normal Distribution

The percentile statistics, ρ_1, ρ_2, ρ_3, ρ_4, for $N(\mu, \sigma^2)$ are

$$\rho_1 = \mu, \quad \rho_2 = 2.5631\sigma, \quad \rho_3 = 1, \quad \rho_4 = 0.52631.$$

Locating (ρ_3, ρ_4) in Figure 4.3–1, we see that three fits are available from Tables D–1, D–2, and D–3. Through the use of these table entries and FindLambdasP for direct computations, we get the three fits GLD_1, GLD_2, and GLD_3 to $N(0,1)$, associated with Tables D–1, D–2, and D–3, respectively. These fits are

$$GLD_1(-0.8584, 0.3967, 1.5540, 15.4770),$$

$$GLD_2(0, 0.5456, 3.3897, 3.3897),$$

$$GLD_3(0, 0.2142, 0.1488, 0.1488),$$

with respective supports

$$[-3.38, 1.66], \quad [-1.83, 1.83], \quad \text{and} \quad [-4.67, 4.67].$$

GLD_1 is asymmetric (even though it has $\rho_3 = 1$ and hence, as measured by the left-right tail-weight is ρ_3-symmetric) because $\lambda_3 \neq \lambda_4$; therefore, it may not be a suitable fit for $N(0,1)$, depending on why $N(0,1)$ is chosen in a particular application. GLD_2 is symmetric but its support is much too confined to be a suitable fit for $N(0, 1)$. GLD_1, GLD_2, GLD_3, and the $N(0, 1)$ p.d.f.s are shown in Figure 4.4–1 and the GLDs are marked by "(1)," "(2),"

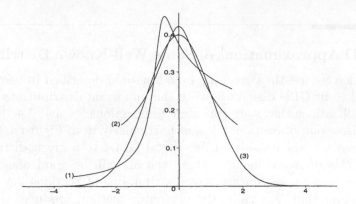

Figure 4.4–1. The GLD_1, GLD_2, and GLD_3 fits to $N(0,1)$ marked
by "(1)," "(2)," and "(3)," respectively; the $N(0,1)$
and GLD_3 p.d.f.s cannot be distinguished.

and "(3)." The $N(0,1)$ p.d.f. cannot be seen as a distinct curve because it
coincides (visually) with GLD_3.

The support of the moment-based GLD fit of $N(0,1)$ obtained in Section
2.4.1 was $[-5.06, 5.06]$ and that of the GBD fit obtained in Section 3.4.1
was $[-10119, 10119]$. By comparison, the support of GLD_3 is more limited;
however, this may not be a problem in many applications (see discussion at
the beginning of Chapter 2) because the $N(0,1)$ tail probability outside of
the GLD_3 support is approximately 3×10^{-6}. We complete our **first check**
for the $N(0,1)$ fits by noting that

$$\sup |\hat{f}_1(x) - f(x)| = 0.1547,$$

$$\sup |\hat{f}_2(x) - f(x)| = 0.08626,$$

$$\sup |\hat{f}_3(x) - f(x)| = 0.0006322,$$

where $\hat{f}_i(x)$ are the GLD_i p.d.f.s and $f(x)$ is the p.d.f. of $N(0,1)$. (See Section
2.4 for an explanation of how $\sup |\hat{f}(x) - f(x)|$ and $\sup |\hat{F}(x) - F(x)|$ are
computed.)

There are perceptible differences between the graphs of the GLD_1 and
GLD_2 d.f.s and the d.f. of $N(0,1)$. However, the graphs of the $N(0,1)$ and
GLD_3 d.f.s appear to be identical and to complete our **second check**, we
note that

$$\sup |\hat{F}_1(x) - F(x)| = 0.04783,$$

$$\sup |\hat{F}_2(x) - F(x)| = 0.03306,$$

$$\sup |\hat{F}_3(x) - F(x)| = 0.0005067,$$

where $\hat{F}_i(x)$ are the GLD$_i$ d.f.s and $F(x)$ is the d.f. of $N(0,1)$.

As indicated in the introductory paragraph of this section, we obtained several fits, with one of the fits, GLD$_3$, clearly superior to the others.

4.4.2 The Uniform Distribution

The values of ρ_1, ρ_2, ρ_3, ρ_4 for the uniform distribution on the interval (a, b) are

$$\rho_1 = \frac{1}{2}(a + b), \quad \rho_2 = \frac{4}{5}(a + b), \quad \rho_3 = 1, \quad \rho_4 = \frac{5}{8}.$$

We can see from Figure 4.3–1 that it is possible to find fits from Tables D–1, D–2, D–3, and D–4. Computations associated with the four entries, when $a = 0$ and $b = 1$, obtained from Tables D–1 through D–4, yield the following fits, respectively.

$$\text{GLD}_1(0.5309 \times 10^{-7}, 1.0000, 1.0000, 154.5288),$$

$$\text{GLD}_2(0.5000, 2.0000, 2.0000, 2.0000),$$

$$\text{GLD}_3(0.5000, 2.0000, 1.0000, 1.0000),$$

$$\text{GLD}_4(-0.2552 \times 10^{-12}, 1.0000, -0.1112 \times 10^{-12}, 1.0000).$$

GLD$_1$ lacks symmetry, as did one of the fits to $N(0,1)$ in Section 4.4.1. GLD$_4$ is not valid because its (λ_3, λ_4) lies in the region excluded for valid GLDs (see Figure 1.2–1). The fits GLD$_2$ and GLD$_3$ are actually a single fit. For either GLD$_2$ or GLD$_3$, the substitution of $\lambda_1, \lambda_2, \lambda_3, \lambda_4$ into $Q(y)$, the inverse distribution function that defines the GLD$(\lambda_1, \lambda_2, \lambda_3, \lambda_4)$ (see (1.2.1)), yields $Q(y) = y$. This not only establishes that GLD$_2$ and GLD$_3$ are the same, but that this common fit is a perfect one that matches the inverse distribution function of the uniform distribution on $(0, 1)$.

Perfect fits for the uniform distribution were also obtained via the GBD in Section 3.4.2 and through moment-based estimation in Section 2.4.2. See the latter part of Section 2.4.2 for a discussion of how to obtain a GLD fit to the general uniform distribution on the interval (a, b) from a fit of the uniform distribution on $(0, 1)$.

4.4.3 The Student's t Distribution

Because of symmetry, $t(\nu)$, the Student's t distribution with ν degrees of freedom, has $\rho_3 = 1$. When $\nu = 1$, $\rho_4 = 0.3249$ and as ν gets large, ρ_4 gets close to 0.5263, the ρ_4 of $N(0, 1)$. To fit $t(1)$, we first determine ρ_1, ρ_2, ρ_3, ρ_4 to obtain

$$\rho_1 = 0, \quad \rho_2 = 6.1554, \quad \rho_3 = 1, \quad \rho_4 = 0.3249.$$

Next, we locate $(\rho_3, \rho_4) = (1, 0.32)$ in Figure 4.3–1 and find out that fits are available through Tables D–1 and D–5. Doing the required computations, we are led to the fits (associated, respectively, with Tables D–1 and D–5),

$$\text{GLD}_1(0, 0.1698, 6.1596, 6.1596),$$

$$\text{GLD}_5(0, -2.0676, -0.8727, -0.8727).$$

Both GLD_1 and GLD_5 are symmetric and GLD_5 has $(-\infty, \infty)$ for its support. The p.d.f. plots of $t(1)$, GLD_1, and GLD_5 are shown in Figure 4.4–2 where the p.d.f.s of $t(1)$ and GLD_5 appear almost identical. To complete our **first check** for the GLD_5 fit, we note that

$$|\hat{f}(x) - f(x)| = 0.005157$$

(for the GLD_1 fit this is 0.1743). The d.f. of $t(1)$ cannot be graphically distinguished from the d.f. of the GLD_5 and

$$|\hat{F}(x) - F(x)| = 0.002376$$

(for the GLD_1 this figure is 0.05086).

To compare this method of fitting $t(\nu)$ with previous ones, we note that the GBD did not provide a fit for any ν (see Section 3.4.3) and the method of moments was not applicable for $\nu \leq 4$ because requisite moments did not exist for such ν (see Section 2.4.3). We were able to obtain the fit

$$\text{GLD}_m(0.1613 \times 10^{-17}, -0.2481, -0.1359, -0.1359)$$

for $t(5)$ through the method of moments in Section 2.4.3. For $t(5)$, the percentile-based method of this chapter produces

$$\rho_1 = 0, \quad \rho_2 = 2.9518, \quad \rho_3 = 1, \quad \rho_4 = 0.4924$$

and the fit (actually, the best of three possible fits)

$$\text{GLD}_p(0, -0.06445, -0.04118, -0.04118).$$

The p.d.f.s of $t(5)$ and GLD_p are visually indistinguishable but there is a perceptible distinction between the p.d.f.s of $t(5)$ and GLD_m (see Figure 2.4–2), indicating that GLD_p is a superior fit. For the GLD_p, we have

$$\sup |\hat{f}(x) - f(x)| = 0.001398 \quad \text{and} \quad \sup |\hat{F}(x) - F(x)| = 0.001550,$$

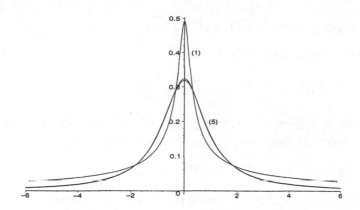

Figure 4.4–2. The GLD_1 and GLD_5 fits to $t(1)$ marked by "(1)" and "(5)," respectively; the $t(1)$ and GLD_5 p.d.f.s are almost identical.

compared to 0.03581 and 0.01488, respectively, for the GLD_m.

For $1 \leq \nu \leq 6$, the percentile-based fitting of $t(\nu)$ provides good fits with support $(-\infty, \infty)$. When $\nu = 7$,

$$\rho_1 = 0, \quad \rho_2 = 2.8298, \quad \rho_3 = 1, \quad \rho_4 = 0.5026$$

and a very good fit is obtained from $GLD(0, 0.02171, 0.01422, 0.01422)$, which has support $[-46.07, 46.07]$ and

$$\sup |\hat{f}(x) - f(x)| = 0.001256, \quad \sup |\hat{F}(x) - F(x)| = 0.001273.$$

As ν increases, the supports of the fits provided by the use of percentiles shrink to $[-4.67, 4.67]$ but the fits remain quite good, as can be seen from the limiting case of $N(0, 1)$ described in Section 4.4.1. The "best" fits for $\nu = 6$, 10, and 30 yield

$$(\lambda_1, \lambda_2, \lambda_3, \lambda_4) = (0, -0.01352, -0.008763, -0.008763), \text{ with}$$
$$\sup |\hat{f}(x) - f(x)| = 0.001324, \text{ and } \sup |\hat{F}(x) - F(x)| = 0.001388;$$

$$(\lambda_1, \lambda_2, \lambda_3, \lambda_4) = (0, 0.08277, 0.05521, 0.05521), \text{ with}$$
$$\sup |\hat{f}(x) - f(x)| = 0.001105 \text{ and } \sup |\hat{F}(x) - F(x)| = 0.001054;$$

$$(\lambda_1, \lambda_2, \lambda_3, \lambda_4) = (0, 0.1721, 0.1179, 0.1179), \text{ with}$$
$$\sup |\hat{f}(x) - f(x)| = 0.0007980 \text{ and } \sup |\hat{F}(x) - F(x)| = 0.0006888.$$

4.4.4   The Exponential Distribution

The d.f. of the exponential distribution is

$$F(x) = \int_0^x f(t)\, dt = 1 - e^{-x/\theta} \ \text{ for } x \geq 0 \tag{4.4.1}$$

and 0 for $x < 0$, where $f(t)$ is the p.d.f. of the distribution (see Section 2.4.4). The percentile function, $Q(x)$, of this distribution is

$$Q(x) = -\theta \ln(1 - x).$$

Since for $0 < p < 1$, π_p, the $100p$-th percentile, is characterized by $\pi_p = Q(p)$, we can easily compute $\pi_{0.1}$, $\pi_{0.25}$, $\pi_{0.5}$, $\pi_{0.75}$, and $\pi_{0.9}$ to obtain the ρ_1, ρ_2, ρ_3, ρ_4 of the exponential distribution with parameter θ. These are

$$\rho_1 = \theta \ln 2, \quad \rho_2 = 2\theta \ln 3, \quad \rho_3 = \frac{\ln 9}{\ln 5} - 1 = 0.3652, \quad \rho_4 = \frac{1}{2}.$$

Since for all values of θ, (ρ_3, ρ_4), rounded to two decimals, is $(0.37, 0.50)$, Figure 4.3–1 indicates the possibility of solutions from Tables D–1 and D–2. When $\theta = 3$, these lead, through the use of **FindLambdasP**, to the two fits

$$\text{GLD}_1(1.0498, 0.1250, 2.9578, 22.5624),$$

$$\text{GLD}_2(5.0180, 0.1967, 5.6153, 0.7407),$$

with respective supports

$$[-6.95, 9.05] \qquad \text{and} \qquad [-0.066, 10.10].$$

Figure 4.4–3 shows the p.d.f.s of GLD_1, GLD_2, and the exponential distribution with $\theta = 3$. It seems from Figure 4.4–3 that both fits have significant differences from the exponential, with GLD_2 (marked by "(2)") being the better of the two fits. To complete our **first check** for GLD_2 we note that

$$\sup |\hat{f}(x) - f(x)| = 0.2653.$$

For our **second check** we compare the d.f.s of the GLD_2 and exponential $(\theta = 3)$ distributions and observe that although they are visibly distinct, they are close to each other and

$$\sup |\hat{F}(x) - F(x)| = 0.03285.$$

In Section 2.4.4 the reasonably good fit

$$\text{GLD}_m(0.02100, -0.0003603, -0.4072 \times 10^{-5}, -0.001076)$$

with support $(-\infty, \infty)$ was obtained through the method of moments, and in Section 3.4.4 a very good fit that could not be visually distinguished from the exponential was obtained through the GBD. Figure 4.4–4 provides a comparison of the GLD_m and GLD_2 fits by showing their p.d.f.s with the p.d.f. of the exponential distribution with $\theta = 3$.

When $\theta = 1$, the better of the two percentile fits is

$$\text{GLD}(5.0180, 0.1967, 5.6153, 0.7407),$$

with support $[-0.022, 3.37]$ and

$$\sup |\hat{f}(x) - f(x)| = 0.7959 \quad \text{and} \quad \sup |\hat{F}(x) - F(x)| = 0.03285.$$

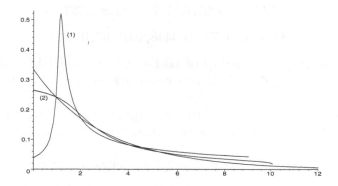

Figure 4.4–3. The p.d.f.s of the exponential distribution (with $\theta = 3$), GLD_1 marked by "(1)," and GLD_2 marked by "(2)".

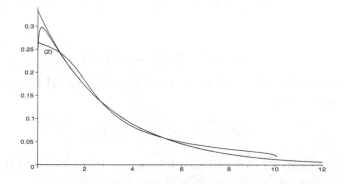

Figure 4.4–4. The p.d.f.s of the exponential distribution (with $\theta = 3$), GLD_2, and the moment-based GLD fit of Section 2.4.4.

4.4.5   The Chi-Square Distribution

In Sections 2.4.5 and 3.4.5 we considered $\chi^2(5)$, the chi-square distribution with $\nu = 5$ degrees of freedom. For this distribution,

$$\rho_1 = 4.3515, \quad \rho_2 = 7.6260, \quad \rho_3 = 0.5611, \quad \rho_4 = 0.5181$$

and from Figure 4.3–1 we see that there are three solutions associated with Tables D–1, D–2, and D–3. Using `FindLambdasP` we get the percentile-based $GLD(\lambda_1, \lambda_2, \lambda_3, \lambda_4)$:

$$GLD_1(2.7300, 0.1198, 2.3638, 18.7648),$$

$$GLD_2(6.5087, 0.1775, 6.0038, 1.3273),$$

$$GLD_3(2.4772, 0.03448, 0.01867, 0.1163).$$

When the graphs of the p.d.f.s of GLD_1, GLD_2, and GLD_3 are compared with that of the $\chi^2(5)$ p.d.f., it becomes obvious that GLD_3 is the superior fit.

The GBD in Section 3.4.5 led to a very good fit (graphically indistinguishable from $\chi^2(5)$) and the method of moments produced the reasonably good fit

$$GLD_m(2.6040, 0.01756, 0.009469, 0.05422)$$

in Section 2.4.5 (see Figure 2.4–4). Figure 4.4–5 shows the p.d.f.s of the $\chi^2(5)$, GLD_3, and the GLD_m fit obtained in Section 2.4.5. The highest rising curve is the moment-based p.d.f., the next highest is the GLD_3 p.d.f., and the lowest curve is the p.d.f. of $\chi^2(5)$. GLD_3 seems to be the better of the two fits; however, both fitted distributions have supports that extend below zero. To complete our **first check** we note that the support of GLD_3 is $[-26.53, 31.48]$ and

$$\sup |\hat{f}(x) - f(x)| = 0.01525.$$

For our **second check** we observe that the d.f.s of GLD_3 and $\chi^2(5)$ are indistinguishable and

$$\sup |\hat{F}(x) - F(x)| = 0.01202.$$

For other values of ν we continue to obtain three distinct GLD fits. The best of these fits for $\nu = 3, 10$, and 30 are, respectively,

$$(\lambda_1, \lambda_2, \lambda_3, \lambda_4) = (0.7445, 0.02524, 0.004756, 0.06524), \text{ with}$$

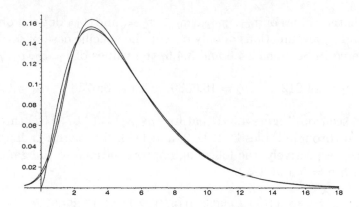

Figure 4.4–5. The $\chi^2(5)$ p.d.f. with its moment-based GLD fit (highest rising curve) and percentile-based GLD fit (next highest).

$$\sup|\hat{f}(x) - f(x)| = 0.05540 \text{ and } \sup|\hat{F}(x) - F(x)| = 0.01556;$$

$$(\lambda_1, \lambda_2, \lambda_3, \lambda_4) = (7.1811, 0.03341, 0.04262, 0.1541), \text{ with}$$
$$\sup|\hat{f}(x) - f(x)| = 0.004830, \text{ and } \sup|\hat{F}(x) - F(x)| = 0.008034;$$

$$(\lambda_1, \lambda_2, \lambda_3, \lambda_4) = (26.8430, 0.02403, 0.07999, 0.1743), \text{ with}$$
$$\sup|\hat{f}(x) - f(x)| = 0.001234 \text{ and } \sup|\hat{F}(x) - F(x)| = 0.004327.$$

4.4.6 The Gamma Distribution

The d.f. of $\Gamma(\alpha, \theta)$, the gamma distribution with parameters α and θ can be expressed as

$$F(x) = \int_0^x f(t)\, dt = 1 - \frac{\Gamma_{x/\theta}(\alpha)}{\Gamma(\alpha)} \quad \text{for } x \ge 0, \tag{4.4.2}$$

(and 0 if $x < 0$) where $f(t)$ is the p.d.f. of the gamma distribution (see Section 2.4.6), and $\Gamma_b(a)$ is the **incomplete gamma function**

$$\Gamma_b(a) = \int_0^b t^{a-1} e^{-t}\, dt.$$

For a discussion of the incomplete gamma function, see Abramowitz and Stegun (1964).

While the values of $\pi_{0.1}$, $\pi_{0.25}$, $\pi_{0.5}$, $\pi_{0.75}$, and $\pi_{0.9}$ depend on both α and θ, ρ_3 and ρ_4 are functions of only α. For the specific case of $\alpha = 5$ and $\theta = 3$ considered in Sections 2.4.6 and 3.4.6, the values of ρ_1, ρ_2, ρ_3, ρ_4 are

$$\rho_1 = 14.0127, \qquad \rho_2 = 16.6830, \qquad \rho_3 = 0.6736, \qquad \rho_4 = 0.5225.$$

We can see from Figure 4.3–1 that for $(\rho_3, \rho_4) = (0.67, 0.52)$ three fits can be obtained through Tables D–1, D–2, and D–3. Through `FindLambdasP` these produce, respectively, the following approximations to the gamma distribution with $\alpha = 5$ and $\theta = 3$:

$$\text{GLD}_1(9.9636, 0.0571, 2.1126, 17.3966),$$

$$\text{GLD}_2(17.4845, 0.0819, 5.6490, 1.7161),$$

$$\text{GLD}_3(10.7717, 0.0223, 0.0426, 0.1541).$$

Upon inspection, we find that GLD_3, with support $[-34.07, 55.61]$, is the best of these fits with

$$\sup |\hat{f}(x) - f(x)| = 0.003220.$$

This takes care of our **first check**. For our **second check** we note that the graphs of the d.f.s of GLD_3 and the gamma distribution with $\alpha = 5$ and $\theta = 3$ look identical and

$$\sup |\hat{F}(x) - F(x)| = 0.008034.$$

In Section 2.4.6 a reasonably good fit

$$\text{GLD}_m(10.7620, 0.01445, 0.02520, 0.09388)$$

was obtained when the $\text{GLD}(\lambda_1, \lambda_2, \lambda_3, \lambda_4)$ parameters were estimated via the method of moments.

In Section 3.4.6, through the GBD approximation, a very good fit was produced. The p.d.f. of this fit was visually indistinguishable from that of the gamma p.d.f. with $\alpha = 5$ and $\theta = 3$. Figure 4.4–6 shows the p.d.f. of the gamma distribution with $\alpha = 5$ and $\theta = 3$, along with the two approximations GLD_m and GLD_3. The curve that rises highest is the GLD_m p.d.f., the next highest is the GLD_3 p.d.f., and the lowest one is the p.d.f. of the gamma distribution. Of the two approximations, GLD_3 may be superior in many applications.

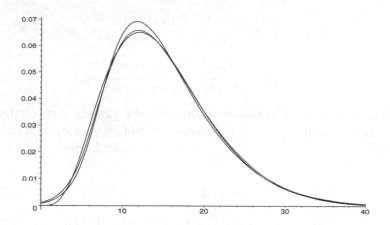

Figure 4.4–6. The p.d.f.s of GLD_m (the highest curve), GLD_3 (the next highest), and the gamma distributions (with $\alpha = 5$, $\theta = 3$).

4.4.7 The Weibull Distribution

The cumulative distribution function for the Weibull distribution with parameters α and β is expressible in closed form as

$$F(x) = \int_0^x f(t)\, dt = 1 - e^{-\alpha x^\beta} \quad \text{for } x \geq 0, \tag{4.4.3}$$

(and 0 if $x < 0$), where $f(t)$ is the Weibull p.d.f. (see Section 2.4.7). From (4.4.3) and the fact that $p = F(\pi_p)$, where π_p is the p-th percentile of this distribution, we have

$$p = 1 - e^{-\alpha \pi_p^\beta}$$

and solving this for π_p we get

$$\pi_p = \left[-\frac{\ln(1 - p)}{\alpha} \right]^{1/\beta}. \tag{4.4.4}$$

This allows us to derive $\pi_{0.1}$, $\pi_{0.25}$, $\pi_{0.5}$, $\pi_{0.75}$, and $\pi_{0.9}$ and subsequently obtain

$$\rho_1 = \left(\frac{\ln 2}{\alpha} \right)^{1/\beta}, \tag{4.4.5}$$

$$\rho_2 = \left(\frac{\ln 10}{\alpha} \right)^{1/\beta} - \left(\frac{\ln(10/9)}{\alpha} \right)^{1/\beta}, \tag{4.4.6}$$

$$\rho_3 = \frac{(\ln 2)^{1/\beta} - (\ln(10/9))^{1/\beta}}{(\ln 10)^{1/\beta} - (\ln 2)^{1/\beta}},\tag{4.4.7}$$

$$\rho_4 = \frac{(2\ln 2)^{1/\beta} - (\ln(4/3))^{1/\beta}}{(\ln 10)^{1/\beta} - (\ln(10/9))^{1/\beta}}.\tag{4.4.8}$$

We can see that as in the situation with the gamma distribution ρ_1 and ρ_2 depend on both distribution parameters, but ρ_3 and ρ_4 depend only on β.

When we take $\alpha = 1$ and $\beta = 5$, as was done in Sections 2.4.7 and 3.4.7, we obtain

$$\rho_1 = 0.9293, \quad \rho_2 = 0.5439, \quad \rho_3 = 1.1567, \quad \rho_4 = 0.52959.$$

Note that since $\rho_3 > 1$ we need to use $1/\rho_3 = 0.8645$ and exchange λ_3 and λ_4. The three solutions available from Tables D–1, D–2, and D–3 and obtained through FindLambdasP are, respectively,

$$\text{GLD}_1(1.0938, 1.8256, 16.3599, 1.7350),$$

$$\text{GLD}_2(0.8858, 2.5734, 2.5933, 4.2158),$$

$$\text{GLD}_3(0.9823, 1.0492, 0.2031, 0.1136).$$

Visual inspection leads us to discard GLD_1 and GLD_2 in favor of GLD_3, whose p.d.f. cannot be distinguished from that of the Weibull p.d.f. with $\alpha = 1$ and $\beta = 5$. To complete our **first check** we observe that the support of GLD_3 is $[0.029, 1.935]$ and

$$\sup |\hat{f}(x) - f(x)| = 0.01914.$$

For our **second check** we note that the graphs of the d.f.s of GLD_3 and the Weibull distribution with $\alpha = 1$ and $\beta = 5$ appear to be identical and

$$\sup |\hat{F}(x) - F(x)| = 0.001932.$$

The moment-based GLD and the GBD approximations obtained in Sections 2.4.7 and 3.4.7 were also good fits to this Weibull distribution (see Figure 3.4–3).

4.4.8 The Lognormal Distribution

The lognormal random variable, W, is defined by $W = e^X$ where X is $N(\mu, \sigma^2)$ (i.e., X is normally distributed with mean μ and variance σ^2). Since $X = \sigma Z + \mu$, where Z is $N(0,1)$, we can define W by

$$W = e^{\sigma Z + \mu} = e^\mu e^{\sigma Z}.$$

For $0 \leq p \leq 1$, π_p, the p-th percentile of W has, by definition, the property

$$p = P(W \leq \pi_p) = P(e^{\mu}e^{\sigma Z} \leq \pi_p) = P(Z \leq \frac{\ln \pi_p - \mu}{\sigma}).$$

If we define z_p to be the $100p$-th percentile of Z (i.e., $p = P(Z \leq z_p)$), we would have

$$z_p = \frac{\ln \pi_p - \mu}{\sigma}$$

and

$$\pi_p = e^{\sigma z_p + \mu}. \tag{4.4.9}$$

Since tabled values of z_p are commonly available, the $\pi_{0.1}$, $\pi_{0.25}$, $\pi_{0.5}$, $\pi_{0.75}$, and $\pi_{0.9}$ needed can be obtained from the corresponding values of z_p and equations (4.1.2) through (4.1.5) can be used to obtain the following formulas for ρ_1, ρ_2, ρ_3, ρ_4:

$$\rho_1 = e^{\mu} \tag{4.4.10}$$

$$\rho_2 = e^{\sigma z_{0.9} + \mu} - e^{-\sigma z_{0.9} + \mu} \tag{4.4.11}$$

$$\rho_3 = \frac{e^{\sigma z_{0.5} + \mu} - e^{-\sigma z_{0.9} + \mu}}{e^{\sigma z_{0.9} + \mu} - e^{\sigma z_{0.5} + \mu}} = \frac{e^{\sigma z_{0.5}} - e^{-\sigma z_{0.9}}}{e^{\sigma z_{0.9}} - e^{\sigma z_{0.5}}} \tag{4.4.12}$$

$$\rho_4 = \frac{e^{\sigma z_{0.75} + \mu} - e^{-\sigma z_{0.75} + \mu}}{e^{\sigma z_{0.9} + \mu} - e^{-\sigma z_{0.9} + \mu}} = \frac{e^{\sigma z_{0.75}} - e^{-\sigma z_{0.75}}}{e^{\sigma z_{0.9}} - e^{-\sigma z_{0.9}}}. \tag{4.4.13}$$

We see from (4.4.12) and (4.4.13) that the expressions for ρ_3 and ρ_4 do not involve μ.

We now consider, as we did in Sections 2.4.8 and 3.4.8, the specific lognormal distribution that has $\mu = 0$ and $\sigma = 1/3$. For these values of μ and σ,

$$\rho_1 = 1, \quad \rho_2 = 0.8806, \quad \rho_3 = 0.6523, \quad \rho_4 = 0.5149$$

and we are led to the following three solutions from Tables D–1, D–2, and D–3, respectively:

$$\text{GLD}_1(0.8001, 1.0811, 2.2099, 17.0513),$$

$$\text{GLD}_2(1.2074, 1.5217, 6.1996, 1.6032),$$

$$\text{GLD}_3(0.8393, 0.2934, 0.02937, 0.1005).$$

By looking at the graphs of the GLD_1 and GLD_2 we determine that they are not suitable approximations to the lognormal distribution under consideration, but GLD_3 provides a very good fit. Computation yields $[-2.57, 4.24]$ as the support of this fit and

$$\sup |\hat{f}(x) - f(x)| = 0.05381$$

completes our **first check**. For our **second check** we note that the graphs of the d.f.s of GLD_3 and the lognormal distribution under consideration appear identical and

$$\sup |\hat{F}(x) - F(x)| = 0.006971.$$

In Section 3.4.8 we found out that the GBD does not provide approximations to lognormal distributions and in Section 2.4.8 we found that

$$GLD_m(0.8451, 0.1085, 0.01017, 0.03422)$$

was a reasonably good fit for the lognormal with $\mu = 0$ and $\sigma = 1/3$ (see Figure 2.4-7). The p.d.f.s of the lognormal with $\mu = 0$ and $\sigma = 1/3$, the GLD_m, and the GLD_3 are shown in Figure 4.4-7. The one that rises highest is the GLD_m p.d.f., the next highest is the GLD_3 p.d.f., and the lowest one is the p.d.f. of the lognormal distribution with $\mu = 0$ and $\sigma = 1/3$. Of these two approximations, GLD_3 appears to be the better one.

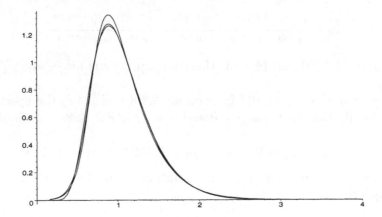

Figure 4.4–7. The p.d.f.s of GLD_m (the highest curve), GLD_3 (the next highest), and the lognormal distributions ($\mu = 0$, $\sigma = 1/3$).

4.4.9 The Beta Distribution

The percentiles, and hence ρ_1, ρ_2, ρ_3, ρ_4, of the beta distribution require numerical integration even when the parameters, β_3 and β_4, are specified. (For the definition of the beta distribution, see Section 2.4.9.) We know from the development of the GBD$(\beta_1, \beta_2, \beta_3, \beta_4)$ in Chapter 3 that the GBD$(\beta_1, \beta_2, \beta_3, \beta_4)$ is a generalization of the beta distribution and, consequently, provides perfect fits for the beta distribution. In Section 2.4.9 we found out that, in general, the (α_3^2, α_4) of the beta distributions lies outside of the range covered by the GLD$(\lambda_1, \lambda_2, \lambda_3, \lambda_4)$ moments, making it impossible to fit a beta distribution through the method of moments. We will see in Section 4.4.17 that the (ρ_3, ρ_4) of the beta distribution are within the (ρ_3, ρ_4)-space of the GLD$(\lambda_1, \lambda_2, \lambda_3, \lambda_4)$, allowing us to approximate beta distributions with GLDs through our percentile-based method.

In the specific case considered in Section 2.4.9 where $\beta_3 = \beta_4 = 1$, the distribution function turns out to be $F(x) = 6x - 6x^2$ for $0 \leq x \leq 1$. Computing ρ_1, ρ_2, ρ_3, ρ_4 we have

$$\rho_1 = \frac{1}{2}, \quad \rho_2 = 0.6084, \quad \rho_3 = 1, \quad \rho_4 = 0.5708.$$

From Tables D–1, D–2, D–3, and D–4 and `FindLambdasP`, we are led to the following four approximations.

$$\text{GLD}_1(0.2231, 1.5068, 1.2612, 22.2424),$$

$$\text{GLD}_2(0.5000, 2.4331, 2.8374, 2.8374),$$

$$\text{GLD}_3(0.5000, 1.9443, 0.4398, 0.4398),$$

$$\text{GLD}_4(0.3819, 1.7030, 0.2054, 0.5859).$$

GLD$_1$ and GLD$_2$ are not good fits but GLD$_3$ and GLD$_4$ are, with GLD$_3$ somewhat superior to GLD$_4$. Figure 4.4–8 shows the p.d.f.s of GLD$_3$, GLD$_4$ and beta distribution with $\beta_3 = \beta_4 = 1$. The GLD$_4$ p.d.f. is higher at the endpoints near 0 and 1, GLD$_3$ gets closer to the x-axis at these points and the beta p.d.f. reaches the x-axis at these points.

The supports of GLD$_3$ and GLD$_4$ are

$$[-0.014, 1.014] \quad \text{and} \quad [-0.20, 0.97],$$

respectively, and the $\lambda_1, \lambda_2, \lambda_3, \lambda_4$ of GLD$_3$ are very close to those of the moment-based GLD$(\lambda_1, \lambda_2, \lambda_3, \lambda_4)$ approximation obtained in Section 2.4.9. To complete our **first check** we note that

$$\sup |\hat{f}_3(x) - f(x)| = 0.009618 \quad \text{and} \quad \sup |\hat{f}_4(x) - f(x)| = 0.1917,$$

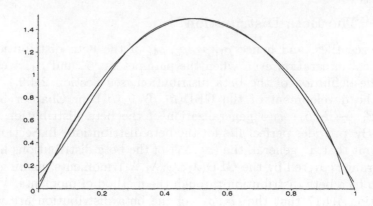

Figure 4.4–8. The p.d.f.s of GLD_3 (highest at 0 and 1), GLD_2 (next
highest), and the beta distributions ($\beta_3 = \beta_4 = 1$).

where $f_3(x)$ and $f_4(x)$ are the p.d.f.s of GLD_3 and GLD_4, respectively. The
graphs of the d.f.s of GLD_3, GLD_4 and the beta distribution with $\beta_3 = \beta_4 =
1$ appear to be identical and

$$\sup |\hat{F}_3(x) - F(x)| = 0.001030 \quad \text{and} \quad \sup |\hat{F}_4(x) - F(x)| = 0.008360,$$

where $F_3(x)$ and $F_4(x)$ are the d.f.s of GLD_3 and GLD_4, respectively. This
takes care of our **second check**.

4.4.10   The Inverse Gaussian Distribution

This distribution, whose p.d.f. is given in Section 2.4.10, does not have a
closed form d.f. or percentile function. We will see in Section 4.4.17 that the
(ρ_3, ρ_4) of the inverse Gaussian distribution are within the (ρ_3, ρ_4)-space of
the GLD, making it possible to fit any inverse Gaussian distribution.

To fit a percentile-based GLD to the inverse Gaussian we specify its parameters μ and λ and use numeric methods to obtain the needed percentiles.
The distribution that we considered in Section 2.4.10 had $\mu = 0.5$ and $\lambda = 6$;
for this distribution, we obtain

$$\pi_{.1} = 0.3339, \ \ \pi_{.25} = 0.3962, \ \ \pi_{.5} = 0.4801, \ \ \pi_{.75} = 0.5821, \ \ \pi_{.9} = 0.6916$$

from which we obtain

$$\rho_1 = 0.4801, \quad \rho_2 = 0.3578, \quad \rho_3 = 0.6915, \quad \rho_4 = 0.5197.$$

Figure 4.3–1 suggests the presence of three solutions associated with Tables D–1, D–2, and D–3 of Appendix D. Through `FindLambdasP` we obtain the fits

$$\text{GLD}_1(0.3934, 2.6872, 2.1017, 16.9322),$$

$$\text{GLD}_2(0.5523, 3.7980, 5.7431, 1.7715),$$

$$\text{GLD}_3(0.4176, 0.9580, 0.04304, 0.1349).$$

Graphic inspection of the three fits reveals that GLD_3 is the best of the three fits. Figure 4.4–9 shows the p.d.f. of the inverse Gaussian distribution with the p.d.f.s of GLD_3 and GLD_m, where GLD_m is the moment-based fit

$$\text{GLD}_m(0.4164, 0.6002, 0.02454, 0.08009)$$

that was obtained in Section 2.4.10. The GLD_m p.d.f. is the highest of the three curves near the center and the GLD_3 is the lowest.

To complete our **first check** we note that the supports of GLD_3 and GLD_m are $[-0.63, 1.46]$ and $[-1.25, 2.08]$, respectively, and for GLD_3

$$\sup |\hat{f}(x) - f(x)| = 0.1143.$$

For our **second check** we observe that the d.f.s of GLD_3 and the lognormal with $\mu = 0.5$ and $\lambda = 6$ are visually indistinguishable and

$$\sup |\hat{F}(x) - F(x)| = 0.006340.$$

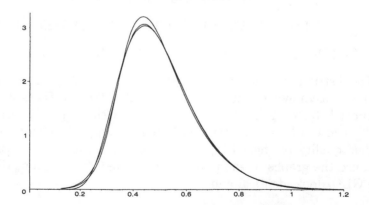

Figure 4.4–9. The p.d.f.s of GLD_m (highest at center), GLD_3 (lowest at center), and the inverse Gaussian ($\mu = 0.5$, $\lambda = 6$).

4.4.11   The Logistic Distribution

The p.d.f. of this distribution is given in Section 2.4.11 and the following d.f. is obtained from it through integration:

$$F(x) = \frac{1}{1 + e^{-(x-\mu)/\sigma}} \quad \text{for} \quad -\infty < x < \infty.$$

The percentiles needed for the computation of ρ_1, ρ_2, ρ_3, ρ_4 are

$$\pi_{.1} = \mu - \sigma \ln 9, \ \pi_{.25} = \mu - \sigma \ln 3, \ \pi_{.5} = \mu, \ \pi_{.75} = \mu + \sigma \ln 3, \ \pi_{.9} = \mu + \sigma \ln 9,$$

and

$$\rho_1 = \mu, \quad \rho_2 = 2\sigma \ln 9, \quad \rho_3 = 1, \quad \rho_4 = \frac{1}{2}.$$

Since ρ_3 and ρ_4 are both independent of the distribution's parameters, if we can fit a specific logistic distribution, we should be able to fit any logistic distribution. For the specific case of $\mu = 0$ and $\sigma = 1$,

$$\rho_1 = 0, \quad \rho_2 = 4.3944, \quad \rho_3 = 1, \quad \rho_4 = \frac{1}{2}.$$

Figure 4.3–1 indicates 4 possible fits from Tables D–1, D–2, D–3, and D–5 of Appendix D; from `FindLambdasP` these fits are, respectively,

$$\text{GLD}_1(-1.1574, 0.2464, 1.8093, 12.3687),$$

$$\text{GLD}_2(0, 0.3078, 3.7094, 3.7094),$$

$$\text{GLD}_3(0, 1.0825 \times 10^{-9}, 1.0825 \times 10^{-9}, 1.0825 \times 10^{-9}),$$

$$\text{GLD}_5(0, -8.45701 \times 10^{-7}, -8.45701 \times 10^{-7}, -8.45701 \times 10^{-7}).$$

The distribution being fitted is symmetric with support the entire real line. GLD_1 is asymmetric with support $[-5.22, 2.90]$, GLD_2 is symmetric with support $[-3.25, 3.25]$, GLD_3 is symmetric with support $[-9.2 \times 10^{10}, 9.2 \times 10^{10}]$, and GLD_5 is symmetric with infinite support in both directions. The higher quality of the GLD_3 and GLD_5 fits is also substantiated visually because the graphs of the p.d.f.s of the fitted distribution, the GLD_3, and the GLD_5 look identical and

$$\sup |\hat{f}_3(x) - f(x)| = 7.9262 \times 10^{-11},$$

$$\sup |\hat{f}_5(x) - f(x)| = 1.0803 \times 10^{-7},$$

where $f_3(x)$ and $f_5(x)$ are the p.d.f.s of GLD$_3$ and GLD$_5$, respectively. Having completed our **first check**, we take up our **second check** by noting that the d.f.s of these distributions also appear identical with

$$\sup |\hat{F}_3(x) - F(x)| = 1.0258 \times 10^{-10},$$

$$\sup |\hat{F}_5(x) - F(x)| = 6.5028 \times 10^{-8},$$

where $F_3(x)$ and $F_5(x)$ are the d.f.s of GLD$_3$ and GLD$_5$, respectively. These are clearly excellent fits. The rather good moment-based fit

$$\text{GLD}_m(0, -0.0003637, -0.0003630, -0.0003630)$$

was obtained in Section 2.4.11 with

$$\sup |\hat{f}(x) - f(x)| = 0.0004191 \text{ and } \sup |\hat{F}(x) - F(x)| = 0.0003541.$$

4.4.12 The Largest Extreme Value Distribution

The largest extreme value distribution, whose p.d.f. is given in Section 2.4.12, has, for $\sigma > 0$,

$$F(x) = e^{-e^{-\frac{x-\mu}{\sigma}}} \quad \text{for} \quad -\infty < x < \infty.$$

From $F(x)$ we can derive

$$\pi_{.1} = \mu - \sigma \ln(\ln 10), \tag{4.4.14}$$

$$\pi_{.25} = \mu - \sigma \ln(\ln 4), \tag{4.4.15}$$

$$\pi_{.5} = \mu - \sigma \ln(\ln 2), \tag{4.4.16}$$

$$\pi_{.75} = \mu - \sigma \ln(\ln 4/3), \tag{4.4.17}$$

$$\pi_{.9} = \mu - \sigma \ln(\ln 10/9), \tag{4.4.18}$$

from which we obtain

$$\rho_1 = \mu - \sigma \ln(\ln 2), \tag{4.4.19}$$

$$\rho_2 = \sigma \left(\ln(\ln 10) - \ln(\ln(10/9)) \right), \tag{4.4.20}$$

$$\rho_3 = \frac{\ln(\ln 10) - \ln(\ln 2)}{\ln(\ln 2) - \ln(\ln 10/9)} \approx 0.6373, \tag{4.4.21}$$

$$\rho_4 = \frac{\ln 2 + \ln(\ln 2) - \ln(\ln 4/3)}{\ln(\ln 10) - \ln(\ln 10/9)} \approx 0.5098. \tag{4.4.22}$$

For the specific case of $\mu = 0$ and $\sigma = 1$ considered in Section 2.4.12, we have

$$\rho_1 = 0.3665, \quad \rho_2 = 3.08440, \quad \rho_3 = 0.6373, \quad \rho_4 = 0.5098$$

and we see from Figure 4.3-1 there should be three fits associated with Tables D–1, D–2, and D–3 of Appendix D. Through `FindLambdasP` we obtain the fits

$$GLD_1(-0.3025, 0.3081, 2.2782, 16.8717),$$

$$GLD_2(1.1584, 0.4276, 6.6331, 1.5199),$$

$$GLD_3(-0.1761, 0.05676, 0.01964, 0.06540).$$

Graphic inspection of these fits reveals that GLD_3 is the best of the three fits. Figure 4.4-10 shows the p.d.f. of the largest extreme value distribution with the p.d.f.s of GLD_3 and GLD_m, where GLD_m is the moment-based fit

$$GLD_m(-0.1857, 0.02107, 0.006696, 0.02326)$$

that was obtained in Section 2.4.12. The p.d.f. of GLD_m rises highest near the center and the p.d.f. of GLD_3 is the lowest near the center.

The observation that GLD_3 has support $[-17.79, 17.44]$ and

$$\sup |\hat{f}(x) - f(x)| = 0.01393,$$

completes our **first check.** For our **second check** we note that the d.f.s of GLD_3 and the distribution being fitted are visually indistinguishable and

$$\sup |\hat{F}(x) - F(x)| = 0.006260.$$

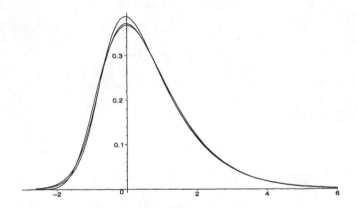

Figure 4.4–10. The p.d.f.s of GLD_m (highest at the center),
GLD_3 (lowest at the center), and the largest
extreme value distribution ($\mu = 0$, $\sigma = 1$).

4.4.13 The Extreme Value Distribution

The p.d.f. of this distribution is given in Section 2.4.13 and its d.f. is

$$F(x) = 1 - e^{-e^{(x-\mu)/\sigma}} \quad \text{for} \quad -\infty < x < \infty.$$

The percentiles needed for the computation of ρ_1, ρ_2, ρ_3, ρ_4 are

$$\pi_{.1} = \mu + \sigma \ln(\ln(10/9)), \tag{4.4.23}$$

$$\pi_{.25} = \mu + \sigma \ln(\ln(4/3)), \tag{4.4.24}$$

$$\pi_{.5} = \mu + \sigma \ln(\ln 2), \tag{4.4.25}$$

$$\pi_{.75} = \mu + \sigma \ln(\ln 4), \tag{4.4.26}$$

$$\pi_{.9} = \mu + \sigma \ln(\ln 10), \tag{4.4.27}$$

and

$$\rho_1 = \mu + \sigma \ln(\ln 2), \tag{4.4.28}$$

$$\rho_2 = \sigma \left(\ln(\ln 10) - \ln(\ln 10/9) \right), \tag{4.4.29}$$

$$\rho_3 = \frac{\ln(\ln 2) - \ln(\ln 10/9)}{\ln(\ln 10) - \ln(\ln 2)} \approx 1.5692, \tag{4.4.30}$$

$$\rho_4 = \frac{\ln 2 + \ln(\ln 2) - \ln(\ln 4/3)}{\ln(\ln 10) - \ln(\ln 10/9)} \approx 0.5098. \tag{4.4.31}$$

Because of the close relationship of this distribution with the largest extreme value distribution, we obtain the same ρ_2 and ρ_4 that we did in Section 4.4.12 and the ρ_3 that we get is the reciprocal of the ρ_3 of the largest extreme value distribution (see (4.4.21) and (4.4.22)). Recall that (ρ_3, ρ_4) and $(1/\rho_3, \rho_4)$ lead to GLD fits with their λ_3 and λ_4 interchanged (see (4.1.11) of Section 4.1). Therefore, we can get a percentile-based GLD fit for the extreme value distribution whenever we are able to fit the largest extreme value distribution with the same parameters and since ρ_3 and ρ_4 are constants, percentile-based fits are available for all possible parameter values of the distribution.

4.4.14 The Double Exponential Distribution

The double exponential distribution, whose p.d.f. is given in Section 2.4.14, has d.f.

$$F(x) = \begin{cases} 1 - \frac{1}{2}e^{-x/\lambda} & \text{if } x \geq 0 \\ \frac{1}{2}e^{x/\lambda} & \text{if } x < 0. \end{cases}$$

The percentiles $\pi_{.1} = -\lambda \ln 5$, $\pi_{.25} = -\lambda \ln 2$, $\pi_{.5} = 0$, $\pi_{.75} = \lambda \ln 1$, and $\pi_{.9} = \lambda \ln 5$ lead to

$$\rho_1 = 0, \quad \rho_2 = 2\lambda \ln 5, \quad \rho_3 = 1, \quad \rho_4 = \ln 2/\ln 5.$$

Since ρ_3 and ρ_4 are both independent of λ if we can fit a specific double exponential, then we should be able to fit all double exponential distributions. Figure 4.3–1 indicates two possible fits from Tables D–1 and D–5 of Appendix D for all possible values of the parameter λ. In the specific case of $\lambda = 1$ considered in Section 2.4.14, from FindLambdasP we obtain the two fits

$$\mathrm{GLD}_1(0, 0.3833, 4.5849, 4.5849),$$

$$\mathrm{GLD}_5(0, -0.7626, -0.3552, -0.3552).$$

The support of the distribution being fitted is $(-\infty, \infty)$ whereas the supports of GLD_1 and GLD_5 are, respectively, $[-2.61, 2.61]$ and $(-\infty, \infty)$. Also, visual inspection indicates that GLD_5 is the better fit. In Section 2.4.14 we obtained the moment-based fit

$$\mathrm{GLD}_m(7.5505 \times 10^{-17}, -0.1192, -0.08020, -0.08020).$$

Figure 4.4–11 shows the p.d.f.s of the double exponential (with $\lambda = 1$), GLD_5, and GLD_m. The double exponential p.d.f. rises to a sharp point at 0 and the higher rising curve at the center is the p.d.f. of GLD_5. The computation

$$\sup |\hat{f}(x) - f(x)| = 0.08034$$

completes our **first check** (note that for GLD_m this figure was 0.1457).

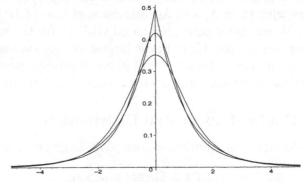

Figure 4.4–11. The p.d.f.s of GLD_m (lowest at center), GLD_5, and the double exponential with $\lambda = 1$ (highest at center).

For our **second check** we observe that the d.f.s of the double exponential and GLD$_5$ appear to be identical and

$$\sup|\hat{f}(x) - f(x)| = 0.01046.$$

For GLD$_m$ this figure was 0.02871.

4.4.15 The F-Distribution

The F-distribution whose p.d.f. is given in Section 2.4.15 does not have a closed-form representation of its d.f. or its percentile function. Thus, we have to use numeric methods to obtain p_1, p_2, p_3, p_4 in specific instances.

When $\nu_1 = 6$ and $\nu_2 = 25$, the case considered in Section 2.4.15, we obtain

$$\rho_1 = 0.9158, \quad \rho_2 = 1.6688, \quad \rho_3 = 0.5058, \quad \rho_4 = 0.5031.$$

The three possible fits suggested by Tables D–1, D–2, and D–3 of Appendix D, respectively, lead to

$$\text{GLD}_1 = (0.6080, 0.5372, 2.5962, 18.77855),$$

$$\text{GLD}_2 = (1.4933, 0.7809, 6.7632, 1.1197),$$

$$\text{GLD}_3 = (0.5290, 0.02885, 0.003070, 0.01930).$$

Visual inspection indicates that GLD$_3$, with support $[-34.13, 35.19]$, is the best of these three fits. Figure 4.4–12 shows the $F(6, 25)$, GLD$_3$, and GLD$_m$ p.d.f.s, where GLD$_m$ is the fit

$$\text{GLD}_m = (0.5898, -0.09063, -0.01095, -0.05314)$$

that was obtained in Section 2.4.15. The curve that rises the highest at the center is the GLD$_m$ p.d.f., the next highest is the GLD$_3$ p.d.f., and the lowest is the p.d.f. of $F(6, 25)$. To complete our **first check**, we note that for GLD$_3$,

$$\sup|\hat{f}(x) - f(x)| = 0.06559.$$

For our **second check**, we observe that the d.f.s of the $F(6, 25)$ and GLD$_3$ distributions cannot be visually distinguished and

$$\sup|\hat{F}(x) - F(x)| = 0.01154.$$

Our discussion of the F-distribution in Section 2.4.15 showed that moment-based fits were possible only for relatively large values of ν_2. The percentile-based approach of this chapter turns out to be far more flexible, allowing

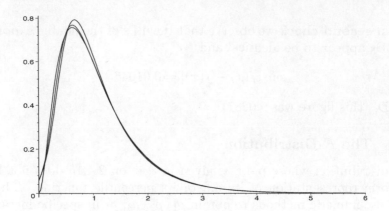

Figure 4.4–12. The p.d.f.s of GLD_m (highest at center),
GLD_3, and $F(6, 25)$ (lowest at center).

us to obtain fits for a variety of choices of ν_1 and ν_2. For example, for
$(\nu_1, \nu_2) = (2, 4)$, $(4, 6)$ and $(6, 12)$ we get, respectively, the following fits:

$$(\lambda_1, \lambda_2, \lambda_3, \lambda_4) = (1.9576 \times 10^{-6}, -0.5000, -7.1117 \times 10^{-7}, -0.5000), \text{ with}$$
$$\sup |\hat{f}(x) - f(x)| = 0.004965 \text{ and } \sup |\hat{F}(x) - F(x)| = 8.4394 \times 10^{-6};$$

$$(\lambda_1, \lambda_2, \lambda_3, \lambda_4) = (0.3244, -0.3426, -0.02427, -0.2969), \text{ with}$$
$$\sup |\hat{f}(x) - f(x)| = 0.1283, \text{ and } \sup |\hat{F}(x) - F(x)| = 0.01386;$$

$$(\lambda_1, \lambda_2, \lambda_3, \lambda_4) = (0.4963, -0.1270, -0.01241, -0.09145), \text{ with}$$
$$\sup |\hat{f}(x) - f(x)| = 0.06865 \text{ and } \sup |\hat{F}(x) - F(x)| = 0.01199.$$

4.4.16   The Pareto Distribution

From the p.d.f. of the Pareto distribution given in Section 2.4.16, through
integration we obtain the d.f.

$$F(x) = 1 - \left(\frac{\lambda}{x}\right)^{\beta} \quad \text{for } x > \lambda$$

and $F(x) = 0$ if $x \leq \lambda$, and the percentile function

$$Q(x) = \frac{\lambda}{(1 - x)^{1/\beta}}.$$

From $Q(x)$ we can compute

$$\rho_1 = \lambda 2^{1/\beta}, \tag{4.4.32}$$

$$\rho_2 = \lambda \left(10^{1/\beta} - (10/9)^{1/\beta}\right), \tag{4.4.33}$$

$$\rho_3 = \frac{1 - (5/9)^{1/\beta}}{5^{1/\beta} - 1}, \tag{4.4.34}$$

$$\rho_4 = \frac{2^{1/\beta} - (2/3)^{1/\beta}}{5^{1/\beta} - (5/9)^{1/\beta}}. \tag{4.4.35}$$

In Sections 2.4.16 and 3.4.16 we were not able to obtain fits to this important distribution (widely used in studies of income) for any of its parameter values because its (α_3^2, α_4) pairs are outside the regions covered by the GLD and EGLD. However, the (ρ_3, ρ_4) of the Pareto distribution depends only on the single parameter β and is within the (ρ_3, ρ_4)-space covered by the tables of Appendix D. For example, when $\lambda = 1$ and $\beta = 5$, we have

$$\rho_1 = 1.1487, \quad \rho_2 = 0.5636, \quad \rho_3 = 0.2921, \quad \rho_4 = 0.4618$$

and we use FindLambdasP to obtain the following fits suggested by Tables D–1, D–2, and D–3, respectively.

$$\text{GLD}_1(1.0842, 1.3832, 3.4873, 23.1347),$$

$$\text{GLD}_2(1.4839, 2.0467, 6.4014, 0.5190),$$

$$\text{GLD}_3(1.0000, -1.0000, -7.3451 \times 10^{-12}, -0.2000).$$

Of these, GLD$_3$, with support $[0.99, 1.97]$, appears to be the best fit and its p.d.f. seems to be indistinguishable from that of the Pareto p.d.f. with $\lambda = 1$ and $\beta = 5$. We complete our **first check** for this fit by computing

$$\sup |\hat{f}(x) - f(x)| = 6.0 \times 10^{-7}.$$

As has generally been the case, in our **second check** the d.f.s of GLD$_3$ and that of the Pareto distribution under consideration also appear to be identical and in this case

$$\sup |\hat{F}(x) - F(x)| = 4.1 \times 10^{-9}.$$

Fits of the Pareto distribution with other choices of λ and β are not only possible but also quite good. The fits for $(\lambda, \beta) = (1, 2)$, $(1, 10)$, and $(1, 20)$

are summarized below.

$$(\lambda_1, \lambda_2, \lambda_3, \lambda_4) = (1.0000, -1.0000, -4.1976 \times 10^{-11}, -0.5000), \text{ with}$$
$$\sup |\hat{f}(x) - f(x)| = 3.8 \times 10^{-8} \text{ and } \sup |\hat{F}(x) - F(x)| = 1.2 \times 10^{-9};$$

$$(\lambda_1, \lambda_2, \lambda_3, \lambda_4) = (1.0000, -1.0000, -2.1159 \times 10^{-7}, -0.1000), \text{ with}$$
$$\sup |\hat{f}(x) - f(x)| = 0.05271, \text{ and } \sup |\hat{F}(x) - F(x)| = 1.1 \times 10^{-5};$$

$$(\lambda_1, \lambda_2, \lambda_3, \lambda_4) = (1.0000, -1.0000, -5.6691 \times 10^{-8}, -0.05000), \text{ with}$$
$$\sup |\hat{f}(x) - f(x)| = 0.05664 \text{ and } \sup |\hat{F}(x) - F(x)| = 6.4 \times 10^{-6}.$$

4.4.17   Summary of Distribution Approximations

It should be clear from Sections 4.4.1 through 4.4.16 that the use of percentiles allows us to approximate a variety of distributions, in many cases yielding better approximations than the moment-based methods of Chapters 2 and 3. It also seems that the (ρ_3, ρ_4) points of the $\text{GLD}(\lambda_1, \lambda_2, \lambda_3, \lambda_4)$ cover a large enough area to provide flexibility in approximating distributions (e.g., the beta, F, Pareto, and some Student's t distributions) that could not be fitted with the $\text{GBD}(\beta_1, \beta_2, \beta_3, \beta_4)$ or with the $\text{GLD}(\lambda_1, \lambda_2, \lambda_3, \lambda_4)$ when moments are used. Figure 4.4–13 charts the location of the (ρ_3, ρ_4) points for the distributions considered in Sections 4.4.1 through 4.4.16. The (ρ_3, ρ_4) points associated with these distributions consist of a single point, a curve, or a region in (ρ_3, ρ_4)-space.

The (ρ_3, ρ_4) points of the uniform, normal, logistic, double exponential, Cauchy (or $t(1)$, the Student's t distribution with one degree of freedom), exponential, and largest extreme value (hence also extreme value) distributions are marked with small rectangles and labeled with "u," "n," "l," "d," "c," "e," and "v," respectively. The first five of these have $\rho_3 = 1$ and are located at the right edge of Figure 4.4–13; the last two are more centrally located.

The (ρ_3, ρ_4) points of Student's t, gamma (this includes the chi-square as a special case), Weibull, lognormal, inverse Gaussian, and Pareto distributions are represented by curves that are labeled with "T," "G," "W," "LN," "IG," and "P," respectively. The curves for the gamma and Weibull intersect at "e," the (ρ_3, ρ_4) of the exponential distribution. The lognormal and inverse Gaussian curves are very close to each other; the higher of the two curves that the label "LN, IG" points to is the curve for the lognormal, the lower one is for the inverse Gaussian. With the exception of the curve for the Student's t and Pareto distributions, all curves extend from the vicinity of $(0,0)$ to a point where $\rho_3 = 1$. The curve for the Pareto distribution also starts

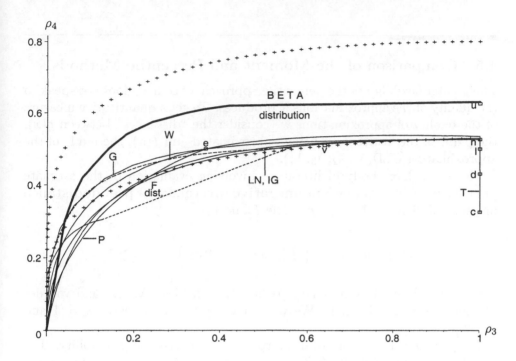

Figure 4.4–13. The (ρ_3, ρ_4) points associated with the distributions considered in Sections 4.4.1 through 4.4.16.

near $(0,0)$ but extends only to the (ρ_3, ρ_4) of the exponential distribution. The curve, actually straight line, of the t distribution connects the points associated with the Cauchy and normal distributions.

The region enclosed by the two curves marked with "+" represents the area covered by the beta distribution. The region enclosed by the dashed lines and marked "F dist." is the area covered by the F-distribution. Except for the region of the F-distribution, the points, curves, and regions of Figure 4.4–13 provide an accurate representation of the (ρ_3, ρ_4) points for each distribution. The (ρ_3, ρ_4) points of the F-distribution are much harder to compute and the region marked "F dist." is only a reasonable approximation of the true (ρ_3, ρ_4)-space of the F-distribution.

The thick curve in Figure 4.4–13 that goes through the beta region is the boundary that separates the points that are within the range of the tables of Appendix D (the points below the curve) from those that are outside this range (the points above the curve). Points that are above this curve are generally also outside of computation range of `FindLambdasP`.

4.5   Comparison of the Moment and Percentile Methods

The greater flexibility of the percentile approach, of course, does not speak to the quality of the approximations it produces. To get a quantitative measure of the quality of approximations we consider the "distances" between $g(x)$, the p.d.f. of the distribution being approximated and $f(x)$, the p.d.f. of the approximating GLD($\lambda_1, \lambda_2, \lambda_3, \lambda_4$).

There is a large body of literature involving evaluation of the estimate $f(x)$ of a p.d.f. $g(x)$, through **nonnegative divergence** or **pseudodistance** measures designated by $D(f, g)$. The L_p-norm,

$$||f(x) - g(x)||_p = \left(\int |f(x) - g(x)| \, dx \right)^{1/p}, \quad p \geq 1 \qquad (4.5.1)$$

is a commonly used form of $D(f, g)$ (see Györfi, Liese, Vajda, and van der Meulen (1998) for details). We will concentrate on the cases $p = 1$ and $p = 2$.

The case $p = 1$ has a natural interpretation in terms of probability. The **overlapping coefficient**, $\Delta(f, g)$, of any two p.d.f.s, $f(x)$ and $g(x)$, is defined as the area that is under **both** p.d.f.s and above the horizontal axis; equivalently, $\Delta(f, g)$ is the area above the horizontal axis and below the function $\min(f(x), g(x))$. (For an introductory discussion of the history and literature of this subject, see Mishra, Shah, and Lefante (1986) and for examples, see Dudewicz and Mishra (1988).) The relationship of the L_1 distance, $||f(x) - g(x)||_1$, to $\Delta(f, g)$ is shown in Figure 4.5–1 where $||f(x) - g(x)||_1$ is the area between the two illustrated p.d.f.s and $\Delta(f, g)$ is the area under $\min(f(x), g(x))$ which is shown in heavy print. It can be seen from Figure 4.5–1 that

$$\Delta(f, g) + ||f(x) - g(x)||_1 = \int_{-\infty}^{\infty} \max(f(x), g(x)) \, dx. \qquad (4.5.2)$$

Hence, the L_1 distance and $\Delta(f, g)$ are two sides of the same coin: the L_1 distance measures the difference, while $\Delta(f, g)$ measures the commonality, of two p.d.f.s

In Point 6 of Section 2.5.1 we noted some of the many statistical tests of the hypothesis that given data comes from a specified d.f., $G(x)$. Many of these tests rely on the sample (empirical) d.f. of the data. However, a number of the tests have an interpretation based on the closeness of an $f(x)$, the estimating p.d.f., to $g(x)$, the p.d.f. being estimated (see Dudewicz and van

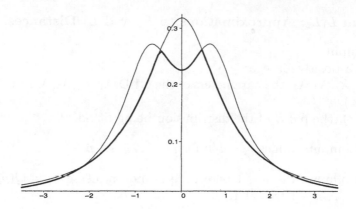

Figure 4.5–1. Relationship between the overlapping coefficient $\Delta(f, g)$ and the L_1 norm, $||f(x) - g(x)||_1$ of any two p.d.f.s $f(x)$ and $g(x)$.

der Meulen (1987) for such an interpretation with the entropy test). Since it is of interest to consider distance measures between p.d.f.s $f(x)$ and $g(x)$, based on $f(x)$ and $g(x)$ themselves (and not their d.f.s), we begin with the definition of L_1 and L_2 norms.

Definition 4.5.1. *The L_1 and L_2 distances between two p.d.f.s are, respectively,*

$$||f - g||_1 = \int_{-\infty}^{\infty} |f(x) - g(x)| \, dx$$

and

$$||f - g||_2 = \left[\int_{-\infty}^{\infty} (f(x) - g(x))^2 \, dx \right]^{1/2}.$$

When there is complete agreement between $f(x)$ and $g(x)$,

$$||f - g||_1 = ||f - g||_2 = 0.$$

Moreover, because they are p.d.f.s, the integrals of f and g are equal to 1 and

$$||f - g||_1 \leq 2.$$

The integrations that lead to $||f-g||_1$ and $||f-g||_2$ must be done numerically because we do not have a closed-form expression for $f(x)$, the p.d.f. of the approximating $GLD(\lambda_1, \lambda_2, \lambda_3, \lambda_4)$. The results, therefore, will be numerical approximations. The following algorithm will produce approximate L_1 and L_2 distances.

Algorithm L_1L_2: Approximations to L_1 and L_2 Distances.

L_1L_2–1. Input
- n, a positive integer ≥ 3.
- $\lambda_1, \lambda_2, \lambda_3, \lambda_4$, the parameters of the GLD($\lambda_1, \lambda_2, \lambda_3, \lambda_4$) fit with p.d.f. $f(x)$.
- $g(x)$, the p.d.f. of the distribution being fitted.

L_1L_2–2. Compute values $p_i = i/n$ for $i = 1, 2, \ldots, n-1$.

L_1L_2–3. Compute the $n-1$ percentile points π_p using $\pi_i = Q(p_i)$ for $i = 1, 2, \ldots, n-1$ (see (1.1.1)).

L_1L_2–4. Compute the $n-1$ y-coordinates of the GLD($\lambda_1, \lambda_2, \lambda_3, \lambda_4$) p.d.f. using

$$y_i = \frac{\lambda_2}{\lambda_3 p_i^{\lambda_3-1} + \lambda_4(1-p_i)^{\lambda_4-1}},$$

for $i = 1, 2, \ldots, n-1$ (see (1.1.3)). The points (π_i, y_i) are on the graph of f.

L_1L_2–5. Compute the $n-1$ values on $g(x)$, the function being fitted, by $Y_i = g(\pi_i)$ for $i = 1, 2, \ldots, n-1$, making sure that when pi_i is outside of the support of $g(x)$, Y_i is assigned a value of zero. The points (π_i, Y_i) are on the graph of g.

L_1L_2–6. Let $\Delta_i = \pi_{i+1} - \pi_i$ for $i = 1, 2, \ldots, n-2$.

L_1L_2–7. Compute the sums

$$S_1 = \sum_{i=1}^{n-2} \Delta_i |y_i - Y_i| \quad \text{and} \quad S_2 = \sum_{i=1}^{n-2} \Delta_i (y_i - Y_i)^2.$$

L_1L_2–8. Compute

$$T_1 = \int_{-\infty}^{\pi_1} g(x)\, dx + \int_{\pi_{n-1}}^{\infty} g(x)\, dx$$

and

$$T_2 = \int_{-\infty}^{\pi_1} g^2(x)\, dx + \int_{\pi_{n-1}}^{\infty} g^2(x)\, dx.$$

L_1L_2–9. $S_1 + T_1$ approximates the L_1 distance between f and g and $\sqrt{S_2 + T_2}$ approximates the L_2 distance between f and g.

In Algorithm L_1L_2, S_1 and S_2 are Riemann sums for the integrals

$$\int_{\pi_1}^{\pi_{n-1}} |f(x) - g(x)|\, dx \quad \text{and} \quad \int_{\pi_1}^{\pi_{n-1}} (f(x) - g(x))^2\, dx, \qquad (4.5.3)$$

respectively. Therefore, S_1 and S_2 converge to the values of these definite integrals as $n \to \infty$. Since the accumulated probability of $f(x)$ on the intervals $(-\infty, \pi_1)$ and (π_{n-1}, ∞) is $2/n$ and $f(x)$ and $g(x)$ are non-negative,

$$\int_{-\infty}^{\pi_1} |\ f(x) - g(x)|\, dx + \int_{\pi_{n-1}}^{\infty} |f(x) - g(x)|\, dx$$

$$\leq \int_{-\infty}^{\pi_1} (f(x) + g(x))\, dx + \int_{\pi_{n-1}}^{\infty} (f(x) + g(x))\, dx$$

$$\leq \frac{2}{n} + T_1. \qquad (4.5.4)$$

Therefore,

$$\left|\ \|f - g\|_1 - (S_1 + T_1)\ \right| \leq \frac{2}{n} \qquad (4.5.5)$$

and

$$\|f - g\|_1 = \lim_{n \to \infty} (S_1 + T_1). \qquad (4.5.6)$$

It can be similarly established that

$$\|f - g\|_2 = \lim_{n \to \infty} \sqrt{S_2 + T_2}, \qquad (4.5.7)$$

justifying the conclusions in Step L_1L_2–9.

   Table 4.5.2 summarizes the L_1 and L_2 distances between the p.d.f.s of the distributions considered in Sections 2.4.1 through 2.4.16 (and also in Sections 4.4.1 through 4.2.16) and their GLD$(\lambda_1, \lambda_2, \lambda_3, \lambda_4)$ approximations that resulted from moment-based and percentile-based estimations. The percentile-based fit is the fit that was deemed visually most appropriate in Sections 4.4.1 through 4.4.16.

   The computations of the L_1 and L_2 distances in Table 4.5–2 were obtained through Algorithm L_1L_2 with a large enough n ($n = 5000$) to have an error bound of 10^{-3}. This assertion is based on the fact that when n was increased from 2500 to 5000, only negligible differences (less than 10^{-3}) were observed in the computed values.

Table 4.5–2. L_1 and L_2 distances between distributions and their moment and percentile based GLD approximations.

Distribution	L_1 distance		L_2 distance	
	Moment Fit	Percentile Fit	Moment Fit	Percentile Fit
Normal: $\mu = 0$, $\sigma^2 = 1$	0.007	0.003	0.003	0.001
Uniform[†]: on interval $(0, 1)$	0	0	0	0
Student's t: $\nu = 1$	‡	0.019	‡	0.005
Student's t: $\nu = 5$	0.066	0.008	0.034	0.002
Chi-Square: $\nu = 5$	0.101	0.060	0.025	0.017
Exponential: $\theta = 3$	0.022	0.154	0.045	0.092
Gamma: $\alpha = 5$, $\theta = 3$	0.060	0.036	0.012	0.007
Weibull: $\alpha = 1$, $\beta = 5$	0.015	0.008	0.012	0.006
Lognormal: $\mu = 0$, $\sigma = 1/3$	0.090	0.042	0.054	0.027
Beta: $\beta_3 = \beta_4 = 1$	‡	0.006	‡	0.014
Inv. Gaussian: $\mu = 0.5$, $\lambda = 6$	0.055	0.008	0.017	0.010
Logistic: $\mu = 0$, $\sigma = 1$	0.004	0.003	0.002	0.002
Largest Ext. Value: $\mu = 0$, $\sigma = 1$	0.071	0.034	0.023	0.013
Extreme Value[‡‡]: $\mu = 0$, $\sigma = 1$	0.071	0.034	0.023	0.013
Double Exponential: $\lambda = 1$	0.123	0.087	0.082	0.038
F: $\nu_1 = 6$, $\nu_2 = 25$	0.068	0.080	0.063	0.063
Pareto: $\lambda = 1$, $\beta = 5$	‡	0.002	‡	0.032

[†]The moment and percentile fits to the uniform distribution were exact.
[‡] Moment fits are not possible.
[‡‡] Figures for this distribution are the same as those for the largest extreme value distribution (see Section 2.4.13 for an explanation).

We see from Table 4.5–2 that with the exception of the exponential case (with $\theta = 3$) and $F(6, 25)$, the percentile fits produce GLD($\lambda_1, \lambda_2, \lambda_3, \lambda_4$) approximations with smaller L_1 and L_2 distances from the p.d.f.s being approximated. Furthermore, the percentile-based estimation of $\lambda_1, \lambda_2, \lambda_3, \lambda_4$ led to good approximations for the Student's t, beta, Pareto and F-distributions when moment-based estimation was not possible.

As was noted in Section 4.1, a percentile-based method of fitting the GLD is important when moments do not exist (or are out of table range), and to avoid the possibly relatively large variability of sample moments of orders 3 and 4. Indeed, the use of sample percentiles in estimation of population quantities such as the mean, variance, skewness, and kurtosis has been popular in statistics for some time due to its robustness. In terms of the sample mean and sample median, one often finds such statements as "... the sample median is less affected by extreme values than the mean. With particularly small sample sizes, the sample median is often a better indicator of central tendency than the mean." (Gibbons (1997), p. 97).

Even for higher order moments such as variance, skewness, and kurtosis, it has been found that better tail weight classification can be found using percentiles rather that the sample kurtosis (Hogg (1972)). In light of a substantial literature along these lines, it is natural to expect the percentile-based method to compare favorably with its moment-based counterpart in the context of fitting a GLD.

We are currently investigating ways in which the relative goodness of these two methods can be quantified. The methods we are studying involve comparisons of the Mean Squared Errors (MSEs) of the estimators of $\lambda_1, \lambda_2, \lambda_3, \lambda_4$ obtained by the two methods, as well as comparing the resulting fitted distributions (using chi-squared discrepancies and Kolmogorov-Smirnov distances, for example). While detailed results are not available, preliminary indications are that the percentile approach, while not uniformly superior to the moment method, does better especially with datasets where moments are misled by high variability of sample moments.

One quantitative comparison may be of some interest. It is known that the sample mean $\bar{X}$ is asymptotically normal with center the population mean and variance σ^2/n, based on a random sample of size n from a population with finite variance σ^2. This follows from the classical Central Limit Theorem (see Theorem 6.3.2 of Dudewicz and Mishra (1988)). It is also known that the 0.5 quantile $Z_{0.5}$ is, in the same setting, asymptotically normal with center the population median $\xi_{0.5}$ and variance

$$\left(4f^2(\xi_{0.5})n\right)^{-1},$$

where f is the population probability density function, which is assumed to be continuous and positive at $\xi_{0.5}$ (see, for example, Theorem 7.4.21 of Dudewicz and Mishra (1988)). In the case of the GLD$(\lambda_1, \lambda_2, \lambda_3, \lambda_4)$ it is known that

$$\sigma^2 = \frac{B - A^2}{\lambda_2^2} \tag{4.5.8}$$

with

$$A = \frac{1}{1 + \lambda_3} - \frac{1}{1 + \lambda_4} \tag{4.5.9}$$

$$B = \frac{1}{1 + 2\lambda_3} + \frac{1}{1 + 2\lambda_4} - 2\beta(1 + \lambda_3, 1 + \lambda_4), \tag{4.5.10}$$

where β designates the beta function. It follows from (16) and the definition of the $\mathrm{GLD}(\lambda_1, \lambda_2, \lambda_3, \lambda_4)$ that

$$f(\xi_{0.5}) = \lambda_2 \left(\lambda_3 (0.5)^{\lambda_3-1} + \lambda_4 (0.5)^{\lambda_4-1} \right)^{-1}.$$

Therefore, the asymptotic distribution of $Z_{0.5}$ has a smaller variance than the asymptotic distribution of $\bar{X}$ if and only if

$$\left(\frac{\lambda_3}{2^{\lambda_3}} + \frac{\lambda_4}{2^{\lambda_4}} \right)^2 <$$
$$\frac{1}{1+2\lambda_3} + \frac{1}{1+2\lambda_4} - 2\beta(1+\lambda_3, 1+\lambda_4) - \left(\frac{1}{1+\lambda_3} - \frac{1}{1+\lambda_4} \right)^2.$$

In Figures 4.5–3 and 4.5–4 the regions of (λ_3, λ_4)-space where this inequality holds are marked by "P" (indicating a lower asymptotic variance, and a potential preference, for the percentile method) and the regions where the reverse inequality holds are marked by "M" (indicating a potential preference for the moment method).

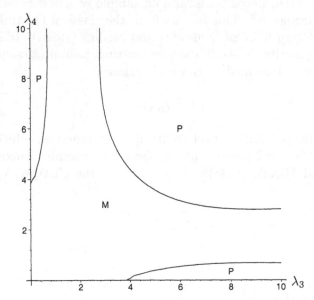

Figure 4.5–3. First quadrant regions of (λ_3, λ_4)-space where variance of $\bar{X}$ is greater than the variance of $Z_{0.5}$ (marked by "P") and less than the variance of $Z_{0.5}$ (marked by "M").

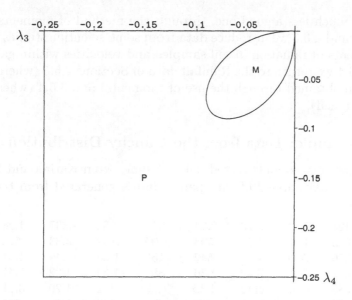

Figure 4.5–4. Third quadrant regions of (λ_3, λ_4)-space where variance of $\bar{X}$ is greater than the variance of $Z_{0.5}$ (marked by "P") and less than the variance of $Z_{0.5}$ (marked by "M").

Since, for both the moment and percentile methods, tabled solutions have so far been confined to the first and third quadrants of (λ_3, λ_4)-space, these are the only (λ_3, λ_4) sets for which we show the status of the inequality. Moreover, in the third quadrant of (λ_3, λ_4)-space, the GLD has its first four moments if and only if $-0.25 \leq \lambda_3, \lambda_4 \leq 0$. Thus, a comparison of the two methods is irrelevant outside of this square. We can see that on a relatively large portion of this square the asymptotic distribution of $Z_{0.5}$ has smaller variance than the asymptotic distribution of $\bar{X}$.

4.6 Examples: GLD Fits of Data via the Method of Percentiles

In this section we apply the method of this chapter to various datasets. In the first application, in Section 4.6.1, we consider data generated from the Cauchy distribution. Since the $\alpha_1, \alpha_2, \alpha_3, \alpha_4$ of the Cauchy distribution do not exist, we do not expect that data generated from the Cauchy could have a fit, much less a good one, through the method of moments. In Sections 4.6.2 and 4.6.3 we introduce data from some scientific studies involving measurements of radiation in soil samples and velocities within galaxies. In Section 4.6.4 we return to the Rainfall data of Sections 2.5.5 (where no GLD fit could be obtained through the use of moments) and 3.5.4 (where a GBD fit was obtained).

4.6.1   Example: Data from the Cauchy Distribution

The data for this example (listed below) comes from Karian and Dudewicz (1999a). It is Example 2 of that paper and is generated from the Cauchy distribution.

1.99	−.424	5.61	−3.13	−2.24	−.014	−3.32	−.837	−1.98	−.120
7.81	−3.13	1.20	1.54	−.594	1.05	.192	−3.83	−.522	.605
.427	.276	.784	−1.30	.542	−.159	−1.66	−2.46	−1.81	−.412
−9.67	6.61	−.598	−3.42	.036	.851	−1.34	−1.22	−1.47	−.592
−.311	3.85	−4.92	−.112	4.22	1.89	−.382	1.20	3.21	−.648
−.523	−.882	.306	−.882	−.635	13.2	.463	−2.60	.281	1.00
−.336	−1.69	−.484	−1.68	−.131	−.166	−.266	.511	−.198	1.55
−1.03	2.15	.495	6.37	−.714	−1.35	−1.55	−4.79	4.36	−1.53
−1.51	−.140	−1.10	−1.87	.095	48.4	−.998	−4.05	−37.9	−.368
5.25	1.09	.274	.684	−.105	20.6	.311	.621	3.28	1.56

We first attempt to obtain fits by using the moment-based methods of Chapters 2 and 3 and compute $\hat{\alpha}_1$, $\hat{\alpha}_2$, $\hat{\alpha}_3$, $\hat{\alpha}_4$ to get

$$\hat{\alpha}_1 = .3464 \qquad \hat{\alpha}_2 = 49.4908 \qquad \hat{\alpha}_3 = 1.8671 \qquad \hat{\alpha}_4 = 31.3916.$$

We note that these computations as well as subsequent ones yield slightly different results from those given in Karian and Dudewicz (1999a) because the computations in Karian and Dudewicz (1999a) are based on the simulated data prior to its truncation to three digits. The (α_3^2, α_4) that we have is well outside the range of the tables in Appendices B and C and also beyond our range of computation, making it impossible to fit a distribution from the EGLD family by the methods discussed in Chapters 2 and 3.

To obtain a percentile-based fit, we compute $\hat{\rho}_1$, $\hat{\rho}_2$, $\hat{\rho}_3$, $\hat{\rho}_4$:

$$\hat{\rho}_1 = -0.1820 \qquad \hat{\rho}_2 = 7.2600 \qquad \hat{\rho}_3 = 0.6632 \qquad \hat{\rho}_4 = 0.2981$$

and obtain two fits from FindLambdasP based on entries from Tables D–1 and D–5, respectively, of Appendix D.

$$GLD_1(-0.3848, 0.1260, 5.2455, 10.2631),$$

$$\text{GLD}_5(-0.2830, -2.4471, -0.9008, -1.0802).$$

GLD_5 turns out to be the superior fit. (This is not surprising since the (λ_3, λ_4) for GLD_5 is from Region 4, assuring us that the support of the resulting fit will be $(-\infty, \infty)$; by contrast, the support of GLD_1 is $[-8.3, 7.6]$.)

Although most of the data is concentrated on the interval $[-6, 6]$, the range of the data is $[-37.9, 48.4]$. A histogram on $[-37.9, 48.4]$ would be so compressed that its main features would not be visible. A slightly distorted histogram of the data (when the 8 of the 100 observations outside of the interval $[-6, 6]$ are ignored) and the GLD_1 and GLD_5 p.d.f.s are shown in Figure 4.6–1 (a) (the p.d.f. of GLD_1 rises higher at the center). Figure 4.6–1 (b) shows the e.d.f. of the data with d.f.s of GLD_1 and GLD_5.

When the data is partitioned into the intervals

$$(-\infty, -3], \quad (-3, -1.5], \quad (-1.5, -.7], \quad (-.7, -.4], \quad (-.4, 0],$$
$$(0, .4], \quad (.4, .7], \quad (.7, 1.5], \quad (1.5, 3], \quad (3, \infty),$$

we obtain observed frequencies of

$$10, \quad 12, \quad 11, \quad 11, \quad 14, \quad 8, \quad 8, \quad 7, \quad 6, \quad 13.$$

and the expected frequencies for these intervals that result from GLD_1 are

$$10.2541, \quad 7.1461, \quad 9.3964, \quad 10.8408, \quad 18.5659,$$
$$8.1299, \quad 4.0927, \quad 7.5986, \quad 8.9770, \quad 14.9985.$$

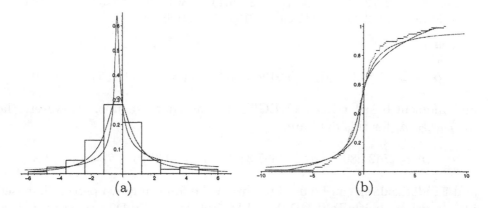

Figure 4.6–1. Histogram of data generated from the Cauchy distribution and the p.d.f.s of the fitted GLD_1 and GLD_5 (a); the e.d.f. of the data with the d.f.s of the fitted GLD_1 and GLD_2 (b).

These lead to the chi-square goodness-of-fit statistic and corresponding p-value of

$$9.7352 \quad \text{and} \quad 0.08310,$$

respectively. For GLD_5, the expected frequencies are

$$10.2644, \quad 9.7906, \quad 14.5337, \quad 8.6285, \quad 12.2154,$$
$$9.8868, \quad 5.4195, \quad 8.9350, \quad 7.4654, \quad 12.8607,$$

and the resulting chi-square statistic and p-value are

$$4.5740 \quad \text{and} \quad 0.4700,$$

justifying our earlier observation that GLD_5 is the better of the two fits.

4.6.2   Data on Radiation in Soil Samples

Florida gypsum and phosphate mine tailings produce radiation in the form of radon 222. A monitoring of these mines in Polk County, reported by Horton (1979) (also see McClave, Dietrich, and Sincich (1997, p. 38)) gave data on these exhalation rates, part of which follows (in increasing order).

178.99	205.84	357.17	393.55	538.37	558.33	599.84
752.89	878.56	880.84	961.40	1096.43	1150.94	1322.76
1426.57	1480.04	1489.86	1572.69	1698.39	1709.79	1774.77
1830.78	1888.22	1977.97	2055.20	2315.52	2367.40	2617.57
2758.84	2770.23	2796.42	2996.49	3017.48	3750.83	3764.96
4132.28	5402.35	6815.69	9139.21	11968.23		

For this data

$$\hat{\alpha}_1 = 2384.8422, \quad \hat{\alpha}_2 = 5.5198 \times 10^6, \quad \hat{\alpha}_3 = 2.3811, \quad \hat{\alpha}_4 = 9.1440$$

and moment-based GLD and EGLD fits are not possible. However, the $\hat{\rho}_1$, $\hat{\rho}_2$, $\hat{\rho}_3$, $\hat{\rho}_4$ for this data are

$$\hat{\rho}_1 = 1742.2800, \quad \hat{\rho}_2 = 4867.3110, \quad \hat{\rho}_3 = 0.3776, \quad \hat{\rho}_4 = 0.3881$$

and `FindLambdasP` can be used to obtain the following two percentile-based fits, associated with Tables D–1 and D–5 of Appendix D, respectively.

$$GLD_1(1377.6122, 0.0001678, 4.0305, 17.2300),$$

$$GLD_5(1314.9599, -0.0008317, -0.2609, -0.6357).$$

The support of GLD_1, $[-4582.25, 7337.47]$, indicates that the fitted GLD_1 distribution extends too far to the left and not far enough to the right. By contrast, the GLD_5 distribution extends indefinitely in both directions and its support is $(-\infty, \infty)$. Visual inspection indicates that GLD_5, in spite of its support, is the better of the two fits. Figure 4.6–2 (a) shows a histogram of the data with the p.d.f. of GLD_5 and Figure 4.6–2 (b) shows the e.d.f. of the data with the d.f. of GLD_5.

To get a chi-square goodness-of-fit statistic for the GLD_5 fit, we partition the data into the intervals

$$[0, 600), \quad [600, 1200), \quad [1200, 1700), \quad [1700, 2100), \quad [2100, 3000), \quad [3000, \infty)$$

and obtain the observed frequencies

$$7, \quad 6, \quad 6, \quad 6, \quad 7, \quad 8$$

and expected frequencies

$$5.3512, 6.9761, 7.1251, 4.6083, 6.3302, 9.6092.$$

The chi-square statistic and its associated p-value for this fit are

$$1.5829 \quad \text{and} \quad 0.2083.$$

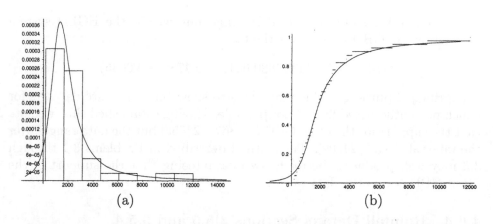

(a)   (b)

Figure 4.6–2. Histogram of radiation data and the p.d.f. of GLD_5 (a); the e.d.f. of the data with the d.f. of GLD_5 (b).

4.6.3 Data on Velocities within Galaxies

In astronomy, the cluster named A1775 is believed to consist of two clusters that are in close proximity. Oegerle, Hill, and Fitchett (1995) gave velocity observations (in kilometers per second) from A1775 (see also McClave, Dietrich, and Sincich (1997, p. 85)) which include the following velocity observations (in increasing order).

18499	18792	18933	19026	19111	19111	19130
19179	19225	19404	19408	19595	19595	19619
19673	19740	19807	19866	20210	20210	20875
21911	21993	22192	22193	22417	22417	22426
22513	22625	22647	22682	22738	22738	22744
22779	22781	22796	22809	22922	22922	23017
23059	23121	23220	23261	23303	23303	23408
23432	24909					

The $\hat{\rho}_1$, $\hat{\rho}_2$, $\hat{\rho}_3$, $\hat{\rho}_4$ for this data are

$$\hat{\rho}_1 = 22355.0000, \quad \hat{\rho}_2 = 4218.0000, \quad \hat{\rho}_3 = 3.5316, \quad \hat{\rho}_4 = 0.7620.$$

The (ρ_3, ρ_4) for this data is not covered by any of the tables of Appendix D and if we locate this point in Figure 4.4–13, we see that it is located outside the region covered by the GLD tables but within the region of the generalized beta distribution of the EGLD discussed in Chapter 3.

The $\hat{\alpha}_1, \hat{\alpha}_2, \hat{\alpha}_3, \hat{\alpha}_4$ for this data,

$$\hat{\alpha}_1 = 21450.0196, \quad \hat{\alpha}_2 = 3.0489 \times 10^6, \quad \hat{\alpha}_3 = -0.2626, \quad \hat{\alpha}_4 = 1.5451,$$

places (α_3^2, α_4) outside of the GLD range but within the EGLD system covered by the GBD which gives the fit

$$\text{GBD}(18675.3862, 4880.6411, -0.4789, -0.6045).$$

The principal purpose of this example is to show that there are datasets for which percentile-based fits are not possible. Having established this, we note that the support for the fitted GBD is [18675, 23556] but the data ranges over the interval [18499, 24909]. The method described in Problems 3.1 through 3.3 may well produce a better fit by guaranteeing that the support of the fitted GBD covers the data range.

4.6.4 Rainfall Data of Sections 2.5.5 and 3.5.4

Rainfall (in inches) at Rochester (X) and Syracuse (Y), New York was given in Table 2.5–6 of Section 2.5.5 where we were not able to find a GLD fit to either X or Y by using moments. In Section 3.5.4 we found GBD fits for both X and Y. Here we reconsider this data with the view of developing percentile-based GLD fits to both X and Y.

We compute the $\hat{\rho}_1$, $\hat{\rho}_2$, $\hat{\rho}_3$, $\hat{\rho}_4$ for X (rainfall in Rochester) to get

$$\hat{\rho}_1 = 0.1900, \qquad \hat{\rho}_2 = 1.4060, \qquad \hat{\rho}_3 = 0.1302, \qquad \hat{\rho}_4 = 0.4339$$

and, using `FindLambdasP`, obtain the two fits

$$\text{GLD}_1(0.08299, 0.4698, 4.3140, 34.6818),$$

$$\text{GLD}_2(1.7684, 0.5682, 4.9390, 0.1055),$$

from the (λ_3, λ_4)-spaces associated with Tables 1 and 2 of Appendix D. The supports of GLD_1 and GLD_2 are, respectively, $[-2.0, 2.2]$ and $[0.0085, 3.53]$. The support of GLD_1 is ill-suited for this data and visual inspection of the GLD_1 p.d.f confirms the view that GLD_2 is the better of the two fits. Figure 4.6–3 (a) shows a histogram of X with the p.d.f. of GLD_2 and the p.d.f of the EGLD fit

$$\text{GBD}(0.06129, 3.0802, -0.7491, 0.5463)$$

obtained in Section 3.5.4. The two p.d.f.s are almost indistinguishable. Figure 4.6–3 (b) shows the e.d.f. of X with the d.f. of GLD_2 (the d.f. of the GBD is not included as it is indistinguishable from that of the GLD_2 distribution).

To check the quality of the GLD_2 fit, we partition X into the intervals

$$(-\infty, 0.07), \quad [0.07, 0.1), \quad [0.1, 0.2), \quad [0.2, 0.45), \quad [0.45, 1.0), \quad [1.0, \infty)$$

(the same intervals used in Section 3.5.4) and obtain observed frequencies of

$$9, \quad 6, \quad 9, \quad 7, \quad 8, \quad 8$$

and expected frequencies of

$$12.8895, \quad 3.9439, \quad 7.1669, \quad 7.5657, \quad 7.3532, \quad 8.0808.$$

This leads to a chi-square statistic and corresponding p-value of

$$2.8145 \quad \text{and} \quad 0.09342,$$

respectively. For comparison, we note that the chi-square statistic and p-value associated with the GBD fit of Section 3.5.4 were

$$1.8663 \quad \text{and} \quad 0.1719,$$

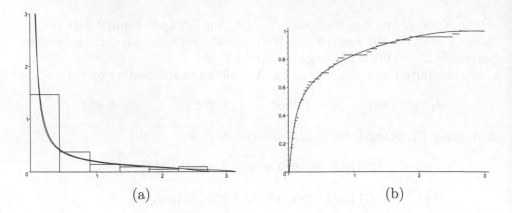

(a)   (b)

Figure 4.6–3. Histogram of rainfall data (X) and the p.d.f.s
of the fitted GLD_2 and GBD (a); the e.d.f. of
the data with the d.f. of the fitted GLD_2 (b).

respectively.

The $\hat{\rho}_1$, $\hat{\rho}_2$, $\hat{\rho}_3$, $\hat{\rho}_4$ for Y are

$$\hat{\rho}_1 = 0.2200, \qquad \hat{\rho}_2 = 1.0100, \qquad \hat{\rho}_3 = 0.2053, \qquad \hat{\rho}_4 = 0.3663.$$

Appendix D indicates the possibilities of fits associated with Tables 1, 2, and
5. These fits, obtained through `FindLambdasP`, are

$$GLD_1(0.1674, 0.6762, 4.8129, 23.8840),$$

$$GLD_2(1.3381, 0.7557, 9.6291, 0.2409),$$

$$GLD_5(0.08251, -4.0872, -0.08545, -0.6986).$$

The supports of GLD_1, GLD_2 and GLD_5 are, respectively,

$$[-1.31, 0.84], \qquad [.015, 2.09], \qquad (-\infty, \infty).$$

The most reasonable of these is the support of GLD_2 and it can also be
confirmed visually that the GLD_2 p.d.f. is the most suitable of the three
fitted GLDs.

Figure 4.6–4 (a) shows a histogram of Y with the p.d.f. of GLD_2 and the
p.d.f. of the EGLD fit

$$GBD(0.07209, 2.1729, -0.5816, 1.4355)$$

that was obtained in Section 3.5.4. The moment-based GBD p.d.f. is marked
with "m" and the p.d.f. of GLD_2 is marked with "p." Figure 4.6–4 (b) depicts

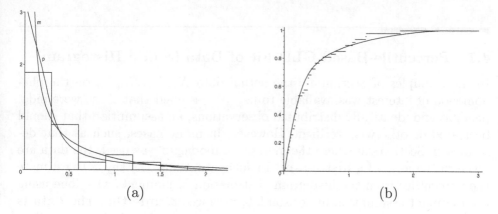

Figure 4.6–4. Histogram of rainfall data (Y) and the p.d.f.s of the fitted GLD_2 and GBD (a); the e.d.f. of the data with the d.f. of the fitted GLD_2 (b).

the e.d.f. of Y with the d.f. of GLD_2 (the d.f. of the GBD fit is not included as it cannot be distinguished from that of the GLD_2).

To obtain a chi-square statistic for the GLD_2 fit we partition Y into the intervals

$$(-\infty, 0.08), \quad [0.08, 0.15), \quad [0.15, 0.22), \quad [0.22, 0.38), \quad [0.38, 0.73), \quad [0.73, \infty)$$

(the same intervals that were used in Section 3.4.5) and determine the observed frequencies

$$7, \quad 7, \quad 9, \quad 8, \quad 8, \quad 8$$

and expected frequencies

$$8.8883, \quad 8.0537, \quad 6.5580, \quad 9.0282, \quad 6.9056, \quad 7.5662.$$

From these we obtain the chi-square statistic and p-value

$$1.1715 \quad \text{and} \quad 0.2791,$$

respectively. For comparison we note that the chi-square statistic and p-value associated with the GBD fit of Section 3.5.4 were

$$3.0371 \quad \text{and} \quad 0.0814,$$

respectively.

4.7   Percentile-Based GLD Fit of Data from a Histogram

In the examples of Section 4.6 the actual data $X_1, X_2, \ldots, X_n$ on the phenomenon of interest was available to us. We assumed that these were independent and identically distributed observations, an assumption that should be tested or otherwise verified. However, in many cases, such as that described in Section 2.6 where the method of moments was used, the data are given in the form of a **histogram**. In such situations, the key is to estimate the percentiles from the histogram. Estimation of moments was done using the midpoint assumption in Section 2.6, here we **assume that the data is uniformly spread in each class to estimate the needed percentiles**.

As an illustration, we consider the data of Section 2.6 (Table 2.6–1). The 10th percentile, $\pi_{0.1}$, is estimated by first observing that the first class, $(0.010, 0.015)$, has a relative frequency of $1/250 = 0.004$; the second class, $(0.015, 0.020)$, has relative frequency $9/250 = 0.036$, and a cumulative relative frequency of $0.004 + 0.036 = 0.04$; the next class, $(0.020, 0.025)$ has a relative frequency of $30/250 = 0.12$ and a cumulative relative frequency of 0.16. The third class is the first one whose cumulative relative frequency exceeds 0.1. For $\pi_{0.1}$, we must reach 0.1: 0.04 from the first two classes and 0.06 from the third, which represents $0.06/0.12 = 0.5$ of the third class. This gives us the estimate $0.020 + 0.5(0.025 - 0.020) = 0.0225$ for $\pi_{0.1}$. In a similar manner we obtain the following percentiles that are needed for the computation of $\hat{\rho}_1, \hat{\rho}_2, \hat{\rho}_3, \hat{\rho}_4$:

$$\pi_{0.1} = 0.0225, \qquad \pi_{0.25} = 0.02756, \qquad \pi_{0.5} = 0.03353,$$
$$\pi_{0.75} = 0.04009, \qquad \pi_{0.9} = 0.04765.$$

From these, using (4.1.2) through (4.1.5), we obtain

$$\hat{\rho}_1 = 0.03353, \quad \hat{\rho}_2 = 0.02515, \quad \hat{\rho}_3 = 0.7819, \quad \hat{\rho}_4 = 0.4982.$$

There are three possible fits for the $(\hat{\rho}_3, \hat{\rho}_4)$ that we have (GLD$_1$, GLD$_2$, and GLD$_5$, from Tables D–1, D–2, and D–5, respectively, of Appendix D). Using `FindLambdasP` we obtain

$$\text{GLD}_1(0.02789, 40.4376, 2.1314, 14.1404),$$

$$\text{GLD}_2(0.03839, 51.1844, 6.8756, 1.9588),$$

$$\text{GLD}_5(0.03151, -1.5821, -0.006655, -0.01124).$$

Visual inspection shows that GLD_5 is the best of these three fits. In Section 2.6 we obtained two fits based on the method of moments. The better of the two fits as measured by the chi-square goodness-of-fit statistic was

$$GLD_m(0.02889, 18.1935, .05744, .1850).$$

Figure 4.7–1 (a) shows the histogram of the data with the p.d.f.s of GLD_5 and GLD_m (the one that rises higher near the center is the p.d.f. of GLD_5. In Figure 4.7–1 (b) we show the e.d.f. of the data with the d.f.s of GLD_5 and GLD_m (the d.f. of GLD_m is slightly lower on the left side of the figure and slightly higher on the right side).

To compare the GLD_5 and GLD_m fits, we note that the chi-square statistic and p-value of 1.6000 and 0.8088 were obtained for the GLD_m fit in Section 2.6. For GLD_5 we use the same class intervals used in Section 2.6 and obtain the observed frequencies

$$10, \quad 30, \quad 44, \quad 58, \quad 45, \quad 29, \quad 17, \quad 9, \quad 8$$

and expected frequencies

$$14.9844, \quad 25.0469, \quad 45.8615, \quad 54.8447, \quad 45.0772,$$
$$29.1585, \quad 16.6068, \quad 8.8906, \quad 9.5295.$$

The chi-square statistic and p-value for the GLD_5 fit are

$$3.1517 \quad \text{and} \quad 0.5328,$$

respectively.

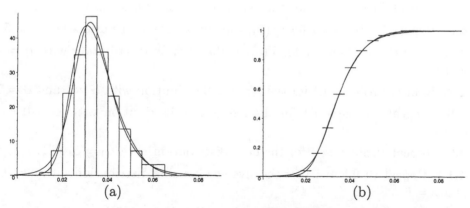

Figure 4.7–1. Histogram of coefficients of friction and the p.d.f.s of the fitted GLD_5 and GLD_m (a); the e.d.f. of the data with the d.f. of the fitted GLD_5 and GLD_m (b).

Problems for Chapter 4

4.1. In Section 4.4.1 we fitted the GLD to the $N(0,1)$ p.d.f. using Algorithm GLD–P of Section 4.3 with $u = 0.1$. Recall that $u = 0.1$ accommodates sample sizes as small as 9 whereas $u = 0.01$ requires sample sizes of at least 99 and, in general, u can be any number strictly between 0 and 0.25. The quality of the fit that we obtained, as measured by the L_1 norm (see Table 4.5–2), was 0.003. The fit and its L_1 norm varies as u varies, hence the L_1 norm of the fit is a function of u, say $L_1(u)$.

    a. Use Algorithm GLD–P to find GLD fits to $N(0,1)$ with $u = 0.01, 0.05, 0.1, 0.2, 0.24$. When several fits are available choose what seems to be the best of the available fits, as was done in Section 4.4.1.

    b. Find $L_1(0.01)$, $L_1(0.05)$, $L_1(0.1)$, $L_1(0.2)$, $L_1(0.24)$ for the fits obtained in part a.

    c. Graph $L_1(u)$ of u in the interval $(0, 0.25)$. What u do you predict will yield the L_1-best GLD approximation to $N(0,1)$? Test your prediction by finding the the fit and its L_1 norm for that u value.

4.2. Repeat Problem 4.1 for $U(0,1)$, the uniform distribution on the interval $(0,1)$.

4.3. Repeat Problem 4.1 for $t(1)$, the Student's t distribution with $\nu = 1$.

4.4. Repeat Problem 4.1 for the exponential distribution with $\theta = 3$.

4.5. Repeat Problem 4.1 for $\chi^2(5)$, the chi-square distribution with $\nu = 5$.

4.6. Repeat Problem 4.1 for $\Gamma(5,3)$, the gamma distribution with $\alpha = 5$ and $\theta = 3$.

4.7. Repeat Problem 4.1 for the Weibull distribution with $\alpha = 1$ and $\beta = 5$.

4.8. Repeat Problem 4.1 for the lognormal distribution with $\mu = 0$ and $\sigma = 1/3$.

4.9. Repeat Problem 4.1 for the beta distribution with $\beta_3 = \beta_4 = 1$.

4.10. Repeat Problem 4.1 for the inverse Gaussian distribution with $\mu = 0.5$ and $\lambda = 6$.

4.11. Repeat Problem 4.1 for the logistic distribution with $\mu = 0$ and $\sigma = 1$.

4.12. Repeat Problem 4.1 for the largest extreme value distribution with $\mu = 0$ and $\sigma = 1$.

4.13. Repeat Problem 4.1 for the double exponential distribution with $\lambda = 1$.

4.14. Repeat Problem 4.1 for $F(6, 25)$, the F-distribution with $\nu_1 = 6$ and $\nu_2 = 25$.

4.15. Repeat Problem 4.1 for the Pareto distribution with $\beta = 5$ and $\lambda = 1$.

4.16. For each of Problems 4.1 through 4.15, find the value of u that achieves

$$\min_u \max_g \; L_1(u, g)$$

where the maximum is taken over all g in the class of distributions under consideration.

4.17. In Section 4.6.1 two fits GLD_1 and GLD_5 were obtained to data that was generated from the Cauchy distribution. By plotting the p.d.f.s of GLD_1 and GLD_5 with the p.d.f. of the Cauchy distribution, make an initial assessment of the "closeness" of the Cauchy p.d.f. to the GLD_1 and GLD_5 p.d.f.s. Next, substantiate your observation quantitatively by computing

$$\sup_x |\hat{f}_1(x) - f(x)| \quad \text{and} \quad \sup_x |\hat{f}_5(x) - f(x)|.$$

Now do graphic and quantitative assessments of the d.f.s of the Cauchy and the fitted GLDs.

4.18. In addition to introducing the terms and concepts of "percentile," "quartile," and "decile" (see the introduction to this chapter), Galton went on to develop "regression" (see Section 10.2 of Dudewicz, Chen, and Taneja (1989)). While regression often uses the assumption of normal distribution for its residuals, a GLD or EGLD distribution for the residuals is also a possibility. Develop the details and in an example contrast the results with those of normal regression. You can use simulated data from a non-normal distribution such as the Cauchy, which has no mean or higher order moments but can be fitted well using the GLD as seen in Section 4.6.1.

4.19. In Example 2.5.4 we fitted each of X and Y from birth weight of twins with a GLD. The fit to Y was good (p-value of 0.24), but that to X was surprisingly poor (p-value of 0.03). Use the method of this chapter to fit X and Y using percentiles, and compare the quality of the fits with those obtained in Chapter 2.

Chapter 5

GLD–2: The Bivariate GLD Distribution

We saw in Chapters 1–4 that the GLD and EGLD are very successful in practice in fitting a variety of datasets. Since data is often bivariate (the simplest multivariate setting), it is desirable to have a way of fitting a **bivariate GLD**, a distribution whose univariate components (not necessarily independent) have univariate GLD (or EGLD) distributions. A bivariate GLD, called GLD–2, was developed by Beckwith and Dudewicz (1996), and its essential features are described in this chapter; for full details, see Beckwith and Dudewicz (1996) and Karian and Dudewicz (1999a). The algorithm developed by Beckwith and Dudewicz (1996) is based on the following procedure.

1. Fit univariate GLDs separately to the components X and Y of the bivariate random variable (X, Y).

2. Use the method R. L. Plackett (1965) proposed for generating bivariate distributions with specified marginals to develop a bivariate distribution with these marginals.

3. Optimize over the infinite set of distributions in the Plackett class to obtain a fit to the actual dataset.

4. Calculate a bivariate plot and various quantitative measures to visually and quantitatively assess the quality of the fit.

Following the brief overview of Section 5.1 this chapter describes in Section 5.2 Plackett's Method and its use to develop the GLD–2. Next, the GLD–2 is used to fit several well-known bivariate distributions with identical and non-identical margins (Sections 5.3 and 5.4), and datasets (Section 5.5). Section 5.6 gives a procedure for generating GLD–2 random variates. We conclude by describing in Section 5.7 some open problems related to the GLD–2, followed by a set of problems for this chapter that will allow students in courses using this book as a text and other readers to test their understanding of the methods and concepts of the GLD–2.

5.1   Overview

Suppose the random variables X and Y have p.d.f.s $f(x)$ and $g(y)$ and d.f.s $F(x)$, and $G(y)$, respectively. The d.f., $H(x,y)$, and p.d.f., $h(x,y)$, formulas for the GLD–2 are, respectively:

$$H(x,y) = \begin{cases} \dfrac{S(x,y) - \sqrt{S^2(x,y) - 4\Psi(\Psi - 1)F(x)G(y)}}{2(\Psi - 1)} & (\Psi \neq 1) \\ F(x)G(y) & (\Psi = 1) \end{cases} \qquad (5.1.1)$$

and

$$h(x,y) = \frac{\Psi f(x)g(y)\left[1 + (\Psi - 1)(F(x) + G(y) - 2F(x)G(y))\right]}{\left(S^2(x,y) - 4\Psi(\Psi - 1)F(x)G(y)\right)^{3/2}}, \qquad (5.1.2)$$

where Ψ is nonnegative and

$$S(x,y) = 1 + (F(x) + G(y))(\Psi - 1).$$

Note that $\Psi = 1$ is the case when X and Y are independent random variables, in which case $H(x,y) = F(x)G(y)$ and $h(x,y) = f(x)g(y)$ for all (x,y).

To illustrate this method we give the final results of fitting the bivariate data which arose in a study of imaging in the brain (Dudewicz, Levy, Lienhart, and Wehrli (1989)). The issue was to fit the measured quantity called AD (the ^1H hydrogen density in a specific portion of brain tissue as related to the average density in the whole brain). In this example, the AD measurements are for the Cortical White Matter Right side (CWR) (designated by X) and the Cortical White Matter Left side (CWL) (designated by Y). Thus, the data are the pairs (X, Y) and the concern of the research was to determine if X and Y were related. The data, from Dudewicz, Levy, Lienhart, and Wehrli (1989, p. 337), is given in Table 5.1–1. The results of applying Algorithm GLD–2 of Section 5.3 to the data of Table 5.1–1, shown in Figures 5.1–2, 5.1–3, and 5.1–4, can be interpreted as follows:

Table 5.1–1. Brain tissue ^1H hydrogen density data.

X	Y	X	Y	X	Y
96.8	99.2	96.8	96.8	95.6	100.8
107.6	102.9	86.5	86.1	87.3	85.7
99.6	98.1	94.5	95.5	89.8	91.2
99.5	103.9	97.0	92.9	88.2	88.8
102.9	109.2	92.2	91.0	89.4	94.1
84.8	82.8	100.3	98.2	88.6	89.9
97.9	95.9	102.5	95.6	87.7	86.4
103.2	100.2	100.6	99.4	88.1	91.0

1. Figure 5.1–2 shows the GLD p.d.f. that is fitted to X (together with the histogram for X) and the fitted d.f. (with the empiric distribution of X). Figure 5.1–3 does the same for Y. These figures indicate that the GLD fits the data well.

2. A two-dimensional scatterplot with median lines is shown in Figure 5.1–4. The counts a, b, c, and d specified in the algorithm of Section 5.3 are shown on the scatterplot.

3. Figure 5.1–4 also shows the three-dimensional GLD–2 p.d.f. that is obtained by using (5.1.2). The "snake shape" of this surface indicates that the GLD–2 can produce a good fit even in the absence of symmetry, a property that is associated with some distributions such as the bivariate normal. For non-convex contours, see McLachlan (1992), p. 41.

We see that X and Y are not independent. In fact, they are highly correlated, and so a tumor in one side might stand out on a visual X and Y comparison. Also, the fits are quite good for each variable. Some software that could be useful in this connection is given in the article by Beckwith and Dudewicz (1996); for additional details, also see Karian and Dudewicz (1999a).

We conclude that the GLD–2 has, in this case, handled well a dataset that might have led researchers seeing the histogram of Figures 5.1–2 and 5.1–3 to wonder: How can we ever fit the pair (X, Y) since X is so non-normal, and the data contours are non-convex? With this motivation, we proceed to develop the GLD–2 in detail.

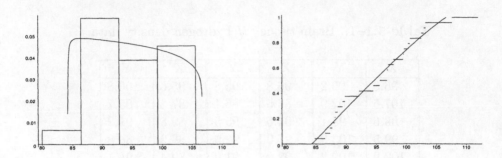

Figure 5.1–2. Histogram of X with its GLD p.d.f.; empirical
distribution with its GLD d.f.

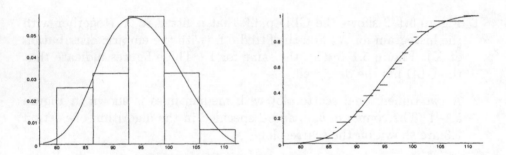

Figure 5.1–3. Histogram of Y with its GLD p.d.f.; empirical
distribution with its GLD d.f.

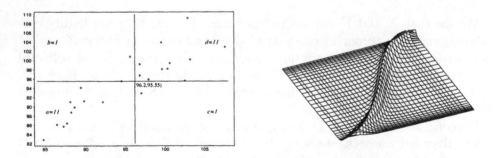

Figure 5.1–4. The (X, Y) scatterplot and the bivariate
GLD–2 p.d.f. for (X, Y).

5.2   Plackett's Method of Bivariate d.f. Construction: The GLD–2

The goal here is to construct a family of bivariate distributions, with given marginals $F_X(x)$ and $G_Y(y)$, such that the family contains the bivariate distributions that correspond to the independent case as well as the upper and lower Fréchet bounds; i.e., the family includes the bivariate d.f.s

$$F(x)G(y), \tag{5.2.1}$$

$$H_0(x,y) = \max\{F(x) + G(y) - 1, 0\}, \tag{5.2.2}$$

and

$$H_1(x,y) = \min\{F(x), G(y)\}. \tag{5.2.3}$$

Fréchet considered convex mixtures of the upper and lower bounds,

$$H(x,y) = \lambda H_0(x,y) + (1 - \lambda)H_1(x,y) \ (0 \le \lambda \le 1), \tag{5.2.4}$$

and Morgenstern (later extended by Farlie and Gumbel, now called the FGM family; see Hutchinson and Lai (1990)) considered

$$H(x,y) = F(x)G(y)[1 + \alpha A(F)B(G)] \tag{5.2.5}$$

where $A(F(x))$ and $B(F(y))$ are functions of x and y that tend to zero as $F(x)$ and $G(y) \to 1$, respectively. Fréchet's mixture certainly contains the upper and lower bounds, and the FGM family contains the independent case, but neither contains all three. Another problem with the FGM family is that the constraints on $A(F)$ and $B(G)$ require that α be small, hence the family only models cases of weak dependence.

In his method, Plackett used the ideas of a two-by-two contingency table. Given a scatterplot of bivariate data, partition the x and y values with a horizontal line and a vertical line, dividing the data into four quadrants. Each quadrant has a probability. Suppose that the true bivariate d.f. of the pair (X, Y) is

$$H(x,y) = P(X \le x, Y \le y). \tag{5.2.6}$$

Then the probabilities that a pair (X, Y) falls into the four quadrants are $H(x,y)$ in the lower left quadrant, $F(x) - H(x,y)$ in the upper left quadrant, $G(y) - H(x,y)$ in the lower right quadrant, and $1 - F(x) - G(y) + H(x,y)$

in the upper right. For example, the probability that (X, Y) is in the upper left quadrant is

$$P((X, Y) \text{ is in the upper left quadrant})$$
$$= P(X \le x, Y > y)$$
$$= P(X \le x, Y \le \infty) - P(X \le x, Y \le y) \qquad (5.2.7)$$
$$= F(x) - H(x, y),$$

where $F(x)$ is the d.f. of X (and we also used $G(y)$ for d.f. of Y). We show the situation in Figure 5.2–1.

Plackett proposed a system that solves

$$\Psi = \frac{H(1 - F - G + H)}{(F - H)(G - H)}, \qquad (5.2.8)$$

and showed that for all $\Psi \in [0, \infty)$ equation (5.2.8) has a unique root H which satisfies $H_0(x, y) \le H(x, y) \le H_1(x, y)$, a condition shown by Fréchet to be necessary for any bivariate distribution. This H is a valid bivariate distribution, and its marginals are $F_X(x) = F(x)$ and $G_Y(y) = G(y)$. Further, it is possible to attain the independent joint distribution $F(x)G(y)$. Note that if we call the probabilities of the four quadrants a, b, c, d as in Figure 5.2–1, then Plackett is proposing a system where, for some value of the constant Ψ,

$$ad = \Psi bc \qquad (5.2.9)$$

for all points (x, y) in the plane. We will see below that this in fact specifies a bivariate d.f. with marginals $F(x)$ and $G(y)$ (though this is not obvious at this point, and the proof will take some effort).

It is unlikely that H, the solution of (5.2.8), will be in a simple form; often numerical calculations will be needed. **To find the solution of Plackett's equation (5.2.8),** we follow the development given by Mardia (1967, 1970) with some additional details. First, rewrite (5.2.8) as a quadratic in H. From (5.2.8) we have the equivalent statements

$$\Psi(F - H)(G - H) = H(1 - F - G + H),$$
$$\Psi(FG - H(F + G) + H^2) = H(1 - F - G) + H^2,$$
$$H^2(1 - \Psi) + H[1 + (F + G)(\Psi - 1)] - \Psi FG = 0.$$

Next, solve the last equation for H and obtain

$$H = \frac{-(1 + (F + G)(\Psi - 1)) \pm \sqrt{(1 + (F + G)(\Psi - 1))^2 + 4(1 - \Psi)\Psi FG}}{2(1 - \Psi)}.$$

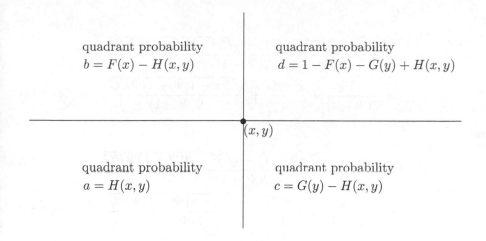

Figure 5.2–1.   The probabilities that (X, Y) falls into the four quadrants determined by a point (x, y).

To simplify, let

$$S = 1 + (F + G)(\Psi - 1), \tag{5.2.10}$$

and obtain

$$H = \frac{-S \pm \sqrt{S^2 + 4(1 - \Psi)\Psi FG}}{2(1 - \Psi)} = \frac{S \pm \sqrt{S^2 - 4(\Psi - 1)\Psi FG}}{2(\Psi - 1)}.$$

For the independent case, let $\Psi = 1$ in (5.2.8); that is,

$$1 = \frac{H(1 - F - G + H)}{(F - H)(G - H)},$$

$$H(1 - F - G + H) = (F - H)(G - H),$$

$$H - HF - HG + H^2 = FG - HG - HF + H^2,$$

$$H = FG. \tag{5.2.11}$$

For $\Psi \neq 1$, Mardia (1967, 1970) showed that **only one root will yield a valid d.f.:**

$$H = \frac{S - \sqrt{S^2 - 4(\Psi - 1)\Psi FG}}{2(\Psi - 1)}. \tag{5.2.12}$$

To see this, label the two roots of the quadratic as

$$\alpha = \frac{S + \sqrt{S^2 - 4(\Psi - 1)\Psi FG}}{2(\Psi - 1)} \quad \text{and} \quad \beta = \frac{S - \sqrt{S^2 - 4(\Psi - 1)\Psi FG}}{2(\Psi - 1)},$$

or

$$\alpha = \frac{1 + (F+G)(\Psi-1) + \sqrt{S^2 - 4(\Psi-1)\Psi FG}}{2(\Psi-1)}$$

$$= \frac{1}{2(\Psi-1)} + \frac{F+G}{2} + \frac{\sqrt{S^2 - 4(\Psi-1)\Psi FG}}{2(\Psi-1)} \qquad (5.2.13)$$

and

$$\beta = \frac{1 + (F+G)(\Psi-1) - \sqrt{S^2 - 4(\Psi-1)\Psi FG}}{2(\Psi-1)}$$

$$= \frac{1}{2(\Psi-1)} + \frac{F+G}{2} - \frac{\sqrt{S^2 - 4(\Psi-1)\Psi FG}}{2(\Psi-1)}. \qquad (5.2.14)$$

Then consider four cases: $\Psi = 0, 0 < \Psi < 1, \Psi = 1$, and $\Psi > 1$. **If $\Psi > 1$,** then the root

$$\alpha > \frac{1}{2(\Psi-1)} + \frac{F+G}{2} > \frac{F+G}{2} > \min(F,G)$$

and hence does not satisfy the Fréchet upper bound (and thus is not a valid bivariate d.f.).

Noting that

$$\alpha\beta = \left(\frac{S + \sqrt{S^2 - 4(\Psi-1)\Psi FG}}{2(\Psi-1)} \right) \left(\frac{S - \sqrt{S^2 - 4(\Psi-1)\Psi FG}}{2(\Psi-1)} \right)$$

$$= \frac{S^2 - S^2 + 4(\Psi-1)\Psi FG}{4(\Psi-1)^2} = \frac{\Psi FG}{\Psi - 1}$$

is negative **for all** $\Psi \in (0,1)$, we conclude that one of the roots α and β is negative (the other positive) in this case. Now for $\Psi \in (0,1), -4(\Psi-1)\Psi FG$ is positive, so

$$\beta = \frac{1}{2(\Psi-1)} + \frac{F+G}{2} - \frac{\sqrt{S^2 - 4(\Psi-1)\Psi FG}}{2(\Psi-1)}$$

$$\geq \frac{1}{2(\Psi-1)} + \frac{F+G}{2} - \frac{\sqrt{S^2}}{2(\Psi-1)}$$

$$= \frac{1}{2(\Psi-1)} + \frac{F+G}{2} - \frac{1 + F(\Psi-1) + G(\Psi-1)}{2(\Psi-1)}$$

$$= \frac{1}{2(\Psi-1)} + \frac{F+G}{2} - \frac{1}{2(\Psi-1)} - \frac{F+G}{2} = 0.$$

Thus for $\Psi \in (0,1)$, we have $\beta \geq 0$, hence α is negative and therefore invalid. **If $\Psi = 1$, we have $H = FG$, the independent case in (5.2.11) above.** From (5.2.8) we see $H = \max(0, F + G - 1)$ if and only if $\Psi = 0$, and $H = \min(F, G)$ if and only if $\Psi = \infty$. These observations together imply that for $\Psi \in [0, \infty)$, **α is not a valid distribution, and also that we can achieve the independent case and the Fréchet bounds.**
What remains to be shown then is that

$$\beta = H(x,y) = \frac{S - \sqrt{S^2 - 4(\Psi - 1)\Psi FG}}{2(\Psi - 1)} \qquad (5.2.15)$$

is a valid bivariate d.f. There are (see p. 123 of Dudewicz and Mishra (1988), e.g.) four criteria to check:

1. $\displaystyle\lim_{x,y \to \infty} H(x,y) = 1$;

2. $\displaystyle\lim_{x \to -\infty} H(x,y) = \lim_{y \to -\infty} H(x,y) = 0$;

3. $\displaystyle\lim_{h \to 0^+} H(x+h, y) = \lim_{h \to 0^+} H(x, y+h) = H(x,y)$; and

4. $H(a,b) + H(a+h, b+k) - H(a+h, b) - H(a, b+k) \geq 0$, for any $h, k > 0$; that is, H assigns positive probability to all rectangles.

We would also like to verify that the given joint d.f. has the desired marginals; that is,

5. $\displaystyle\lim_{y \to \infty} H(x,y) = F(x)$ and $\displaystyle\lim_{x \to \infty} H(x,y) = G(y)$.

To establish criterion 1, note that $\lim_{x \to \infty} F(x) = \lim_{y \to \infty} G(y) = 1$, so by (5.2.11)

$$\lim_{x,y \to \infty} S = \lim_{F,G \to 1} 1 + (F+G)(\Psi - 1) = 2\Psi - 1,$$

hence, from (5.2.12), when $\Psi \neq 1$ (the case $\Psi = 1$ follows from (5.2.11))

$$\begin{aligned}
\lim_{x,y \to \infty} H(x,y) &= \lim_{F,G \to 1} \frac{S - \sqrt{S^2 - 4(\Psi - 1)\Psi FG}}{2(\Psi - 1)} \\
&= \frac{2\Psi - 1 - \sqrt{(2\Psi - 1)^2 - 4(\Psi - 1)\Psi}}{2(\Psi - 1)} \\
&= \frac{2\Psi - 1 - \sqrt{4\Psi^2 - 4\Psi + 1 - 4\Psi^2 + 4\Psi}}{2(\Psi - 1)} \\
&= \frac{2\Psi - 2}{2(\Psi - 1)} = 1.
\end{aligned}$$

Criterion 2 is similar: one notes that

$$\lim_{x \to -\infty} F(x) = 0,$$

hence

$$\lim_{x \to -\infty} S = \lim_{F \to 0} 1 + (F + G)(\Psi - 1) = 1 + G(\Psi - 1)$$

and note that $1 + G(\Psi - 1) \geq 0$, so

$$\begin{aligned}
\lim_{x \to -\infty} H(x, y) &= \lim_{F \to 0} \frac{S - \sqrt{S^2 - 4(\Psi - 1)\Psi F G}}{2(\Psi - 1)} \\
&= \frac{1 + G(\Psi - 1) - \sqrt{[1 + G(\Psi - 1)]^2}}{2(\Psi - 1)} \\
&= \frac{1 + G(\Psi - 1) - |1 + G(\Psi - 1)|}{2(\Psi - 1)} = 0;
\end{aligned}$$

by symmetry, $\lim_{y \to -\infty} H(x, y) = 0$.

To verify criteria 3 and 4, it is sufficient to show that

$$h(x, y) \equiv \frac{\partial^2 H}{\partial F \partial G} \geq 0$$

for all real numbers x and y. Below at (5.2.19) we find that

$$h(x, y) = \frac{\Psi f g[1 + (\Psi - 1)(F + G - 2FG)]}{(S^2 - 4\Psi(\Psi - 1)FG)^{3/2}},$$

which is (since the other terms are nonnegative) ≥ 0 if and only if

$$1 + (\Psi - 1)(F + G - 2FG) \geq 0. \tag{5.2.16}$$

Now consider three cases: $0 \leq \Psi < 1$, $\Psi = 1$, and $\Psi > 1$.

Clearly (5.2.16) is satisfied if $\Psi = 1$. If $0 \leq \Psi < 1$, then note the left-hand side of (5.2.16) can be bounded as follows:

$$1 + 2FG - F - G \leq 1 + (\Psi - 1)(F + G - 2FG) < 1,$$

since using calculus we find $F + G - 2FG$ achieves its minimum of 0 at $F = 0, G = 1$ or $F = G = 0$ and its maximum of 1 at $F = 0, G = 1$ or $F = 1, G = 0$. It follows that for $0 \leq \Psi < 1$ we have

$$0 \leq 1 + (\Psi - 1)(F + G - 2FG) < 1.$$

In the case of $\Psi > 1, 1 + (\Psi - 1)(F + G - 2FG) > 1 + F + G - 2FG$. As $1 + F + G - 2FG$ achieves its minimum of 1 at $F = G = 0$ or $F = G = 1$ and its maximum of 2 at $F = 0, G = 1$ or $F = 1, G = 0$, when $\Psi > 1$ we have

$$1 + (\Psi - 1)(F + G - 2FG) \geq 1.$$

Thus, H is a valid bivariate d.f. for all $\Psi \geq 0$ and is given by

$$H = \begin{cases} \dfrac{S - \sqrt{S^2 - 4\Psi(\Psi - 1)FG}}{2(\Psi - 1)} & (\Psi \neq 1) \\ FG & (\Psi = 1). \end{cases} \qquad (5.2.17)$$

To see that the marginals are as in 5, note that since $\lim_{x \to \infty} F(x) = 1$, by (5.2.10)

$$\lim_{x \to \infty} S = \lim_{F \to 1} 1 + (F + G)(\Psi - 1) = \Psi + G(\Psi - 1),$$

thus (see (5.2.14))

$$\begin{aligned} \lim_{x \to \infty} H(x, y) = \lim_{F \to 1} H(x, y) &= \lim_{F \to 1} \frac{S - \sqrt{S^2 - 4(\Psi - 1)\Psi FG}}{2(\Psi - 1)} \\ &= \frac{\Psi + G(\Psi - 1) - \sqrt{[\Psi + G(\Psi - 1)]^2 - 4(\Psi - 1)\Psi G}}{2(\Psi - 1)} \\ &= \frac{\Psi + G(\Psi - 1) - \sqrt{\Psi^2 - 2G\Psi(\Psi - 1) + G^2(\Psi - 1)^2}}{2(\Psi - 1)} \\ &= \frac{\Psi + G(\Psi - 1) - \sqrt{[\Psi - G(\Psi - 1)]^2}}{2(\Psi - 1)} \\ &= \frac{\Psi + G(\Psi - 1) - \Psi + G(\Psi - 1)}{2(\Psi - 1)} \\ &= \frac{2G(\Psi - 1)}{2(\Psi - 1)} = G \end{aligned}$$

and by symmetry $\lim_{y \to \infty} H(x, y) = F$.

Mardia also derived **the p.d.f. of H** as follows. Given a d.f. $H(x, y)$, the p.d.f. $h(x, y)$ is given by

$$\begin{aligned} h(x, y) &= \frac{\partial}{\partial y}\left(\frac{\partial H}{\partial x}\right) = \frac{\partial}{\partial y}\left(\frac{\partial H}{\partial F}\frac{dF}{dy}\right) = \frac{\partial}{\partial y}\left(\frac{\partial H}{\partial F}f\right) \\ &= \frac{\partial}{\partial F}\left(\frac{\partial H}{\partial y}\right)f = \frac{\partial}{\partial F}\left(\frac{\partial H}{\partial G}\frac{dG}{dy}\right) = fg\frac{\partial^2 H}{\partial F \partial G}, \qquad (5.2.18) \end{aligned}$$

with the derivative interchanges permissible due to the continuity of $H(x, y)$, $F(x)$, and $G(y)$. Making substitutions to simplify notation:

$$D = (S^2 - 4\Psi(\Psi - 1)FG)^{\frac{3}{2}}, p = \Psi,$$

and noting that (5.2.10) implies $\frac{\partial S}{\partial F} \equiv \frac{\partial S}{\partial G} = \Psi - 1 = p - 1$, from (5.2.14) when $\Psi \neq 1$, we find

$$\frac{\partial H}{\partial F} = \frac{p-1}{2(p-1)} - \frac{1}{2(p-1)}$$

$$\times \left[\frac{1}{2}(S^2 - 4p(p-1)FG)^{-\frac{1}{2}} \times (2S(p-1) - 4p(p-1)G) \right]$$

$$= \frac{1}{2} - \frac{1}{2}(S^2 - 4p(p-1)FG)^{-\frac{1}{2}} \times (S - 2pG).$$

Thus,

$$\frac{\partial^2 H}{\partial F \partial G} = -\frac{1}{2} \left[\begin{array}{l} -\frac{1}{2}(S^2 - 4p(p-1)FG)^{-\frac{3}{2}}(2S(p-1) - 4p(p-1)F) \\ \times(S - 2pG) + (S^2 - 4p(p-1)FG)^{-\frac{1}{2}}((p-1) - 2p) \end{array} \right]$$

$$= -\frac{1}{2} \left[\begin{array}{l} -\frac{1}{2}(S^2 - 4p(p-1)FG)^{-\frac{3}{2}}2(p-1)(S - 2pF) \\ \times(S - 2pG) - (S^2 - 4p(p-1)FG)^{-\frac{1}{2}} \times (p+1) \end{array} \right]$$

$$= \frac{-1}{2D} \left[(p-1)(S - 2pF)(S - 2pG) + (p+1)(S^2 - 4p(p-1)FG) \right]$$

$$= \frac{p}{D} \left[S^2 - p(SF + SG) + SF + SG - 2(p-1)FG \right]$$

$$= \frac{p}{D} \left[S^2 - (p-1)(SF + SG - 2FG) \right]$$

$$= \frac{p}{D} \left[S(1 + (p-1)(F+G) - (p-1)(F+G)) - 2(p-1)FG \right]$$

$$= \frac{p}{D} \left[S - 2(p-1)FG \right]$$

$$= \frac{p}{D} \left[1 + (p-1)(F+G) - 2(p-1)FG \right]$$

$$= \frac{p}{D} \left[1 + (p-1)(F+G-2FG) \right]$$

$$= \frac{\Psi \left[1 + (\Psi-1)(F+G-2FG) \right]}{(S^2 - 4\Psi(\Psi-1)FG)^{\frac{3}{2}}}.$$

Hence,

$$h = \frac{\Psi fg[1 + (\Psi-1)(F+G-2FG)]}{(S^2 - 4\Psi(\Psi-1)FG)^{\frac{3}{2}}}. \tag{5.2.19}$$

One sees from (5.2.14) and (5.2.10) that this also holds for $\Psi = 1$, i.e., for all Ψ.

The results derived above are summarized in the following theorem.

Theorem 5.2.20. *Let X be a r.v. with d.f. $F(x)$, and let Y be a r.v. with d.f. $G(y)$. Then for each value of $\Psi \neq 1$ from $[0, \infty)$ the bivariate function (5.2.15)*

$$\beta = H(x, y) = \frac{S - \sqrt{S^2 - 4\Psi(\Psi - 1)FG}}{2(\Psi - 1)},$$

where $S = 1 + (F + G)(\Psi - 1)$ (see (5.2.10)), is a bivariate d.f. with the marginals $F(x)$ and $G(y)$. When $\Psi = 1$, we have

$$H(x, y) = F(x)G(y),$$

the case of independent components X and Y. In all cases, H is the unique solution of the equation (see (5.2.8))

$$\Psi = \frac{H(1 - F - G + H)}{(F - H)(G - H)}.$$

When $\Psi = 0$,

$$H = \max(0, F + G - 1),$$

and when $\Psi = \infty$,

$$H = \min(F, G),$$

showing that the Fréchet bounds can be attained. The p.d.f. of $H(x, y)$ is (see (5.2.19))

$$h = \frac{\Psi fg[1 + (\Psi - 1)(F + G - 2FG)]}{(S^2 - 4\Psi(\Psi - 1)FG)^{\frac{3}{2}}}.$$

Given then that H is a valid distribution, the question arises: **What value of Ψ is appropriate?** Plackett gives two examples of how one might answer this. In an example using the Bivariate Normal he equates the two d.f.s at the univariate medians, and given a contingency table proposes a consistent estimator Ψ^+. The standard (zero means and unit variances) **Bivariate Normal** is given by its p.d.f.

$$N(x, y | \rho) = \frac{\exp\left(\frac{-(x^2 - 2\rho xy + y^2)}{2(1 - \rho^2)}\right)}{2\pi\sqrt{1 - \rho^2}}, \quad |\rho| \leq 1; \tag{5.2.21}$$

the corresponding d.f. is given by

$$M(x,y|\rho) = \int_{-\infty}^{x} \int_{-\infty}^{y} \frac{\exp\left(\frac{-(u^2-2\rho uv + v^2)}{2(1-\rho^2)}\right)}{2\pi\sqrt{1-\rho^2}} \, du \, dv. \qquad (5.2.22)$$

Thus, at $x = y = 0$ (the "median vector") the d.f. equals

$$M(0,0|\rho) = \int_{-\infty}^{0} \int_{-\infty}^{0} \frac{\exp(\frac{-(u^2-2\rho uv + v^2)}{2(1-\rho^2)})}{2\pi\sqrt{1-\rho^2}} \, du \, dv = \frac{\cos^{-1}(-\rho)}{2\pi} \qquad (5.2.23)$$

and since $x = y = 0$, $F = G = \frac{1}{2}$, and $S = 1 + (F + G)(\Psi - 1) = \Psi$ (see (5.2.14)) at the medians the d.f. is given by

$$H(0,0) = \frac{S - \sqrt{S^2 - 4\Psi(\Psi-1)FG}}{2(\Psi-1)} = \frac{\Psi - \sqrt{\Psi^2 - \Psi^2 + \Psi}}{2(\Psi-1)} = \frac{\Psi - \sqrt{\Psi}}{2(\Psi-1)}$$

$$= \frac{\sqrt{\Psi}}{2(1+\sqrt{\Psi})} \times \frac{(\Psi - \sqrt{\Psi})(1+\sqrt{\Psi})}{\sqrt{\Psi}(\Psi-1)} = \frac{\sqrt{\Psi}}{2(1+\sqrt{\Psi})}. \qquad (5.2.24)$$

Equating (5.2.23) and (5.2.24), we have

$$\frac{\cos^{-1}(-\rho)}{2\pi} = \frac{\sqrt{\Psi}}{2(1+\sqrt{\Psi})}.$$

We now solve for Ψ to obtain

$$\Psi = \left(\frac{\cos^{-1}(-\rho)}{\cos^{-1}(-\rho) - \pi}\right)^2. \qquad (5.2.25)$$

Thus with the Ψ of (5.2.25) the Bivariate Normal and H will agree (at least) at this point.

Plackett also suggested **an estimator of** Ψ useful for fitting datasets: Divide the joint distribution into four quadrants, using lines $x = h$ and $y = k$ for some constants h, k. Then count the number of (x, y) points in each quadrant; let a, b, c, d denote the counts; that is,

$$a(x \le h, y \le k), b(x \le h, y > k), c(x > h, y \le k), \text{ and } d(x > h, y > k).$$

Then $\Psi^+ = \frac{ad}{bc}$ is Plackett's estimator. One may motivate this estimator as follows: Since $ad = bc$ for the d.f. $H(x, y)$ for any pair (x, y) (see 5.2.9) by the derivation of the class of d.f.s H, it is always true that $\Psi = ad/(bc)$. It is then reasonable to estimate Ψ using the counts a, b, c, d from the data.

Plackett states that Ψ^+ is asymptotically normal, and that the variance is given by $V(\Psi^+) = (\Psi)^2[1/a + 1/b + 1/c + 1/d]$ where a, b, c, and d are as given in Figure 5.2–1. Mardia (1967) showed that the variance is minimized if h and k are chosen as the respective medians.

With the above groundwork, we are in a position to develop a bivariate GLD in Section 5.3. There (5.2.17) and (5.2.19) are taken with f, g, F, and G chosen as univariate GLDs; to fit a known bivariate distribution, we fit the marginals $f(x)$ and $g(y)$ each with a univariate GLD, while for datasets with unknown distributions the GLDs can be fitted with some appropriate method and substituted. Additionally, we can specify any arbitrary pair of marginals in (5.2.17) and (5.2.19) to investigate the possible curve shapes. In the sections that follow we develop these methods and discuss several examples.

5.3 Fitting the GLD–2 to Well-Known Bivariate Distributions

We now introduce the new bivariate GLD–2. The d.f. is (see Theorem 5.2.20) the solution of Plackett's equation

$$\Psi = \frac{H(1 - F - G + H)}{(F - H)(G - H)},$$

i.e.,

$$H(x,y) = \begin{cases} \dfrac{S - \sqrt{S^2 - 4\Psi(\Psi - 1)FG}}{2(\Psi - 1)} & (\Psi \neq 1) \\[2mm] FG & (\Psi = 1), \end{cases} \qquad (5.3.1)$$

with

$$S = S(x,y) = 1 + (F(x) + G(y))(\Psi - 1),$$

where $F = F(x)$ and $G = G(y)$ are GLD marginal d.f.s and $\Psi \in [0, \infty)$ is a measure of association between the marginals. The p.d.f. $h(x,y)$ is

$$h(x,y) = \frac{\Psi f(x)g(y)[1 + (\Psi - 1)(F(x) + G(y) - 2F(x)G(y))]}{(S^2(x,y) - 4\Psi(\Psi - 1)F(x)G(y))^{\frac{3}{2}}},$$

where $f(x)$ and $g(x)$ are (possibly different) GLD marginal p.d.f.s.

In order to develop a bivariate GLD using Plackett's method Beckwith and Dudewicz (1996) proposed the following algorithm for approximating any specified p.d.f. $f(x, y)$:

Algorithm GLD–BV: Bivariate Approximation to $f(x, y)$ with a GLD–2.

GLD–BV–1. Specify the bivariate p.d.f. $f(x, y)$ to be fitted and its marginals $f_1(x)$ and $f_2(y)$.

GLD–BV–2. Fit both marginals with GLDs, using some suitable method (such as the Method of Moments, or if the moments do not exist the Method of Percentiles).

GLD–BV–3. Graph the marginals $f_1(x), f_2(y)$ and their GLD fits to verify the quality of the univariate fits.

GLD–BV–4. Evaluate the GLD–2 distribution numerically on a grid of $n \times n$ values of x and y for a broad range of Ψ values.

GLD–BV–5. Choose Ψ optimizing the fit with some criterion, yielding $\Psi = \Psi_0$.

GLD–BV–6. Graph the true distribution $f(x, y)$ and the GLD–2 with $\Psi = \Psi_0$ to visualize the quality of the fit.

In the remainder of this section we consider four bivariate distributions: the **Bivariate Normal, Gumbel's Bivariate Exponential**, the **Bivariate Cauchy**, and **Kibble's Bivariate Gamma**, each with univariate marginals of the same name. The marginals are fitted with the Method of Moments (except for the Cauchy, which has no moments, in which case the Method of Percentiles is used). Hutchinson and Lai (1990) is an excellent reference on bivariate distributions that is broad and encyclopedic.

5.3.1   The Bivariate Normal (BVN) Distribution

Here the standard BVN of (5.2.21) is used, hence we have $N(0, 1)$ marginals, correlation coefficient ρ, and moments $\mu = 0, \sigma^2 = 1, \alpha_3 = 0$, and $\alpha_4 = 3$. From Section 2.4.1 we find the fitted GLD has $\lambda = (0, 0.1975, 0.1349, 0.1349)$, and is a good univariate fit.

Having the univariate fits, the appropriate value of Ψ for each ρ is needed next. Use of Algorithm GLD–BV of Section 5.3 proceeds as follows:

- Set n, the number of grid values for each marginal.

- Calculate equally spaced x and y values, and the corresponding ρ value. Also calculate $f(x), f(y)$, and store these values in arrays.

- Find the value of Ψ (to three decimal places) which minimizes the L_1 error.

- Output the results for each $\rho = -0.9, -0.8, \cdots, 0.9$.

Since (see Beckwith and Dudewicz (1996)) there is a unique value of Ψ that minimizes the L_1 error, our program (the listing is in Appendix A) asks the user for N, the maximum number of iterations ($10 \leq N \leq 40$ is usually sufficient). The L_1 error is calculated for $\Psi = 0, 1, 2, \ldots, N$ (for brevity, we designate this succession of values of Ψ by $\Psi = 0(1)N$) and the least value Ψ_1 is kept. The initial error should be large for small Ψ, decrease near the minimum, and then increase again. So the error is calculated from $\Psi_1 - 1$ to $\Psi_1 + 1$ in increments of 0.001 until a three decimal place "best" $\Psi = \Psi_0$ is found.

At each iteration, four other measures of error are calculated: the smallest ratio and the largest ratio of the estimated $f(x, y)$ to the true $f(x, y)$, and the smallest and largest absolute differences of estimated and true $f(x, y)$. Also, the estimated volume of the two joint distributions is calculated to ensure accuracy of the L_1 error (if the volumes aren't near one, the process is suspect). The truncation required (at ± 4 in this case) will lead to error, as will the choice of Riemann partition size. These errors are summarized in Table 5.3–1 (excluding $\min |f - h|$, which was $< 10^{-5}$ for all cases). See Figure 5.3–2[1] for a graph of $\Psi = 3(0.1)5$ (i.e., for $\Psi = 3, 3.1, 3.2, \ldots, 5$), with the corresponding errors, from fitting the Bivariate Normal with $\rho = 0.5$, and optimal $\Psi = 3.653$.

With this estimate of Ψ, a visual check is important. See Figures 5.3–3 through 5.3–8 for graphs of the GLD–2, the true BVN, and the error

$$f(x, y|\rho) - h(x, y|\Psi),$$

both as 3-D density graphs and 2-D contour graphs. The values of ρ used are $-0.8, -0.5, 0, 0.2, 0.5$, and 0.9. The BVN seems to work well for most of these examples, the method is quite efficient, and most of the error seems due to the univariate fits. Good references on graphic visualization of bivariate distributions are the paper of Johnson, Wang, and Ramberg (1984) and the book of Johnson (1987).

---

[1]Some of the figures of this and the next two sections (specifically, Figures 5.3–2 through 5.3–12, 5.3–14, 5.3–15, 5.3–17 through 5.3–20, 5.4–2, 5.4–3, 5.4–5, and 5.5–1 through 5.5–6) are reprinted with permission from Beckwith and Dudewicz (1996), and are copyright ©1996 by American Sciences Press, Inc., 20 Cross Road, Syracuse, New York 13224-2104.

Table 5.3–1. Optimal Ψ and Corresponding Error Estimates.

Distribution	h = GLD–2	Max $\|h - f\|$	Max h/f	Min h/f	L_1 Error
BVN, $\rho = -0.8$	Ψ = 0.083	0.0611	1772332.764	0.003	0.17694
BVN, $\rho = -0.5$	Ψ = 0.274	0.0135	8739.460	0.006	0.08661
BVN, $\rho = 0$	Ψ = 1.000	0.0023	1.101	0.330	0.00927
BVN, $\rho = 0.2$	Ψ = 1.624	0.0038	10.942	0.040	0.03159
BVN, $\rho = 0.5$	Ψ = 3.653	0.0119	8727.850	0.006	0.08661
BVN, $\rho = 0.9$	Ψ = 28.395	0.0721	7721885.406	0.003	0.22283
BVE, $\theta = 0$	Ψ = 0.989	0.3031	1.004	0.695	0.01919
BVE, $\theta = 0.5$	Ψ = 0.477	0.1705	130.895	0.666	0.08628
BVE, $\theta = 0.5$	Ψ = 0.278	0.0934	15793.256	0.723	0.10465
BVC of (5.3.4)	Ψ = 1.553	0.0512	8.093	0.097	0.32887
BVC, Independent	Ψ = 1.000	0.0041	1.084	0.958	0.02019
BVG, $\rho = 0.2$	Ψ = 1.611	0.0338	3.894	0.109	0.09931
BVG, $\rho = 0.4$	Ψ = 2.605	0.0458	62.180	0.053	0.12313
BVG, $\rho = 0.6$	Ψ = 4.614	0.0731	25062.656	0.038	0.15694
BVG, $\rho = 0.8$	Ψ = 11.004	0.1298	1204440.624	0.039	0.21297

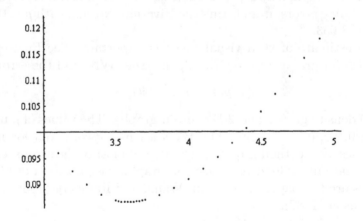

Figure 5.3–2. L_1 error (vertical axis) vs. Ψ (horizontal axis) for the BVN with $\rho = 0.5$.

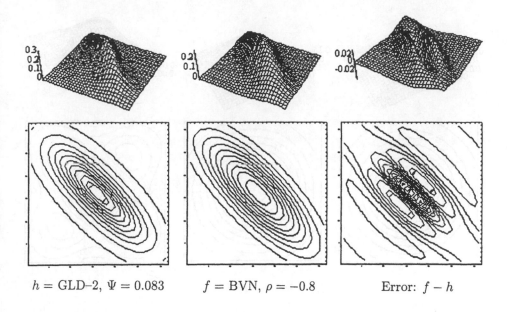

$h = \text{GLD–2}, \ \Psi = 0.083$ $f = \text{BVN}, \ \rho = -0.8$ Error: $f - h$

Figure 5.3–3. Bivariate normal ($\rho = -0.8$) and GLD–2.

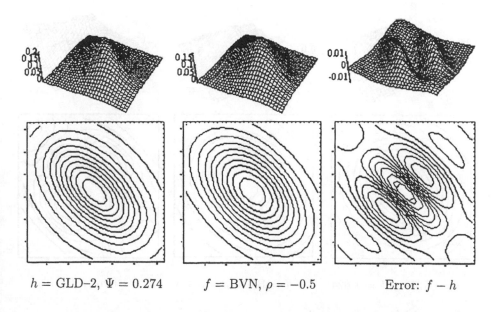

$h = \text{GLD–2}, \ \Psi = 0.274$ $f = \text{BVN}, \ \rho = -0.5$ Error: $f - h$

Figure 5.3–4. Bivariate normal ($\rho = -0.5$) and GLD–2.

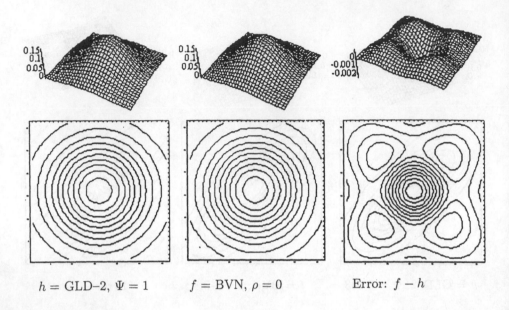

$h = \text{GLD–2}, \Psi = 1$ $f = \text{BVN}, \rho = 0$ Error: $f - h$

Figure 5.3–5. Bivariate normal ($\rho = 0$) and GLD–2.

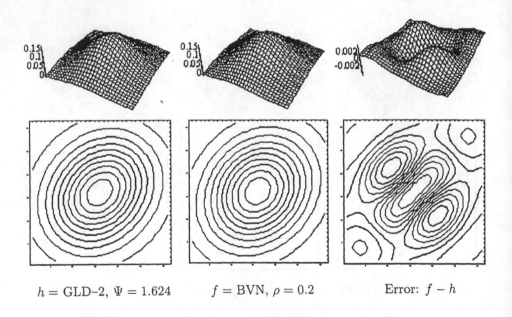

$h = \text{GLD–2}, \Psi = 1.624$ $f = \text{BVN}, \rho = 0.2$ Error: $f - h$

Figure 5.3–6. Bivariate normal ($\rho = 0.2$) and GLD–2.

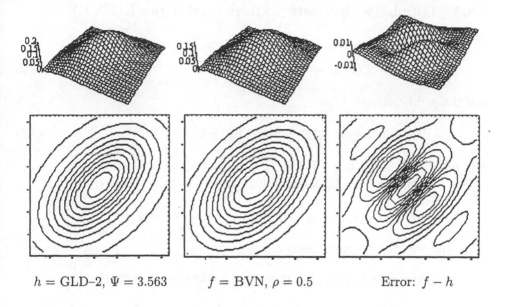

$h = $ GLD–2, $\Psi = 3.563$ $f = $ BVN, $\rho = 0.5$ Error: $f - h$

Figure 5.3–7. Bivariate normal ($\rho = 0.5$) and GLD–2.

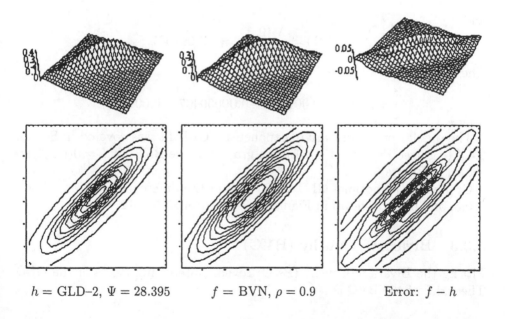

$h = $ GLD–2, $\Psi = 28.395$ $f = $ BVN, $\rho = 0.9$ Error: $f - h$

Figure 5.3–8. Bivariate normal ($\rho = 0.9$) and GLD–2.

5.3.2   Gumbel's Bivariate Exponential Type I (BVE)

Here the d.f. is

$$F(x,y) = 1 - e^{-x} - e^{-y} + e^{-(x+y+\theta xy)} \tag{5.3.2}$$

and the p.d.f. is

$$f(x,y) = [(1 + \theta x)(1 + \theta y) - \theta]e^{-(x+y+\theta xy)}, \tag{5.3.3}$$

where $0 \le \theta \le 1$ and $x, y \ge 0$ (see Hutchinson and Lai (1990)).

The marginals are each standard Exponential, that is Exp[1], hence

$$\mu = 1, \quad \sigma^2 = 1, \quad \alpha_3 = 2, \quad \alpha_4 = 9.$$

From the tables (also from Section 2.4.4),

$$\lambda(0, 1) = (-0.993, -0.001081, -0.00000407, -0.001076),$$

so by Step 5 of Algorithm GLD–M in Section 2.3.2, we have

$$\lambda_1(1, 1) = \lambda_1(0, 1)1 + 1 = -0.993 + 1 = 0.007$$

and

$$\lambda_2(1, 1) = \frac{\lambda_2(0, 1)}{1} = -0.001081.$$

Therefore,

$$\lambda = (0.007, -0.001081, -0.00000407, -0.001076).$$

The quality of the univariate exponential GLD fit was covered in Section 2.4.4 (e.g., see Figure 2.4–3). Eleven graphs of the BVE for $\theta = 0(0.1)1$ are shown in Figure 5.3–9.

Graphs of the optimal GLD–2, BVE, and the difference function for $\theta = 0, 0.5$, and 0.9 are shown in Figures 5.3–10 through 5.3–12.

5.3.3   Bivariate Cauchy (BVC)

Fitting the Bivariate Cauchy (BVC) reveals interesting facets of the GLD. The p.d.f. of the BVC is

$$f(x,y) = \frac{1}{2\pi}(1 + x^2 + y^2)^{(-3/2)}, \quad -\infty < x,\ y < \infty, \tag{5.3.4}$$

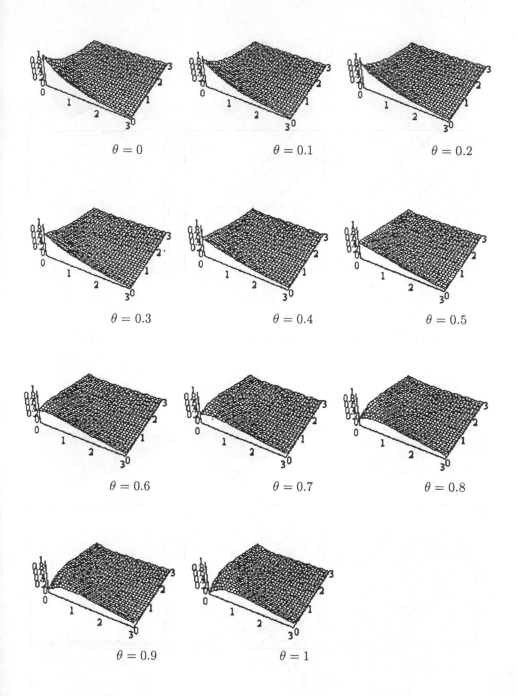

Figure 5.3–9. Graphs of the bivariate exponential p.d.f.

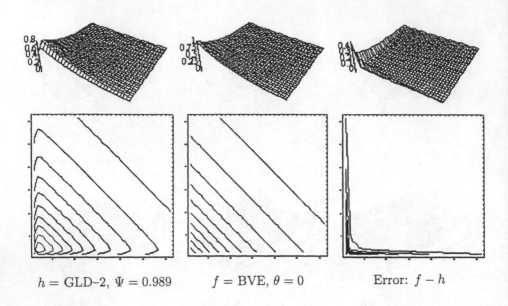

$h = $ GLD–2, $\Psi = 0.989$ $f = $ BVE, $\theta = 0$ Error: $f - h$

Figure 5.3–10. Bivariate exponential ($\theta = 0$) and GLD–2.

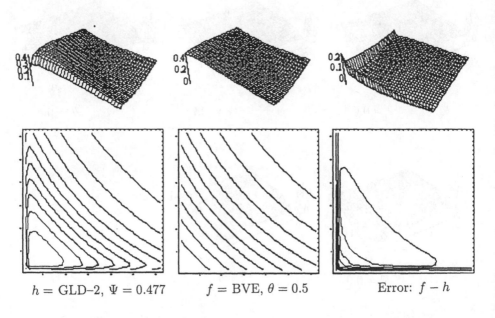

$h = $ GLD–2, $\Psi = 0.477$ $f = $ BVE, $\theta = 0.5$ Error: $f - h$

Figure 5.3–11. Bivariate exponential ($\theta = 0.5$) and GLD–2.

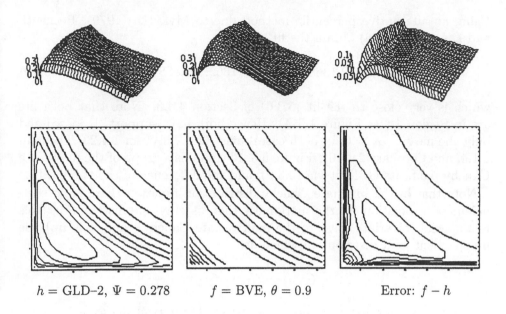

| $h = $ GLD–2, $\Psi = 0.278$ | $f = $ BVE, $\theta = 0.9$ | Error: $f - h$ |

Figure 5.3–12. Bivariate exponential ($\theta = 0.9$) and GLD–2.

with marginals

$$f_1(x) = f_2(x) = \frac{1}{\pi(1 + x^2)} \quad -\infty < x, \ y < \infty. \qquad (5.3.5)$$

Since the Cauchy distribution has no expected values, the Method of Moments cannot be used in fitting the GLD to the Cauchy. With the Method of Percentiles, noting that the Cauchy is $t(1)$, the Student's t with 1 degree of freedom, from Section 4.4.3 we have the fit with

$$\lambda = (0, -2.0676, -0.8727, -0.8727). \qquad (5.3.6)$$

The percentile function for $t(1)$ is available in closed form: $F(x) = p$ where $0 < p < 1$ yields

$$p = \frac{1}{\pi}\left(\tan^{-1}(x) + \frac{\pi}{2}\right) = \frac{1}{\pi}\tan^{-1}(x) + \frac{1}{2}, \qquad (5.3.7)$$

from which one finds

$$Q(p) = \tan\left(\pi(p - \frac{1}{2})\right). \qquad (5.3.8)$$

Using an alternative percentile method due to Mykytka (1979), Beckwith and Dudewicz (1996) obtain the fit with

$$\lambda = (0, -2.136833, -0.88871662, -0.88871662), \qquad (5.3.9)$$

which is very close to the fit (5.3.6) of Section 4.4.3. Note that both fits are in Region 4 (see Figure 1.3–1). Hence both have support $(-\infty, \infty)$ and only the means ($\alpha_1 = 0$, in both cases) exist (see Theorem 1.4.23, Theorem 2.1.4, and Corollary 2.1.10). Figure 5.3–13 contains a graph of the univariate Cauchy p.d.f., its GLD fit via (5.3.9), and the difference of the two p.d.f.s.

Note that in the bivariate Cauchy distribution function there is no coefficient of association for the marginals: there is only one distribution (this is a special (bivariate) case of the multivariate Cauchy), and it is **not** the independent joint distribution which has

$$f(x, y) = f_1(x) \cdot f_2(y).$$

Using Algorithm GLD–BV with f the BVC of (5.3.3) yields Figure 5.3–14. For f the independent joint distribution, we obtain Figure 5.3–15. For each, the smallest L_1 error occurred at or near $\Psi = 1$, which yields an independent bivariate p.d.f. The errors max $| f - h |$ for the BVC and the independent joint distribution were very low, 0.0512 and 0.0041, respectively; the L_1 error using the BVC was 0.32887 (the largest of all cases) whereas the L_1 error for the independent joint distribution was 0.02019, indicating a very good fit. Note that we used $\Psi = 1$ in Figure 5.3–15 (due to theoretical implications of the symmetry, although Algorithm GLD–BV yields $\Psi = 1.5534$ for minimum L_1 error).

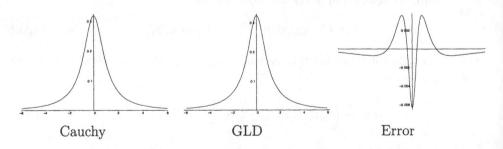

Cauchy   GLD   Error

Figure 5.3–13.   The univariate Cauchy p.d.f., its GLD
approximation, and approximation error.

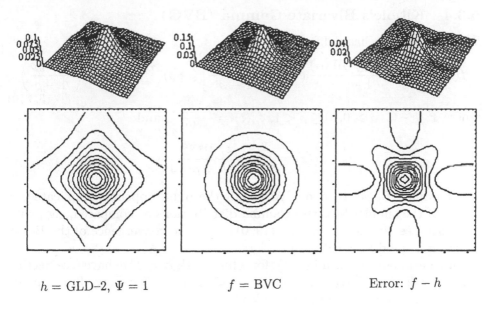

$h = \text{GLD–2}, \Psi = 1$ $f = \text{BVC}$ Error: $f - h$

Figure 5.3–14. Bivariate Cauchy of (5.3.4).

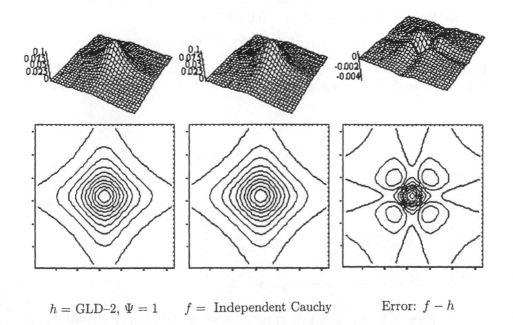

$h = \text{GLD–2}, \Psi = 1$ $f = \text{Independent Cauchy}$ Error: $f - h$

Figure 5.3–15. Bivariate independent Cauchy.

5.3.4   Kibble's Bivariate Gamma (BVG)

This distribution has p.d.f.

$$f(x,y) = \frac{f_\alpha(x) f_\alpha(y) \Gamma(\alpha)}{1-\rho} \exp\left[-\frac{\rho(x+y)}{1-\rho} \right] (xy\rho)^{-(a-1)/2} I_{\alpha-1}\left(\frac{2\sqrt{xy\rho}}{1-\rho} \right)$$

$$(5.3.10)$$

where $x, y \geq 0$, $\alpha > 0$, $0 \leq \rho < 1$, $f_\alpha(t) = \frac{t^{\alpha-1} e^{-t}}{\Gamma(\alpha)}$, and

$$I_k(z) = \sum_{r=0}^{\infty} \frac{(z/2)^{k+2r}}{r! \Gamma(k+r+1)}$$

is a modified Bessel function. Although the p.d.f. appears complicated, it is one of the simpler bivariate Gamma distributions as compared, e.g., with Jensen's (see Hutchison and Lai (1990)), and the convergence of the Bessel function is good for small values of ρ.

The marginals are each $\Gamma(\alpha, 1)$, for which with $\alpha = 2$ we have (see Section 2.4.6) $\mu = 2$, $\sigma^2 = 2$, $\alpha_3 = \sqrt{2}$, and $\alpha_4 = 6$. From the Ramberg, Tadikamalla, Dudewicz, and Mykytka (1979) tables

$$\lambda(0,1) = (-0.782, 0.0379, 0.005603, 0.0365),$$

so by Step 5 of Algorithm GLD-M of Section 2.3.2, $\lambda_1(2,2) = \lambda_1(0,1)\sqrt{2} + 2 = 0.8941$, $\lambda_2(2,2) = \lambda_2(0,1)/\sqrt{2} = 0.02680$, and

$$\lambda = (0.894085, 0.0267993, 0.005603, 0.0365).$$

Figure 5.3–16 contains graphs of $\Gamma(2,1)$, along with its GLD fit. See Figures 5.3–17 through 5.3–20 for graphs of the GLD–2, Bivariate Gamma, and error.

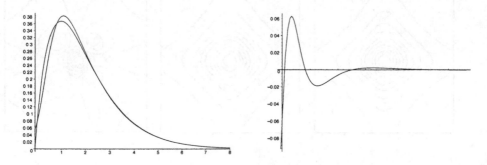

Figure 5.3–16. The $\Gamma(2,1)$ p.d.f. with its GLD approximation, and the error of the GLD approximation.

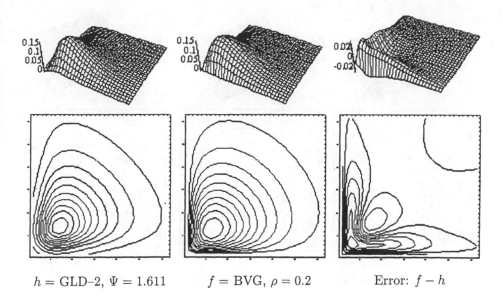

$h = $ GLD–2, $\Psi = 1.611$ $f = $ BVG, $\rho = 0.2$ Error: $f - h$

Figure 5.3–17. Bivariate gamma ($\rho = 0.2$) and GLD–2.

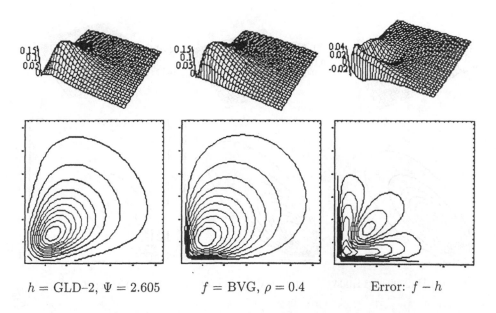

$h = $ GLD–2, $\Psi = 2.605$ $f = $ BVG, $\rho = 0.4$ Error: $f - h$

Figure 5.3–18. Bivariate gamma ($\rho = 0.4$) and GLD–2.

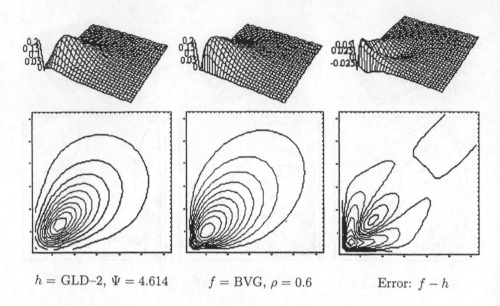

$h = \text{GLD–2}, \Psi = 4.614$ $f = \text{BVG}, \rho = 0.6$ Error: $f - h$

Figure 5.3–19. Bivariate gamma ($\rho = 0.6$) and GLD–2.

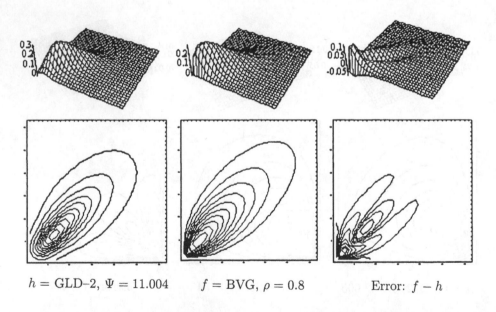

$h = \text{GLD–2}, \Psi = 11.004$ $f = \text{BVG}, \rho = 0.8$ Error: $f - h$

Figure 5.3–20. Bivariate gamma ($\rho = 0.8$) and GLD–2.

5.4 GLD–2 Fits: Distributions with Non-Identical Marginals

In the examples of Section 5.3, known bivariate distributions whose marginals are identical were considered. We now consider bivariate distributions with different marginals to examine the variety of shapes that can be achieved by the GLD–2.

5.4.1 Bivariate Gamma BVG with Non-Identical Marginals

First, consider X to be the $\Gamma(2,1)$ r.v. and Y the $\Gamma(4,1)$ r.v., both fitted by GLDs. $\Gamma(2,1)$ was fitted in Section 5.3.4 via

$$\lambda = (0.894085, 0.0267993, 0.005603, 0.036)$$

and $\Gamma(4,1)$ can be fitted by a univariate GLD with

$$\lambda = (2.656, 0.0421, 0.0194, 0.08235).$$

See Figure 5.4–1 for graphs of the univariate GLD fits, and Figure 5.4–2 for contour graphs of the fitted GLD–2 with $\Psi = 0.05(0.115)0.95$ (negative correlation) and $\Psi = 1(3)13$ (zero and positive correlation). The contour graphs in Figure 5.4–2 are truncated to $0 \leq x, y \leq 10$. (Note the contour heights between graphs are **not** the same.)

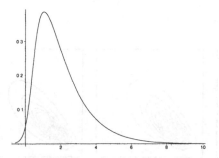 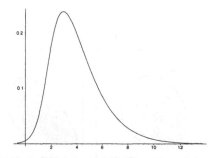

Figure 5.4–1. Univariate GLD fits to $\Gamma(2,1)$ (on the left) and $\Gamma(4,1)$ (on the right).

5.4.2 Bivariate with Normal and Cauchy Marginals

As our second example with non-identical marginals, suppose X is Cauchy and Y is $N(0,1)$. Each of these distributions was used previously in Section

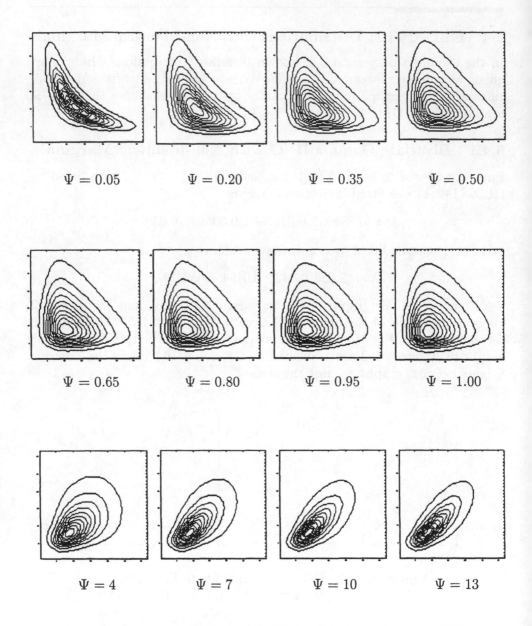

$\Psi = 0.05$ $\Psi = 0.20$ $\Psi = 0.35$ $\Psi = 0.50$

$\Psi = 0.65$ $\Psi = 0.80$ $\Psi = 0.95$ $\Psi = 1.00$

$\Psi = 4$ $\Psi = 7$ $\Psi = 10$ $\Psi = 13$

Figure 5.4–2. GLD–2 with different gamma marginals (X
is $\Gamma(2, 1)$ and Y is $\Gamma(4, 1)$) for 12 values
of Ψ (graphs show $0 < x, y < 6$).

5.3 in fitting their respective bivariate p.d.f.s. Figure 5.4–3 shows several contour graphs, again with $\Psi = 0.05(0.15)0.95$ and $\Psi = 1(3)13$.

5.4.3 Bivariate with Gamma and "Backwards Gamma" Marginals

The final example of the non-identical marginals case involves a non-standard marginal. To reverse the shape of one of the marginals, one can switch λ_3 and λ_4; this will skew the distribution exactly opposite to the original. In the case of Gamma, however, the p.d.f. is non-negative only for $x \geq 0$, whereas this new "Backwards Gamma" is defined for x both negative and positive. To create a p.d.f. more similar to $\Gamma(4,1)$ with a reversed shape, we add an offset of 5 to λ_1 to shift the p.d.f. The new parameters are then

$$\lambda = (7.656, 0.0421, 0.08235, 0.0194).$$

See Figure 5.4–4 for a graph of the GLD "Backwards Gamma" and Figure 5.4–5 for contour graphs of the GLD–2 for 12 values of Ψ.

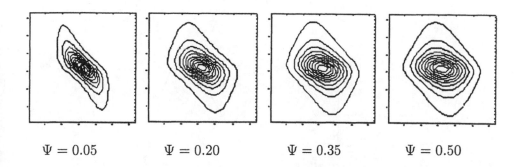

$\Psi = 0.05$ $\Psi = 0.20$ $\Psi = 0.35$ $\Psi = 0.50$

Figure 5.4–3. The GLD–2 for one Cauchy marginal (X) and one normal marginal (Y) for $\Psi = 0.05(0.15)0.95$ and $1(3)12$ (continued).

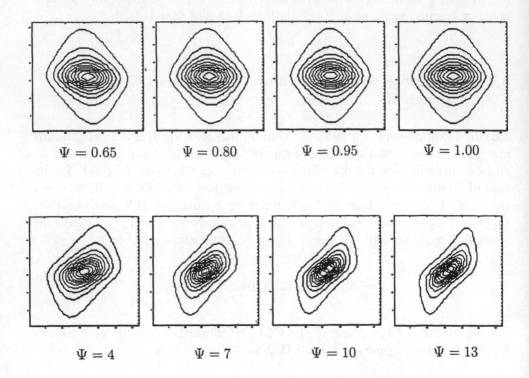

$\Psi = 0.65$ $\Psi = 0.80$ $\Psi = 0.95$ $\Psi = 1.00$

$\Psi = 4$ $\Psi = 7$ $\Psi = 10$ $\Psi = 13$

Figure 5.4–3. The GLD–2 for one Cauchy marginal
(X) and one normal marginal (Y) for
$\Psi = 0.05(0.15)0.95$ and $1(3)12$ (concluded).

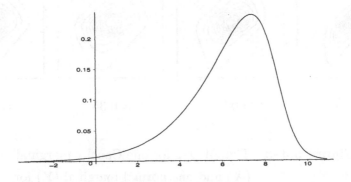

Figure 5.4–4. The univariate GLD for the "backwards $\Gamma(4,1)$".

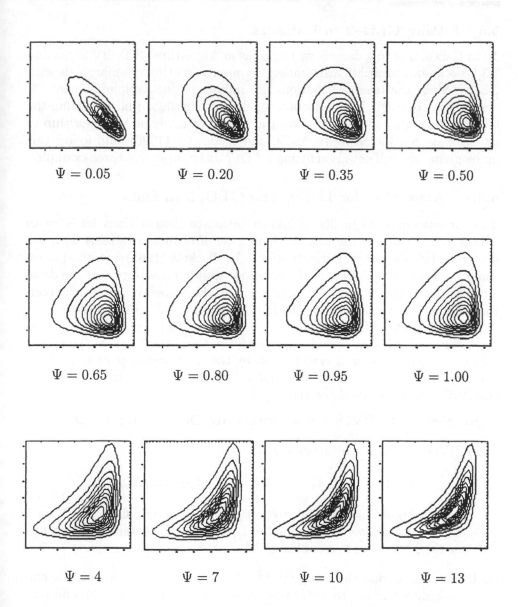

$\Psi = 0.05$ $\Psi = 0.20$ $\Psi = 0.35$ $\Psi = 0.50$

$\Psi = 0.65$ $\Psi = 0.80$ $\Psi = 0.95$ $\Psi = 1.00$

$\Psi = 4$ $\Psi = 7$ $\Psi = 10$ $\Psi = 13$

Figure 5.4–5. The GLD–2 with X as the $\Gamma(2, 1)$ and Y as the "backwards $\Gamma(4, 1)$" for $\Psi = 0.05(0.15)0.95$ and $1(3)12$.

5.5   Fitting GLD–2 to Datasets

To fit the GLD–2 to a dataset we begin, as in Algorithm GLD–BV in Section 5.3 for a known bivariate distribution, by approximating the marginals with GLDs. Since the true distributions are unknown, the sample moments or percentiles are used. In this section we develop an algorithm for fitting the GLD–2 to a dataset; next, we show the results of applying the algorithm to six datasets considered by Beckwith and Dudewicz (1996), and follow this up by giving the full details of fitting GLD–2 distributions in three examples.

5.5.1   Algorithm for Fitting the GLD–2 to Data

The success obtained in fitting known bivariate distributions leads us to expect good results in the more difficult case of bivariate datasets with unknown marginals and joint distributions. In Plackett (1965) a method of estimating Ψ was proposed for this situation: graph a scatterplot of the data, along with the two lines at $x = \tilde{x}$ and $y = \tilde{y}$, the respective medians; count the number of points in each of the four resulting quadrants, denoting these four counts in the sets $\{x \leq \tilde{x}, y \leq \tilde{y}\}$, $\{x \leq \tilde{x}, y > \tilde{y}\}$, $\{x > \tilde{x}, y \leq \tilde{y}\}$, and $\{x > \tilde{x}, y > \tilde{y}\}$, by a, b, c, and d, respectively; then use $\Psi^+ = (a \cdot d)/(b \cdot c)$ to estimate Ψ (this estimator is motivated by the relationship given in (5.2.9)).

Beckwith and Dudewicz (1996) proposed the following algorithm for fitting the GLD–2 to bivariate datasets:

Algorithm GLD–BVD: Fitting Bivariate Data with a GLD–2.

GLD–BVD–1. Given a bivariate dataset

$$Z_1, Z_2, \ldots, Z_n = (X_1, Y_1), (X_2, Y_2), \ldots, (X_n, Y_n),$$

with unknown bivariate p.d.f $f(x, y)$, fit both marginals X and Y with GLDs, using some suitable method (such as the Method of Moments, or the Method of Percentiles).

GLD–BVD–2. Graph the marginal d.f.s $F(x)$ and $F(y)$ along with the empiric d.f.s (e.d.f.s) to verify the quality of the univariate fits, and perform other checks of the univariate fits as discussed in previous chapters.

GLD–BVD–3. Graph a scatterplot of the (X,Y) data with the lines $x = \tilde{x}$ and $y = \tilde{y}$, i.e., at the respective medians.

GLD–BVD–4. Count the number of points in each of the four resulting quadrants, labeling them a, b, c, and d, as in Figure 5.2–1.

GLD–BVD–5. Calculate $\Psi^+ = (a \cdot d)/(b \cdot c)$, and use this value of Ψ in (5.3.1) with the GLD marginal d.f.s (or in $h(x, y)$ which follows (5.3.1)).

GLD–BVD–6. Graph the resulting GLD–2 and a 3-D scatterplot using height values from the GLD–2 to visualize the quality of the fit, and perform quantitative checks of the fit as well.

Algorithm GLD–BVD was applied by Beckwith and Dudewicz (1996) to three datasets involving imaging data in the normal brain (see Dudewicz, Levy, Lienhart, and Wehri (1989)) and three datasets from Johnson and Wichern's (1992) text. As it is instructive to see the wide range of applications, of p.d.f. shapes and contours, and of values of Ψ, we provide some of the details of these six applications and then give new examples with full details. For each of the datasets, the univariate GLD fits were calculated using the method of moments and the empiric d.f.s were graphed for a visual check of the fit. The resulting GLD p.d.f. was also graphed. Then the data were graphed in a bivariate scatterplot along with the horizontal and vertical lines at the respective medians to determine the values a, b, c, and d. The respective counts were labeled in each quadrant, and the median lines intersect at $(\tilde{x}, \tilde{y})$. With the resulting Ψ^+, the data were graphed in a 3-D scatterplot with heights calculated from the GLD–2 p.d.f. Also graphed was the resulting GLD–2, both as a 3-D density plot and a 2-D contour plot. Figures 5.5–1 through 5.5–6 give the bivariate graphs for each of the six datasets.

The *first dataset* is from national track records for 55 countries; the first variable is the time (in seconds) of the 100-meter run, and the second is the time (in seconds) of the 400-meter run. The computed sample correlation coefficient is $r = 0.834692$; this modestly high correlation results in a large Ψ^+ of 162.5. The *second dataset* compares the effect of two levels of carbon dioxide on an anesthetic, halothane, in sleeping dogs; here $r = 0.718082$ and $\Psi^+ = 14$. The *third dataset* compares the GPAs and GMAT scores for 31 students admitted to a graduate school of business; here $r = 0.00409619$, and (since this is close to zero), $\Psi^+ = 1.469$ is close to 1.

The fourth, fifth, and sixth datasets are Magnetic Resonance Imaging (MRI) brain scan data. Several measures were taken from MRI scans on approximately 40 patients over a period of time (the "Screened" data was used).

The *first MRI dataset* (our fourth dataset) is from the Putamen section, and the two variables are T2 left and right; here $r = 0.797282$ and $\Psi^+ = 64$.

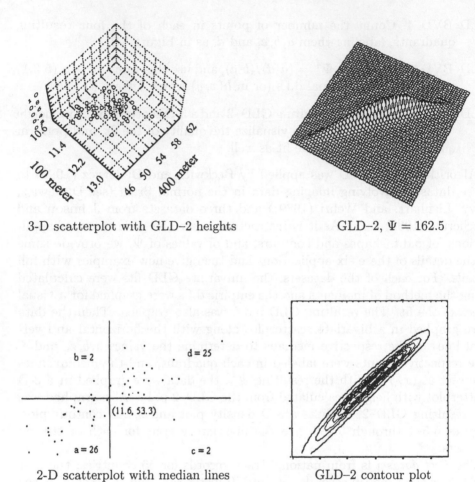

3-D scatterplot with GLD-2 heights   GLD-2, $\Psi = 162.5$

2-D scatterplot with median lines   GLD-2 contour plot

Figure 5.5–1. Track records (in seconds) for 55 countries,
100 meter vs. 400 meter.

The *second MRI dataset* is AD data, with the first variable Cortical White
Right, and the second Caudate Right; here $r = 0.783512$ and $\Psi^+ = 23.333$.
The *third MRI dataset* is Putamen Right, T2 vs. AD; here $r = -0.603856$
and (since r is negative and somewhat large in absolute value) $\Psi^+ = 0.0222$
is close to zero.

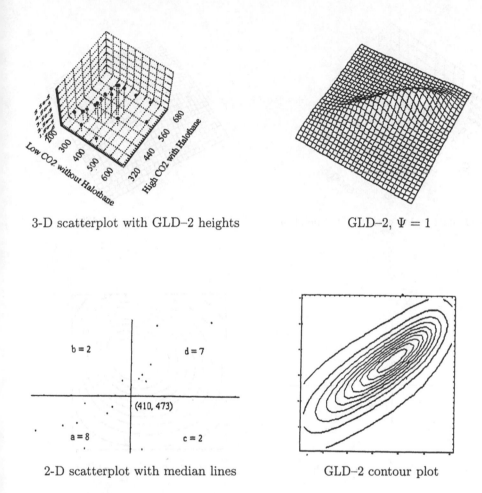

3-D scatterplot with GLD–2 heights

GLD–2, $\Psi = 1$

2-D scatterplot with median lines

GLD–2 contour plot

Figure 5.5–2. Comparison of the effect of two levels
of carbon dioxide in sleeping dogs.

It is difficult to judge the quality of the fits when the distributions are unknown, as we do not have the error values like $|h(x, y) - f(x, y)|$ that were available in the known-distribution case, or the known distribution, to compare with. Given this lack of information, however, the graphs presented here do seem to be evidence of fits that are reasonable.

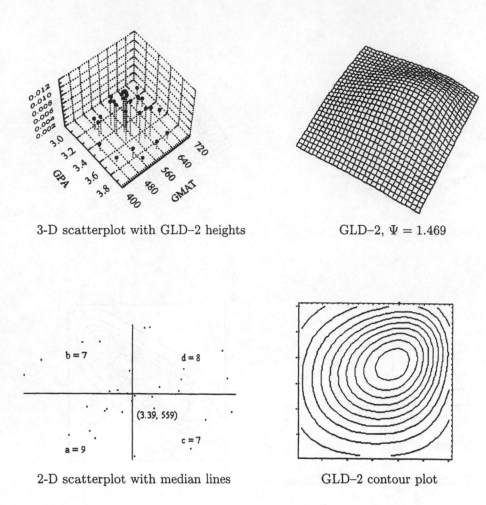

3-D scatterplot with GLD–2 heights   GLD–2, $\Psi = 1.469$

2-D scatterplot with median lines   GLD–2 contour plot

Figure 5.5–3. Comparison of GPA and GMAT scores of students admitted to graduate business schools.

Up to this point in Chapter 5 we have emphasized **visual assessment of goodness-of-fit**. In the univariate case in Section 2.5.1, we noted that in addition to the "eyeball test," there were many statistical tests of the hypothesis that the data come from the fitted distribution, and that it was important to perform at least one of these, lest one's assessment be overly

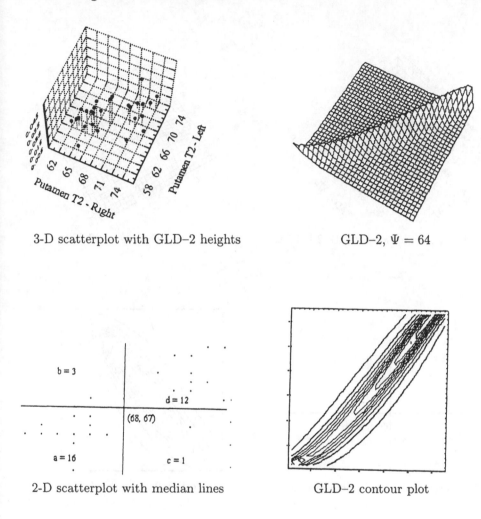

3-D scatterplot with GLD–2 heights

GLD–2, $\Psi = 64$

2-D scatterplot with median lines

GLD–2 contour plot

Figure 5.5–4. Comparison of the T2 left and T2 right
variables from Putamen sections.

subjective. In the bivariate case, we lack the key result of Point 6 of Section
2.5.1 (Theorem 2.5.1) that one can transform to univariate uniformity. There
is, in fact, no such transformation in two or more dimensions. This limits the
number of quantitative assessments available; this is an important problem
in theoretical statistics which is currently under investigation.

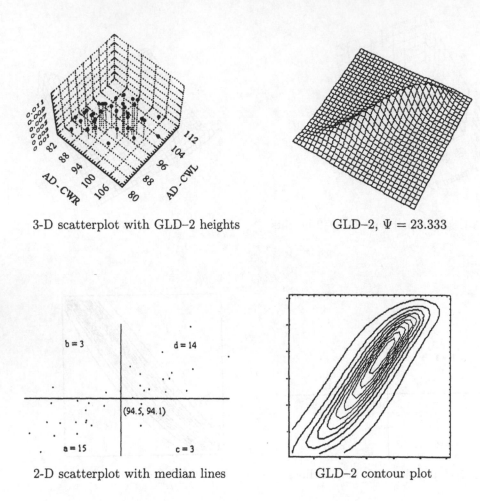

3-D scatterplot with GLD–2 heights   GLD–2, $\Psi = 23.333$

2-D scatterplot with median lines   GLD–2 contour plot

Figure 5.5–5. AD of CWR and CWL variables in MRI data.

One test that is available is the **chi-square test of the hypothesis that the data** $(X_1, Y_1), (X_2, Y_2), \ldots, (X_n, Y_n)$ **are independent observations from the fitted distribution** $H(x, y)$. To conduct this test, we divide the plane into k non-overlapping and exhaustive cells and obtain the probability that H assigns to each cell. For any rectangular area such as that shown in Figure 5.5–7, the probability can be found by computing (for $a < b$, $c < d$)

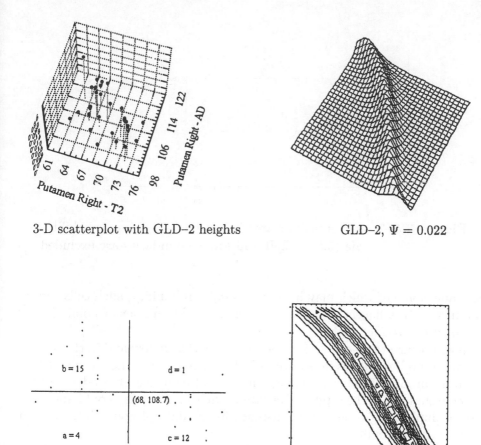

3-D scatterplot with GLD–2 heights   GLD–2, $\Psi = 0.022$

2-D scatterplot with median lines   GLD–2 contour plot

Figure 5.5–6. Putamen right, T2 vs. AD.

$$P(a < x \leq b, c < Y \leq d) = H(b,d) - H(a,d) - H(b,c) + H(a,c). \quad (5.5.1)$$

This formula holds even if a, b, c, d take on values such as $\pm\infty$, making it convenient to divide the plane into rectangular cells. As one desires the expected number of points in each cell to be at least 5, one can partition each of X and Y into cells, say 4 cells each. This yields $4 \times 4 = 16$ cells in

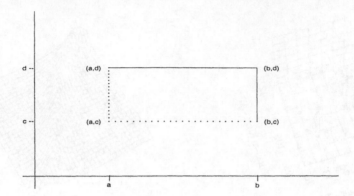

Figure 5.5-7. A rectangular area in the plane and its probability
via (5.5.1). Left and lower boundaries are excluded.

the plane, some of which may have very small probability; such cells can be combined with adjacent cells to raise the expected number of points in the combined cell to 5 or higher.

We will return to assessment in the examples that follow and in the discussion of research problems in Section 5.6. We have already shown the GLD-2 fits to a number of datasets, along with visual evidence of goodness-of-fit, plus contours which the typical bivariate models in use today could not come close to duplicating (due to the existence in the data of non-elliptical, even non-convex, contours).

In two of the following three examples we examine in greater detail the quantitative assessment of goodness-of-fit via the chi-square test as discussed above.

5.5.2 Example: Human Twin Data of Section 2.5.4

This dataset of birth weights of twins was considered in Section 2.5.4 (see Table 2.5-3) where both of its components were fitted with GLDs through the method of moments. The fits for X and Y that were obtained were, respectively,

$$\text{GLD}_x(5.5872, 0.2266, 0.2089, 0.1762)$$

and

$$\text{GLD}_y(5.3904, 0.2293, 0.1884, 0.1807).$$

A histogram of X with the fitted GLD_x and a plot of the e.d.f. of X with the d.f. of GLD_x, given in Figure 2.5-4 indicate that GLD_x fits X reasonably well. Similar graphic comparisons in Figure 2.5–5 establish that GLD$_y$ is also a reasonably good fit to Y. The qualities of the two fits were further substantiated through chi-square tests that yielded p-values of 0.03175 for GLD$_x$ and 0.2450 for GLD$_y$.

The scatterplot of Figure 5.5–8 (a) shows

$$a = 49, \qquad b = 13, \qquad c = 13, \qquad d = 48$$

from which we obtain $\Psi^+ = 13.9172$. The GLD–2 that results from this is shown in Figure 5.5–8 (b).

To perform a chi-square goodness-of-fit test we partition the support of the GLD–2 into rectangular cells designated by

$$I_x \times I_y = \{(x, y) \mid x \in I_x \text{ and } y \in I_y\},$$

where I_x and I_y are intervals of finite or infinite length. Since the rectangular cells are of the type shown in Figure 5.5–7, we can use (5.5.1) to compute the probabilities, and hence the expected frequencies for each cell. Table 5.5–9 gives the cells used in this case as well as the observed and expected frequencies associated with each cell.

The chi-square statistic and p-value for this test are 9.3116 and 0.02542. Note that to compute the chi-square statistic and p-value we use the chi-square distribution with 3 degrees of freedom because we have 13 cells and 9 estimated parameters: $\lambda_1, \lambda_2, \lambda_3, \lambda_4$ for each of the GLD$_x$ and GLD$_y$ fits and Ψ^+, hence degrees of freedom is $13 - 9 - 1 = 3$.

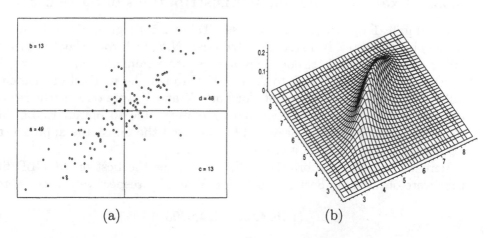

(a)   (b)

Figure 5.5–8. The scatterplot of the birth weights of twins (a) and the fitted GLD–2 (b).

Table 5.5–9. The cells, expected frequencies, and observed frequencies for the twin data and its fitted GLD–2.

Cell	Expected Frequency	Observed Frequency
$(-\infty, 4] \times (-\infty, 4.2]$	7.5625	11
$(4, \infty) \times (-\infty, 4.2]$	10.7254	7
$(-\infty, 4.5] \times (4.2, 5]$	8.4201	9
$(4.5, 5.5] \times (4.2, 5]$	13.8320	13
$(5.5, \infty) \times (4.2, 5]$	5.1798	8
$(-\infty, 5.25] \times (5, 5.75]$	11.1951	8
$(5.25, 5.75] \times (5, 5.75]$	9.4963	9
$(5.75, \infty) \times (5, 5.75]$	11.3240	13
$(-\infty, 5.75], (5.75, 6.5]$	7.5109	8
$(5.75, 6.5] \times (5.75, 6.5]$	11.0697	9
$(6.5, \infty) \times (5.75, 6.5]$	7.5785	12
$(-\infty, 6.75] \times (6.5, \infty)$	9.3363	7
$(6.75, \infty) \times (6.5, \infty)$	9.7695	9

5.5.3   Example: The Rainfall Distributions of Section 2.5.5

The rainfall data of Section 2.5.5 (see Table 2.5–6) gives the rainfall (in inches) in Rochester, N.Y. (X) and Syracuse, N.Y. (Y) from May to October 1998, on days when both locations had positive rainfall. Our attempt, in Section 2.5.5, to fit a GLD to X and Y through the method of moments failed because the (α_3^2, α_4) points for both X and Y were outside the region covered by the GLD; in Section 3.5.4 we were able to obtain EGLD fits to both X and Y; and in Section 4.6.4 we used the percentile approach to obtain two GLD fits to X and three fits to Y.

Here we consider the bivariate (X, Y) and use the best of the GLD fits that were obtained in Section 4.6.4. For X and Y, respectively, these were

$$\text{GLD}_x(1.7684, 0.5682, 4.9390, 0.1055)$$

and

$$\text{GLD}_y(1.3381, 0.7557, 9.6291, 0.2409).$$

Figure 4.6–3 (a) shows a histogram of X with the p.d.f.s of GLD_x and the GBD fit of Section 3.5.4. In Figure 4.6–3 (b) we see the e.d.f. of X with the d.f. of GLD_x. Figure 4.6–4 gives similar illustrations for Y. These are reasonably good fits, an assertion that is supported by p-values of 0.09342 and 0.2791 for GLD_x and GLD_y, respectively.

From the scatterplot of (X, Y) in Figure 5.5–10 (a) we see that

$$a = 19, \qquad b = 5, \qquad c = 5, \qquad d = 18$$

from which we obtain $\Psi^+ = 13.68$. The GLD–2 fit to this bivariate data is shown in Figure 5.5–10 (b).

A chi-square goodness-of-fit test is not available in this case because there are 47 data points and 9 estimated parameters. Hence, we would need to partition the support of the fitted GLD–2 into a minimum of 11 cells to have $11 - 9 - 1 = 1$ be positive, forcing some of the expected frequencies to fall below 5. In such a situation the test statistic will not have an approximate chi-square distribution. One could obtain the actual null distribution by Monte Carlo methods and then complete a test. This would also be true if degrees of freedom were zero or negative.

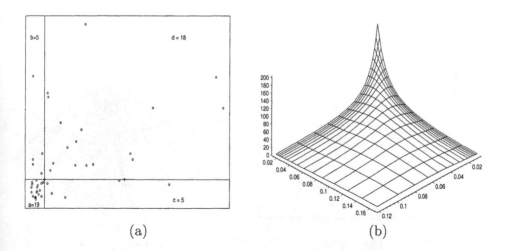

(a)   (b)

Figure 5.5–10. The scatterplot of the rainfall data (a)
and the fitted GLD–2 (b).

5.5.4   Example: The Tree Stand Data of Section 3.5.5

Two variables, DBH representing the tree diameter at breast height (in inches) and H representing the tree height (in feet), were considered in Section 3.5.5, and GLD (moment-based) and GBD fits for both variables were obtained. To obtain a GLD–2 for the bivariate (DBH, H), we use the GLD fits of Section 3.5.5 that were

$$\text{GLD}_{DBH}(4.7744, 0.08911, 0.06257, 0.3056)$$

and

$$\text{GLD}_H(82.0495, 0.01442, 0.6212, 0.02459)$$

for DBH and H, respectively. We note that p-values of 0.09696 and 0.3776 were computed for the GLD_{DBH} and GLD_H fits, respectively, in Section 3.5.5.

The (DBH, H) scatterplot is given in Figure 5.5–11 (a) with $a = 42$, $b = 3$, $c = 4$, and $d = 40$, giving us $\Psi^+ = 140$. The GLD–2 fit to (DBH,H) is shown in Figure 5.5–11 (b).

With the notation used earlier, a partitioning of the plane into 13 cells and associated expected and observed frequencies are shown in Table 5.5–12. We have degrees of freedom $13 - 9 - 1 = 3$ (since there are 13 cells, 9 fitted parameters, and we lose 1 for the chi-square even if we fit no parameters). The chi-square statistic and p-value that result from this test are 11.0670 and 0.01137, respectively.

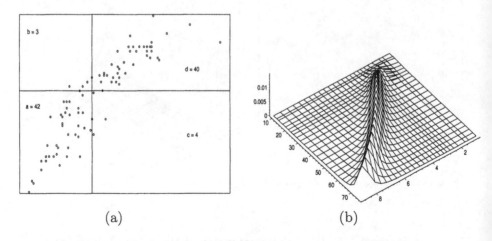

(a)   (b)

Figure 5.5–11. The scatterplot of the tree stand data (a)
and the fitted GLD–2 (b).

Table 5.5–12. The cells, expected frequencies, and observed frequencies for the tree stand data and its fitted GLD–2.

Cell	Expected Frequency	Observed Frequency
$(-\infty, \infty) \times (-\infty, 26]$	6.1395	5
$(-\infty, 4] \times (26, 39]$	6.0533	8
$(4, \infty) \times (26, 39]$	6.0464	5
$(\infty, 5] \times (39, 50]$	6.6154	8
$(5, \infty) \times (39, 50]$	6.8840	9
$(-\infty, 6] \times (50, 60]$	6.9502	8
$(6, \infty) \times (50, 60]$	7.3736	3
$(-\infty, 7.25] \times (60, 69]$	8.0411	8
$(7.25, \infty), (60, 69]$	6.0959	8
$(-\infty, 8.5) \times (69, 78]$	7.3022	5
$(8.5, \infty) \times (69, 78]$	7.2650	9
$(-\infty, 10.5] \times (78, \infty)$	6.5677	10
$(10.5, \infty) \times (78, \infty)$	7.6502	3

5.6 GLD–2 Random Variate Generation

The generation of r.v.s for Monte Carlo studies is one of the important applications of fitted distributions. In previous chapters we developed methods for generating GLD random variates (Section 1.5) and EGLD random variates (Section 2.6). Here we describe a method for generating random variates from a GLD–2. Design of experiments using the GLD–2 is also possible (generalizing Section 2.7).

Recall that the GLD–2 is specified by its distribution function $H(x,y)$ given in (5.1.1) and its p.d.f. $h(x,y)$ given in (5.1.2). These involve a univariate GLD $F(x)$, a univariate GLD $G(y)$, and a nonnegative constant Ψ. The $F(x)$, $G(y)$, and Ψ may be chosen to match a particular theoretical model distribution, as noted in Sections 5.3 and 5.4, or to fit a dataset through Algorithm GLD–BVD of Section 5.5. Since "bivariate inverse functions" are not available, the generation of r.v.s from a GLD–2 involves considerations that did not arise in our previous discussions of the univariate cases in Sections 1.5 and 3.6. The three theorems that follow enable us to overcome the non-existence of bivariate inverse functions.

Theorem 5.6.1. *If* $\mathbf{Z} = (X, Y)$ *is a r.v. with any bivariate distribution, then its distribution function* $H(x, y)$ *may be written as*

$$H(x, y) = F(x)G(y \mid x) \tag{5.6.2}$$

where $F(x)$ *and* $G(y)$ *are the distribution functions of* X *and* Y, *respectively, and*

$$G(y \mid x) = P(Y \leq y \mid X \leq x) \tag{5.6.3}$$

is the conditional distribution function of Y *given that* $X \leq x$.

The proof of Theorem 5.6.1 is a simple application of the definitions of conditional probability and distribution function (in the bivariate case). The following is a direct consequence of Theorem 5.6.1.

Theorem 5.6.4. *Suppose* $\mathbf{Z} = (X, Y)$ *is a r.v. with a GLD–2 distribution,* X_1 *is generated from* $F(x)$ *through the method of Section 1.5, and* Y_1 *is generated from the p.d.f.*

$$g(y \mid x_1) = \frac{h(x_1, y)}{f(x_1)}. \tag{5.6.5}$$

Then the pair (X_1, Y_1) *has the same distribution as* (X, Y).

Since the $g(y \mid x_1)$ in (5.6.5) is the conditional p.d.f. of Y given that $X = x_1$, Theorem 5.6.4 provides us with a method of generating a GLD–2 r.v. (X, Y) if we can generate Y from the p.d.f. $g(y \mid x_1)$. This is non-trivial, since $g(y \mid x_1)$ is not a GLD distribution. However, the following method can be used because we have explicit expressions for both the numerator and denominator of (5.6.5).

Theorem 5.6.6. *Let* $Y_1 = G^{-1}(U \mid x_1)$ *where* U *is a uniform r.v. on* $(0, 1)$. *Then* Y_1 *has p.d.f.* $g(y \mid x_1)$ *described in (5.6.5) and is the solution of the equation*

$$G(y_1 \mid x_1) = U. \tag{5.6.7}$$

To obtain a solution to (5.6.7) numerical integration can be used to find y_1 such that

$$\int_{-\infty}^{y_1} \frac{h(x_1 \mid y)}{f(x_1)} \, dy = U,$$

i.e., such that

$$\int_{-\infty}^{y_1} h(x_1 \mid y) \, dy = U f(x_1). \tag{5.6.8}$$

Example: The program GLD2RAND for generating GLD–2 random variates (written in Maple) is included in Section A.5 of Appendix A. The first two parameters of the program are the $\lambda_1, \lambda_2, \lambda_3, \lambda_4$ of the two marginal GLDs, each entered as a list; the third parameter is the value of Ψ; and the fourth parameter is the number of observations that are to be generated.

For a specific illustration, we fit the GLD–2 to the bivariate normal with marginals that are $N(0,1)$ and $N(100, 225)$ with $\rho = 0.5$. The GLD fits to $N(0,1)$ and $N(100, 225)$ are, respectively,

$$\text{GLD}_X(0, 0.1975, 0.1349, 0.1349)$$

and

$$\text{GLD}_Y(100, .01316, .1349, .1349).$$

GLD_X is the fit that was obtained in Section 2.4.1. We know from Table 5.3–1 that $\Psi = 3.653$ when $\rho = 0.5$. Thus, we invoke GLD2RAND with the following sequence of Maple commands:

```
> LambdaX := [0,0.1975,0.1349,0.1349];
> LambdaY := [100,.01316,.1349,.1349];
> GLD2RAND(LambdaX, LambdaY, 3.653, 200);
```

to generate 200 random observations from the GLD–2 that has GLD_X and GLD_Y for its marginals and $\Psi = 3.653$. In Table 5.6–1 we give the 200 (X, Y) pairs that are generated and note that when the means and variances of the X and Y components and the correlation coefficient are computed we get

$$\bar{X} = 0.01100, \quad \bar{Y} = 99.4321, \quad s_X^2 = 0.8573, \quad s_Y^2 = 219.9745, \quad r = 0.4035.$$

To obtain a quantitative assessment of the quality of this generator, we perform a chi-square goodness-of-fit test on the data of Table 5.6–1 by partitioning the plane into 16 rectangles. These rectangular cells along with their observed and expected frequencies shown in Table 5.6–2 produce a chi-square statistic of 10.1753 and a p-value of 0.1175.

Table 5.6–1. Random sample generated through GLD2RAND.

X	Y	X	Y	X	Y	X	Y	X	Y
−.644	95.90	−.951	96.24	−.008	117.3	−1.46	87.50	−2.06	99.53
1.00	104.3	−.061	110.0	.857	109.0	−.811	85.76	−.136	107.2
−1.11	79.16	1.05	88.49	.257	106.9	−.830	88.89	1.25	119.1
−.938	95.23	−.549	123.4	−.439	102.5	−1.82	88.84	.479	109.5
1.45	121.9	1.18	106.9	.752	113.9	.898	98.30	−.394	95.90
−.144	106.5	−1.04	88.79	−.156	104.6	−.151	105.2	−.343	89.72
.797	103.6	.871	110.8	−.455	89.81	−.197	119.1	.351	103.4
−.750	93.42	−1.81	84.39	−.717	90.97	−.075	90.65	.617	101.7
−.976	112.8	−1.16	92.58	.700	110.5	1.77	113.4	−.264	105.3
.020	99.50	1.15	101.1	.209	115.5	.087	109.1	−1.14	85.28
.099	104.8	.335	122.8	.904	125.6	.748	131.8	−2.44	77.43
.250	84.78	−.444	86.17	−.345	110.6	.298	112.4	−.985	82.95
−1.02	85.22	.166	110.0	−1.29	98.00	1.20	95.31	.546	89.96
−.359	100.6	.277	84.71	−1.56	98.15	−.640	92.79	.445	94.80
.134	109.6	−.762	108.5	1.07	85.22	−1.18	112.9	.608	100.7
.824	83.38	.253	102.4	.227	85.70	.630	112.7	−.924	38.98
−2.38	80.74	−.097	85.03	−.033	90.83	−1.26	77.55	.971	107.9
−1.37	84.06	.345	101.2	2.44	120.6	1.01	112.2	.076	91.81
1.29	84.13	−.319	102.6	1.29	111.6	.864	89.72	.461	98.07
.327	71.78	−.333	90.42	.751	113.8	.182	100.5	−.048	100.1
−.400	97.81	−.621	88.13	1.19	128.9	.615	124.8	−.887	101.6
.147	98.30	1.56	107.9	−.100	103.9	1.00	115.2	1.00	100.6
−1.09	86.96	−2.31	95.62	.748	77.89	−1.34	102.9	−.887	86.29
.426	85.58	−2.12	79.84	.650	112.3	−.756	93.09	1.16	93.09
.566	102.9	−1.45	106.0	−.252	98.52	−.606	90.38	1.38	87.50
−.128	108.0	1.08	107.8	.136	120.3	−.087	96.36	.366	128.0
−2.17	68.89	−1.47	85.28	.818	105.7	1.19	102.5	.907	127.0
1.34	103.0	1.32	92.07	.057	128.4	1.45	111.2	.021	113.7
1.53	112.4	−.526	102.0	.213	91.01	−1.71	93.34	−.145	74.41
−1.09	87.40	−.357	61.71	−.455	85.88	−.075	96.93	.540	102.1
.287	107.2	−.605	86.74	.704	122.1	−.135	93.95	−.274	113.3
.937	101.1	−.209	74.73	.594	117.3	.955	117.7	−.601	78.33
.291	87.02	−.545	126.4	−.024	95.27	.030	79.93	−.334	80.02
1.70	119.3	.691	95.43	.475	114.8	−.596	115.3	.612	114.4
.644	113.4	.696	128.9	.182	106.7	−.371	83.38	−1.45	89.33
−.688	105.5	−.214	81.97	.099	117.0	.344	105.7	.936	81.08
.106	93.99	−.960	85.34	−1.35	107.7	1.54	90.28	.614	134.8
−.425	60.69	−.729	74.89	1.32	84.00	1.29	115.2	.678	96.28
.109	100.3	−.635	95.07	.301	102.6	1.85	94.92	.386	121.3
−.559	99.72	−.028	89.96	−1.63	111.6	−.678	76.44	1.07	103.4

Table 5.6–2. The cells, expected frequencies, and observed
frequencies for the data of Table 5.6–1.

Cell	Expected Frequency	Observed Frequency
$(-\infty, -1] \times (-\infty, 89.5]$	16.7091	17
$(-1, -0.25] \times (-\infty, 89.5]$	15.6534	17
$(-0.25, 0.25] \times (-\infty, 89.5]$	8.3016	7
$(0.25, \infty) \times (-\infty, 89.5]$	7.7286	12
$(-\infty, -0.7] \times (89.5, 100]$	13.9775	11
$(-0.7, 0] \times (89.5, 100]$	15.7682	18
$(0, 0.7] \times (89.5, 100]$	13.6189	10
$(0.7, \infty) \times (89.5, 100]$	8.2427	7
$(-\infty, -0.7] \times (100, 110.5]$	8.2427	5
$-0.7, 0] \times (100, 110.5]$	13.6189	14
$(0, 0.7] \times (100, 110.5]$	15.7682	19
$(0.7, \infty) \times (100, 110.5]$	13.9775	15
$(-\infty, -0.25] \times (110.5, \infty)$	7.7286	8
$(-0.25, 0.25] \times (110.5, \infty)$	8.3016	7
$(0.25, 1] \times (110.5, \infty)$	15.6534	21
$(1, \infty] \times (110.5, \infty)$	16.7091	12

5.7 Conclusions and Research Problems Regarding GLD–2

The GLD is a versatile family for distribution fitting. In this Chapter 5 we
have seen a bivariate version, the GLD–2. From the examples given, using
a variety of marginals and correlations as well as a variety of datasets, it
appears that the GLD–2 fits both some bivariate known distributions and
some bivariate datasets well. We are currently reinvestigating an extension of
the EGLD to a bivariate version, so that an even broader range of probability
distributions arising in practice can be fitted.

Some historical, as well as research frontiers, comments may be useful to
the reader seeking to put the material on bivariate distributions into per-
spective. **First**, since any bivariate d.f. $F(x, y)$ is a function of the marginals
$F_1(x)$ and $F_2(y)$ called a **copula**, $F(x, y) = C(F_1(x), F_2(y))$, it seems unlikely that a bivariate GLD (by any generalization to two dimensions) would
have a closed form for its d.f. since its univariate form is specified by its per-

centile function. This would also seem to apply to the multivariate (three or more variates) cases. Note that Johnson and Kotz (1973) proposed a non-closed-form multivariate extension of Tukey's original (ungeneralized) lambda distribution.

Second, the method R.L. Plackett (1965) proposed for generating bivariate distributions from specified marginals, used by Beckwith and Dudewicz (1996) to yield the GLD–2, rarely generates a closed-form expression for d.f.s, but this is not a problem as computer calculation of the properties of the resulting distribution is readily available.

Third, Plackett (1981, p. 151) has noted that the method of Plackett (1965) for generating bivariate distributions was in fact introduced in 1913 in Pearson (1913) and Pearson and Heron (1913).

Fourth, due to the lack of a transformation from (X, Y) into a bivariate uniform distribution on the unit square (as noted in Section 5.5, in the paragraph before that containing equation (5.5.1)), a Kolmogorov-Smirnov type test based on the statistic

$$D_n = \sup_{(x,y)} |F_n(x,y) - H(x,y)| \qquad (5.7.1)$$

does not reduce to testing bivariate uniformity; hence, critical values will be functions of the fitted H (when we act as if it is the true d.f. of (X, Y)) as well as of the sample size n. This means we would have to calculate the critical values anew in each application. (Above, $F_n(x, y)$ is the bivariate empiric d.f. .) In addition, the *sup* will require care to compute (but will be possible since it should occur at or just before a jump). While $F_n(x, y)$ is called the **empirical process** and there is a great deal of research on it (e.g., see Csörgő and Szyszkowicz (1994) and its references; on p. 97 these authors seem to indicate computational methods will be needed for the critical values, and in any case they do not seem to have reduced their theory to practice). We are not aware of this concluding in a (non-asymptotic) goodness-of-fit test, even if, e.g., H is bivariate normal (in which case one could calculate the necessary tables, which might also depend on the correlation coefficient). There appear to be many interesting and important research problems in this area (and some of them might be approached via computer simulation in terms of the tables needed for implementation, or via algorithms to calculate the needed critical values anew each time one tests using D_n of (5.6.1)).

Fifth, from the graphs in Section 5.5, it will be clear that the GLD–2 can have non-elliptic contours, in fact even non-convex contours. This is of importance in applications, as it allows modeling of the (widespread) situations where these sorts of contours arise. Work on multivariate models is still in rapid development; some references one might consult for some of these aspects (elliptically contoured distributions, multivariate models with given marginals, etc.) include Dall'Aglio, Kotz, and Salinetti (1991), Anderson, Fang, and Olkin (1994), and Hayakawa, Aoshima, Shimizu, and Taneja (1995, 1996, 1997, 1998).

Problems for Chapter 5

5.1 Let X and Y be random variables with marginal p.d.f.s that are uniform on $(0, 1)$. Find and provide a contour plot of $H(x, y)$ with marginals the same as those of X and Y, but for which (5.2.9) holds for $\Psi = 4$.

5.2 Do Problem 5.1 with $\Psi = 0.25$.

5.3 Do Problem 5.1 for each of the following values of Ψ: 0.01, 0.1, 1, 10, 100, ∞.

5.4 Suppose (X, Y) has a bivariate d.f. $F(x, y)$ which is one discussed in Hutchinson and Lai (1990) but not considered in Section 5.3. Develop a GLD–2 fit $H(x, y)$, plot contours, and consider the error $f - h$. How good a fit to (X, Y) can be obtained using the GLD–2?

5.5 In Section 5.5.3, a GLD–2 fit was obtained to a set of data on rainfall (X, Y) at two cities. Test the hypothesis that the data comes from the fitted distribution. [Hint: Split the plane into 10 rectangles with frequencies of at least 4.0 per rectangle. Compute a chi-square discrepancy measure. To assess its significance, we cannot use a chi-square distribution, since the degrees of freedom would be $10 - 9 - 1 = 0$. However, one can find the approximate null distribution by Monte Carlo simulation. Do so and find the p-value of the computed test statistic. At level of significance 0.01 do you accept or reject the null hypothesis that the data $(X_1, Y_1), (X_2, Y_2), \ldots, (X_{47}, Y_{47})$ comes from the fitted GLD–2?]

5.6 Generate 25 points from the GLD–2 $h(x, y)$ fitted to the rainfall data of Section 5.5.3. Fit a GLD–2 $h^*(x, y)$ to this set of simulated data. Compare this GLD–2 to that with p.d.f. plotted in Figure 5.5–10 (b). In particular, find

$$\sup_{(x,y)} |h(x, y) - h^*(x, y)|,$$

and plot $h^*(x, y)$ and $e(x, y) = |h(x, y) - h^*(x, y)|$.

Chapter 6

The Generalized Bootstrap (GB) and Monte Carlo (MC) Methods

In Chapter 1 we noted the key role of statistical models, in particular of probability distributions, in modern human endeavor, and the recent explosive developments in **fitting distributions to data**. In Chapters 1 through 5, through the use of the EGLD, we have provided modern, cutting-edge methods for this area.

The fitted distributions are often used in a **Monte Carlo Simulation (MC) problem solution**, where one samples from the fitted distribution using random numbers — recall that we saw how to accomplish such sampling in Section 1.5 (GLD Random Variate Generation), Section 3.6 (EGLD Random Variate Generation), and Section 5.6 (GLD–2 Random Variate Generation). The resulting samples from the fitted distribution can then be used to assess the performance of the simulated system, making it possible to optimize the system before it is built or modified (we saw in Section 2.7 how design of experiments could be used in this process with a fitted GLD). For some examples of MC in optimization, vehicle routing, and inventory management see, e.g., Golden, Assad, and Zanakis (1984), Golden and Eiselt (1992), and Dudewicz (1997), respectively. For uses in quality control, see Dudewicz (1999, p. 44–84).

An alternative approach in use as early as 1967, that did not gain wide use until it was given the name **bootstrap** by B. Efron in 1979, after which interest in it increased markedly, does not fit a distribution and then sample from it. Rather, **it draws its samples at random with replacement from the data at hand.**

The Bootstrap Method has attracted much interest since it is simple and seems to yield solutions to problems. However, the method can behave badly if the sample size is not large, and most of the theory supporting it is developed as the sample size becomes infinite. For example, if we have 50 years of data on rainfall in an area and we are designing a levee, the bootstrap method will never yield observations more extreme than the most extreme in the dataset of size 50. This can easily lead to design of a levee that will not withstand the 100-year flood (a typical design criterion in the U.S.) or the 1000-year flood (a design criterion in Europe) with high probability. For more details, see Dudewicz (1992) and Dudewicz and Mishra (1988, Section 5.6). For these reasons, **we will not detail the Bootstrap Method further here; it is a method fraught with danger of seriously inadequate results.**

We do detail the **Generalized Bootstrap** which was introduced formally in Dudewicz (1992) (and discussed earlier in Karian and Dudewicz (1991, Section 6.6)).

6.1 The Generalized Bootstrap (GB) Method

The essence of the **Generalized Bootstrap (GB) Method** is to fit an EGLD to the available data, and then take samples from the fitted distribution (and work with them as the BM does with its samples from the data itself). This method has been shown to do better than the BM when the number of data points is not very large, and do as well as the BM when the number of data points is large. Sun and Müller–Schwarze (1996) have an excellent exposition with real-data examples. More recent examples in **data mining**, including interactive computer systems, consumer purchases, crop damage, and pathogens, are considered by Dudewicz and Karian (1999b).

Suppose that we are interested in

$$\theta = \theta(F), \tag{6.1.1}$$

some function of the distribution $F(\cdot)$, but $F(\cdot)$ is unknown (if it were known we could simply calculate $\theta(F)$). However, we have a random sample $X_1, X_2, \ldots, X_n$ from $F(\cdot)$. How can we use this sample to estimate $\theta(F)$?

The Generalized Bootstrap (GB) approaches the problem as follows. Suppose that one would typically estimate θ by $\hat{\theta} = \hat{\theta}(X_1, X_2, \ldots, X_n)$. Then, instead, proceed as follows:

GB–1. Estimate F by $\hat{F}$. (The estimating $\hat{F}$ should always be one that has the properties known for the true F. For example, $\hat{F}$ should be continuous if it is known that F is continuous. Fitting with the Generalized Lambda Distribution or with the empiric p.d.f. will often be appropriate.)

GB–2. Independently generate N random samples of size n from $\hat{F}$.

From

$\qquad Y_1, Y_2, \ldots, Y_n$ estimate $\theta(F)$, calling the estimate $\hat{\theta}_1$.

From

$\qquad Y_{n+1}, Y_{n+2}, \ldots, Y_{2n}$ estimate $\theta(F)$, calling the estimate $\hat{\theta}_2$.

$\qquad \vdots$

From

$\qquad Y_{(N-1)n+1}, \ldots, Y_{Nn}$ estimate $\theta(F)$, calling the estimate $\hat{\theta}_N$.

GB–3. Use the sample $\hat{\theta}_1, \hat{\theta}_2, \ldots, \hat{\theta}_N$ to estimate $\hat{\theta}$. For example, one may calculate

$$\bar{\theta} = \left(\frac{1}{N} \right) \sum_{i=1}^{N} \hat{\theta}_i \qquad (6.1.2)$$

and

$$\hat{\sigma}^2 = \left(\frac{1}{N-1} \right) \sum_{i=1}^{N} (\hat{\theta}_i - \bar{\theta})^2, \qquad (6.1.3)$$

which give the sample mean and sample variance, respectively, of the GB estimators found in GB–2. Then, assuming approximate normality, we have

$$\mathrm{Var}(\bar{\theta}) = \mathrm{Var}(\hat{\theta}_i)/N \approx \hat{\sigma}^2/N, \qquad (6.1.4)$$

and **an approximate $100(1 - \alpha)\%$ confidence interval for θ is**

$$\hat{\theta} \pm \Phi^{-1}(1 - \alpha/2)\hat{\sigma}. \qquad (6.1.5)$$

The GB method uses the original estimator based on the original sample. However, the GB method uses the samples generated in step GB–2 to arrive at an estimate of the variability of the original estimator. The method of (6.1.5) is called the **"standard method"** (e.g., see Chernick (1999, p. 54)), and is appropriate if $\hat{\theta}$ has an approximate normal distribution.

A widely used alternative to the standard method is the **"percentile method,"** which uses the upper and lower $\alpha/2$ percentiles of the GB sample estimators as the confidence interval. Specifically, **the percentile method proceeds as follows:**

PM–1. Use the Generalized Bootstrap GB (Steps GB–1 and GB–2) to obtain N estimates $\hat{\theta}_1, \hat{\theta}_2, \ldots, \hat{\theta}_N$.

PM–2. Place the N estimates $\hat{\theta}_1, \hat{\theta}_2, \ldots, \hat{\theta}_N$ from step PM–1 in increasing numerical order, obtaining

$$\hat{\theta}_{(1)} \leq \hat{\theta}_{(2)} \leq \cdots \leq \hat{\theta}_{(N)}, \tag{6.1.6}$$

the order statistics, where $\hat{\theta}_{(1)}$ is the smallest of $\hat{\theta}_{(1)} \ldots \hat{\theta}_{(N)}$, $\hat{\theta}_{(2)}$ is the second smallest of $\hat{\theta}_{(1)} \ldots \hat{\theta}_{(N)}$, etc., and $\hat{\theta}_{(N)}$ is the largest of $\hat{\theta}_{(1)} \ldots \hat{\theta}_{(N)}$.

PM–3. The percentile method's (approximate) $100(1 - \alpha)\%$ confidence interval for θ is

$$(\hat{\theta}_{(a)}, \hat{\theta}_{(b)}) \tag{6.1.7}$$

where

$$a = \lfloor N\frac{\alpha}{2} \rfloor, \quad b = \lceil N(1 - \frac{\alpha}{2}) \rceil. \tag{6.1.8}$$

(That is, the lower $100\alpha/2\%$, and upper $100\alpha/2\%$, of the samples are deleted. Note that $\lfloor \cdot \rfloor$ rounds down and $\lceil \cdot \rceil$ rounds up; e.g., $\lfloor 24.8 \rfloor = 24$, $\lceil 76.3 \rceil = 77$.)

Example: We have the $n = 60$ observations $X_1, X_2, \ldots, X_{60}$ given in Table 6.1–1. We wish to obtain a 95% (approximate) confidence interval for p, the probability that a value at least 7.30 will be found. (This data was considered, in another context, in Dudewicz and Karian (1999b).)

Table 6.1–1. Data ($n = 60$ data points).

3.88	5.31	3.26	3.65	6.78	3.31	6.09	3.42	3.45	4.83
4.01	4.37	3.34	4.20	2.23	3.17	3.20	3.12	5.50	5.45
4.28	4.77	4.49	3.74	4.59	3.64	3.16	4.66	6.44	2.04
3.87	3.37	2.35	4.29	6.25	2.77	3.73	7.28	4.22	4.48
3.54	3.79	3.78	2.28	4.09	5.02	3.37	3.00	2.75	4.12
4.86	4.17	5.94	5.14	3.29	3.79	3.59	4.79	3.18	4.84

Using the Generalized Bootstrap, in Step GB–1 we fit a GLD using the methods of Chapter 2. From the data we find

$$\hat{\alpha}_1 = 4.1053, \quad \hat{\alpha}_2 = 1.2495, \quad \hat{\alpha}_3 = 0.6828, \quad \hat{\alpha}_4 = 3.2998$$

and the fit is

$$\lambda_1 = 3.2530, \quad \lambda_2 = 0.1549, \quad \lambda_3 = 0.04033, \quad \lambda_4 = 0.2059$$

or the fit
$$GLD(3.2530, 0.1549, 0.04033, 0.2059).$$

This fit, with a histogram of the $n = 60$ data points, is shown in Figure 6.1–2.

Following step GB–2, we generate $N = 500$ random samples of size $n = 60$ from the fitted GLD $(3.2530, 0.1549, 0.04033, 0.2059)$. (For details of the random number generator used, see Karian and Dudewicz (1999b, p. 133), where it is called URN41. It is a generator with high-quality test results and an astronomic period 2^{62}. Also see Dudewicz and Karian (1999b, p. 262).) Samples $1, 2, 3, 4, 5, 6$ and 500 are shown in Table 6.1–3. From these we calculate the estimates of p, probability of values at least 7.30, as

$$\hat{p}_1 = \hat{p}_2 = \hat{p}_3 = \hat{p}_4 = 0, \quad \hat{p}_5 = 0.0333, \quad \hat{p}_6 = 0, \ldots, \hat{p}_{500} = 0.$$

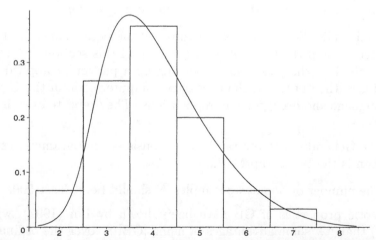

Figure 6.1–2. Data in the form of a histogram, and fitted GLD$(3.2530, 0.1549, 0.04033, 0.2059)$.

Note that in Table 6.1–3 values at or above 7.30 are highlighted for the reader's ease. The only such values, 8.15 and 7.76, are in the fifth sample, hence $\hat{p_5} = 2/60 = .0333$.

The resulting $\hat{p}_1, \hat{p}_2, \ldots, \hat{p}_{500}$ are shown in Table 6.1–4 where the entry in the i-th row and j-th column is $\hat{p}_{10(i-1)+j}$. The order statistics (6.1.6) of the entries of Table 6.1–4 are shown in Table 6.1–5; the latter are needed for the percentile method of obtaining an approximate confidence interval (step PM–3).

For an approximate 95% confidence interval, we have

$$N\alpha/2 = 12.5 \quad \text{and} \quad N(1 - \alpha/2) = 487.5;$$

hence, our interval for p is

$$\left(\hat{p}_{(12)}, \hat{p}_{(488)}\right) = (0, 0.0333).$$

We know (see Dudewicz and Karian (1999b, p. 262)) that the fitted GLD actually has $p = 0.0056$, so this interval covers the true p of the fitted GLD. (As the distribution from which the original data was drawn is not known, we are not in a position to conclude with certainty whether its p is also covered by the interval. This is the case with all "bootstrap" methods, which allow us to make inferences from data but cannot guarantee correctness of those inferences.)

The following theorem, discussed in Dudewicz (1992), implies that by proper use of the GB one will obtain better results than with the BM.

Theorem 6.1.9. *The Bootstrap Method is a special case of the Generalized Bootstrap (namely, the BM takes the empiric d.f. in step GB–1).*

Sun and Müller-Schwarze (1996) compared the performances of the Bootstrap Method and the Generalized Bootstrap of this section (as well as another method — the jackknife). Focusing on a particular applied problem and dataset, they provided the necessary computer code in the C programming language and drew general implications. They reported two important conclusions.

1. The Generalized Bootstrap is more consistent in parameter estimation than is the Bootstrap Method.

2. The number of bootstrap samples N should be at least 500.

Asymptotic properties of GB have been shown by Lin (1997), who notes (p. 302) that "... some advantages of using [GB] ... over the ordinary bootstrap [BM] have been found ... ." For further discussion and examples, see Dudewicz (1992).

Table 6.1–3. Samples 1 through 6 and 500, each of size 60, from the fitted GLD. Highlighted observations are at least 7.30.

$i = 1$	$i = 2$	$i = 3$	$i = 4$	$i = 5$	$i = 6$	$\cdots$	$i = 500$
3.30	3.02	6.03	5.14	2.73	3.67	$\cdots$	3.25
5.53	4.01	2.87	5.84	3.36	4.25	$\cdots$	4.25
2.90	3.73	5.13	5.56	6.52	3.74	$\cdots$	4.66
5.01	3.45	3.54	5.70	2.48	2.87	$\cdots$	6.08
5.78	2.73	2.84	3.49	3.91	5.91	$\cdots$	4.81
4.94	5.35	3.49	4.33	4.12	3.60	$\cdots$	5.20
3.56	4.29	4.42	4.16	3.99	2.47	$\cdots$	5.73
5.64	3.82	4.75	3.85	3.06	4.77	$\cdots$	4.52
5.15	2.66	2.65	3.62	3.78	5.43	$\cdots$	4.00
3.02	3.81	4.72	5.09	2.96	3.59	$\cdots$	4.91
3.48	4.34	6.29	3.71	3.17	4.31	$\cdots$	3.18
5.10	3.73	4.17	6.35	5.51	5.85	$\cdots$	2.51
4.33	3.21	1.99	4.66	6.94	4.51	$\cdots$	4.37
2.81	3.26	3.68	3.02	5.69	4.18	$\cdots$	3.28
2.36	2.21	3.01	5.26	4.37	4.42	$\cdots$	3.22
5.10	5.49	2.54	6.35	3.71	5.14	$\cdots$	5.65
4.81	4.13	4.33	3.08	3.78	2.82	$\cdots$	3.49
2.46	4.21	2.67	5.50	3.56	2.49	$\cdots$	4.21
5.11	4.77	3.20	2.44	2.84	4.95	$\cdots$	4.03
2.94	4.36	2.42	4.05	4.43	4.29	$\cdots$	3.48
5.29	4.75	3.95	2.30	4.49	3.52	$\cdots$	4.01
4.18	7.21	6.23	3.39	3.83	4.43	$\cdots$	3.78
2.65	4.13	3.56	3.86	3.59	2.83	$\cdots$	3.39
4.44	3.03	2.87	3.79	3.21	4.43	$\cdots$	4.31
5.44	3.64	4.09	3.42	5.13	1.84	$\cdots$	2.76
3.81	5.60	3.12	3.15	3.56	4.03	$\cdots$	4.81
2.77	6.59	4.49	6.19	5.21	3.46	$\cdots$	4.82
3.76	3.71	4.85	3.41	3.39	3.48	$\cdots$	6.15
3.76	4.11	3.17	2.57	3.05	3.03	$\cdots$	3.08
5.55	3.52	3.25	4.07	4.48	5.71	$\cdots$	4.30
3.12	4.35	2.98	3.52	3.78	3.75	$\cdots$	2.89
3.83	4.97	3.39	3.36	4.28	6.03	$\cdots$	5.79
4.99	4.64	4.47	3.72	4.16	3.16	$\cdots$	3.48
3.18	4.75	4.83	4.97	4.36	4.20	$\cdots$	3.47
2.70	4.46	4.11	5.73	**8.15**	3.42	$\cdots$	3.48
3.55	3.66	2.87	5.56	3.25	3.12	$\cdots$	4.94
2.70	5.07	4.95	4.33	6.48	5.27	$\cdots$	4.04
1.95	3.50	4.06	4.25	4.83	3.12	$\cdots$	3.15
2.99	2.93	4.98	3.67	3.22	4.77	$\cdots$	3.27
3.44	3.64	4.51	3.08	3.75	5.03	$\cdots$	4.19
2.98	6.56	2.97	6.17	3.68	5.20	$\cdots$	2.94
3.12	6.32	6.68	3.52	3.57	4.32	$\cdots$	4.83
3.78	3.88	6.24	4.27	3.27	4.05	$\cdots$	4.75
5.71	4.58	5.11	2.67	3.30	4.20	$\cdots$	5.25
4.84	3.26	4.58	3.60	3.28	5.91	$\cdots$	4.64
5.74	3.63	3.78	4.77	4.36	4.02	$\cdots$	4.02
3.45	3.37	6.48	5.02	4.79	5.73	$\cdots$	2.94
3.05	4.69	4.41	2.67	3.73	4.81	$\cdots$	3.25
3.55	6.11	4.30	3.86	5.44	3.97	$\cdots$	4.79
3.32	2.44	3.49	3.72	4.23	6.26	$\cdots$	3.30
4.47	4.93	2.23	3.99	4.31	6.34	$\cdots$	4.11
4.16	4.02	6.29	3.04	4.91	4.52	$\cdots$	5.64
2.34	2.51	4.64	5.02	3.72	3.86	$\cdots$	3.52
5.08	3.17	6.22	4.36	3.95	5.89	$\cdots$	5.13
2.95	4.91	3.52	3.49	3.43	4.16	$\cdots$	4.83
3.30	5.94	2.87	2.93	**7.76**	5.29	$\cdots$	4.15
3.22	3.12	2.42	4.07	3.22	4.81	$\cdots$	4.24
6.48	4.67	2.36	5.19	3.20	3.45	$\cdots$	3.67
4.50	3.62	2.27	6.10	4.07	3.39	$\cdots$	4.59
2.94	2.77	3.89	5.36	2.30	3.37	$\cdots$	3.85

Table 6.1–4. The 500 GB estimates $\hat{p}_1, \hat{p}_2, \ldots, \hat{p}_{500}$.

0	0	0	0	.0333	0	.0333	0	0	0
0	0	0	.0500	0	.0167	0	0	.0500	0
.0167	.0167	0	0	.0167	0	0	0	0	.0167
0	0	0	0	.0167	0	0	.0333	0	.0167
0	.0167	0	.0333	.0167	0	.0167	0	.0167	.0167
.0167	.0167	0	0	.0167	.0167	.0333	0	0	0
0	0	.0167	.0667	0	.0167	.0167	0	.0167	.0167
.0333	.0333	.0167	.0167	0	.0167	0	.0167	0	0
0	.0167	0	0	0	0	0	0	0	0
0	.0333	.0167	0	.0167	.0167	0	0	0	0
0	0	0	.0667	.0500	.0167	0	0	0	.0500
.0333	0	0	0	0	.0167	0	.0167	.0500	0
0	.0167	0	0	.0500	0	0	.0167	0	0
.0333	0	0	0	0	0	0	.0167	0	.0167
0	0	0	0	.0167	0	.0167	0	0	0
0	0	0	0	0	0	.0167	.0500	.0167	0
.0167	.0167	0	0	0	.0333	0	.0167	0	0
0	0	0	0	0	0	.0167	0	0	0
.0167	0	0	0	.0333	.0333	0	0	0	.0167
.0167	0	0	0	0	0	0	.0167	0	0
0	.0333	.0333	0	.0167	.0500	0	0	.0167	.0167
.0167	0	0	0	0	.0333	0	.0333	0	0
.0333	.0333	.0167	0	.0167	0	0	0	0	0
.0167	0	.0333	·.0167	.0167	.0333	.0333	.0333	0	0
0	0	.0167	.0167	0	.0333	.0333	.0167	0	.0333
0	.0333	0	0	0	.0167	0	.0167	0	0
0	.0167	0	.0333	0	0	0	0	0	.0167
0	.0167	0	.0167	0	.0167	0	0	.0333	0
.0167	0	0	0	.0167	0	0	.0167	.0333	0
0	0	0	0	.0167	.0167	0	0	.0167	0
.0167	.0333	0	.0333	.0167	0	0	0	0	0
0	0	0	.0167	.0167	0	.0167	.0333	.0167	0
0	0	.0333	0	.0167	.0167	.0333	.0167	.0167	0
.0167	.0167	.0167	.0167	.0167	0	.0167	0	0	0
0	0	0	0	.0167	.0167	0	.0167	0	0
0	0	0	0	.0333	0	0	.0167	0	.0167
0	0	0	.0333	0	.0333	.0167	.0167	0	.0333
.0333	0	.0167	0	.0167	0	.0167	0	.0333	0
0	.0167	.0167	.0167	0	.0333	0	0	0	.0167
.0167	.0167	0	.0333	0	0	0	.0167	0	0
0	0	.0167	.0167	.0333	.0167	.0333	0	0	0
0	.0167	.0167	0	0	0	0	.0333	0	0
0	.0167	0	.0167	0	.0167	0	.0167	.0167	0
.0167	0	0	.0333	.0167	.0167	0	0	0	0
0	.0167	.0167	0	0	0	0	0	0	.0167
0	0	.0167	.0167	0	.0167	0	.0500	0	0
0	.0167	0	0	.0167	0	0	.0167	0	0
.0167	0	.0167	.0167	0	.0167	0	.0167	0	0
.0167	0	0	.0167	.0167	0	0	.0167	0	0
0	0	0	.0167	0	0	.0167	.0167	0	0

Table 6.1–5. The order statistics $\hat{p}_{(1)} \leq \hat{p}_{(2)} \leq \cdots \leq \hat{p}_{(500)}$.

0	0	0	0	0	0	0	0	0	0
0	0	0	0	0	0	0	0	0	0
0	0	0	0	0	0	0	0	0	0
0	0	0	0	0	0	0	0	0	0
0	0	0	0	0	0	0	0	0	0
0	0	0	0	0	0	0	0	0	0
0	0	0	0	0	0	0	0	0	0
0	0	0	0	0	0	0	0	0	0
0	0	0	0	0	0	0	0	0	0
0	0	0	0	0	0	0	0	0	0
0	0	0	0	0	0	0	0	0	0
0	0	0	0	0	0	0	0	0	0
0	0	0	0	0	0	0	0	0	0
0	0	0	0	0	0	0	0	0	0
0	0	0	0	0	0	0	0	0	0
0	0	0	0	0	0	0	0	0	0
0	0	0	0	0	0	0	0	0	0
0	0	0	0	0	0	0	0	0	0
0	0	0	0	0	0	0	0	0	0
0	0	0	0	0	0	0	0	0	0
0	0	0	0	0	0	0	0	0	0
0	0	0	0	0	0	0	0	0	0
0	0	0	0	0	0	0	0	0	0
0	0	0	0	0	0	0	0	0	0
0	0	0	0	0	0	0	0	0	0
0	0	0	0	0	0	0	0	0	0
0	0	0	0	0	0	0	0	0	0
0	0	0	0	0	0	0	0	0	0
0	0	0	0	0	0	0	0	0	0
0	0	0	0	0	0	0	0	0	0
0	0	0	0	0	0	0	0	0	0
0	0	0	0	0	0	0	0	0	0
0	0	0	0	0	0	0	0	0	0
0	0	0	0	0	0	0	0	0	0
0	0	0	0	0	0	0	0	0	0
0	0	.0167	.0167	.0167	.0167	.0167	.0167	.0167	.0167
.0167	.0167	.0167	.0167	.0167	.0167	.0167	.0167	.0167	.0167
.0167	.0167	.0167	.0167	.0167	.0167	.0167	.0167	.0167	.0167
.0167	.0167	.0167	.0167	.0167	.0167	.0167	.0167	.0167	.0167
.0167	.0167	.0167	.0167	.0167	.0167	.0167	.0167	.0167	.0167
.0167	.0167	.0167	.0167	.0167	.0167	.0167	.0167	.0167	.0167
.0167	.0167	.0167	.0167	.0167	.0167	.0167	.0167	.0167	.0167
.0167	.0167	.0167	.0167	.0167	.0167	.0167	.0167	.0167	.0167
.0167	.0167	.0167	.0167	.0167	.0167	.0167	.0167	.0167	.0167
.0167	.0167	.0167	.0167	.0167	.0167	.0167	.0167	.0167	.0167
.0167	.0167	.0167	.0167	.0167	.0167	.0167	.0167	.0167	.0167
.0167	.0167	.0167	.0167	.0167	.0167	.0167	.0167	.0167	.0167
.0167	.0167	.0167	.0167	.0167	.0167	.0167	.0167	.0167	.0167
.0167	.0167	.0167	.0167	.0167	.0167	.0167	.0167	.0167	.0167
.0167	.0167	.0333	.0333	.0333	.0333	.0333	.0333	.0333	.0333
.0333	.0333	.0333	.0333	.0333	.0333	.0333	.0333	.0333	.0333
.0333	.0333	.0333	.0333	.0333	.0333	.0333	.0333	.0333	.0333
.0333	.0333	.0333	.0333	.0333	.0333	.0333	.0333	.0333	.0333
.0333	.0333	.0333	.0333	.0333	.0333	.0333	.0333	.0333	.0500
.0500	.0500	.0500	.0500	.0500	.0500	.0500	.0500	.0667	.0667

6.2   Comparisons of the GB and BM Methods

Since the BM has defects that are not generally well-known, in addition to the comments already made in Section 6.1, we detail here some of the recent results relevant to the comparison of GB and BM. These comparisons, in our opinion, make GB the method of choice when one intends to use MC methods.

In one study of distributional inference, Kroese (1994, pp. 41–42) stated, "A disadvantage of the Bootstrap procedure is that discrete distributions are generated." This is because the data is resampled by the BM, so one never sees any but the same data points one started with when the BM is used. Indeed, criticism of the BM has been growing in recent years (Young (1994)).

Meanwhile, recent theoretical work (e.g., Lin (1997)) has shown that methods like the GB, which beat the BM for small to moderate sample sizes, also give good asymptotic performance. In this light, it is not surprising that the most recent comprehensive book on bootstrap methods states, ". . . the generalized bootstrap. . . [is] a promising alternative to the. . . bootstrap since it has the advantage of taking account of the fact that the data are continuous but it does not seem to suffer the drawbacks. . . "(Chernick (1999, p. 109)).

Problems for Chapter 6

6.1 Suppose we are faced with the question: What is the probability that a value of a random variable X greater than 6.30 will be encountered? We have available a random sample of data of size 20, namely

| 3.88 | 4.01 | 4.28 | 3.87 | 3.54 | 4.86 | 5.31 | 4.37 | 4.77 | 3.37 |
| 3.79 | 4.17 | 3.26 | 3.34 | 4.49 | 2.35 | 3.78 | 5.94 | 3.65 | 4.20 |

(this is part of the data considered in Dudewicz and Karian (1999b, p. 260) in another context). Use the GB method to answer this question. [Hint: Fit an EGLD to the data. Then either (a) find the tail probability above 6.30 in the fitted distribution, or (b) take a sample of size 10,000 from the fitted EGLD and estimate the desired probability by $Y/10{,}000$ where Y is the number of the 10,000 values which are 6.30 or larger.]

6.2 Use the BM to attempt to answer the question of Problem 6.1. [Hint: Sample 10,000 from the data at random with replacement, and estimate by $Z/10,000$ where Z is the number of the 10,000 values which are 6.30 or larger.]

6.3 In Problem 6.1, find an approximate 95% confidence interval for the desired probability p as follows. Do the sampling of 10,000 values 1,000 times. This yields 1,000 estimates of p, say $p_1, \ldots, p_{1000}$. Discard the 25 smallest of the p_is and also discard the 25 largest, leaving the middle 95% of the estimates. Claim this is a 95% confidence interval for p.

6.4 As in Problem 6.3, but using the BM.

6.5 Comparing the results of Problem 6.1 with those of Problem 6.2, and of Problem 6.3 with those of Problem 6.4, given an assessment of the GB in comparison with the BM.

Appendix A

Programs for Fitting the GLD, GBD, and GLD–2

This Appendix consists of the compendium of programs that deal with various computations discussed in the text.

1. Section A.1 lists programs that are of general utilty. These consist of

   Histogram: for plotting data histograms;

   Percentile: for obtaining sample percentiles;

   GLDPDFplot: for graphing a GLD p.d.f.;

   GLDDFplot: for graphing the d.f. of a GLD.

2. Section A.2 lists programs associated with fitting of GLDs with the method of moments described in Chapter 2. These consist of

   FindAlphas: for finding the α_1, α_2, α_3, α_4 of a dataset;

   FindLambdasM: for determining λ_1, λ_2, λ_3, λ_4 from α_1, α_2, α_3, α_4.

3. The single program of Section A.3 is **FindBetasM** and it is used for computing β_1, β_2, β_3, β_4 from α_1, α_2, α_3, α_4.

4. Section A.4 contains programs associated with the percentile-based method discussed in Chapter 4. These are

   FindRhos: for the determination of ρ_1, ρ_2, ρ_3, ρ_4 for a dataset;

   FindLambdasP: for determining λ_1, λ_2, λ_3, λ_4 from ρ_1, ρ_2, ρ_3, ρ_4.

5. Section A.5 deals with GLD–2 fits and contains

   PlackettPsi: to compute Plackett's Ψ from bivariate data;

   GLD2PDFplot: for graphing the p.d.f. of a GLD–2;

GLD2RAND: for generating random samples from a fitted GLD–2; and

P.301ff: the program (listing starts on page 301) written in the C programming language for obtaining GLD–2 approximations.

With the exception of the last program in Section A.5 (which is written in the programming language C), all programs are written in *Maple* for execution within the *Maple* computing environment. There are advantages and disadvantages associated with any choice of programming language or environment. The principal disadvantages of *Maple* are

1. *Maple* is not as universally available as FORTRAN or C. Consequently, some individuals, including expert programmers, will have to assume the expense (in money and time) of acquiring *Maple* and getting acclimated to it.

2. The primary strength of *Maple* is in symbolic mathematical manipulations, not in numeric computations. This makes numeric computations in *Maple* considerably slower than they would be in FORTRAN or C.

The advantages of using *Maple* are

1. Access to a host of built-in symbolic manipulation algorithms that enable users to perform differentiations, symbolic integrations, computations of Jacobians, solving differential equations, etc.

2. The ability to set almost arbitrary levels of accuracy, including precisions that exceed the computer's hardware floating-point capabilities. Use of high levels of accuracy slows program executions; however, in many instances this is a very worthwhile trade-off.

3. Access to numeric, symbolic and **graphic** computation in a single environment. The ability to produce images such as those given in figures throughout this book greatly enhances one's ability to solve equations associated with GLD, GBD, and GLD–2 fits.

The programs listed in the following sections can be obtained from the CRC Press website through

http://www.crcpress.com

follwed by a click on "Electronic Download and Update Page" (in the leeft column), in turn followed by a click on the title of this text ("Fitting Statistical Distributions ...") that is provided in a pop-up list. The last click is on "Download 2885.zip."

A.1 General Utility Programs

```
####################################################
#                                                  #
#   Procedure for Graphing a Data Histogram        #
#                                                  #
####################################################
# FUNCTION:   Histogram
# PURPOSE:    Plot a histogram of data
# ARGUMENTS:  L -- list of data values
#             R -- (optional) range of data values of the form
#                  a..b (if not present the max and min of the
#                  data will be used).
#             I -- (optional) number of intervals (if not present
#                  8 intervals will be used).

Histogram := proc()
local YYY, YY, Y, n, l, u, v, i, j, k, p, t, U, V, count,
      truemin, truemax, nint;

    YY := sort(evalf(args[1]));
    n := nops(YY);
    if nargs = 1 then
        truemin := YY[1]; truemax := YY[n]; nint := 8
    else
        truemin := lhs(args[2]);
        truemax := rhs(args[2]);
        nint := args[3]
    fi;
    count := 0;
    for i to n do
        if YY[i] < truemin or truemax < YY[i] then
            count := count + 1
        fi
    od;
    if 0 < count then print("WARNING: There are", count,
        " data points out of plot range.")
    fi;
    j := 1;
    YYY := array(1 .. n);
    u := array(1 .. nint + 1);
    v := array(1 .. nint + 1);
```

```
for i to n do
    if truemin <= YY[i] and YY[i] <= truemax then
        YYY[j] := YY[i]; j := j + 1
    fi
od;
n := j - 1;
Y := [seq(YYY[k], k = 1 .. n)];
p := array(1 .. 8*nint);
l := (truemax - truemin)/nint;
u[1] := truemin;
v[1] := 0;
for i from 2 to nint + 1 do
    u[i] := u[i - 1] + l; v[i] := 0
od;
if u[nint + 1] < Y[n] then u[nint + 1] := Y[n] fi;
i := 2;
for j to n do
    if Y[j] <= u[i] then v[i - 1] := v[i - 1] + 1
    else i := i + 1; j := j - 1
    fi
od;
t := 1;
for i to nint do
    U[t] := u[i];
    V[t] := 0;
    t := t + 1;
    U[t] := U[t - 1];
    V[t] := v[i];
    t := t + 1;
    U[t] := U[t - 1] + 1;
    V[t] := v[i];
    t := t + 1;
    U[t] := U[t - 1];
    V[t] := 0;
    t := t + 1
od;
for i to t - 1 do
    p[2*i - 1] := U[i]; p[2*i] := V[i]/(l*n)
od;
plot([seq([p[2*i - 1], p[2*i]], i = 1 .. t - 1)],
    color = black)
end:
```

```
##################################################
#                                                #
#    Procedure to Determine Data Percentiles     #
#                                                #
##################################################
# FUNCTION:   Percentile
# PURPOSE:    Compute sample percentiles
# ARGUMENTS:  L -- list of data values
#             p -- probability for the percentile

Percentile:=proc(L::list(numeric), p::numeric)
    local n, LL, F, r, ab:
    description "The 100*p-th percentile of the list L":

    if p<0 or p>1 then
        ERROR("Second argument must be between 0 and 1") fi:
    n := nops(L):
    if p < 1/(n+1) or p > n/(n+1) then
        ERROR("Percentile does not exist") fi:
    LL:=sort(L):    F:=convert((n+1)*p, fraction):
    r:=trunc(F):    ab:=F-r:
    LL[r]+ ab*(LL[r+1]-LL[r])
end:




##################################################
#                                                #
#     Procedure for Graphing a GLD p.d.f.        #
#                                                #
##################################################
# FUNCTION:   GLDPDFplot
# PURPOSE:    Graph the p.d.f. of a GLD distribution
# ARGUMENTS:  L -- list of lambda values.


GLDPDFplot := proc(L::list)

local P1, P2, P, X, f, i, n, Points;

n := 101;
P:=array(1..n);
for i from 1 to n do
```

```
    if i <= 5 then P[i] := i*0.0005
        elif i <= 10 then P[i] := 0.0025+(i-5)*0.001
        elif i <= 20 then P[i] := 0.0075+(i-10)*0.002
        elif i <= 30 then P[i] := 0.0275+(i-20)*0.01
        elif i <= 67 then P[i] := 0.1275+(i-30)*0.02
        elif i <= 77 then P[i] := 0.8675+(i-67)*0.01
        elif i <= 87 then P[i] := 0.9675+(i-77)*0.002
        elif i <= 97 then P[i] := 0.9875+(i-87)*0.001
        else P[i] := 0.9975+(i-97)*0.0005
    fi
od;

X := [seq(L[1]+ ( P[i]^L[3] - (1-P[i])^L[4] )/L[2], i=1..n)]:
f := [seq(L[2]/(L[3]*P[i]^(L[3]-1)+L[4]*(1-P[i])^(L[4]-1)), i=1..n)]:
Points:= [seq( [X[i], f[i]], i= 1..n)]:
plot(Points,color=black):
end:


###############################################
#                                             #
#      Procedure for Graphing a GLD d.f.      #
#                                             #
###############################################
# FUNCTION:  GLDCDFplot
# PURPOSE:   Graph the d.f. of a GLD distribution
# ARGUMENTS: L -- list of lambda values.

GLDCDFplot := proc(L::list)

local P, X, i, n, Points;

n := 101;
P:=array(1..n);
for i from 1 to n do
    if i <= 5 then P[i] := i*0.0005
        elif i <= 10 then P[i] := 0.0025+(i-5)*0.001
        elif i <= 20 then P[i] := 0.0075+(i-10)*0.002
        elif i <= 30 then P[i] := 0.0275+(i-20)*0.01
        elif i <= 67 then P[i] := 0.1275+(i-30)*0.02
        elif i <= 77 then P[i] := 0.8675+(i-67)*0.01
        elif i <= 87 then P[i] := 0.9675+(i-77)*0.002
        elif i <= 97 then P[i] := 0.9875+(i-87)*0.001
        else P[i] := 0.9975+(i-97)*0.0005
    fi
```

```
od;

X := [seq(L[1]+ ( P[i]^L[3] - (1-P[i])^L[4] )/L[2], i=1..n)]:
Points:= [seq( [X[i], P[i]], i= 1..n)]:
plot(Points,color=black):
end:
```

A.2   GLD Parameter Estimation: Method of Moments

```
####################################################
#                                                  #
#    Procedure to Determine Alphas from Data    #
#                                                  #
####################################################
# FUNCTION:  FindAlphas
# PURPOSE:   Compute First 4 sample moments
# ARGUMENTS: DL -- list of data.

FindAlphas := proc (DL::list)

local n, i, AH1, AH2, AH3, AH4, S;

n:=nops(DL);
AH1 := convert(DL, '+')/n;
S:= [seq((DL[i]-AH1)^2, i=1..n)];
AH2 := convert(S, '+')/n;
S:= [seq((DL[i]-AH1)^3, i=1..n)];
AH3 := convert(S, '+')/(n*AH2^(3/2));
S:= [seq((DL[i]-AH1)^4, i=1..n)];
AH4 := convert(S, '+')/(n*AH2^2);

evalf([AH1, AH2, AH3, AH4]);
end:
```

```
##################################################
#                                                #
#      Procedure to Determine Lambdas from       #
#                 Sample Moments                 #
#                                                #
##################################################
# FUNCTION:   FindLambdasM
# PURPOSE:    Newton's method for GLD Approx. via moments
# ARGUMENTS:  AH -- list of sample alphas;
#             I3, I4 -- Initial approx. of lambda3 and lambda4
#             IterCount -- (optional) Max. No. of Iterations
#                          (if not present it is set to 10).

FindLambdasM := proc(AH:: list, I3::numeric, I4::numeric)

local A, B, C, D1, D2, D3, Alph1, Alph2, Alph3, Alph4, F,
      AHat1, AHat2, AHat3, AHat4, V, J, err3, err4, iterations,
      Fk, Jk, Y, Eq3, Eq4, L1, L2, L3, L4, a, b, temp, lambda1,
      lambda2, lambda3, lambda4, IterCount, L, FirstL, SecondL;

Digits:=20;
if nargs=3 then IterCount := 10 else IterCount := args[4] fi;
with(linalg, vector, matrix, jacobian, linsolve):

AHat1:=0; AHat2:=1;
AHat3:=evalf(AH[3]); AHat4:=evalf(AH[4]);
L3:=I3; L4:=I4;

A:=1/(1+lambda3)-1/(1+lambda4);
B:=1/(1+2*lambda3)+1/(1+2*lambda4)-2*Beta(1+lambda3,1+lambda4);
C:=1/(1+3*lambda3)-1/(1+3*lambda4)-3*Beta(1+2*lambda3,1+lambda4)
   +3*Beta(1+lambda3,1+2*lambda4);
D1:=1/(1+4*lambda3)+1/(1+4*lambda4)+6*Beta(1+2*lambda3,1+2*lambda4);
D2:=-4*Beta(1+3*lambda3,1+lambda4)-4*Beta(1+lambda3,1+3*lambda4);
D3:=D1+D2;

Alph1:=lambda1+A/lambda2;
Alph2:=abs(B-A^2)/lambda2^2;
Alph3:=(C-3*A*B+2*A^3)/abs((B-A^2)^(3/2));
Alph4:=(D3-4*A*C+6*B*A^2-3*A^4)/((B-A^2)^2);

Eq3 := Alph3-AHat3;
Eq4 := Alph4-AHat4;
```

```
F:= vector([Eq3, Eq4]);
V:=vector([lambda3, lambda4]);
J:= jacobian(F, V);
err3 := 1; err4 := 1; iterations:=1;

while (((err3 > 0.00001) or
        (err4 > 0.00001)) and (iterations <= IterCount)) do
# if iterations=IterCount then lprint('WARNING: iterations=',
     iterations, 'ERRORS: ',err3, err4) fi;
  Fk:=vector([evalf(subs({lambda3=L3, lambda4=L4},-Eq3)),
              evalf(subs({lambda3=L3, lambda4=L4},-Eq4))]);
  Jk:= matrix([
       [subs({lambda3=L3, lambda4=L4}, J[1,1]),
        subs({lambda3=L3, lambda4=L4}, J[1,2])],
       [subs({lambda3=L3, lambda4=L4}, J[2,1]),
        subs({lambda3=L3, lambda4=L4}, J[2,2])]
      ]);
  Y:= linsolve(Jk,Fk);
  L3:= L3+Y[1];
  L4:= L4+Y[2];
  err3:= evalf(abs(subs({lambda3=L3, lambda4=L4}, Eq3)));
  err4:= evalf(abs(subs({lambda3=L3, lambda4=L4}, Eq4)));
  if iterations=IterCount then lprint('WARNING: iterations=',
     iterations, 'ERRORS: ',err3, err4) fi;
  iterations := iterations+1
od;

a:=evalf(subs({lambda3=L3, lambda4=L4},A));
b:=evalf(subs({lambda3=L3, lambda4=L4},B));

L2:=abs(sqrt((b-a^2)/AHat2));
L1 := AHat1 -a/L2;
FirstL:=[L1,L2,L3,L4];

if L3 < 0 then SecondL:=[-FirstL[1],FirstL[2],
    FirstL[4],FirstL[3]] else SecondL:= FirstL fi;
if evalf(AH[3]) < 0 then L:=[-SecondL[1],SecondL[2],
    SecondL[4],SecondL[3]] else L:=SecondL fi;

[L[1]*sqrt(AH[2])+AH[1], L[2]/sqrt(AH[2]),L[3],L[4]];
evalf(%,10);
end:
```

A.3   GBD Parameter Estimation: Method of Moments

```
##################################################
#                                                #
#       Procedure to Determine Betas from        #
#                 Sample Moments                 #
#                                                #
##################################################
# FUNCTION:  FindBetasM
# PURPOSE:   Newton's method for GBD approx. via moments
# ARGUMENTS: AH -- list of sample alphas;
#            I3, I4 -- Initial approx. of beta 3 and beta 4
#            IterCount -- (optional) Max. No. of Iterations
#                         (if not present it is set to 10).

FindBetasM := proc(AH:: list, I3::numeric, I4::numeric)

local Alph1, Alph2, Alph3, Alph4, F, AHat1, AHat2,
      AHat3, AHat4, V, J, err3, err4, iterations,
      Alph4Num, Alph4Den, B2, B3, B4, B5,
      Fk, Jk, Y, Eq3, Eq4, Bet1, Bet2, Bet3, Bet4,
      temp, beta1, beta2, beta3, beta4, IterCount;

Digits:=20;
if nargs=3 then IterCount := 10 else IterCount := args[4] fi;
with(linalg, vector, matrix, jacobian, linsolve):
AHat1:=AH[1];  AHat2:=AH[2];  AHat4:=AH[4];
if AH[3] < 0 then AHat3 := -AH[3] else AHat3 := AH[3] fi;
Bet3:=I3; Bet4:=I4;

B2 := beta3+beta4+2; B3 := B2+1; B4 := B2+2; B5 := B2+3;
Alph1 := beta1+(beta2*(beta3+1))/B2;
Alph2 := (beta2^2*(beta3+1)*(beta4+1))/(B2^2*B3);
Alph3 := 2*((beta4-beta3)*sqrt(B3))/(B4*sqrt((beta3+1)*(beta4+1)));
Alph4Num:=3*B3*(beta3*beta4*B2+3*beta3^2+5*beta3+3*beta4^2+5*beta4+4)
Alph4Den := B4*B5*(beta3+1)*(beta4+1);
Alph4 := Alph4Num/Alph4Den;

Eq3 := Alph3-AHat3;
Eq4 := Alph4-AHat4;

F:= vector([Eq3, Eq4]);
```

```
V:=vector([beta3, beta4]);
J:= jacobian(F, V);
err3 := 1; err4 := 1; iterations:=0;
while (((err3 > 0.0000001) or
        (err4 > 0.0000001)) and (iterations < IterCount+1)) do
  if iterations=IterCount then lprint('WARNING:
     iterations=', iterations, 'ERRORS: ',err3, err4) fi;
  Fk:=vector([evalf(subs({beta3=Bet3, beta4=Bet4},-Eq3)),
              evalf(subs({beta3=Bet3, beta4=Bet4},-Eq4))]);
  Jk:= matrix([
      [subs({beta3=Bet3, beta4=Bet4}, J[1,1]),
       subs({beta3=Bet3, beta4=Bet4}, J[1,2])],
      [subs({beta3=Bet3, beta4=Bet4}, J[2,1]),
       subs({beta3=Bet3, beta4=Bet4}, J[2,2])]
     ]);
  Y:= linsolve(Jk,Fk);
  Bet3:= Bet3+Y[1];
  Bet4:= Bet4+Y[2];
  err3:= evalf(abs(subs({beta3=Bet3, beta4=Bet4}, Eq3)));
  err4:= evalf(abs(subs({beta3=Bet3, beta4=Bet4}, Eq4)));
  iterations := iterations+1
od;

if AH[3] < 0 then temp := Bet3; Bet3 := Bet4; Bet4 := temp fi;
Bet2:=(Bet3+Bet4+2)*sqrt( ((Bet3+Bet4+3)*AHat2)/((Bet3+1)*(Bet4+1)) );
Bet1 := AHat1- (Bet2*(Bet3+1))/(Bet3+Bet4+2);
evalf([Bet1, Bet2, Bet3, Bet4], 10);
end:
```

A.4 GLD Parameter Estimation: Method of Percentiles

```
###################################################
#                                                 #
#     Procedure to Determine Rhos from Data       #
#                                                 #
###################################################
# FUNCTION:  FindRhos
# PURPOSE:   Compute Rho values for a sample
# ARGUMENTS: DL -- list of data
#            u -- (optional-if not present, assumed to be 0.1)

FindRhos := proc (DL::list)
    local P10, P25, P50, P75, P90,
          Rho2, Rho3, Rho4, u;

if nargs=1 then u:=0.1 else u:=args[2] fi:

P10:=Percentile(DL,u):
P25:=Percentile(DL,0.25):
P50:=Percentile(DL,0.5);
P75:=Percentile(DL,0.75):
P90:=Percentile(DL,1-u):

Rho2:=P90-P10:
Rho3:=(P50-P10)/(P90-P50):
Rho4:=(P75-P25)/Rho2:
[P50,Rho2,Rho3,Rho4]
end:


###################################################
#                                                 #
#   Procedure to Determine Lambdas from Rhos      #
#                                                 #
###################################################
# FUNCTION:  FindLambdasP
# PURPOSE:   Newton's method for GLD (percentiles)
# ARGUMENTS: RH -- list of rho values
#            I3, I4 -- Initial approx. of lambda3 and lambda4
#            u -- (optional; if not present assumed 0.1)

FindLambdasP := proc(RH:: list, I3::numeric, I4::numeric)
```

```
local u, IterCount, RHat1, RHat2, RHat3, RHat4, L3, L4,
     lambda1, lambda2, lambda3, lambda4, Rho1, Rho2,
     Rho3, Rho4, Eq3, Eq4, F, V, J, Fk, Jk, err3, err4,
     Y, L1, L2, iterations, temp;

Digits:=30;
if nargs=4 then u:=args[4] else u:=0.1 fi;
IterCount := 10;
with(linalg, vector, matrix, jacobian, linsolve):
RHat1:=RH[1];   RHat2:=RH[2];   RHat4:=RH[4];
if RH[3] > 1 then RHat3 := 1/RH[3] else RHat3 := RH[3] fi;
L3:=I3; L4:=I4;

Rho1:=lambda1+(0.5^lambda3-0.5^lambda4)/lambda2;
Rho2:= ((1-u)^lambda3-u^lambda4+(1-u)^lambda4-u^lambda3)/lambda2;
Rho3:=((1-u)^lambda4-u^lambda3+0.5^lambda3-0.5^lambda4)/
      ((1-u)^lambda3-u^lambda4+0.5^lambda4-0.5^lambda3);
Rho4:=(0.75^lambda3-0.25^lambda4+0.75^lambda4-0.25^lambda3)/
      ((1-u)^lambda3-u^lambda4+(1-u)^lambda4-u^lambda3);

Eq3 := Rho3-RHat3; Eq4 := Rho4-RHat4;

F:= vector([Eq3, Eq4]);
V:=vector([lambda3, lambda4]);
J:= jacobian(F, V);
err3 := 1; err4 := 1; iterations:=0;
while (((err3 > 0.000001) or
        (err4 > 0.000001)) and (iterations < IterCount+1)) do
  if iterations=IterCount then lprint('WARNING:
    iterations=', iterations, 'ERRORS: ',err3, err4) fi;
  Fk:=vector([evalf(subs({lambda3=L3, lambda4=L4},-Eq3)),
             evalf(subs({lambda3=L3, lambda4=L4},-Eq4))]);
  Jk:= matrix([
       [subs({lambda3=L3, lambda4=L4}, J[1,1]),
        subs({lambda3=L3, lambda4=L4}, J[1,2])],
       [subs({lambda3=L3, lambda4=L4}, J[2,1]),
        subs({lambda3=L3, lambda4=L4}, J[2,2])]
      ]);
  Y:= linsolve(Jk,Fk);
  L3:= L3+Y[1];
  L4:= L4+Y[2];
  err3:= evalf(abs(subs({lambda3=L3, lambda4=L4}, Eq3)));
  err4:= evalf(abs(subs({lambda3=L3, lambda4=L4}, Eq4)));
```

```
    iterations := iterations+1
od;

if RH[3] > 1 then temp:=L3; L3:=L4; L4:=temp fi;
L2:=((1-u)^L3-u^L4+(1-u)^L4-u^L3)/RHat2;
L1:=RHat1-(0.5^L3-0.5^L4)/L2;

evalf([L1, L2, L3, L4],10);
end:
```

A.5   Programs for GLD–2 Fitting

```
##################################################
#                                                #
#      Procedure to Determine No. of Points in   #
#           Each Quadrant and Plackett's Psi      #
#                                                #
##################################################
# FUNCTION:  PlackettPsi
# PURPOSE:   Compute a, b, c, d, and Plackett's Psi
# ARGUMENTS: X, Y -- X and Y data values (as lists)

PlackettPsi := proc(X::list, Y::list)

local n, i, C1, C2, C3, C4, MX, MY;

n := nops(X);
C1 := 0; C2 := 0; C3 := 0; C4 := 0;
MX := Percentile(X,0.5);
MY := Percentile(Y,0.5);
for i from 1 to n do
    if X[i] >= MX and Y[i] >= MY then C1 := C1+1
        elif X[i] >= MX and Y[i] < MY then C4 := C4+1
        elif X[i] <= MX and Y[i] > MY then C2 := C2+1
        else C3 := C3+1
    fi
od;
[C1, C2,C3,C4,(C1*C3)/(C2*C4)]
end:
```

```
##################################################
#                                                #
#  Procedure for a 3-D Graph of a Fitted GLD-2   #
#                                                #
##################################################
# FUNCTION:   GLD2PDFplot
# PURPOSE:    Produce a 3-D graph of a fitted GLD-2
# ARGUMENTS: Psi -- Plackett's Psi
#            L1, L2 -- Lambda values for the X and Y
#                      GLD fits
GLD2PDFplot := proc(Psi::numeric, L1::list, L2::list)

local A, P1, P2, P, XX, fX, YX, fY, i,
      j, S, PL, n, top, bottom, PlotSet;

with(plots,[display, spacecurve]);
n := 25;
P1 := .001,.005,.012,.003,.02,.03,.042,.055,.075,
      .1,.13,.165,.205,.25,.3,.35,.4,.45,.5;
P2 := seq(1-P1[19-i],i=1..18);
P := [P1,P2];

XX:=[seq(L1[1]+ ( P[i]^L1[3] - (1-P[i])^L1[4] )/L1[2], i=1..n)]:
fX:=[seq(L1[2]/(L1[3]*P[i]^(L1[3]-1)+L1[4]*(1-P[i])^(L1[4]-1)),i=1..n)]
YX:=[seq(L2[1]+ ( P[i]^L2[3] - (1-P[i])^L2[4] )/L2[2], i=1..n)]:
fY:=[seq(L2[2]/(L2[3]*P[i]^(L2[3]-1)+L2[4]*(1-P[i])^(L2[4]-1)),i=1..n)]

if Psi = 1 then
    for i from 1 to n do
        for j from 1 to n do
            A[i,j] := [XX[i], YX[j], fX[i]*fY[j]];
        od
    od
    else
    for i from 1 to n do
      for j from 1 to n do
        top := Psi*fX[i]*fY[j]*(1+(Psi-1)*(P[i]+P[j]-2*P[i]*P[j]));
        S := 1+(P[i]+P[j])*(Psi-1);
        bottom := (S^2-4*Psi*(Psi-1)*P[i]*P[j])^(3/2);
        A[i,j] := [XX[i], YX[j], top/bottom]
      od
    od
fi;
```

```
PlotSet := {};
for i from 1 to n do
    PL := spacecurve([seq(A[i,j], j=1..n)], color=black);
    PlotSet := PlotSet union {PL};
od;
for j from 1 to n do
    PL := spacecurve([seq(A[i,j], i=1..n)], color=black);
    PlotSet := PlotSet union {PL};
od;

display(PlotSet);
end:
}


####################################################
#                                                  #
# Procedure for Generating GLD-2 Random Samples #
#                                                  #
####################################################
# FUNCTION:   GLD2RAND
# PURPOSE:    Generate random samples from GLD-2
# ARGUMENTS: LX, LY -- Lambdas (as lists) of the X
#                      and Y components
#            Psi -- Plackett's Psi
#            N -- Size of sample to be generated

GLD2RAND := proc(LX::list, LY::list, psi::numeric, N::posint)
    local y, l1, l2, l3, l4, FInv, GInv, f, g, R,
          gInt, k, r, u, x, Fx, fx, Gy, X, i, Delta,
          gy, S, hNum, hDen, h, gPDF, j, iter;

l1+(y^l3-(1-y)^l4)/l2;
FInv:=subs({l1=LX[1],l2=LX[2],l3=LX[3],l4=LX[4]},%);
GInv:=subs({l1=LY[1],l2=LY[2],l3=LY[3],l4=LY[4]},%%);
l2/(l3*y^(l3-1)+l4*(1-y)^(l4-1));
f:=simplify(subs({l1=LX[1],l2=LX[2],l3=LX[3],l4=LX[4]},%));
g:=simplify(subs({l1=LY[1],l2=LY[2],l3=LY[3],l4=LY[4]},%%));
iter:=1000;
R:=array(1..N):  gInt:=array(1..iter);
    Gy:=[seq(i*(1.0/iter), i=0..iter)]:
    X:=[seq(subs(y=Gy[i],GInv), i=1..nops(Gy))]:
    Delta:=[seq(X[i+1]-X[i], i=1..nops(Gy)-1)]:
    gy:=[seq(subs(y=Gy[i],g), i=1..nops(Gy))]:
```

```
for k from 1 to N do
    r:=evalf(rand()/10^12):
    u:=evalf(rand()/10^12):
    x:=subs(y=r,FInv):
    Fx:=r; fx:=subs(y=r,f):
    S:=[seq(1+(Fx+Gy[i])*(psi-1), i=1..nops(Gy))]:
    hNum:=[seq(psi*fx*gy[i]*(1+(psi-1)*(Fx+Gy[i]-2*Fx*Gy[i])),
            i=1..nops(Gy))]:
    hDen:=[seq((S[i]^2-4*psi*(psi-1)*Fx*Gy[i])^(3/2),
            i=1..nops(Gy))]:
    h:=[seq(hNum[i]/hDen[i], i=1..nops(Gy))]:
    gPDF:=[seq(h[i]/fx, i=1..nops(Gy))]:
    gInt[1]:=Delta[1]*gPDF[1];
    for i from 2 to iter do
        gInt[i]:=gInt[i-1]+Delta[i]*gPDF[i];
    od:
    j:=1:
    while gInt[j] < u do j:=j+1 od:
    R[k]:=[x,(X[j-1]+X[j])/2]
od:
[seq(R[i], i=1..N)]:

end:




/* ************************************************************ */
/* This C program, reproduced with permission from Beckwith    */
/* and Dudewicz (1996), copyright (c) 1996 by American Sciences */
/* Press, Inc., 20 Cross Road, Syracuse, New York 13224-2104,   */
/* calculates the value of Psi (to 3 decimal places) which      */
/* results in the smallest L1 error in fitting the GLD-2        */
/* to the Bivariate Normal distribution, for rho= -0.8(0.1)0.9. */
/* By changing the values of Lambda in function Table, and      */
/* the formula for the BVN p.d.f. in function bvn, other        */
/* distributions can be fitted also.                           */
/* ************************************************************ */

#include <math.h>
#include <stdio.h>
#include <stdlib.h>
#include <conio.h>
#define Pi 3.14159265
```

```
/* function prototypes */
   double f (double x, double a, double b, double c, double d);
   double R (double x, double a, double b, double c, double d);
   double bvn (double x, double y);
   double h (double Psi, double fx, double fy, double px, double py);
   int table (void);
   double get_Psi (int places, int steps, double old_pos);

/* global variables */
   double min, sup, inf, best_bvnvol, best_gld_vol, Rho;
   double best-max_ratio, best_min_ratio, E[20][20], Lx[5];
   double px[500], py[500], Ly[5], fx[500], fy[500], zx[500], zy[500]
   int numpoints;

/* functions */
double f(double x, double a, double b, double c, double d) {

    /* GLD p.d.f. */

    return (b/(c * pow(x, c- 1) + d * pow (1-x, d-1)));
}

double R(double x, double a, double b, double c, double d) {

    /* GLD R(p) */

    return (a + (pow (x, c) - pow (1-x, d)) / b);
}

double bvn (double x, double y) {

    /* Bivariate Normal (standard) p.d.f. */

    double e, c;

    e = (x*x - 2*Rho*x*y + y*y) / (2 - 2*Rho*Rho);
    c = 2 * Pi * pow(1 - Rho*Rho, 0.5);
    return (exp(-e)/c);
}

double h (double Psi, double fx, double fy, double px, double py) {

    /* GLD-2 p.d.f., with Plackett's h(x,y) and GLD marginals */
```

```
      double num, den, S;

      S = 1 + (px + py) * (Psi- 1);
      num = Psi*fx*fy*(1 + (Psi- 1) * (px + py - 2 * px * py));
      den = pow(S*S - 4 * Psi * (Psi- 1) * px * py, 1.5);
      if (den == 0)
         den = 0.0000000 1;
      return(num/den);
}

int table (void) {

   /* Create arrays of:
      equally spaced x and y values between their minimums and
      maximums, the corresponding p-values for each x, y, and
      the GLD p.d.f. values.       */

   double x_range, y_range, error, a, b, h;
   double zxmax=4, zxmin=-4, zymax=4, zymin=-4;
   int i, count, top;

   printf ("Enter the number of pairs: ");
   scanf ("%d", &numpoints);

   printf ("Enter the largest integer value of Psi: ");
   scanf ("%d", &top);

   /* Lambda values for univariate normal p.d.f.'s */
   Lx[1] = 0;
   Lx[2] = 0.1974513695;
   Lx[3] = 0.13491245465;
   Lx[4] = Lx[3];
   Ly[1] = 0;
   Ly[2] = 0.1974513695;
   Ly[31 = 0.13491245465;
   Ly[4] = Ly[3];

   x_range = zxmax - zxmin;
   y_range = zymax - zymin;

   /* Bisection method to find the p-values for equally spaced x */

   for (i = 0; i <= numpoints; ++i) {
      zx[i] = zxmin + x_range * i / numpoints;
```

```
        error = 1;
        count = 0;
        if (Lx[2] < 0)
            a = 0.0001;    /* 0^(-L) = infinity */
        else
            a= 0;
  b = 1;
  while (error> .0000000001) {
        h= (b+ a) / 2;
        if(count > 50)
            error = 0;
        if ((R(a, Lx[1], Lx[2], Lx[3], Lx[4]) - zx[i])*\
            (R(h, Lx[1], Lx[21, Lx[3], Lx[4]) -.zxli]) > 0)
            a= h;
        else
            b = h;
        error = fabs(R(h, Lx[1], Lx[2], Lx[3], Lx[4]) - zx[i]);
        ++count;
        }

px[i] = h;
}


/* Bisection method to find the p-values for equally spaced y */

for (i 0; i <= numpoints; ++i) {

        zy[i] zymin + y_range * i/numpoints;
        error = 1;
        count = 0;
        if (Ly[2] < 0)
            a - 0.000 1; /* 0^(-L) infinity */
        else
            a = 0;
        b = 1;
        while (error> .0000000001) {
            h = (b+a)/2;
        if(count > 50)
            error = 0;
        if ((R(a, Ly[1], Ly[2], Ly[3], Ly[4]) - zy[ij)*\
            (R(h, Ly[1], Ly[2], Ly[3], Ly[4]) - zy[i]) > 0)
            a= h;
        else
            b = h;
```

```
            error = fabs(R(h, Ly[1], Ly[2], Ly[3], Ly[4]) - zy[i]);
            ++count;
            }
        py[i] = h;
        }

    for (i = 1; i <= numpoints; ++i) {
        fx[i]=f(px[i], Lx[1], Lx[2], Lx[3], Lx[4]);
        fy[i]=f(py[i], Ly[1], Ly[2], Ly[3], Ly[4]);
    }
return (top);
}

double get_Psi (int places, int steps, double old_pos) {

    /* Find the Psi which corresponds to the smallest L1 error
       volume. This is a four-step procedure, finding Psi as an
       integer, then within 0.1, 0.01, and finally 0.001  */

    double pos, norm_vol, gld-vol, min_ratio, max_ratio;
    double delta, error, Psi, new_pos, t1, t2, term, temp;
    int i, j, count;

    min = 5000; sup = 0; inf = 1;
    for (count = 0; count <= steps; ++count) {
        max_ratio = -1;
        min_ratio = 1000;
        error = 0;
        norm_vol = 0;
        gld_vol = 0;
        if (places == 0)
            Psi = count;
        else
            Psi = old_pos - 1 / pow(10, places - 1)
                    + count pow(10, places));
        for (i = 1; i < numpoints; ++i)
            for (j = 1; j < numpoints; ++j) {
                t1 = bvn(zx[i], zy[j]);
                t2 = h(Psi, fx[i], fy[i], px[i], py[j]);
                delta = (zx[i+ 1] - zx[i])*(zy[j+ 1] - zy[j]);
                norm_vol += t1 * delta;
                gld_vol += t2 * delta;
                term = fabs(t1 - t2);
                error += term * delta;
```

```
            if (term > sup)
                sup = term;
            if (term < inf)
                inf = term;
            if (fabs(t1) > .0000000001) {
                temp = t2 / t1;
                if (temp > max_ratio)
                    max_ratio = temp;
                if (temp < min_ratio)
                    min_ratio = temp;
            }
        }
        printf ("Rho=%.2f Psi=%.3f L1error=%.8f\n",
            Rho, Psi, error);
        if (error < min) {
            min = error;
            best_bvn_vol = norm_vol;
            best-gld_vol = gld_vol;
            best_max_ratio max_ratio;
            best_min_ratio min_ratio;
            new_pos = Psi;
        }
    }
    return (new-pos);
}

void main (void) {
    int count, r, t, last;
    double Psi;

    clrscr();
    last = table();

    for (r = -8; r < 10; ++r) {
        Rho= r / 10.0;
        t = r + 8;

        Psi = get_Psi(0,last,0); /* Find Psi between 0 and
                                    specified maximum    */
        Psi = get_Psi (1,20,Psi); /* Find Psi to 0.1      */
        Psi = get-Psi (2,20,Psi); /* Find Psi to 0.01     */
        Psi = get_Psi (3,20,Psi); /* Find Psi to 0.001    */

      printf ("Minimum L1 at Psi=%.3f, error=%.8f\n",Psi,min);
```

```
    printf ("Smallest error: %.8f,largest: %.8f\n",inf,sup);
    printf ("Rho = %f\n", Rho);
    printf ("At best psi, volume of BVN = %f, volume of GLD = %f\n",
            best_bvn_vol, best_gld_vol);

    E[t][1] = Rho; /* Rho for BVN */
    E[t][2] = Psi; /* Psi value */
    E[t][3] = min; /* Smallest L1 volume */
    E[t][4] = sup; /* Largest abs. difference between p.d.f.s */
    E[t][5] = inf; /* Smallest abs. difference between p.d.f.s */
    E[t][6] = best_bvn_vol;    /* BVN vol at optimal Psi */
    E[t][7] = best_gld_vol;    /* GLD-2 volume at optimal Psi */
    E[t][8] = best_min_ratio;  /* smallest ratio of p.d.f.'s */
    E[t][9] = best_max_ratio;  /* largest ratio of p.d.f.'s */
  }
  printf ("Rho   Psi   err vol max err min error");
  printf (" min ratio max ratio   BVN vol   GLD vol\n");
  printf ("---- ----- ------- ------- ----------");
  printf ("-------- ---------- ------- -------\n");

  for (t = 0; t < 18; ++t)
     printf("%4.1f%6.3f%.5f%.4f %.8f %.3f %11.3f%8.3f %8.3f\n",
         E[t][1], E[t][2], E[t][3], E[t][4], E[t][5], E[t][8],
         E[t][9], E[t][6], E[t][7]);
  }
```

Appendix B

Table B–1 for GLD Fits: Method of Moments

For specified α_3 and α_4, Table B–1 gives the $\lambda_1(0,1)$, $\lambda_2(0,1)$, λ_3, and λ_4 for GLD fits in the shaded region of Figure 3.3–5. In order to give a sufficient number of significant digits, superscripts are used to designate factors of $\frac{1}{10}$. Thus, an entry of a^s designates $a \times 10^s$. For example, for (α_3, α_4) close to $(0.15, 4.1)$, the table gives

$$(\lambda_1(0,1),\ \lambda_2(0,1),\ \lambda_3,\ \lambda_4) = (-0.7268^1, 0.1603^1, 0.8378^2, 0.9564^2).$$

The proper interpretation of these table entries is

$$(\lambda_1(0,1),\ \lambda_2(0,1),\ \lambda_3,\ \lambda_4) = (-0.07268, 0.01603, 0.008378, 0.009564).$$

With few exceptions, Table B–1 provides values of $\lambda_1, \lambda_2, \lambda_3, \lambda_4$ for which

$$\max_{1 \leq i \leq 4} |\alpha_i - \hat{\alpha}_i| < 10^{-5}.$$

The exceptions occur when very small changes in λ_3 or λ_4 cause large variations in α_3 and α_4 — a situation that arises when λ_3 or λ_4 gets close to 0. When $|\lambda_3| < 10^{-2}$ or $|\lambda_4| < 10^{-2}$, we generally have

$$\max_{1 \leq i \leq 4} |\alpha_i - \hat{\alpha}_i| < 10^{-3}.$$

In the rare instances where $|\lambda_3| < 10^{-4}$ or $|\lambda_4| < 10^{-4}$, we can only be assured of

$$\max_{1 \leq i \leq 4} |\alpha_i - \hat{\alpha}_i| < 10^{-2}.$$

The entries of Table B–1 are from "The Extended Generalized Lambda Distribution (EGLD) System for Fitting Distributions to Data with Moments, II: Tables" by E.J. Dudewicz and Z.A. Karian, *American Journal of Mathematical and Management Sciences*, V. 16, 3 & 4 (1996), pp. 287–307, copyright ©1996 by American Sciences Press, Inc., 20 Cross Road, Syracuse, New York 13224-2104. Reprinted with permission.

	$\hat{\alpha}_3 = 0.00$					$\hat{\alpha}_3 = 0.05$		
$\hat{\alpha}_4$	$\lambda_1(0,1)$	$\lambda_2(0,1)$	λ_3	λ_4	$\hat{\alpha}_4$	$\lambda_1(0,1)$	$\lambda_2(0,1)$	λ_3	λ_4
1.8	0	0.5773	1.0000	1.0000	1.9	-1.4668	0.2991	0.2295^1	0.8558
1.9	0	0.5360	0.7315	0.7315	2.0	-1.2292	0.3122	0.5052^1	0.7602
2.0	0	0.4952	0.5843	0.5843	2.1	-1.0128	0.3230	0.7990^1	0.6698
2.1	0	0.4563	0.4839	0.4839	2.2	-0.8022	0.3314	0.1128	0.5802
2.2	0	0.4197	0.4092	0.4092	2.3	-0.5873	0.3361	0.1502	0.4881
2.3	0	0.3854	0.3507	0.3507	2.4	-0.3748	0.3328	0.1876	0.3941
2.4	0	0.3533	0.3032	0.3032	2.5	-0.2200	0.3164	0.2046	0.3148
2.5	0	0.3232	0.2637	0.2637	2.6	-0.1431	0.2924	0.1973	0.2605
2.6	0	0.2950	0.2303	0.2303	2.7	-0.1046	0.2673	0.1809	0.2212
2.7	0	0.2684	0.2016	0.2016	2.8	-0.8253^1	0.2429	0.1625	0.1903
2.8	0	0.2434	0.1765	0.1765	2.9	-0.6818^1	0.2196	0.1445	0.1645
2.9	0	0.2198	0.1545	0.1545	3.0	-0.5835^1	0.1975	0.1276	0.1425
3.0	0	0.1974	0.1349	0.1349	3.1	-0.5110^1	0.1765	0.1119	0.1232
3.1	0	0.1763	0.1174	0.1174	3.2	-0.4554^1	0.1565	0.9745^1	0.1061
3.2	0	0.1563	0.1016	0.1016	3.3	-0.4115^1	0.1375	0.8412^1	0.9081^1
3.3	0	0.1372	0.8724^1	0.8724^1	3.4	-0.3758^1	0.1194	0.7182^1	0.7700^1
3.4	0	0.1191	0.7418^1	0.7418^1	3.5	-0.3463^1	0.1021	0.6045^1	0.6444^1
3.5	0	0.1018	0.6222^1	0.6222^1	3.6	-0.3215^1	0.8562^1	0.4993^1	0.5297^1
3.6	0	0.8527^1	0.5122^1	0.5122^1	3.7	-0.3003^1	0.6986^1	0.4015^1	0.4243^1
3.7	0	0.6951^1	0.4106^1	0.4106^1	3.8	-0.2820^1	0.5478^1	0.3105^1	0.3270^1
3.8	0	0.5442^1	0.3166^1	0.3166^1	3.9	-0.2661^1	0.4033^1	0.2256^1	0.2369^1
3.9	0	0.3996^1	0.2291^1	0.2291^1	4.0	-0.2518^1	0.2646^1	0.1463^1	0.1531^1
4.0	0	0.2610^1	0.1476^1	0.1476^1	4.1	-0.2395^1	0.1315^1	0.7184^2	0.7504^2
4.1	0	0.1279^1	0.7140^2	0.7140^2	4.2	-0.2284^1	0.3585^3	0.1936^3	0.2018^3
4.2	0	-0.6579^3	-0.3630^3	-0.3630^3	4.3	-0.2183^1	-0.1195^1	-0.6386^2	-0.6643^2
4.3	0	-0.1231^1	-0.6707^2	-0.6707^2	4.4	-0.2091^1	-0.2381^1	-0.1259^1	-0.1308^1
4.4	0	-0.2416^1	-0.1302^1	-0.1302^1	4.5	-0.2010^1	-0.3523^1	-0.1845^1	-0.1913^1
4.5	0	-0.3558^1	-0.1897^1	-0.1897^1	4.6	-0.1935^1	-0.4625^1	-0.2400^1	-0.2485^1
4.6	0	-0.4660^1	-0.2460^1	-0.2460^1	4.7	-0.1867^1	-0.5689^1	-0.2926^1	-0.3026^1
4.7	0	-0.5723^1	-0.2993^1	-0.2993^1	4.8	-0.1803^1	-0.6717^1	-0.3425^1	-0.3538^1
4.9	0	-0.7744^1	-0.3978^1	-0.3978^1	5.0	-0.1691^1	-0.8672^1	-0.4351^1	-0.4485^1
5.1	0	-0.9636^1	-0.4868^1	-0.4868^1	5.2	-0.1593^1	-0.1050	-0.5193^1	-0.5343^1
5.3	0	-0.1141	-0.5678^1	-0.5678^1	5.4	-0.1509^1	-0.1223	-0.5961^1	-0.6124^1
5.5	0	-0.1308	-0.6419^1	-0.6419^1	5.6	-0.1434^1	-0.1385	-0.6666^1	-0.6839^1
5.7	0	-0.1466	-0.7098^1	-0.7098^1	5.8	-0.1368^1	-0.1538	-0.7315^1	-0.7495^1
5.9	0	-0.1614	-0.7725^1	-0.7725^1	6.0	-0.1309^1	-0.1683	-0.7914^1	-0.8101^1
6.1	0	-0.1755	-0.8304^1	-0.8304^1	6.2	-0.1256^1	-0.1820	-0.8471^1	-0.8662^1
6.3	0	-0.1889	-0.8842^1	-0.8842^1	6.4	-0.1208^1	-0.1950	-0.8988^1	-0.9183^1
6.5	0	-0.2016	-0.9342^1	-0.9342^1	6.6	-0.1165^1	-0.2074	-0.9470^1	-0.9668^1
6.7	0	-0.2137	-0.9809^1	-0.9809^1	6.8	-0.1125^1	-0.2192	-0.9921^1	-0.1012
7.0	0	-0.2307	-0.1045	-0.1045	7.1	-0.1072^1	-0.2359	-0.1055	-0.1075
7.3	0	-0.2466	-0.1104	-0.1104	7.4	-0.1024^1	-0.2515	-0.1112	-0.1132
7.6	0	-0.2615	-0.1158	-0.1158	7.7	-0.9820^2	-0.2661	-0.1164	-0.1184
7.9	0	-0.2755	-0.1207	-0.1207	8.0	-0.9442^2	-0.2798	-0.1212	-0.1232
8.2	0	-0.2887	-0.1253	-0.1253	8.3	-0.9101^2	-0.2927	-0.1256	-0.1277
8.7	0	-0.3089	-0.1321	-0.1321	8.8	-0.8600^2	-0.3126	-0.1323	-0.1343
9.2	0	-0.3274	-0.1382	-0.1382	9.3	-0.8169^2	-0.3307	-0.1383	-0.1403
9.7	0	-0.3443	-0.1436	-0.1436	9.8	-0.7795^2	-0.3474	-0.1436	-0.1456
10.2	0	-0.3599	-0.1485	-0.1485	10.3	-0.7465^2	-0.3627	-0.1484	-0.1503
10.7	0	-0.3743	-0.1529	-0.1529	10.8	-0.7173^2	-0.3768	-0.1527	-0.1547
11.5	0	-0.3951	-0.1592	-0.1592	11.6	-0.6769^2	-0.3974	-0.1589	-0.1608
12.3	0	-0.4138	-0.1646	-0.1646	12.4	-0.6428^2	-0.4158	-0.1643	-0.1661
13.1	0	-0.4305	-0.1694	-0.1694	13.2	-0.6134^2	-0.4323	-0.1690	-0.1708
13.9	0	-0.4457	-0.1736	-0.1736	14.0	-0.5879^2	-0.4473	-0.1732	-0.1750
14.7	0	-0.4595	-0.1774	-0.1774	14.8	-0.5656^2	-0.4610	-0.1769	-0.1787

	$\hat{\alpha}_3 = 0.10$					$\hat{\alpha}_3 = 0.15$			
$\hat{\alpha}_4$	$\lambda_1(0,1)$	$\lambda_2(0,1)$	λ_3	λ_4	$\hat{\alpha}_4$	$\lambda_1(0,1)$	$\lambda_2(0,1)$	λ_3	λ_4
1.9	-1.5049	0.2911	0.1562^1	0.8296	1.9	-1.5589	0.2826	0.7231^2	0.8107
2.0	-1.2708	0.3028	0.4121^1	0.7373	2.0	-1.3231	0.2934	0.3145^1	0.7203
2.1	-1.0645	0.3117	0.6711^1	0.6520	2.1	-1.1221	0.3010	0.5487^1	0.6387
2.2	-0.8721	0.3177	0.9412^1	0.5700	2.2	-0.9405	0.3056	0.7814^1	0.5622
2.3	-0.6878	0.3199	0.1219	0.4897	2.3	-0.7722	0.3066	0.1009	0.4891
2.4	-0.5150	0.3163	0.1477	0.4116	2.4	-0.6175	0.3031	0.1215	0.4194
2.5	-0.3700	0.3049	0.1646	0.3407	2.5	-0.4830	0.2939	0.1366	0.3553
2.6	-0.2687	0.2863	0.1678	0.2831	2.6	-0.3760	0.2791	0.1435	0.2994
2.7	-0.2047	0.2644	0.1607	0.2385	2.7	-0.2984	0.2604	0.1421	0.2533
2.8	-0.1640	0.2417	0.1486	0.2033	2.8	-0.2436	0.2398	0.1350	0.2156
2.9	-0.1366	0.2193	0.1346	0.1746	2.9	-0.2049	0.2187	0.1249	0.1846
3.0	-0.1170	0.1977	0.1205	0.1503	3.0	-0.1766	0.1980	0.1135	0.1586
3.1	-0.1026	0.1770	0.1067	0.1294	3.1	-0.1550	0.1778	0.1017	0.1362
3.2	-0.9153^1	0.1572	0.9365^1	0.1111	3.2	-0.1384	0.1584	0.9006^1	0.1167
3.3	-0.8271^1	0.1383	0.8135^1	0.9490^1	3.3	-0.1251	0.1397	0.7885^1	0.9958^1
3.4	-0.7555^1	0.1203	0.6984^1	0.8034^1	3.4	-0.1143	0.1219	0.6815^1	0.8428^1
3.5	-0.6961^1	0.1031	0.5908^1	0.6719^1	3.5	-0.1053	0.1048	0.5803^1	0.7052^1
3.6	-0.6461^1	0.8666^1	0.4903^1	0.5523^1	3.6	-0.9772^1	0.8838^1	0.4848^1	0.5806^1
3.7	-0.6034^1	0.7092^1	0.3964^1	0.4429^1	3.7	-0.9124^1	0.7268^1	0.3947^1	0.4669^1
3.8	-0.5665^1	0.5585^1	0.3085^1	0.3422^1	3.8	-0.8564^1	0.5764^1	0.3100^1	0.3627^1
3.9	-0.5346^1	0.4141^1	0.2261^1	0.2493^1	3.9	-0.8082^1	0.4320^1	0.2301^1	0.2667^1
4.0	-0.5058^1	0.2755^1	0.1488^1	0.1631^1	4.0	-0.7649^1	0.2935^1	0.1548^1	0.1780^1
4.1	-0.4811^1	0.1423^1	0.7606^2	0.8302^2	4.1	-0.7268^1	0.1603^1	0.8378^2	0.9564^2
4.2	-0.4588^1	0.1432^2	0.7577^3	0.8235^3	4.2	-0.6927^1	0.3218^2	0.1667^2	0.1891^2
4.3	-0.4385^1	-0.1089^1	-0.5703^2	-0.6175^2	4.3	-0.6618^1	-0.9116^2	-0.4682^2	-0.5279^2
4.4	-0.4202^1	-0.2275^1	-0.1181^1	-0.1274^1	4.4	-0.6344^1	-0.2099^1	-0.1069^1	-0.1199^1
4.5	-0.4036^1	-0.3418^1	-0.1758^1	-0.1891^1	4.5	-0.6093^1	-0.3245^1	-0.1639^1	-0.1830^1
4.6	-0.3883^1	-0.4522^1	-0.2306^1	-0.2474^1	4.6	-0.5863^1	-0.4350^1	-0.2180^1	-0.2424^1
4.7	-0.3746^1	-0.5587^1	-0.2826^1	-0.3023^1	4.7	-0.5653^1	-0.5417^1	-0.2694^1	-0.2984^1
4.8	-0.3619^1	-0.6616^1	-0.3320^1	-0.3543^1	4.8	-0.5460^1	-0.6448^1	-0.3184^1	-0.3513^1
5.0	-0.3392^1	-0.8574^1	-0.4238^1	-0.4504^1	5.0	-0.5116^1	-0.8410^1	-0.4095^1	-0.4489^1
5.2	-0.3197^1	-0.1041	-0.5074^1	-0.5373^1	5.2	-0.4820^1	-0.1025	-0.4927^1	-0.5371^1
5.4	-0.3027^1	-0.1213	-0.5838^1	-0.6162^1	5.4	-0.4562^1	-0.1198	-0.5688^1	-0.6171^1
5.6	-0.2877^1	-0.1376	-0.6540^1	-0.6884^1	5.6	-0.4336^1	-0.1361	-0.6388^1	-0.6902^1
5.8	-0.2744^1	-0.1529	-0.7187^1	-0.7547^1	5.8	-0.4135^1	-0.1514	-0.7033^1	-0.7571^1
6.0	-0.2625^1	-0.1674	-0.7785^1	-0.8157^1	6.0	-0.3955^1	-0.1660	-0.7631^1	-0.8188^1
6.2	-0.2519^1	-0.1812	-0.8340^1	-0.8722^1	6.2	-0.3794^1	-0.1798	-0.8186^1	-0.8758^1
6.4	-0.2423^1	-0.1942	-0.8857^1	-0.9246^1	6.4	-0.3649^1	-0.1928	-0.8704^1	-0.9286^1
6.6	-0.2335^1	-0.2066	-0.9339^1	-0.9734^1	6.6	-0.3517^1	-0.2053	-0.9187^1	-0.9778^1
6.8	-0.2255^1	-0.2184	-0.9790^1	-0.1019	6.8	-0.3396^1	-0.2171	-0.9639^1	-0.1024
7.1	-0.2148^1	-0.2351	-0.1042	-0.1082	7.1	-0.3234^1	-0.2339	-0.1027	-0.1087
7.4	-0.2053^1	-0.2507	-0.1099	-0.1139	7.4	-0.3090^1	-0.2495	-0.1084	-0.1145
7.7	-0.1968^1	-0.2653	-0.1151	-0.1192	7.7	-0.2962^1	-0.2641	-0.1136	-0.1197
8.0	-0.1892^1	-0.2791	-0.1199	-0.1240	8.0	-0.2848^1	-0.2779	-0.1185	-0.1246
8.3	-0.1824^1	-0.2920	-0.1244	-0.1284	8.3	-0.2744^1	-0.2908	-0.1229	-0.1290
8.8	-0.1723^1	-0.3119	-0.1311	-0.1351	8.8	-0.2593^1	-0.3108	-0.1297	-0.1357
9.3	-0.1637^1	-0.3301	-0.1370	-0.1411	9.3	-0.2462^1	-0.3290	-0.1357	-0.1417
9.8	-0.1562^1	-0.3467	-0.1424	-0.1464	9.8	-0.2349^1	-0.3457	-0.1411	-0.1470
10.3	-0.1495^1	-0.3621	-0.1472	-0.1511	10.3	-0.2249^1	-0.3611	-0.1459	-0.1518
10.8	-0.1437^1	-0.3763	-0.1516	-0.1554	10.8	-0.2161^1	-0.3753	-0.1503	-0.1561
11.6	-0.1356^1	-0.3969	-0.1578	-0.1616	11.6	-0.2039^1	-0.3959	-0.1565	-0.1622
12.4	-0.1287^1	-0.4153	-0.1632	-0.1669	12.4	-0.1936^1	-0.4144	-0.1620	-0.1676
13.2	-0.1229^1	-0.4318	-0.1679	-0.1716	13.2	-0.1847^1	-0.4310	-0.1668	-0.1722
14.0	-0.1177^1	-0.4468	-0.1721	-0.1757	14.0	-0.1770^1	-0.4460	-0.1710	-0.1764
14.8	-0.1133^1	-0.4605	-0.1759	-0.1794	14.8	-0.1703^1	-0.4597	-0.1748	-0.1801

	$\hat{\alpha}_3 = 0.20$					$\hat{\alpha}_3 = 0.25$			
$\hat{\alpha}_4$	$\lambda_1(0,1)$	$\lambda_2(0,1)$	λ_3	λ_4	$\hat{\alpha}_4$	$\lambda_1(0,1)$	$\lambda_2(0,1)$	λ_3	λ_4
2.0	-1.3871	0.2841	0.2124^1	0.7090	2.0	-1.4649	0.2748	0.1050^1	0.7034
2.1	-1.1865	0.2909	0.4302^1	0.6298	2.1	-1.2592	0.2814	0.3133^1	0.6252
2.2	-1.0104	0.2947	0.6384^1	0.5571	2.2	-1.0840	0.2847	0.5060^1	0.5548
2.3	-0.8509	0.2952	0.8359^1	0.4888	2.3	-0.9287	0.2850	0.6845^1	0.4898
2.4	-0.7062	0.2919	0.1013	0.4246	2.4	-0.7893	0.2820	0.8436^1	0.4294
2.5	-0.5783	0.2841	0.1151	0.3652	2.5	-0.6654	0.2753	0.9730^1	0.3734
2.6	-0.4710	0.2718	0.1233	0.3120	2.6	-0.5584	0.2649	0.1062	0.3226
2.7	-0.3862	0.2558	0.1253	0.2661	2.7	-0.4696	0.2512	0.1103	0.2776
2.8	-0.3217	0.2374	0.1221	0.2273	2.8	-0.3983	0.2349	0.1099	0.2385
2.9	-0.2734	0.2180	0.1153	0.1948	2.9	-0.3423	0.2172	0.1059	0.2050
3.0	-0.2368	0.1983	0.1065	0.1672	3.0	-0.2985	0.1987	0.9957^1	0.1763
3.1	-0.2087	0.1789	0.9673^1	0.1435	3.1	-0.2640	0.1802	0.9173^1	0.1515
3.2	-0.1865	0.1599	0.8659^1	0.1230	3.2	-0.2364	0.1618	0.8313^1	0.1300
3.3	-0.1687	0.1416	0.7651^1	0.1049	3.3	-0.2140	0.1440	0.7423^1	0.1110
3.4	-0.1542	0.1240	0.6668^1	0.8888^1	3.4	-0.1956	0.1266	0.6532^1	0.9419^1
3.5	-0.1420	0.1070	0.5722^1	0.7449^1	3.5	-0.1802	0.1099	0.5658^1	0.7916^1
3.6	-0.1318	0.9075^1	0.4819^1	0.6150^1	3.6	-0.1672	0.9375^1	0.4810^1	0.6560^1
3.7	-0.1230	0.7512^1	0.3960^1	0.4968^1	3.7	-0.1560	0.7821^1	0.3995^1	0.5331^1
3.8	-0.1154	0.6012^1	0.3145^1	0.3888^1	3.8	-0.1463	0.6326^1	0.3214^1	0.4210^1
3.9	-0.1088	0.4570^1	0.2372^1	0.2896^1	3.9	-0.1379	0.4888^1	0.2469^1	0.3182^1
4.0	-0.1030	0.3185^1	0.1640^1	0.1980^1	4.0	-0.1305	0.3504^1	0.1760^1	0.2235^1
4.1	-0.9786^1	0.1853^1	0.9467^2	0.1132^1	4.1	-0.1239	0.2172^1	0.1084^1	0.1360^1
4.2	-0.9323^1	0.5707^2	0.2894^2	0.3429^2	4.2	-0.1180	0.8889^2	0.4408^2	0.5467^2
4.3	-0.8910^1	-0.6640^2	-0.3342^2	-0.3929^2	4.3	-0.1127	-0.3477^2	-0.1713^2	-0.2104^2
4.4	-0.8535^1	-0.1854^1	-0.9261^2	-0.1081^1	4.4	-0.1080	-0.1541^1	-0.7542^2	-0.9178^2
4.5	-0.8195^1	-0.3002^1	-0.1489^1	-0.1727^1	4.5	-0.1036	-0.2691^1	-0.1309^1	-0.1580^1
4.6	-0.7886^1	-0.4110^1	-0.2024^1	-0.2334^1	4.6	-0.9964^1	-0.3801^1	-0.1838^1	-0.2202^1
4.7	-0.7600^1	-0.5179^1	-0.2533^1	-0.2906^1	4.7	-0.9602^1	-0.4875^1	-0.2343^1	-0.2787^1
4.8	-0.7339^1	-0.6213^1	-0.3019^1	-0.3445^1	4.8	-0.9270^1	-0.5912^1	-0.2825^1	-0.3340^1
4.9	-0.7097^1	-0.7213^1	-0.3482^1	-0.3956^1	4.9	-0.8962^1	-0.6916^1	-0.3285^1	-0.3861^1
5.1	-0.6668^1	-0.9118^1	-0.4347^1	-0.4900^1	5.1	-0.8415^1	-0.8828^1	-0.4147^1	-0.4824^1
5.3	-0.6294^1	-0.1091	-0.5139^1	-0.5753^1	5.3	-0.7941^1	-0.1062	-0.4937^1	-0.5693^1
5.5	-0.5968^1	-0.1259	-0.5867^1	-0.6528^1	5.5	-0.7526^1	-0.1231	-0.5665^1	-0.6482^1
5.7	-0.5680^1	-0.1418	-0.6539^1	-0.7237^1	5.7	-0.7161^1	-0.1391	-0.6336^1	-0.7202^1
5.9	-0.5424^1	-0.1568	-0.7160^1	-0.7888^1	5.9	-0.6836^1	-0.1541	-0.6959^1	-0.7862^1
6.1	-0.5195^1	-0.1710	-0.7737^1	-0.8487^1	6.1	-0.6545^1	-0.1684	-0.7537^1	-0.8470^1
6.3	-0.4989^1	-0.1844	-0.8274^1	-0.9041^1	6.3	-0.6284^1	-0.1819	-0.8075^1	-0.9031^1
6.5	-0.4802^1	-0.1972	-0.8775^1	-0.9556^1	6.5	-0.6047^1	-0.1948	-0.8578^1	-0.9552^1
6.7	-0.4632^1	-0.2094	-0.9243^1	-0.1004	6.7	-0.5832^1	-0.2070	-0.9049^1	-0.1004
6.9	-0.4477^1	-0.2210	-0.9683^1	-0.1048	6.9	-0.5635^1	-0.2187	-0.9491^1	-0.1049
7.2	-0.4267^1	-0.2374	-0.1029	-0.1110	7.2	-0.5370^1	-0.2351	-0.1010	-0.1111
7.5	-0.4082^1	-0.2528	-0.1085	-0.1166	7.5	-0.5135^1	-0.2506	-0.1067	-0.1168
7.8	-0.3916^1	-0.2672	-0.1137	-0.1218	7.8	-0.4925^1	-0.2650	-0.1118	-0.1220
8.1	-0.3767^1	-0.2807	-0.1184	-0.1265	8.1	-0.4737^1	-0.2786	-0.1166	-0.1268
8.4	-0.3633^1	-0.2934	-0.1228	-0.1309	8.4	-0.4567^1	-0.2914	-0.1210	-0.1312
8.9	-0.3435^1	-0.3131	-0.1294	-0.1375	8.9	-0.4318^1	-0.3111	-0.1277	-0.1378
9.4	-0.3265^1	-0.3310	-0.1353	-0.1433	9.4	-0.4103^1	-0.3291	-0.1337	-0.1437
9.9	-0.3116^1	-0.3475	-0.1406	-0.1485	9.9	-0.3915^1	-0.3457	-0.1390	-0.1489
10.4	-0.2986^1	-0.3627	-0.1454	-0.1532	10.4	-0.3751^1	-0.3609	-0.1438	-0.1536
10.9	-0.2870^1	-0.3767	-0.1497	-0.1575	10.9	-0.3605^1	-0.3750	-0.1482	-0.1579
11.7	-0.2710^1	-0.3971	-0.1559	-0.1635	11.7	-0.3403^1	-0.3954	-0.1544	-0.1639
12.5	-0.2574^1	-0.4153	-0.1613	-0.1687	12.5	-0.3232^1	-0.4138	-0.1599	-0.1692
13.3	-0.2457^1	-0.4317	-0.1660	-0.1733	13.3	-0.3085^1	-0.4302	-0.1647	-0.1738
14.1	-0.2356^1	-0.4466	-0.1703	-0.1774	14.1	-0.2957^1	-0.4452	-0.1689	-0.1779
14.9	-0.2267^1	-0.4602	-0.1740	-0.1811	14.9	-0.2845^1	-0.4588	-0.1727	-0.1816

$$\hat{\alpha}_3 = 0.30$$

$\hat{\alpha}_4$	$\lambda_1(0,1)$	$\lambda_2(0,1)$	λ_3	λ_4
2.0	-1.5501	0.2660	-0.1001^4	0.7019
2.1	-1.3428	0.2721	0.1956^1	0.6249
2.2	-1.1637	0.2755	0.3797^1	0.5556
2.3	-1.0087	0.2759	0.5467^1	0.4927
2.4	-0.8712	0.2732	0.6949^1	0.4348
2.5	-0.7491	0.2675	0.8187^1	0.3813
2.6	-0.6421	0.2586	0.9108^1	0.3324
2.7	-0.5505	0.2467	0.9658^1	0.2884
2.8	-0.4743	0.2323	0.9830^1	0.2495
2.9	-0.4121	0.2162	0.9671^1	0.2155
3.0	-0.3618	0.1991	0.9255^1	0.1859
3.1	-0.3213	0.1816	0.8661^1	0.1602
3.2	-0.2884	0.1640	0.7957^1	0.1377
3.3	-0.2614	0.1467	0.7191^1	0.1179
3.4	-0.2390	0.1298	0.6398^1	0.1003
3.5	-0.2202	0.1133	0.5600^1	0.8457^1
3.6	-0.2042	0.9733^1	0.4813^1	0.7043^1
3.7	-0.1905	0.8192^1	0.4046^1	0.5763^1
3.8	-0.1786	0.6705^1	0.3303^1	0.4597^1
3.9	-0.1682	0.5272^1	0.2588^1	0.3530^1
4.0	-0.1591	0.3891^1	0.1902^1	0.2549^1
4.1	-0.1510	0.2559^1	0.1245^1	0.1643^1
4.2	-0.1438	0.1275^1	0.6176^2	0.8035^2
4.3	-0.1381	0.7885^3	0.3799^3	0.4890^3
4.4	-0.1314	-0.1158^1	-0.5555^2	-0.7058^2
4.5	-0.1260	-0.2312^1	-0.1103^1	-0.1387^1
4.6	-0.1212	-0.3427^1	-0.1626^1	-0.2026^1
4.7	-0.1167	-0.4504^1	-0.2126^1	-0.2627^1
4.8	-0.1127	-0.5545^1	-0.2604^1	-0.3193^1
4.9	-0.1089	-0.6553^1	-0.3062^1	-0.3728^1
5.1	-0.1022	-0.8474^1	-0.3920^1	-0.4713^1
5.3	-0.9637^1	-0.1028	-0.4709^1	-0.5600^1
5.5	-0.9129^1	-0.1198	-0.5437^1	-0.6404^1
5.7	-0.8682^1	-0.1358	-0.6110^1	-0.7137^1
5.9	-0.8285^1	-0.1509	-0.6734^1	-0.7809^1
6.1	-0.7930^1	-0.1653	-0.7314^1	-0.8426^1
6.3	-0.7611^1	-0.1789	-0.7855^1	-0.8997^1
6.5	-0.7322^1	-0.1918	-0.8361^1	-0.9525^1
6.7	-0.7059^1	-0.2041	-0.8834^1	-0.1002
6.9	-0.6819^1	-0.2158	-0.9279^1	-0.1047
7.2	-0.6496^1	-0.2324	-0.9897^1	-0.1111
7.5	-0.6210^1	-0.2479	-0.1046	-0.1168
7.8	-0.5955^1	-0.2624	-0.1098	-0.1221
8.1	-0.5726^1	-0.2760	-0.1147	-0.1269
8.4	-0.5519^1	-0.2889	-0.1191	-0.1313
8.9	-0.5216^1	-0.3087	-0.1258	-0.1380
9.4	-0.4955^1	-0.3269	-0.1319	-0.1439
9.9	-0.4728^1	-0.3435	-0.1372	-0.1492
10.4	-0.4528^1	-0.3588	-0.1421	-0.1539
10.9	-0.4351^1	-0.3729	-0.1465	-0.1582
11.7	-0.4106^1	-0.3935	-0.1528	-0.1643
12.5	-0.3899^1	-0.4118	-0.1583	-0.1695
13.3	-0.3721^1	-0.4284	-0.1632	-0.1742
14.1	-0.3566^1	-0.4434	-0.1675	-0.1783
14.9	-0.3431^1	-0.4570	-0.1713	-0.1819

$$\hat{\alpha}_3 = 0.35$$

$\hat{\alpha}_4$	$\lambda_1(0,1)$	$\lambda_2(0,1)$	λ_3	λ_4
2.1	-1.4414	0.2630	0.7342^2	0.6297
2.2	-1.2523	0.2667	0.2556^1	0.5599
2.3	-1.0934	0.2674	0.4168^1	0.4979
2.4	-0.9548	0.2653	0.5590^1	0.4415
2.5	-0.8324	0.2604	0.6797^1	0.3897
2.6	-0.7245	0.2527	0.7745^1	0.3423
2.7	-0.6307	0.2424	0.8394^1	0.2992
2.8	-0.5504	0.2297	0.8727^1	0.2606
2.9	-0.4830	0.2153	0.8761^1	0.2263
3.0	-0.4271	0.1996	0.8543^1	0.1961
3.1	-0.3809	0.1832	0.8131^1	0.1695
3.2	-0.3428	0.1665	0.7581^1	0.1462
3.3	-0.3112	0.1498	0.6944^1	0.1255
3.4	-0.2847	0.1333	0.6254^1	0.1072
3.5	-0.2623	0.1171	0.5540^1	0.9077^1
3.6	-0.2432	0.1014	0.4819^1	0.7602^1
3.7	-0.2268	0.8619^1	0.4104^1	0.6267^1
3.8	-0.2126	0.7144^1	0.3403^1	0.5053^1
3.9	-0.2001	0.5718^1	0.2722^1	0.3943^1
4.0	-0.1892	0.4341^1	0.2062^1	0.2925^1
4.1	-0.1794	0.3011^1	0.1427^1	0.1986^1
4.2	-0.1707	0.1727^1	0.8158^2	0.1116^1
4.3	-0.1629	0.4870^2	0.2292^2	0.3090^2
4.4	-0.1559	-0.7105^2	-0.3332^2	-0.4431^2
4.5	-0.1495	-0.1868^1	-0.8725^2	-0.1146^1
4.6	-0.1437	-0.2986^1	-0.1389^1	-0.1804^1
4.7	-0.1383	-0.4067^1	-0.1884^1	-0.2423^1
4.8	-0.1334	-0.5113^1	-0.2358^1	-0.3004^1
4.9	-0.1289	-0.6125^1	-0.2813^1	-0.3553^1
5.0	-0.1247	-0.7106^1	-0.3249^1	-0.4072^1
5.2	-0.1173	-0.8977^1	-0.4070^1	-0.5029^1
5.4	-0.1108	-0.1074	-0.4827^1	-0.5893^1
5.6	-0.1051	-0.1240	-0.5528^1	-0.6677^1
5.8	-0.1001	-0.1396	-0.6178^1	-0.7392^1
6.0	-0.9563^1	-0.1545	-0.6782^1	-0.8048^1
6.2	-0.9162^1	-0.1685	-0.7345^1	-0.8652^1
6.4	-0.8801^1	-0.1818	-0.7871^1	-0.9210^1
6.6	-0.8473^1	-0.1945	-0.8364^1	-0.9727^1
6.8	-0.8175^1	-0.2066	-0.8826^1	-0.1021
7.0	-0.7902^1	-0.2181	-0.9261^1	-0.1066
7.3	-0.7534^1	-0.2344	-0.9866^1	-0.1128
7.6	-0.7207^1	-0.2497	-0.1042	-0.1184
7.9	-0.6916^1	-0.2640	-0.1093	-0.1236
8.2	-0.6654^1	-0.2774	-0.1141	-0.1283
8.5	-0.6417^1	-0.2901	-0.1185	-0.1327
9.0	-0.6068^1	-0.3097	-0.1251	-0.1393
9.5	-0.5768^1	-0.3276	-0.1310	-0.1451
10.0	-0.5506^1	-0.3440	-0.1364	-0.1503
10.5	-0.5276^1	-0.3592	-0.1412	-0.1550
11.0	-0.5072^1	-0.3732	-0.1456	-0.1592
11.8	-0.4789^1	-0.3935	-0.1518	-0.1652
12.6	-0.4549^1	-0.4117	-0.1573	-0.1704
13.4	-0.4343^1	-0.4282	-0.1621	-0.1750
14.2	-0.4164^1	-0.4430	-0.1664	-0.1790
15.0	-0.4007^1	-0.4566	-0.1703	-0.1827

	$\hat{\alpha}_3 = 0.40$					$\hat{\alpha}_3 = 0.45$			
$\hat{\alpha}_4$	$\lambda_1(0,1)$	$\lambda_2(0,1)$	λ_3	λ_4	$\hat{\alpha}_4$	$\lambda_1(0,1)$	$\lambda_2(0,1)$	λ_3	λ_4
2.2	-1.3537	0.2583	0.1295^{1}	0.5683	2.3	-1.2899	0.2519	0.1629^{1}	0.5173
2.3	-1.1858	0.2595	0.2904^{1}	0.5058	2.4	-1.1383	0.2511	0.3051^{1}	0.4608
2.4	-1.0429	0.2580	0.4305^{1}	0.4500	2.5	-1.0083	0.2479	0.4266^{1}	0.4101
2.5	-0.9180	0.2539	0.5504^{1}	0.3991	2.6	-0.8946	0.2423	0.5277^{1}	0.3641
2.6	-0.8081	0.2473	0.6481^{1}	0.3527	2.7	-0.7947	0.2346	0.6070^{1}	0.3221
2.7	-0.7115	0.2384	0.7207^{1}	0.3103	2.8	-0.7071	0.2248	0.6633^{1}	0.2840
2.8	-0.6277	0.2272	0.7666^{1}	0.2720	2.9	-0.6308	0.2133	0.6963^{1}	0.2495
2.9	-0.5557	0.2143	0.7861^{1}	0.2376	3.0	-0.5650	0.2004	0.7072^{1}	0.2184
3.0	-0.4947	0.2000	0.7817^{1}	0.2069	3.1	-0.5086	0.1863	0.6986^{1}	0.1906
3.1	-0.4433	0.1848	0.7574^{1}	0.1797	3.2	-0.4605	0.1716	0.6737^{1}	0.1657
3.2	-0.4001	0.1690	0.7177^{1}	0.1555	3.3	-0.4196	0.1565	0.6364^{1}	0.1435
3.3	-0.3638	0.1530	0.6671^{1}	0.1341	3.4	-0.3846	0.1412	0.5898^{1}	0.1236
3.4	-0.3331	0.1371	0.6091^{1}	0.1149	3.5	-0.3546	0.1260	0.5370^{1}	0.1058
3.5	-0.3070	0.1214	0.5467^{1}	0.9782^{1}	3.6	-0.3288	0.1110	0.4801^{1}	0.8969^{1}
3.6	-0.2846	0.1060	0.4818^{1}	0.8242^{1}	3.7	-0.3065	0.9623^{1}	0.4209^{1}	0.7514^{1}
3.7	-0.2654	0.9099^{1}	0.4162^{1}	0.6849^{1}	3.8	-0.2870	0.8183^{1}	0.3608^{1}	0.6192^{1}
3.8	-0.2486	0.7639^{1}	0.3508^{1}	0.5583^{1}	3.9	-0.2699	0.6781^{1}	0.3007^{1}	0.4986^{1}
3.9	-0.2339	0.6223^{1}	0.2864^{1}	0.4427^{1}	4.0	-0.2548	0.5419^{1}	0.2412^{1}	0.3880^{1}
4.0	-0.2210	0.4852^{1}	0.2235^{1}	0.3368^{1}	4.1	-0.2414	0.4098^{1}	0.1828^{1}	0.2864^{1}
4.1	-0.2095	0.3525^{1}	0.1623^{1}	0.2391^{1}	4.2	-0.2294	0.2818^{1}	0.1258^{1}	0.1925^{1}
4.2	-0.1992	0.2242^{1}	0.1031^{1}	0.1489^{1}	4.3	-0.2186	0.1577^{1}	0.7045^{2}	0.1055^{1}
4.3	-0.1900	0.1002^{1}	0.4597^{2}	0.6521^{2}	4.4	-0.2090	0.3763^{2}	0.1680^{2}	0.2470^{2}
4.4	-0.1816	-0.1984^{2}	-0.9089^{3}	-0.1269^{2}	4.5	-0.2001	-0.7867^{2}	-0.3508^{2}	-0.5068^{2}
4.5	-0.1741	-0.1358^{1}	-0.6204^{2}	-0.8534^{2}	4.6	-0.1921	-0.1912^{1}	-0.8511^{2}	-0.1211^{1}
4.6	-0.1672	-0.2480^{1}	-0.1130^{1}	-0.1533^{1}	4.7	-0.1848	-0.3002^{1}	-0.1334^{1}	-0.1871^{1}
4.7	-0.1610	-0.3566^{1}	-0.1619^{1}	-0.2171^{1}	4.8	-0.1781	-0.4058^{1}	-0.1799^{1}	-0.2491^{1}
4.8	-0.1552	-0.4617^{1}	-0.2089^{1}	-0.2771^{1}	4.9	-0.1719	-0.5081^{1}	-0.2247^{1}	-0.3075^{1}
4.9	-0.1499	-0.5634^{1}	-0.2541^{1}	-0.3336^{1}	5.0	-0.1662	-0.6072^{1}	-0.2678^{1}	-0.3625^{1}
5.0	-0.1450	-0.6620^{1}	-0.2975^{1}	-0.3870^{1}	5.1	-0.1609	-0.7034^{1}	-0.3094^{1}	-0.4145^{1}
5.1	-0.1404	-0.7576^{1}	-0.3392^{1}	-0.4375^{1}	5.2	-0.1560	-0.7966^{1}	-0.3494^{1}	-0.4638^{1}
5.3	-0.1322	-0.9401^{1}	-0.4179^{1}	-0.5307^{1}	5.4	-0.1472	-0.9750^{1}	-0.4251^{1}	-0.5549^{1}
5.5	-0.1251	-0.1112	-0.4908^{1}	-0.6150^{1}	5.6	-0.1394	-0.1143	-0.4955^{1}	-0.6373^{1}
5.7	-0.1189	-0.1274	-0.5585^{1}	-0.6916^{1}	5.8	-0.1326	-0.1302	-0.5610^{1}	-0.7124^{1}
5.9	-0.1133	-0.1428	-0.6214^{1}	-0.7616^{1}	6.0	-0.1265	-0.1453	-0.6221^{1}	-0.7810^{1}
6.1	-0.1084	-0.1573	-0.6801^{1}	-0.8258^{1}	6.2	-0.1211	-0.1595	-0.6791^{1}	-0.8440^{1}
6.3	-0.1039	-0.1711	-0.7349^{1}	-0.8850^{1}	6.4	-0.1162	-0.1731	-0.7325^{1}	-0.9021^{1}
6.5	-0.9989^{1}	-0.1842	-0.7861^{1}	-0.9397^{1}	6.6	-0.1118	-0.1859	-0.7826^{1}	-0.9559^{1}
6.7	-0.9624^{1}	-0.1967	-0.8342^{1}	-0.9905^{1}	6.8	-0.1078	-0.1982	-0.8297^{1}	-0.1006
6.9	-0.9290^{1}	-0.2085	-0.8794^{1}	-0.1038	7.0	-0.1041	-0.2099	-0.8740^{1}	-0.1052
7.1	-0.8985^{1}	-0.2199	-0.9220^{1}	-0.1082	7.2	-0.1007	-0.2211	-0.9157^{1}	-0.1096
7.4	-0.8573^{1}	-0.2359	-0.9814^{1}	-0.1143	7.5	-0.9618^{1}	-0.2369	-0.9741^{1}	-0.1156
7.7	-0.8207^{1}	-0.2510	-0.1036	-0.1198	7.8	-0.9212^{1}	-0.2517	-0.1028	-0.1211
8.0	-0.7879^{1}	-0.2651	-0.1086	-0.1249	8.1	-0.8849^{1}	-0.2657	-0.1077	-0.1261
8.3	-0.7584^{1}	-0.2783	-0.1133	-0.1296	8.4	-0.8522^{1}	-0.2788	-0.1124	-0.1307
8.6	-0.7317^{1}	-0.2909	-0.1176	-0.1339	8.7	-0.8225^{1}	-0.2912	-0.1166	-0.1349
9.1	-0.6925^{1}	-0.3102	-0.1242	-0.1404	9.2	-0.7788^{1}	-0.3103	-0.1231	-0.1413
9.6	-0.6586^{1}	-0.3279	-0.1301	-0.1461	9.7	-0.7411^{1}	-0.3278	-0.1290	-0.1470
10.1	-0.6290^{1}	-0.3442	-0.1354	-0.1513	10.2	-0.7080^{1}	-0.3439	-0.1342	-0.1521
10.6	-0.6029^{1}	-0.3592	-0.1402	-0.1559	10.7	-0.6789^{1}	-0.3588	-0.1390	-0.1567
11.1	-0.5798^{1}	-0.3730	-0.1445	-0.1601	11.2	-0.6531^{1}	-0.3726	-0.1433	-0.1608
11.9	-0.5477^{1}	-0.3932	-0.1507	-0.1660	12.0	-0.6172^{1}	-0.3926	-0.1495	-0.1667
12.7	-0.5205^{1}	-0.4113	-0.1562	-0.1712	12.8	-0.5867^{1}	-0.4106	-0.1550	-0.1718
13.5	-0.4970^{1}	-0.4276	-0.1610	-0.1757	13.6	-0.5605^{1}	-0.4267	-0.1598	-0.1763
14.3	-0.4767^{1}	-0.4424	-0.1653	-0.1797	14.4	-0.5376^{1}	-0.4414	-0.1640	-0.1803
15.1	-0.4588^{1}	-0.4559	-0.1691	-0.1833	15.2	-0.5176^{1}	-0.4548	-0.1679	-0.1839

	$\hat{\alpha}_3 = 0.50$					$\hat{\alpha}_3 = 0.55$			
$\hat{\alpha}_4$	$\lambda_1(0,1)$	$\lambda_2(0,1)$	λ_3	λ_4	$\hat{\alpha}_4$	$\lambda_1(0,1)$	$\lambda_2(0,1)$	λ_3	λ_4
2.3	-1.4124	0.2443	0.2873^2	0.5334	2.4	-1.3697	0.2379	0.4465^2	0.4931
2.4	-1.2450	0.2445	0.1784^1	0.4748	2.5	-1.2153	0.2366	0.1798^1	0.4393
2.5	-1.1060	0.2422	0.3045^1	0.4233	2.6	-1.0862	0.2331	0.2925^1	0.3920
2.6	-0.9864	0.2376	0.4102^1	0.3770	2.7	-0.9748	0.2276	0.3854^1	0.3495
2.7	-0.8818	0.2310	0.4961^1	0.3350	2.8	-0.8774	0.2202	0.4593^1	0.3109
2.8	-0.7898	0.2225	0.5614^1	0.2969	2.9	-0.7918	0.2113	0.5143^1	0.2758
2.9	-0.7091	0.2123	0.6060^1	0.2621	3.0	-0.7165	0.2009	0.5508^1	0.2440
3.0	-0.6387	0.2007	0.6304^1	0.2307	3.1	-0.6504	0.1893	0.5696^1	0.2150
3.1	-0.5775	0.1878	0.6361^1	0.2023	3.2	-0.5927	0.1767	0.5723^1	0.1889
3.2	-0.5246	0.1742	0.6254^1	0.1768	3.3	-0.5424	0.1634	0.5611^1	0.1652
3.3	-0.4790	0.1600	0.6013^1	0.1539	3.4	-0.4986	0.1497	0.5383^1	0.1438
3.4	-0.4396	0.1454	0.5665^1	0.1332	3.5	-0.4605	0.1358	0.5063^1	0.1245
3.5	-0.4057	0.1308	0.5239^1	0.1146	3.6	-0.4272	0.1217	0.4673^1	0.1070
3.6	-0.3762	0.1163	0.4756^1	0.9786^1	3.7	-0.3981	0.1078	0.4233^1	0.9111^1
3.7	-0.3506	0.1019	0.4237^1	0.8266^1	3.8	-0.3725	0.9394^1	0.3757^1	0.7666^1
3.8	-0.3282	0.8771^1	0.3695^1	0.6885^1	3.9	-0.3499	0.8034^1	0.3260^1	0.6347^1
3.9	-0.3084	0.7387^1	0.3142^1	0.5624^1	4.0	-0.3299	0.6702^1	0.2750^1	0.5139^1
4.0	-0.2910	0.6038^1	0.2587^1	0.4469^1	4.1	-0.3121	0.5400^1	0.2236^1	0.4028^1
4.1	-0.2755	0.4726^1	0.2035^1	0.3409^1	4.2	-0.2963	0.4134^1	0.1723^1	0.3006^1
4.2	-0.2617	0.3450^1	0.1491^1	0.2429^1	4.3	-0.2820	0.2900^1	0.1215^1	0.2059^1
4.3	-0.2492	0.2212^1	0.9582^2	0.1523^1	4.4	-0.2691	0.1700^1	0.7150^2	0.1181^1
4.4	-0.2380	0.1011^1	0.4385^2	0.6817^2	4.5	-0.2574	0.5354^2	0.2258^2	0.3644^2
4.5	-0.2278	-0.1539^2	-0.6683^3	-0.1018^2	4.6	-0.2468	-0.5954^2	-0.2515^2	-0.3975^2
4.6	-0.2186	-0.1283^1	-0.5570^2	-0.8335^2	4.7	-0.2371	-0.1693^1	-0.7160^2	-0.1110^1
4.7	-0.2101	-0.2377^1	-0.1032^1	-0.1518^1	4.8	-0.2282	-0.2758^1	-0.1167^1	-0.1778^1
4.8	-0.2023	-0.3438^1	-0.1491^1	-0.2161^1	4.9	-0.2200	-0.3792^1	-0.1604^1	-0.2405^1
4.9	-0.1952	-0.4466^1	-0.1934^1	-0.2765^1	5.0	-0.2124	-0.4795^1	-0.2028^1	-0.2996^1
5.0	-0.1886	-0.5463^1	-0.2362^1	-0.3335^1	5.1	-0.2055	-0.5769^1	-0.2438^1	-0.3553^1
5.1	-0.1825	-0.6431^1	-0.2775^1	-0.3872^1	5.2	-0.1990	-0.6714^1	-0.2835^1	-0.4080^1
5.2	-0.1769	-0.7370^1	-0.3174^1	-0.4381^1	5.3	-0.1930	-0.7632^1	-0.3218^1	-0.4579^1
5.4	-0.1667	-0.9167^1	-0.3931^1	-0.5320^1	5.5	-0.1821	-0.9392^1	-0.3948^1	-0.5501^1
5.6	-0.1578	-0.1086	-0.4635^1	-0.6169^1	5.7	-0.1726	-0.1105	-0.4631^1	-0.6335^1
5.8	-0.1500	-0.1246	-0.5293^1	-0.6941^1	5.9	-0.1642	-0.1263	-0.5269^1	-0.7094^1
6.0	-0.1430	-0.1398	-0.5907^1	-0.7645^1	6.1	-0.1567	-0.1412	-0.5867^1	-0.7787^1
6.2	-0.1368	-0.1542	-0.6482^1	-0.8291^1	6.3	-0.1500	-0.1553	-0.6428^1	-0.8424^1
6.4	-0.1312	-0.1679	-0.7020^1	-0.8886^1	6.5	-0.1440	-0.1688	-0.6954^1	-0.9011^1
6.6	-0.1261	-0.1809	-0.7526^1	-0.9437^1	6.7	-0.1385	-0.1816	-0.7449^1	-0.9554^1
6.8	-0.1216	-0.1932	-0.8001^1	-0.9947^1	6.9	-0.1335	-0.1938	-0.7916^1	-0.1006
7.0	-0.1174	-0.2050	-0.8449^1	-0.1042	7.1	-0.1290	-0.2054	-0.8356^1	-0.1053
7.2	-0.1135	-0.2163	-0.8872^1	-0.1087	7.3	-0.1248	-0.2165	-0.8771^1	-0.1097
7.5	-0.1083	-0.2323	-0.9462^1	-0.1148	7.6	-0.1192	-0.2323	-0.9353^1	-0.1157
7.8	-0.1037	-0.2472	-0.1001	-0.1204	7.9	-0.1141	-0.2471	-0.9891^1	-0.1212
8.1	-0.9957^1	-0.2613	-0.1051	-0.1255	8.2	-0.1096	-0.2610	-0.1039	-0.1263
8.4	-0.9585^1	-0.2745	-0.1098	-0.1301	8.5	-0.1056	-0.2741	-0.1085	-0.1309
8.7	-0.9248^1	-0.2870	-0.1141	-0.1345	8.8	-0.1019	-0.2865	-0.1128	-0.1352
9.2	-0.8752^1	-0.3063	-0.1207	-0.1410	9.3	-0.9647^1	-0.3056	-0.1194	-0.1416
9.7	-0.8324^1	-0.3240	-0.1266	-0.1467	9.8	-0.9178^1	-0.3231	-0.1252	-0.1474
10.2	-0.7950^1	-0.3402	-0.1319	-0.1519	10.3	-0.8768^1	-0.3392	-0.1305	-0.1525
10.7	-0.7620^1	-0.3552	-0.1368	-0.1565	10.8	-0.8407^1	-0.3541	-0.1354	-0.1571
11.2	-0.7328^1	-0.3691	-0.1412	-0.1607	11.3	-0.8086^1	-0.3679	-0.1398	-0.1612
12.0	-0.6922^1	-0.3892	-0.1475	-0.1666	12.1	-0.7641^1	-0.3879	-0.1460	-0.1671
12.8	-0.6578^1	-0.4073	-0.1530	-0.1718	12.9	-0.7263^1	-0.4059	-0.1515	-0.1722
13.6	-0.6282^1	-0.4236	-0.1579	-0.1763	13.7	-0.6938^1	-0.4221	-0.1564	-0.1768
14.4	-0.6024^1	-0.4384	-0.1622	-0.1803	14.5	-0.6654^1	-0.4369	-0.1608	-0.1808
15.2	-0.5798^1	-0.4519	-0.1661	-0.1839	15.3	-0.6406^1	-0.4503	-0.1647	-0.1843

$\hat{\alpha}_3 = 0.60$

$\hat{\alpha}_4$	$\lambda_1(0,1)$	$\lambda_2(0,1)$	λ_3	λ_4
2.5	-1.3428	0.2310	0.4715^2	0.4596
2.6	-1.1981	0.2286	0.1706^1	0.4098
2.7	-1.0766	0.2242	0.2723^1	0.3659
2.8	-0.9718	0.2180	0.3551^1	0.3265
2.9	-0.8801	0.2102	0.4200^1	0.2908
3.0	-0.7993	0.2010	0.4675^1	0.2583
3.1	-0.7281	0.1905	0.4984^1	0.2288
3.2	-0.6654	0.1791	0.5137^1	0.2020
3.3	-0.6103	0.1668	0.5151^1	0.1776
3.4	-0.5619	0.1540	0.5042^1	0.1554
3.5	-0.5195	0.1408	0.4833^1	0.1354
3.6	-0.4822	0.1274	0.4541^1	0.1171
3.7	-0.4493	0.1139	0.4187^1	0.1005
3.8	-0.4204	0.1004	0.3785^1	0.8541^1
3.9	-0.3948	0.8714^1	0.3350^1	0.7160^1
4.0	-0.3720	0.7404^1	0.2893^1	0.5894^1
4.1	-0.3517	0.6119^1	0.2422^1	0.4731^1
4.2	-0.3336	0.4863^1	0.1945^1	0.3660^1
4.3	-0.3173	0.3637^1	0.1467^1	0.2669^1
4.4	-0.3026	0.2442^1	0.9911^2	0.1750^1
4.5	-0.2892	0.1278^1	0.5215^2	0.8965^2
4.6	-0.2770	0.1491^2	0.6111^3	0.1025^2
4.7	-0.2659	-0.9531^2	-0.3916^2	-0.6425^2
4.8	-0.2558	-0.2022^1	-0.8328^2	-0.1339^1
4.9	-0.2464	-0.3061^1	-0.1262^1	-0.1992^1
5.0	-0.2378	-0.4069^1	-0.1680^1	-0.2606^1
5.1	-0.2299	-0.5048^1	-0.2085^1	-0.3186^1
5.2	-0.2225	-0.6001^1	-0.2479^1	-0.3733^1
5.3	-0.2157	-0.6926^1	-0.2861^1	-0.4250^1
5.4	-0.2093	-0.7825^1	-0.3230^1	-0.4740^1
5.6	-0.1977	-0.9549^1	-0.3936^1	-0.5647^1
5.8	-0.1876	-0.1118	-0.4597^1	-0.6469^1
6.0	-0.1786	-0.1273	-0.5218^1	-0.7217^1
6.2	-0.1706	-0.1419	-0.5801^1	-0.7902^1
6.4	-0.1634	-0.1559	-0.6349^1	-0.8531^1
6.6	-0.1569	-0.1691	-0.6865^1	-0.9111^1
6.8	-0.1510	-0.1817	-0.7351^1	-0.9648^1
7.0	-0.1457	-0.1938	-0.7809^1	-0.1015
7.2	-0.1408	-0.2053	-0.8242^1	-0.1061
7.4	-0.1363	-0.2163	-0.8652^1	-0.1105
7.7	-0.1302	-0.2319	-0.9226^1	-0.1165
8.0	-0.1247	-0.2465	-0.9758^1	-0.1219
8.3	-0.1198	-0.2603	-0.1025	-0.1269
8.6	-0.1154	-0.2732	-0.1071	-0.1315
8.9	-0.1115	-0.2855	-0.1114	-0.1358
9.4	-0.1056	-0.3045	-0.1179	-0.1422
9.9	-0.1005	-0.3219	-0.1237	-0.1479
10.4	-0.9601^1	-0.3379	-0.1290	-0.1529
10.9	-0.9208^1	-0.3526	-0.1338	-0.1575
11.4	-0.8859^1	-0.3663	-0.1382	-0.1616
12.2	-0.8373^1	-0.3863	-0.1445	-0.1675
13.0	-0.7961^1	-0.4042	-0.1500	-0.1726
13.8	-0.7605^1	-0.4204	-0.1549	-0.1771
14.6	-0.7296^1	-0.4350	-0.1592	-0.1811
15.4	-0.7024^1	-0.4484	-0.1631	-0.1846

$\hat{\alpha}_3 = 0.65$

$\hat{\alpha}_4$	$\lambda_1(0,1)$	$\lambda_2(0,1)$	λ_3	λ_4
2.6	-1.3292	0.2240	0.3908^2	0.4318
2.7	-1.1915	0.2208	0.1533^1	0.3853
2.8	-1.0759	0.2157	0.2465^1	0.3443
2.9	-0.9761	0.2090	0.3215^1	0.3075
3.0	-0.8887	0.2010	0.3797^1	0.2742
3.1	-0.8117	0.1916	0.4220^1	0.2438
3.2	-0.7436	0.1813	0.4493^1	0.2162
3.3	-0.6834	0.1700	0.4627^1	0.1911
3.4	-0.6302	0.1581	0.4637^1	0.1682
3.5	-0.5832	0.1457	0.4539^1	0.1473
3.6	-0.5417	0.1330	0.4350^1	0.1283
3.7	-0.5049	0.1201	0.4088^1	0.1110
3.8	-0.4723	0.1071	0.3767^1	0.9516^1
3.9	-0.4434	0.9417^1	0.3402^1	0.8068^1
4.0	-0.4177	0.8136^1	0.3004^1	0.6740^1
4.1	-0.3946	0.6873^1	0.2584^1	0.5519^1
4.2	-0.3740	0.5632^1	0.2148^1	0.4395^1
4.3	-0.3555	0.4418^1	0.1706^1	0.3356^1
4.4	-0.3387	0.3229^1	0.1259^1	0.2393^1
4.5	-0.3235	0.2070^1	0.8137^2	0.1499^1
4.6	-0.3097	0.9398^2	0.3719^2	0.6659^2
4.7	-0.2970	-0.1607^2	-0.6394^3	-0.1116^2
4.8	-0.2855	-0.1232^1	-0.4921^2	-0.8391^2
4.9	-0.2749	-0.2274^1	-0.9114^2	-0.1521^1
5.0	-0.2651	-0.3287^1	-0.1321^1	-0.2162^1
5.1	-0.2560	-0.4273^1	-0.1721^1	-0.2766^1
5.2	-0.2477	-0.5231^1	-0.2110^1	-0.3335^1
5.3	-0.2399	-0.6163^1	-0.2488^1	-0.3873^1
5.4	-0.2327	-0.7069^1	-0.2855^1	-0.4383^1
5.5	-0.2259	-0.7950^1	-0.3212^1	-0.4866^1
5.7	-0.2137	-0.9643^1	-0.3895^1	-0.5761^1
5.9	-0.2029	-0.1125	-0.4537^1	-0.6572^1
6.1	-0.1933	-0.1277	-0.5142^1	-0.7312^1
6.3	-0.1848	-0.1421	-0.5711^1	-0.7989^1
6.5	-0.1771	-0.1558	-0.6248^1	-0.8612^1
6.7	-0.1701	-0.1689	-0.6754^1	-0.9187^1
6.9	-0.1638	-0.1814	-0.7231^1	-0.9720^1
7.1	-0.1581	-0.1933	-0.7682^1	-0.1021
7.3	-0.1528	-0.2046	-0.8109^1	-0.1068
7.5	-0.1480	-0.2155	-0.8513^1	-0.1111
7.8	-0.1414	-0.2309	-0.9081^1	-0.1170
8.1	-0.1355	-0.2454	-0.9607^1	-0.1225
8.4	-0.1303	-0.2591	-0.1010	-0.1274
8.7	-0.1255	-0.2720	-0.1055	-0.1320
9.0	-0.1212	-0.2841	-0.1098	-0.1362
9.5	-0.1149	-0.3030	-0.1162	-0.1426
10.0	-0.1093	-0.3202	-0.1221	-0.1482
10.5	-0.1045	-0.3362	-0.1273	-0.1533
11.0	-0.1003	-0.3508	-0.1321	-0.1578
11.5	-0.9646^1	-0.3645	-0.1365	-0.1620
12.3	-0.9120^1	-0.3843	-0.1428	-0.1678
13.1	-0.8672^1	-0.4022	-0.1483	-0.1729
13.9	-0.8286^1	-0.4183	-0.1532	-0.1774
14.7	-0.7950^1	-0.4329	-0.1576	-0.1813
15.5	-0.7655^1	-0.4462	-0.1615	-0.1849

| | $\hat{\alpha}_3 = 0.70$ | | | | | $\hat{\alpha}_3 = 0.75$ | | | |
$\hat{\alpha}_4$	$\lambda_1(0,1)$	$\lambda_2(0,1)$	λ_3	λ_4	$\hat{\alpha}_4$	$\lambda_1(0,1)$	$\lambda_2(0,1)$	λ_3	λ_4
2.7	-1.3272	0.2170	0.2251^2	0.4090	2.9	-1.2057	0.2061	0.1012^1	0.3487
2.8	-1.1943	0.2132	0.1298^1	0.3651	3.0	-1.0968	0.2003	0.1835^1	0.3119
2.9	-1.0829	0.2077	0.2165^1	0.3264	3.1	-1.0031	0.1932	0.2489^1	0.2790
3.0	-0.9867	0.2008	0.2858^1	0.2918	3.2	-0.9212	0.1850	0.2993^1	0.2492
3.1	-0.9025	0.1926	0.3393^1	0.2604	3.3	-0.8488	0.1758	0.3360^1	0.2221
3.2	-0.8283	0.1833	0.3781^1	0.2319	3.4	-0.7846	0.1658	0.3602^1	0.1974
3.3	-0.7625	0.1731	0.4033^1	0.2059	3.5	-0.7275	0.1552	0.3729^1	0.1748
3.4	-0.7041	0.1621	0.4159^1	0.1821	3.6	-0.6765	0.1440	0.3753^1	0.1542
3.5	-0.6523	0.1505	0.4173^1	0.1604	3.7	-0.6310	0.1324	0.3689^1	0.1352
3.6	-0.6062	0.1386	0.4091^1	0.1406	3.8	-0.5904	0.1206	0.3548^1	0.1179
3.7	-0.5653	0.1263	0.3925^1	0.1225	3.9	-0.5541	0.1086	0.3343^1	0.1019
3.8	-0.5288	0.1138	0.3691^1	0.1060	4.0	-0.5215	0.9654^1	0.3087^1	0.8730^1
3.9	-0.4963	0.1014	0.3403^1	0.9077^1	4.1	-0.4922	0.8453^1	0.2789^1	0.7381^1
4.0	-0.4673	0.8889^1	0.3072^1	0.7683^1	4.2	-0.4659	0.7260^1	0.2460^1	0.6137^1
4.1	-0.4413	0.7654^1	0.2710^1	0.6401^1	4.3	-0.4421	0.6080^1	0.2106^1	0.4988^1
4.2	-0.4179	0.6434^1	0.2324^1	0.5219^1	4.4	-0.4206	0.4918^1	0.1736^1	0.3923^1
4.3	-0.3969	0.5234^1	0.1922^1	0.4126^1	4.5	-0.4011	0.3778^1	0.1355^1	0.2936^1
4.4	-0.3779	0.4058^1	0.1511^1	0.3116^1	4.6	-0.3833	0.2659^1	0.9663^2	0.2016^1
4.5	-0.3607	0.2905^1	0.1094^1	0.2177^1	4.7	-0.3671	0.1564^1	0.5749^2	0.1159^1
4.6	-0.3450	0.1779^1	0.6767^2	0.1303^1	4.8	-0.3523	0.4939^2	0.1833^2	0.3582^2
4.7	-0.3306	0.6799^2	0.2607^2	0.4872^2	4.9	-0.3386	-0.5509^2	-0.2061^2	-0.3916^2
4.8	-0.3175	-0.3918^2	-0.1512^2	-0.2751^2	5.0	-0.3261	-0.1570^1	-0.5915^2	-0.1095^1
4.9	-0.3055	-0.1436^1	-0.5574^2	-0.9894^2	5.1	-0.3145	-0.2564^1	-0.9715^2	-0.1756^1
5.0	-0.2944	-0.2453^1	-0.9565^2	-0.1660^1	5.2	-0.3038	-0.3533^1	-0.1345^1	-0.2379^1
5.1	-0.2842	-0.3443^1	-0.1348^1	-0.2291^1	5.3	-0.2939	-0.4476^1	-0.1711^1	-0.2966^1
5.2	-0.2747	-0.4407^1	-0.1730^1	-0.2885^1	5.4	-0.2846	-0.5396^1	-0.2070^1	-0.3521^1
5.3	-0.2659	-0.5345^1	-0.2103^1	-0.3447^1	5.5	-0.2760	-0.6291^1	-0.2420^1	-0.4046^1
5.4	-0.2577	-0.6258^1	-0.2467^1	-0.3978^1	5.6	-0.2680	-0.7164^1	-0.2762^1	-0.4544^1
5.5	-0.2501	-0.7147^1	-0.2822^1	-0.4481^1	5.7	-0.2605	-0.8014^1	-0.3096^1	-0.5017^1
5.6	-0.2429	-0.8012^1	-0.3167^1	-0.4958^1	5.8	-0.2534	-0.8842^1	-0.3421^1	-0.5467^1
5.8	-0.2300	-0.9675^1	-0.3828^1	-0.5843^1	6.0	-0.2405	-0.1044	-0.4046^1	-0.6303^1
6.0	-0.2186	-0.1125	-0.4453^1	-0.6647^1	6.2	-0.2291	-0.1195	-0.4638^1	-0.7064^1
6.2	-0.2084	-0.1275	-0.5043^1	-0.7380^1	6.4	-0.2189	-0.1339	-0.5198^1	-0.7761^1
6.4	-0.1993	-0.1417	-0.5599^1	-0.8051^1	6.6	-0.2097	-0.1476	-0.5728^1	-0.8401^1
6.6	-0.1911	-0.1553	-0.6125^1	-0.8669^1	6.8	-0.2014	-0.1607	-0.6229^1	-0.8992^1
6.8	-0.1837	-0.1682	-0.6622^1	-0.9240^1	7.0	-0.1939	-0.1732	-0.6705^1	-0.9538^1
7.0	-0.1769	-0.1805	-0.7091^1	-0.9769^1	7.2	-0.1870	-0.1851	-0.7155^1	-0.1005
7.2	-0.1708	-0.1923	-0.7536^1	-0.1026	7.4	-0.1807	-0.1964	-0.7582^1	-0.1052
7.4	-0.1651	-0.2035	-0.7957^1	-0.1072	7.6	-0.1750	-0.2073	-0.7988^1	-0.1096
7.6	-0.1599	-0.2143	-0.8357^1	-0.1115	7.8	-0.1696	-0.2178	-0.8373^1	-0.1137
7.9	-0.1529	-0.2296	-0.8919^1	-0.1174	8.1	-0.1624	-0.2326	-0.8917^1	-0.1195
8.2	-0.1466	-0.2440	-0.9440^1	-0.1228	8.4	-0.1559	-0.2466	-0.9422^1	-0.1247
8.5	-0.1409	-0.2575	-0.9925^1	-0.1278	8.7	-0.1500	-0.2598	-0.9893^1	-0.1295
8.8	-0.1358	-0.2703	-0.1038	-0.1323	9.0	-0.1447	-0.2723	-0.1033	-0.1339
9.1	-0.1312	-0.2824	-0.1080	-0.1365	9.3	-0.1399	-0.2841	-0.1074	-0.1380
9.6	-0.1243	-0.3011	-0.1145	-0.1429	9.8	-0.1327	-0.3024	-0.1138	-0.1442
10.1	-0.1184	-0.3183	-0.1203	-0.1485	10.3	-0.1265	-0.3192	-0.1194	-0.1497
10.6	-0.1132	-0.3341	-0.1255	-0.1535	10.8	-0.1211	-0.3347	-0.1246	-0.1546
11.1	-0.1086	-0.3487	-0.1303	-0.1581	11.3	-0.1162	-0.3490	-0.1293	-0.1590
11.6	-0.1045	-0.3623	-0.1347	-0.1622	11.8	-0.1119	-0.3624	-0.1337	-0.1631
12.4	-0.9883^1	-0.3821	-0.1410	-0.1680	12.6	-0.1059	-0.3818	-0.1399	-0.1688
13.2	-0.9399^1	-0.3999	-0.1466	-0.1731	13.4	-0.1008	-0.3993	-0.1453	-0.1738
14.0	-0.8981^1	-0.4159	-0.1515	-0.1775	14.2	-0.9641^1	-0.4152	-0.1502	-0.1782
14.8	-0.8618^1	-0.4305	-0.1559	-0.1815	15.0	-0.9256^1	-0.4295	-0.1546	-0.1821
15.6	-0.8299^1	-0.4438	-0.1598	-0.1850	15.8	-0.8917^1	-0.4427	-0.1585	-0.1856

$\hat{\alpha}_3 = 0.80$

$\hat{\alpha}_4$	$\lambda_1(0,1)$	$\lambda_2(0,1)$	λ_3	λ_4
3.0	-1.2251	0.1995	0.6849^2	0.3356
3.1	-1.1174	0.1936	0.1481^1	0.3002
3.2	-1.0250	0.1864	0.2111^1	0.2687
3.3	-0.9444	0.1783	0.2597^1	0.2402
3.4	-0.8732	0.1693	0.2955^1	0.2143
3.5	-0.8099	0.1595	0.3195^1	0.1907
3.6	-0.7535	0.1492	0.3330^1	0.1691
3.7	-0.7030	0.1383	0.3369^1	0.1492
3.8	-0.6578	0.1272	0.3325^1	0.1310
3.9	-0.6172	0.1158	0.3210^1	0.1143
4.0	-0.5807	0.1042	0.3035^1	0.9887^1
4.1	-0.5479	0.9260^1	0.2810^1	0.8468^1
4.2	-0.5183	0.8100^1	0.2544^1	0.7157^1
4.3	-0.4915	0.6947^1	0.2246^1	0.5945^1
4.4	-0.4672	0.5806^1	0.1924^1	0.4822^1
4.5	-0.4452	0.4679^1	0.1584^1	0.3780^1
4.6	-0.4252	0.3573^1	0.1231^1	0.2812^1
4.7	-0.4069	0.2485^1	0.8698^2	0.1909^1
4.8	-0.3901	0.1420^1	0.5035^2	0.1066^1
4.9	-0.3747	0.3770^2	0.1352^2	0.2770^2
5.0	-0.3605	-0.6425^2	-0.2326^2	-0.4627^2
5.1	-0.3474	-0.1638^1	-0.5981^2	-0.1157^1
5.2	-0.3353	-0.2610^1	-0.9599^2	-0.1811^1
5.3	-0.3241	-0.3558^1	-0.1317^1	-0.2427^1
5.4	-0.3137	-0.4483^1	-0.1667^1	-0.3009^1
5.5	-0.3040	-0.5385^1	-0.2012^1	-0.3559^1
5.6	-0.2949	-0.6265^1	-0.2349^1	-0.4080^1
5.7	-0.2865	-0.7122^1	-0.2680^1	-0.4574^1
5.8	-0.2785	-0.7958^1	-0.3003^1	-0.5044^1
5.9	-0.2711	-0.8773^1	-0.3318^1	-0.5491^1
6.1	-0.2575	-0.1034	-0.3925^1	-0.6322^1
6.3	-0.2454	-0.1184	-0.4502^1	-0.7080^1
6.5	-0.2346	-0.1326	-0.5050^1	-0.7773^1
6.7	-0.2248	-0.1461	-0.5570^1	-0.8411^1
6.9	-0.2160	-0.1591	-0.6063^1	-0.9000^1
7.1	-0.2080	-0.1714	-0.6531^1	-0.9545^1
7.3	-0.2007	-0.1832	-0.6975^1	-0.1005
7.5	-0.1940	-0.1945	-0.7398^1	-0.1052
7.7	-0.1878	-0.2053	-0.7799^1	-0.1096
7.9	-0.1821	-0.2156	-0.8181^1	-0.1138
8.2	-0.1744	-0.2304	-0.8720^1	-0.1195
8.5	-0.1674	-0.2443	-0.9222^1	-0.1247
8.8	-0.1611	-0.2574	-0.9691^1	-0.1295
9.1	-0.1555	-0.2698	-0.1013	-0.1339
9.4	-0.1503	-0.2815	-0.1054	-0.1380
9.9	-0.1426	-0.2997	-0.1117	-0.1442
10.4	-0.1360	-0.3165	-0.1174	-0.1497
10.9	-0.1301	-0.3319	-0.1226	-0.1546
11.4	-0.1250	-0.3462	-0.1273	-0.1591
11.9	-0.1203	-0.3595	-0.1316	-0.1631
12.7	-0.1139	-0.3790	-0.1379	-0.1688
13.5	-0.1084	-0.3964	-0.1434	-0.1738
14.3	-0.1037	-0.4122	-0.1483	-0.1782
15.1	-0.9953^1	-0.4266	-0.1527	-0.1821
15.9	-0.9589^1	-0.4398	-0.1566	-0.1856

$\hat{\alpha}_3 = 0.85$

$\hat{\alpha}_4$	$\lambda_1(0,1)$	$\lambda_2(0,1)$	λ_3	λ_4
3.1	-1.2524	0.1935	0.3194^2	0.3254
3.2	-1.1444	0.1875	0.1104^1	0.2912
3.3	-1.0523	0.1804	0.1724^1	0.2607
3.4	-0.9721	0.1724	0.2205^1	0.2332
3.5	-0.9013	0.1635	0.2563^1	0.2083
3.6	-0.8384	0.1541	0.2809^1	0.1855
3.7	-0.7822	0.1440	0.2956^1	0.1647
3.8	-0.7318	0.1336	0.3014^1	0.1455
3.9	-0.6866	0.1228	0.2994^1	0.1279
4.0	-0.6458	0.1118	0.2905^1	0.1117
4.1	-0.6090	0.1007	0.2759^1	0.9668^1
4.2	-0.5758	0.8946^1	0.2565^1	0.8286^1
4.3	-0.5457	0.7825^1	0.2330^1	0.7006^1
4.4	-0.5184	0.6710^1	0.2064^1	0.5821^1
4.5	-0.4935	0.5605^1	0.1772^1	0.4720^1
4.6	-0.4709	0.4512^1	0.1461^1	0.3696^1
4.7	-0.4503	0.3438^1	0.1136^1	0.2744^1
4.8	-0.4313	0.2380^1	0.8002^2	0.1854^1
4.9	-0.4140	0.1342^1	0.4581^2	0.1022^1
5.0	-0.3979	0.3243^2	0.1122^2	0.2417^2
5.1	-0.3832	-0.6705^2	-0.2346^2	-0.4896^2
5.2	-0.3696	-0.1645^1	-0.5808^2	-0.1178^1
5.3	-0.3569	-0.2596^1	-0.9250^2	-0.1826^1
5.4	-0.3452	-0.3525^1	-0.1265^1	-0.2437^1
5.5	-0.3343	-0.4432^1	-0.1601^1	-0.3015^1
5.6	-0.3241	-0.5317^1	-0.1932^1	-0.3562^1
5.7	-0.3145	-0.6181^1	-0.2257^1	-0.4080^1
5.8	-0.3056	-0.7024^1	-0.2576^1	-0.4572^1
5.9	-0.2973	-0.7847^1	-0.2888^1	-0.5039^1
6.0	-0.2894	-0.8649^1	-0.3194^1	-0.5484^1
6.2	-0.2751	-0.1020	-0.3784^1	-0.6313^1
6.4	-0.2623	-0.1167	-0.4348^1	-0.7068^1
6.6	-0.2508	-0.1308	-0.4884^1	-0.7761^1
6.8	-0.2404	-0.1442	-0.5394^1	-0.8398^1
7.0	-0.2311	-0.1570	-0.5879^1	-0.8986^1
7.2	-0.2226	-0.1692	-0.6340^1	-0.9531^1
7.4	-0.2148	-0.1809	-0.6779^1	-0.1004
7.6	-0.2076	-0.1921	-0.7196^1	-0.1051
7.8	-0.2011	-0.2028	-0.7594^1	-0.1095
8.0	-0.1950	-0.2131	-0.7973^1	-0.1136
8.3	-0.1867	-0.2277	-0.8508^1	-0.1193
8.6	-0.1793	-0.2415	-0.9008^1	-0.1246
8.9	-0.1726	-0.2546	-0.9475^1	-0.1294
9.2	-0.1666	-0.2669	-0.9912^1	-0.1338
9.5	-0.1610	-0.2786	-0.1032	-0.1379
10.0	-0.1529	-0.2967	-0.1095	-0.1441
10.5	-0.1457	-0.3134	-0.1152	-0.1496
11.0	-0.1395	-0.3289	-0.1204	-0.1545
11.5	-0.1339	-0.3431	-0.1252	-0.1590
12.0	-0.1290	-0.3564	-0.1295	-0.1630
12.8	-0.1221	-0.3758	-0.1358	-0.1687
13.6	-0.1162	-0.3933	-0.1413	-0.1737
14.4	-0.1111	-0.4091	-0.1462	-0.1781
15.2	-0.1067	-0.4234	-0.1507	-0.1820
16.0	-0.1028	-0.4366	-0.1546	-0.1855

	$\hat{\alpha}_3 = 0.90$						$\hat{\alpha}_3 = 0.95$		
$\hat{\alpha}_4$	$\lambda_1(0,1)$	$\lambda_2(0,1)$	λ_3	λ_4	$\hat{\alpha}_4$	$\lambda_1(0,1)$	$\lambda_2(0,1)$	λ_3	λ_4
3.3	-1.1782	0.1821	0.7038^2	0.2846	3.4	-1.2193	0.1773	0.2760^2	0.2803
3.4	-1.0851	0.1751	0.1328^1	0.2548	3.5	-1.1237	0.1704	0.9165^2	0.2509
3.5	-1.0043	0.1672	0.1813^1	0.2281	3.6	-1.0415	0.1628	0.1416^1	0.2247
3.6	-0.9333	0.1586	0.2179^1	0.2039	3.7	-0.9695	0.1544	0.1798^1	0.2010
3.7	-0.8702	0.1494	0.2438^1	0.1818	3.8	-0.9056	0.1455	0.2075^1	0.1794
3.8	-0.8138	0.1397	0.2602^1	0.1615	3.9	-0.8485	0.1362	0.2261^1	0.1596
3.9	-0.7632	0.1296	0.2681^1	0.1429	4.0	-0.7972	0.1264	0.2365^1	0.1414
4.0	-0.7175	0.1192	0.2686^1	0.1258	4.1	-0.7509	0.1164	0.2397^1	0.1246
4.1	-0.6764	0.1086	0.2626^1	0.1099	4.2	-0.7091	0.1062	0.2366^1	0.1091
4.2	-0.6391	0.9788^1	0.2510^1	0.9533^1	4.3	-0.6711	0.9579^1	0.2280^1	0.9477^1
4.3	-0.6053	0.8706^1	0.2346^1	0.8180^1	4.4	-0.6366	0.8533^1	0.2147^1	0.8148^1
4.4	-0.5746	0.7622^1	0.2142^1	0.6926^1	4.5	-0.6051	0.7486^1	0.1974^1	0.6913^1
4.5	-0.5466	0.6543^1	0.1906^1	0.5761^1	4.6	-0.5764	0.6441^1	0.1768^1	0.5765^1
4.6	-0.5211	0.5472^1	0.1644^1	0.4678^1	4.7	-0.5502	0.5402^1	0.1536^1	0.4695^1
4.7	-0.4978	0.4412^1	0.1361^1	0.3669^1	4.8	-0.5261	0.4373^1	0.1281^1	0.3697^1
4.8	-0.4765	0.3367^1	0.1063^1	0.2729^1	4.9	-0.5041	0.3358^1	0.1009^1	0.2766^1
4.9	-0.4569	0.2338^1	0.7531^2	0.1849^1	5.0	-0.4838	0.2355^1	0.7243^2	0.1894^1
5.0	-0.4389	0.1327^1	0.4348^2	0.1025^1	5.1	-0.4650	0.1369^1	0.4294^2	0.1076^1
5.1	-0.4223	0.3339^2	0.1111^2	0.2526^2	5.2	-0.4477	0.3999^2	0.1276^2	0.3075^2
5.2	-0.4069	-0.6388^2	-0.2154^2	-0.4735^2	5.3	-0.4317	-0.5511^2	-0.1786^2	-0.4151^2
5.3	-0.3927	-0.1591^1	-0.5428^2	-0.1157^1	5.4	-0.4168	-0.1483^1	-0.4873^2	-0.1096^1
5.4	-0.3794	-0.2523^1	-0.8694^2	-0.1801^1	5.5	-0.4029	-0.2397^1	-0.7967^2	-0.1738^1
5.5	-0.3671	-0.3435^1	-0.1194^1	-0.2410^1	5.6	-0.3900	-0.3292^1	-0.1106^1	-0.2346^1
5.6	-0.3557	-0.4325^1	-0.1515^1	-0.2985^1	5.7	-0.3780	-0.4165^1	-0.1412^1	-0.2919^1
5.7	-0.3450	-0.5194^1	-0.1833^1	-0.3530^1	5.8	-0.3668	-0.5020^1	-0.1716^1	-0.3463^1
5.8	-0.3350	-0.6044^1	-0.2145^1	-0.4046^1	5.9	-0.3562	-0.5856^1	-0.2017^1	-0.3980^1
5.9	-0.3256	-0.6847^1	-0.2453^1	-0.4537^1	6.0	-0.3464	-0.6673^1	-0.2313^1	-0.4470^1
6.0	-0.3168	-0.7684^1	-0.2755^1	-0.5003^1	6.1	-0.3371	-0.7471^1	-0.2605^1	-0.4937^1
6.1	-0.3085	-0.8475^1	-0.3051^1	-0.5448^1	6.2	-0.3283	-0.8251^1	-0.2892^1	-0.5381^1
6.2	-0.3007	-0.9247^1	-0.3341^1	-0.5871^1	6.3	-0.3201	-0.9014^1	-0.3174^1	-0.5805^1
6.4	-0.2863	-0.1074	-0.3904^1	-0.6661^1	6.5	-0.3049	-0.1049	-0.3721^1	-0.6597^1
6.6	-0.2735	-0.1216	-0.4441^1	-0.7385^1	6.7	-0.3737^1	-0.7977	-0.2443	-0.2609
6.8	-0.2620	-0.1352	-0.4953^1	-0.8049^1	6.9	-0.2792	-0.1324	-0.4749^1	-0.7987^1
7.0	-0.2516	-0.1481	-0.5442^1	-0.8661^1	7.1	-0.2681	-0.1452	-0.5229^1	-0.8601^1
7.2	-0.2421	-0.1605	-0.5908^1	-0.9227^1	7.3	-0.2580	-0.1575	-0.5688^1	-0.9170^1
7.4	-0.2334	-0.1724	-0.6352^1	-0.9753^1	7.5	-0.2488	-0.1693	-0.6126^1	-0.9697^1
7.6	-0.2255	-0.1837	-0.6775^1	-0.1024	7.7	-0.2404	-0.1806	-0.6545^1	-0.1019
7.8	-0.2182	-0.1946	-0.7179^1	-0.1070	7.9	-0.2327	-0.1914	-0.6944^1	-0.1065
8.0	-0.2115	-0.2051	-0.7563^1	-0.1113	8.1	-0.2255	-0.2017	-0.7326^1	-0.1108
8.2	-0.2053	-0.2151	-0.7931^1	-0.1153	8.3	-0.2189	-0.2117	-0.7691^1	-0.1148
8.5	-0.1968	-0.2294	-0.8451^1	-0.1208	8.6	-0.2099	-0.2259	-0.8209^1	-0.1204
8.8	-0.1892	-0.2428	-0.8937^1	-0.1259	8.9	-0.2017	-0.2394	-0.8694^1	-0.1255
9.1	-0.1823	-0.2556	-0.9393^1	-0.1306	9.2	-0.1944	-0.2520	-0.9148^1	-0.1302
9.4	-0.1760	-0.2676	-0.9820^1	-0.1349	9.5	-0.1877	-0.2641	-0.9575^1	-0.1345
9.7	-0.1703	-0.2791	-0.1022	-0.1389	9.8	-0.1816	-0.2755	-0.9977^1	-0.1386
10.2	-0.1618	-0.2969	-0.1084	-0.1450	10.3	-0.1725	-0.2932	-0.1060	-0.1446
10.7	-0.1544	-0.3133	-0.1140	-0.1504	10.8	-0.1646	-0.3096	-0.1116	-0.1501
11.2	-0.1478	-0.3284	-0.1191	-0.1552	11.3	-0.1577	-0.3247	-0.1167	-0.1549
11.7	-0.1421	-0.3425	-0.1238	-0.1596	11.8	-0.1515	-0.3388	-0.1214	-0.1593
12.2	-0.1369	-0.3555	-0.1281	-0.1636	12.3	-0.1460	-0.3518	-0.1257	-0.1633
13.0	-0.1297	-0.3747	-0.1343	-0.1692	13.1	-0.1383	-0.3710	-0.1320	-0.1690
13.8	-0.1235	-0.3919	-0.1398	-0.1742	13.9	-0.1317	-0.3882	-0.1375	-0.1739
14.6	-0.1181	-0.4075	-0.1447	-0.1786	14.7	-0.1260	-0.4038	-0.1424	-0.1783
15.4	-0.1135	-0.4217	-0.1491	-0.1824	15.5	-0.1210	-0.4180	-0.1469	-0.1822
16.2	-0.1094	-0.4347	-0.1530	-0.1858	16.3	-0.1166	-0.4311	-0.1509	-0.1856

$\hat{\alpha}_3 = 1.00$

$\hat{\alpha}_4$	$\lambda_1(0,1)$	$\lambda_2(0,1)$	λ_3	λ_4
3.6	-1.1690	0.1664	0.4829^2	0.2490
3.7	-1.0840	0.1590	0.1009^1	0.2230
3.8	-1.0101	0.1509	0.1414^1	0.1996
3.9	-0.9448	0.1423	0.1715^1	0.1783
4.0	-0.8865	0.1333	0.1927^1	0.1588
4.1	-0.8342	0.1239	0.2058^1	0.1409
4.2	-0.7869	0.1142	0.2119^1	0.1244
4.3	-0.7441	0.1044	0.2119^1	0.1091
4.4	-0.7053	0.9434^1	0.2064^1	0.9500^1
4.5	-0.6699	0.8424^1	0.1962^1	0.8187^1
4.6	-0.6376	0.7411^1	0.1820^1	0.6967^1
4.7	-0.6080	0.6399^1	0.1645^1	0.5830^1
4.8	-0.5809	0.5392^1	0.1442^1	0.4770^1
4.9	-0.5560	0.4392^1	0.1215^1	0.3780^1
5.0	-0.5331	0.3404^1	0.9698^2	0.2855^1
5.1	-0.5120	0.2428^1	0.7098^2	0.1987^1
5.2	-0.4925	0.1466^1	0.4383^2	0.1172^1
5.3	-0.4744	0.5192^2	0.1584^2	0.4061^2
5.4	-0.4577	-0.4110^2	-0.1276^2	-0.3149^2
5.5	-0.4421	-0.1324^1	-0.4176^2	-0.9946^2
5.6	-0.4276	-0.2219^1	-0.7097^2	-0.1636^1
5.7	-0.4141	-0.3096^1	-0.1002^1	-0.2243^1
5.8	-0.4014	-0.3956^1	-0.1295^1	-0.2818^1
5.9	-0.3896	-0.4797^1	-0.1586^1	-0.3363^1
6.0	-0.3785	-0.5619^1	-0.1874^1	-0.3880^1
6.1	-0.3681	-0.6424^1	-0.2159^1	-0.4372^1
6.2	-0.3583	-0.7212^1	-0.2441^1	-0.4840^1
6.3	-0.3491	-0.7982^1	-0.2718^1	-0.5286^1
6.4	-0.3404	-0.8734^1	-0.2992^1	-0.5711^1
6.5	-0.3322	-0.9470^1	-0.3260^1	-0.6117^1
6.7	-0.3170	-0.1089	-0.3783^1	-0.6877^1
6.9	-0.3034	-0.1226	-0.4285^1	-0.7574^1
7.1	-0.2910	-0.1356	-0.4768^1	-0.8217^1
7.3	-0.2799	-0.1480	-0.5230^1	-0.8810^1
7.5	-0.2697	-0.1600	-0.5672^1	-0.9360^1
7.7	-0.2603	-0.1714	-0.6095^1	-0.9872^1
7.9	-0.2517	-0.1824	-0.6500^1	-0.1035
8.1	-0.2438	-0.1929	-0.6888^1	-0.1080
8.3	-0.2365	-0.2030	-0.7258^1	-0.1121
8.5	-0.2297	-0.2128	-0.7613^1	-0.1161
8.8	-0.2205	-0.2267	-0.8118^1	-0.1215
9.1	-0.2121	-0.2398	-0.8591^1	-0.1265
9.4	-0.2045	-0.2522	-0.9035^1	-0.1311
9.7	-0.1976	-0.2640	-0.9454^1	-0.1354
10.0	-0.1913	-0.2752	-0.9848^1	-0.1393
10.5	-0.1819	-0.2927	-0.1046	-0.1453
11.0	-0.1737	-0.3087	-0.1101	-0.1506
11.5	-0.1665	-0.3236	-0.1152	-0.1554
12.0	-0.1600	-0.3375	-0.1198	-0.1598
12.5	-0.1543	-0.3504	-0.1241	-0.1637
13.3	-0.1462	-0.3693	-0.1303	-0.1693
14.1	-0.1393	-0.3863	-0.1358	-0.1742
14.9	-0.1334	-0.4018	-0.1407	-0.1785
15.7	-0.1281	-0.4158	-0.1451	-0.1824
16.5	-0.1235	-0.4288	-0.1491	-0.1858

$\hat{\alpha}_3 = 1.05$

$\hat{\alpha}_4$	$\lambda_1(0,1)$	$\lambda_2(0,1)$	λ_3	λ_4
3.7	-1.2220	0.1631	0.1786^3	0.2491
3.8	-1.1327	0.1559	0.5815^2	0.2230
3.9	-1.0559	0.1481	0.1019^1	0.1997
4.0	-0.9883	0.1398	0.1351^1	0.1785
4.1	-0.9283	0.1311	0.1593^1	0.1592
4.2	-0.8744	0.1220	0.1755^1	0.1415
4.3	-0.8259	0.1127	0.1847^1	0.1251
4.4	-0.7819	0.1032	0.1878^1	0.1100
4.5	-0.7419	0.9349^1	0.1856^1	0.9599^1
4.6	-0.7054	0.8374^1	0.1786^1	0.8297^1
4.7	-0.6720	0.7394^1	0.1676^1	0.7086^1
4.8	-0.6415	0.6413^1	0.1531^1	0.5957^1
4.9	-0.6134	0.5436^1	0.1357^1	0.4902^1
5.0	-0.5875	0.4465^1	0.1159^1	0.3916^1
5.1	-0.5637	0.3502^1	0.9400^2	0.2992^1
5.2	-0.5418	0.2552^1	0.7055^2	0.2128^1
5.3	-0.5214	0.1613^1	0.4577^2	0.1314^1
5.4	-0.5026	0.6882^2	0.1998^2	0.5483^2
5.5	-0.4850	-0.2218^2	-0.6571^3	-0.1730^2
5.6	-0.4687	-0.1116^1	-0.3367^2	-0.8536^2
5.7	-0.4535	-0.1994^1	-0.6113^2	-0.1497^1
5.8	-0.4393	-0.2855^1	-0.8880^2	-0.2105^1
5.9	-0.4260	-0.3700^1	-0.1166^1	-0.2682^1
6.0	-0.4135	-0.4528^1	-0.1443^1	-0.3229^1
6.1	-0.4019	-0.5337^1	-0.1719^1	-0.3748^1
6.2	-0.3909	-0.6131^1	-0.1993^1	-0.4242^1
6.3	-0.3806	-0.6908^1	-0.2264^1	-0.4712^1
6.4	-0.3708	-0.7668^1	-0.2532^1	-0.5160^1
6.5	-0.3616	-0.8411^1	-0.2796^1	-0.5588^1
6.6	-0.3529	-0.9139^1	-0.3057^1	-0.5996^1
6.8	-0.3369	-0.1055	-0.3566^1	-0.6761^1
7.0	-0.3224	-0.1190	-0.4057^1	-0.7463^1
7.2	-0.3094	-0.1319	-0.4530^1	-0.8110^1
7.4	-0.2975	-0.1443	-0.4985^1	-0.8707^1
7.6	-0.2867	-0.1561	-0.5421^1	-0.9262^1
7.8	-0.2768	-0.1675	-0.5839^1	-0.9777^1
8.0	-0.2677	-0.1784	-0.6240^1	-0.1026
8.2	-0.2593	-0.1889	-0.6624^1	-0.1071
8.4	-0.2515	-0.1990	-0.6992^1	-0.1113
8.6	-0.2443	-0.2087	-0.7346^1	-0.1152
8.9	-0.2344	-0.2225	-0.7848^1	-0.1207
9.2	-0.2255	-0.2356	-0.8320^1	-0.1258
9.5	-0.2174	-0.2480	-0.8765^1	-0.1304
9.8	-0.2101	-0.2598	-0.9183^1	-0.1347
10.1	-0.2034	-0.2710	-0.9578^1	-0.1387
10.6	-0.1934	-0.2884	-0.1019	-0.1447
11.1	-0.1846	-0.3045	-0.1074	-0.1501
11.6	-0.1769	-0.3194	-0.1125	-0.1549
12.1	-0.1701	-0.3332	-0.1172	-0.1593
12.6	-0.1639	-0.3461	-0.1215	-0.1632
13.4	-0.1553	-0.3651	-0.1278	-0.1689
14.2	-0.1480	-0.3821	-0.1333	-0.1738
15.0	-0.1416	-0.3976	-0.1383	-0.1782
15.8	-0.1361	-0.4117	-0.1428	-0.1820
16.6	-0.1312	-0.4247	-0.1468	-0.1855

$$\hat{\alpha}_3 = 1.10$$

$\hat{\alpha}_4$	$\lambda_1(0,1)$	$\lambda_2(0,1)$	λ_3	λ_4
3.9	-1.1889	0.1534	0.1235^2	0.2248
4.0	-1.1077	0.1458	0.6040^2	0.2013
4.1	-1.0369	0.1378	0.9740^2	0.1800
4.2	-0.9744	0.1294	0.1251^1	0.1607
4.3	-0.9185	0.1207	0.1448^1	0.1430
4.4	-0.8682	0.1117	0.1574^1	0.1267
4.5	-0.8226	0.1025	0.1639^1	0.1117
4.6	-0.7812	0.9320^1	0.1650^1	0.9774^1
4.7	-0.7434	0.8377^1	0.1613^1	0.8478^1
4.8	-0.7088	0.7428^1	0.1535^1	0.7270^1
4.9	-0.6771	0.6478^1	0.1422^1	0.6144^1
5.0	-0.6479	0.5530^1	0.1278^1	0.5091^1
5.1	-0.6210	0.4586^1	0.1108^1	0.4106^1
5.2	-0.5962	0.3650^1	0.9166^2	0.3182^1
5.3	-0.5733	0.2724^1	0.7076^2	0.2317^1
5.4	-0.5520	0.1808^1	0.4839^2	0.1502^1
5.5	-0.5323	0.9038^2	0.2484^2	0.7342^2
5.6	-0.5136	0.7603^6	0.2140^{10}	0.6045^{10}
5.7	-0.4968	-0.8629^2	-0.2479^2	-0.6726^2
5.8	-0.4808	-0.1724^1	-0.5046^2	-0.1319^1
5.9	-0.4659	-0.2570^1	-0.7648^2	-0.1930^1
6.0	-0.4519	-0.3401^1	-0.1027^1	-0.2510^1
6.1	-0.4388	-0.4214^1	-0.1290^1	-0.3060^1
6.2	-0.4264	-0.5013^1	-0.1553^1	-0.3583^1
6.3	-0.4148	-0.5796^1	-0.1816^1	-0.4081^1
6.4	-0.4039	-0.6561^1	-0.2076^1	-0.4553^1
6.5	-0.3936	-0.7312^1	-0.2335^1	-0.5005^1
6.6	-0.3839	-0.8047^1	-0.2590^1	-0.5436^1
6.7	-0.3747	-0.8767^1	-0.2843^1	-0.5848^1
6.8	-0.3660	-0.9471^1	-0.3092^1	-0.6242^1
7.0	-0.3499	-0.1084	-0.3580^1	-0.6980^1
7.2	-0.3353	-0.1215	-0.4051^1	-0.7660^1
7.4	-0.3221	-0.1340	-0.4506^1	-0.8287^1
7.6	-0.3101	-0.1460	-0.4944^1	-0.8867^1
7.8	-0.2991	-0.1576	-0.5366^1	-0.9407^1
8.0	-0.2890	-0.1687	-0.5771^1	-0.9909^1
8.2	-0.2797	-0.1793	-0.6159^1	-0.1038
8.4	-0.2711	-0.1896	-0.6533^1	-0.1082
8.6	-0.2632	-0.1994	-0.6891^1	-0.1123
8.8	-0.2558	-0.2089	-0.0724^1	-0.1162
9.1	-0.2456	-0.2224	-0.7726^1	-0.1216
9.4	-0.2365	-0.2353	-0.8188^1	-0.1265
9.7	-0.2282	-0.2475	-0.8624^1	-0.1311
10.0	-0.2206	-0.2590	-0.9035^1	-0.1353
10.3	-0.2136	-0.2700	-0.9423^1	-0.1392
10.8	-0.2033	-0.2872	-0.1002	-0.1451
11.3	-0.1942	-0.3030	-0.1057	-0.1504
11.8	-0.1862	-0.3177	-0.1108	-0.1552
12.3	-0.1791	-0.3314	-0.1154	-0.1595
12.8	-0.1727	-0.3441	-0.1197	-0.1635
13.6	-0.1637	-0.3628	-0.1259	-0.1691
14.4	-0.1561	-0.3797	-0.1314	-0.1739
15.2	-0.1494	-0.3951	-0.1364	-0.1783
16.0	-0.1436	-0.4091	-0.1408	-0.1821
16.8	-0.1384	-0.4219	-0.1449	-0.1855

$$\hat{\alpha}_3 = 1.15$$

$\hat{\alpha}_4$	$\lambda_1(0,1)$	$\lambda_2(0,1)$	λ_3	λ_4
4.1	-1.1669	0.1442	0.1565^2	0.2045
4.2	-1.0915	0.1365	0.5746^2	0.1829
4.3	-1.0255	0.1284	0.8937^2	0.1635
4.4	-0.9669	0.1199	0.1130^1	0.1457
4.5	-0.9143	0.1113	0.1293^1	0.1293
4.6	-0.8668	0.1024	0.1395^1	0.1143
4.7	-0.8237	0.9341^1	0.1440^1	0.1003
4.8	-0.7843	0.8430^1	0.1438^1	0.8730^1
4.9	-0.7483	0.7512^1	0.1393^1	0.7521^1
5.0	-0.7152	0.6591^1	0.1312^1	0.6393^1
5.1	-0.6848	0.5671^1	0.1199^1	0.5338^1
5.2	-0.6567	0.4754^1	0.1058^1	0.4351^1
5.3	-0.6307	0.3843^1	0.8948^2	0.3424^1
5.4	-0.6068	0.2940^1	0.7120^2	0.2555^1
5.5	-0.5845	0.2046^1	0.5129^2	0.1736^1
5.6	-0.5637	0.1163^1	0.3005^2	0.9643^2
5.7	-0.5445	0.2912^2	0.7733^3	0.2364^2
5.8	-0.5265	-0.5674^2	-0.1544^2	-0.4513^2
5.9	-0.5096	-0.1412^1	-0.3927^2	-0.1102^1
6.0	-0.4939	-0.2244^1	-0.6361^2	-0.1718^1
6.1	-0.4792	-0.3059^1	-0.8824^2	-0.2302^1
6.2	-0.4653	-0.3861^1	-0.1131^1	-0.2857^1
6.3	-0.4523	-0.4647^1	-0.1381^1	-0.3384^1
6.4	-0.4401	-0.5419^1	-0.1631^1	-0.3886^1
6.5	-0.4286	-0.6175^1	-0.1881^1	-0.4364^1
6.6	-0.4177	-0.6917^1	-0.2129^1	-0.4820^1
6.7	-0.4074	-0.7643^1	-0.2375^1	-0.5255^1
6.8	-0.3976	-0.8355^1	-0.2619^1	-0.5671^1
6.9	-0.3884	-0.9053^1	-0.2861^1	-0.6069^1
7.0	-0.3796	-0.9736^1	-0.3100^1	-0.6450^1
7.2	-0.3634	-0.1106	-0.3567^1	-0.7166^1
7.4	-0.3487	-0.1233	-0.4020^1	-0.7825^1
7.6	-0.3353	-0.1355	-0.4459^1	-0.8435^1
7.8	-0.3231	-0.1473	-0.4882^1	-0.9001^1
8.0	-0.3119	-0.1585	-0.5290^1	-0.9527^1
8.2	-0.3017	-0.1693	-0.5682^1	-0.1002
8.4	-0.2922	-0.1797	-0.6060^1	-0.1048
8.6	-0.2834	-0.1898	-0.6423^1	-0.1091
8.8	-0.2752	-0.1994	-0.6772^1	-0.1131
9.0	-0.2676	-0.2087	-0.7108^1	-0.1169
9.3	-0.2572	-0.2220	-0.7588^1	-0.1222
9.6	-0.2478	-0.2346	-0.8041^1	-0.1271
9.9	-0.2392	-0.2465	-0.8468^1	-0.1316
10.2	-0.2314	-0.2579	-0.8872^1	-0.1357
10.5	-0.2242	-0.2687	-0.9254^1	-0.1396
11.0	-0.2135	-0.2856	-0.9847^1	-0.1454
11.5	-0.2041	-0.3012	-0.1039	-0.1507
12.0	-0.1958	-0.3157	-0.1089	-0.1554
12.5	-0.1883	-0.3292	-0.1135	-0.1597
13.0	-0.1817	-0.3419	-0.1177	-0.1636
13.8	-0.1723	-0.3604	-0.1239	-0.1691
14.6	-0.1643	-0.3771	-0.1294	-0.1740
15.4	-0.1574	-0.3923	-0.1344	-0.1783
16.2	-0.1513	-0.4062	-0.1388	-0.1821
17.0	-0.1459	-0.4190	-0.1429	-0.1855

$\hat{\alpha}_3 = 1.20$

$\hat{\alpha}_4$	$\lambda_1(0,1)$	$\lambda_2(0,1)$	λ_3	λ_4
4.3	-1.1536	0.1356	0.1393^2	0.1874
4.4	-1.0828	0.1278	0.5100^2	0.1675
4.5	-1.0205	0.1197	0.7912^2	0.1494
4.6	-0.9650	0.1114	0.9969^2	0.1329
4.7	-0.9150	0.1028	0.1137^1	0.1177
4.8	-0.8697	0.9410^1	0.1221^1	0.1036
4.9	-0.8284	0.8529^1	0.1255^1	0.9057^1
5.0	-0.7907	0.7641^1	0.1245^1	0.7841^1
5.1	-0.7560	0.6749^1	0.1197^1	0.6706^1
5.2	-0.7241	0.5856^1	0.1115^1	0.5645^1
5.3	-0.6947	0.4964^1	0.1005^1	0.4651^1
5.4	-0.6675	0.4077^1	0.8705^2	0.3718^1
5.5	-0.6422	0.3196^1	0.7145^2	0.2841^1
5.6	-0.6189	0.2325^1	0.5411^2	0.2017^1
5.7	-0.5971	0.1462^1	0.3525^2	0.1239^1
5.8	-0.5768	0.6087^2	0.1515^2	0.5050^2
5.9	-0.5579	-0.2326^2	-0.5954^3	-0.1890^2
6.0	-0.5402	-0.1061^1	-0.2786^2	-0.8456^2
6.1	-0.5236	-0.1878^1	-0.5041^2	-0.1468^1
6.2	-0.5080	-0.2681^1	-0.7344^2	-0.2058^1
6.3	-0.4934	-0.3469^1	-0.9680^2	-0.2619^1
6.4	-0.4797	-0.4244^1	-0.1204^1	-0.3152^1
6.5	-0.4667	-0.5005^1	-0.1441^1	-0.3659^1
6.6	-0.4545	-0.5752^1	-0.1679^1	-0.4143^1
6.7	-0.4430	-0.6484^1	-0.1917^1	-0.4604^1
6.8	-0.4321	-0.7203^1	-0.2153^1	-0.5045^1
6.9	-0.4218	-0.7907^1	-0.2389^1	-0.5466^1
7.0	-0.4120	-0.8598^1	-0.2622^1	-0.5869^1
7.1	-0.4027	-0.9275^1	-0.2853^1	-0.6255^1
7.2	-0.3939	-0.9939^1	-0.3082^1	-0.6625^1
7.4	-0.3775	-0.1123	-0.3531^1	-0.7320^1
7.6	-0.3626	-0.1247	-0.3967^1	-0.7962^1
7.8	-0.3490	-0.1365	-0.4390^1	-0.8557^1
8.0	-0.3366	-0.1480	-0.4799^1	-0.9109^1
8.2	-0.3252	-0.1590	-0.5194^1	-0.9624^1
8.4	-0.3147	-0.1695	-0.5575^1	-0.1010
8.6	-0.3050	-0.1797	-0.5942^1	-0.1055
8.8	-0.2960	-0.1895	-0.6296^1	-0.1098
9.0	-0.2877	-0.1990	-0.6637^1	-0.1137
9.2	-0.2799	-0.2081	-0.6965^1	-0.1175
9.5	-0.2692	-0.2211	-0.7435^1	-0.1227
9.8	-0.2594	-0.2335	-0.7879^1	-0.1275
10.1	-0.2506	-0.2452	-0.8299^1	-0.1319
10.4	-0.2425	-0.2564	-0.8697^1	-0.1360
10.7	-0.2351	-0.2671	-0.9073^1	-0.1398
11.2	-0.2238	-0.2840	-0.9670^1	-0.1457
11.7	-0.2140	-0.2995	-0.1021	-0.1509
12.2	-0.2054	-0.3138	-0.1070	-0.1556
12.7	-0.1977	-0.3271	-0.1116	-0.1598
13.2	-0.1908	-0.3396	-0.1158	-0.1637
14.0	-0.1811	-0.3579	-0.1219	-0.1692
14.8	-0.1727	-0.3745	-0.1274	-0.1740
15.6	-0.1655	-0.3895	-0.1323	-0.1783
16.4	-0.1591	-0.4033	-0.1368	-0.1821
17.2	-0.1535	-0.4160	-0.1408	-0.1855

$\hat{\alpha}_3 = 1.25$

$\hat{\alpha}_4$	$\lambda_1(0,1)$	$\lambda_2(0,1)$	λ_3	λ_4
4.5	-1.1479	0.1277	0.8589^3	0.1730
4.6	-1.0805	0.1199	0.4219^2	0.1545
4.7	-1.0211	0.1119	0.6755^2	0.1376
4.8	-0.9679	0.1036	0.8599^2	0.1221
4.9	-0.9199	0.9524^1	0.9844^2	0.1079
5.0	-0.8763	0.8672^1	0.1057^1	0.9463^1
5.1	-0.8365	0.7812^1	0.1084^1	0.8233^1
5.2	-0.8000	0.6947^1	0.1071^1	0.7086^1
5.3	-0.7664	0.6080^1	0.1023^1	0.6013^1
5.4	-0.7355	0.5214^1	0.9440^2	0.5009^1
5.5	-0.7068	0.4351^1	0.8391^2	0.4066^1
5.6	-0.6802	0.3492^1	0.7114^2	0.3180^1
5.7	-0.6556	0.2642^1	0.5643^2	0.2347^1
5.8	-0.6327	0.1798^1	0.4005^2	0.1560^1
5.9	-0.6113	0.9635^2	0.2227^2	0.8178^2
6.0	-0.5913	-0.1391^2	-0.3323^3	-0.1156^2
6.1	-0.5726	-0.6741^2	-0.1658^2	-0.5490^2
6.2	-0.5550	-0.1476^1	-0.3728^2	-0.1179^1
6.3	-0.5386	-0.2265^1	-0.5858^2	-0.1777^1
6.4	-0.5231	-0.3041^1	-0.8036^2	-0.2345^1
6.5	-0.5086	-0.3805^1	-0.1025^1	-0.2885^1
6.6	-0.4949	-0.4556^1	-0.1249^1	-0.3400^1
6.7	-0.4820	-0.5293^1	-0.1474^1	-0.3890^1
6.8	-0.4697	-0.6017^1	-0.1701^1	-0.4358^1
6.9	-0.4582	-0.6728^1	-0.1927^1	-0.4805^1
7.0	-0.4472	-0.7425^1	-0.2153^1	-0.5232^1
7.1	-0.4368	-0.8109^1	-0.2377^1	-0.5640^1
7.2	-0.4270	-0.8780^1	-0.2601^1	-0.6032^1
7.3	-0.4176	-0.9438^1	-0.2822^1	-0.6407^1
7.4	-0.4087	-0.1008	-0.3041^1	-0.6767^1
7.6	-0.3921	-0.1134	-0.3473^1	-0.7445^1
7.8	-0.3770	-0.1254	-0.3893^1	-0.8072^1
8.0	-0.3633	-0.1370	-0.4302^1	-0.8654^1
8.2	-0.3506	-0.1482	-0.4697^1	-0.9195^1
8.4	-0.3390	-0.1590	-0.5081^1	-0.9699^1
8.6	-0.3282	-0.1693	-0.5451^1	-0.1017
8.8	-0.3183	-0.1793	-0.5808^1	-0.1061
9.0	-0.3091	-0.1889	-0.6154^1	-0.1103
9.2	-0.3005	-0.1981	-0.6486^1	-0.1142
9.4	-0.2925	-0.2071	-0.6807^1	-0.1179
9.7	-0.2815	-0.2199	-0.7268^1	-0.1230
10.0	-0.2714	-0.2320	-0.7703^1	-0.1277
10.3	-0.2623	-0.2436	-0.8117^1	-0.1321
10.6	-0.2540	-0.2546	-0.8508^1	-0.1362
10.9	-0.2463	-0.2651	-0.8880^1	-0.1400
11.4	-0.2348	-0.2816	-0.9458^1	-0.1457
11.9	-0.2246	-0.2968	-0.9990^1	-0.1509
12.4	-0.2157	-0.3110	-0.1048	-0.1555
12.9	-0.2076	-0.3242	-0.1093	-0.1597
13.4	-0.2004	-0.3365	-0.1136	-0.1636
14.2	-0.1903	-0.3547	-0.1197	-0.1691
15.0	-0.1816	-0.3712	-0.1252	-0.1739
15.8	-0.1740	-0.3862	-0.1301	-0.1781
16.6	-0.1673	-0.3999	-0.1345	-0.1819
17.4	-0.1614	-0.4125	-0.1386	-0.1853

$\hat{\alpha}_3 = 1.30$

$\hat{\alpha}_4$	$\lambda_1(0,1)$	$\lambda_2(0,1)$	λ_3	λ_4
4.7	-1.1487	0.1206	0.6892^4	0.1610
4.8	-1.0839	0.1129	0.3178^2	0.1435
4.9	-1.0265	0.1049	0.5525^2	0.1276
5.0	-0.9751	0.9679^1	0.7227^2	0.1130
5.1	-0.9286	0.8855^1	0.8373^2	0.9954^1
5.2	-0.8863	0.8022^1	0.9037^2	0.8702^1
5.3	-0.8476	0.7184^1	0.9278^2	0.7536^1
5.4	-0.8121	0.6342^1	0.9149^2	0.6447^1
5.5	-0.7793	0.5501^1	0.8696^2	0.5428^1
5.6	-0.7490	0.4660^1	0.7960^2	0.4472^1
5.7	-0.7209	0.3824^1	0.6977^2	0.3573^1
5.8	-0.6948	0.2992^1	0.5782^2	0.2726^1
5.9	-0.6706	0.2169^1	0.4405^2	0.1930^1
6.0	-0.6480	0.1353^1	0.2870^2	0.1176^1
6.1	-0.6268	0.5450^2	0.1202^2	0.4638^2
6.2	-0.6071	-0.2528^2	-0.5774^3	-0.2108^2
6.3	-0.5885	-0.1040^1	-0.2450^2	-0.8504^2
6.4	-0.5711	-0.1816^1	-0.4399^2	-0.1457^1
6.5	-0.5547	-0.2581^1	-0.6411^2	-0.2034^1
6.6	-0.5393	-0.3333^1	-0.8469^2	-0.2583^1
6.7	-0.5247	-0.4074^1	-0.1057^1	-0.3106^1
6.8	-0.5110	-0.4802^1	-0.1269^1	-0.3604^1
6.9	-0.4980	-0.5517^1	-0.1483^1	-0.4080^1
7.0	-0.4858	-0.6219^1	-0.1698^1	-0.4533^1
7.1	-0.4741	-0.6910^1	-0.1914^1	-0.4968^1
7.2	-0.4631	-0.7587^1	-0.2129^1	-0.5383^1
7.3	-0.4527	-0.8252^1	-0.2344^1	-0.5781^1
7.4	-0.4427	-0.8905^1	-0.2558^1	-0.6162^1
7.5	-0.4332	-0.9546^1	-0.2770^1	-0.6528^1
7.6	-0.4242	-0.1017	-0.2980^1	-0.6880^1
7.8	-0.4074	-0.1140	-0.3395^1	-0.7543^1
8.0	-0.3921	-0.1257	-0.3801^1	-0.8156^1
8.2	-0.3780	-0.1371	-0.4195^1	-0.8727^1
8.4	-0.3651	-0.1480	-0.4579^1	-0.9258^1
8.6	-0.3532	-0.1585	-0.4951^1	-0.9753^1
8.8	-0.3422	-0.1686	-0.5311^1	-0.1022
9.0	-0.3321	-0.1784	-0.5659^1	-0.1065
9.2	-0.3226	-0.1878	-0.5996^1	-0.1106
9.4	-0.3138	-0.1969	-0.6321^1	-0.1145
9.6	-0.3056	-0.2057	-0.6636^1	-0.1181
9.9	-0.2942	-0.2183	-0.7087^1	-0.1232
10.2	-0.2839	-0.2303	-0.7515^1	-0.1279
10.5	-0.2744	-0.2417	-0.7922^1	-0.1322
10.8	-0.2658	-0.2525	-0.8308^1	-0.1362
11.1	-0.2578	-0.2629	-0.8675^1	-0.1400
11.6	-0.2459	-0.2791	-0.9247^1	-0.1457
12.1	-0.2354	-0.2942	-0.9774^1	-0.1508
12.6	-0.2261	-0.3082	-0.1026	-0.1554
13.1	-0.2177	-0.3213	-0.1071	-0.1596
13.6	-0.2102	-0.3335	-0.1113	-0.1634
14.4	-0.1996	-0.3516	-0.1174	-0.1689
15.2	-0.1905	-0.3679	-0.1229	-0.1737
16.0	-0.1826	-0.3828	-0.1278	-0.1779
16.8	-0.1757	-0.3965	-0.1323	-0.1817
17.6	-0.1695	-0.4090	-0.1363	-0.1851

$\hat{\alpha}_3 = 1.35$

$\hat{\alpha}_4$	$\lambda_1(0,1)$	$\lambda_2(0,1)$	λ_3	λ_4
5.0	-1.0925	0.1066	0.2023^2	0.1344
5.1	-1.0365	0.9874^1	0.4254^2	0.1193
5.2	-0.9863	0.9076^1	0.5874^2	0.1054
5.3	-0.9409	0.8269^1	0.6972^2	0.9257^1
5.4	-0.8995	0.7456^1	0.7615^2	0.8064^1
5.5	-0.8616	0.6640^1	0.7863^2	0.6953^1
5.6	-0.8267	0.5821^1	0.7763^2	0.5913^1
5.7	-0.7945	0.5004^1	0.7359^2	0.4938^1
5.8	-0.7647	0.4188^1	0.6689^2	0.4023^1
5.9	-0.7370	0.3377^1	0.5786^2	0.3161^1
6.0	-0.7113	0.2573^1	0.4683^2	0.2350^1
6.1	-0.6873	0.1774^1	0.3404^2	0.1583^1
6.2	-0.6648	0.9825^2	0.1976^2	0.8577^2
6.3	-0.6438	0.1993^2	0.4177^3	0.1704^2
6.4	-0.6242	-0.5370^2	-0.1246^2	-0.4801^2
6.5	-0.6056	-0.1337^1	-0.3006^2	-0.1099^1
6.6	-0.5883	-0.2088^1	-0.4834^2	-0.1685^1
6.7	-0.5719	-0.2830^1	-0.6728^2	-0.2244^1
6.8	-0.5565	-0.3560^1	-0.8671^2	-0.2776^1
6.9	-0.5419	-0.4278^1	-0.1065^1	-0.3284^1
6.9	-0.5419	-0.4279^1	-0.1066^1	-0.3284^1
7.0	-0.5280	-0.4987^1	-0.1267^1	-0.3769^1
7.1	-0.5151	-0.5680^1	-0.1470^1	-0.4230^1
7.2	-0.5027	-0.6363^1	-0.1674^1	-0.4672^1
7.3	-0.4910	-0.7034^1	-0.1880^1	-0.5096^1
7.4	-0.4798	-0.7693^1	-0.2085^1	-0.5501^1
7.5	-0.4692	-0.8341^1	-0.2291^1	-0.5890^1
7.6	-0.4592	-0.8977^1	-0.2495^1	-0.6263^1
7.7	-0.4496	-0.9602^1	-0.2699^1	-0.6621^1
7.8	-0.4404	-0.1021	-0.2901^1	-0.6965^1
7.9	-0.4317	-0.1082	-0.3101^1	-0.7296^1
8.1	-0.4154	-0.1199	-0.3497^1	-0.7921^1
8.3	-0.4004	-0.1312	-0.3883^1	-0.8502^1
8.5	-0.3867	-0.1420	-0.4260^1	-0.9043^1
8.7	-0.3740	-0.1525	-0.4626^1	-0.9548^1
8.9	-0.3623	-0.1627	-0.4982^1	-0.1002
9.1	-0.3515	-0.1724	-0.5327^1	-0.1046
9.3	-0.3414	-0.1818	-0.5661^1	-0.1088
9.5	-0.3320	-0.1909	-0.5985^1	-0.1127
9.7	-0.3232	-0.1997	-0.6298^1	-0.1164
9.9	-0.3150	-0.2082	-0.6601^1	-0.1199
10.2	-0.3037	-0.2204	-0.7036^1	-0.1248
10.5	-0.2933	-0.2320	-0.7450^1	-0.1293
10.8	-0.2839	-0.2430	-0.7844^1	-0.1335
11.1	-0.2752	-0.2536	-0.8219^1	-0.1374
11.4	-0.2672	-0.2637	-0.8576^1	-0.1411
11.9	-0.2551	-0.2795	-0.9133^1	-0.1466
12.4	-0.2445	-0.2941	-0.9648^1	-0.1516
12.9	-0.2350	-0.3078	-0.1012	-0.1561
13.4	-0.2265	-0.3206	-0.1057	-0.1602
13.9	-0.2188	-0.3326	-0.1098	-0.1639
14.7	-0.2080	-0.3503	-0.1158	-0.1693
15.5	-0.1987	-0.3664	-0.1212	-0.1740
16.3	-0.1906	-0.3810	-0.1261	-0.1782
17.1	-0.1834	-0.3945	-0.1305	-0.1819
17.9	-0.1770	-0.4068	-0.1345	-0.1853

$$\hat{\alpha}_3 = 1.40 \qquad\qquad\qquad\qquad \hat{\alpha}_3 = 1.45$$

$\hat{\alpha}_4$	$\lambda_1(0,1)$	$\lambda_2(0,1)$	λ_3	λ_4	$\hat{\alpha}_4$	$\lambda_1(0,1)$	$\lambda_2(0,1)$	λ_3	λ_4
5.2	-1.1058	0.1011	0.7869^3	0.1268	5.5	-1.0691	0.8871^1	0.1642^2	0.1068
5.3	-1.0507	0.9335^1	0.2960^2	0.1124	5.6	-1.0199	0.8105^1	0.3245^2	0.9396^1
5.4	-1.0012	0.8553^1	0.4548^2	0.9909^1	5.7	-0.9754	0.7333^1	0.4364^2	0.8210^1
5.5	-0.9565	0.7764^1	0.5638^2	0.8680^1	5.8	-0.9348	0.6559^1	0.5068^2	0.7108^1
5.6	-0.9157	0.6970^1	0.6297^2	0.7537^1	5.9	-0.8976	0.5784^1	0.5408^2	0.6078^1
5.7	-0.8783	0.6175^1	0.6579^2	0.6470^1	6.0	-0.8634	0.5009^1	0.5432^2	0.5114^1
5.8	-0.8438	0.5379^1	0.6531^2	0.5471^1	6.1	-0.8317	0.4236^1	0.5177^2	0.4208^1
5.9	-0.8119	0.4584^1	0.6194^2	0.4534^1	6.2	-0.8022	0.3467^1	0.4677^2	0.3356^1
6.0	-0.7824	0.3792^1	0.5603^2	0.3653^1	6.3	-0.7749	0.2703^1	0.3964^2	0.2553^1
6.1	-0.7550	0.3006^1	0.4791^2	0.2824^1	6.4	-0.7494	0.1944^1	0.3063^2	0.1794^1
6.2	-0.7294	0.2225^1	0.3785^2	0.2041^1	6.5	-0.7255	0.1192^1	0.1998^2	0.1076^1
6.3	-0.7056	0.1450^1	0.2611^2	0.1300^1	6.6	-0.7031	0.4469^2	0.7911^3	0.3948^2
6.4	-0.6832	0.6823^2	0.1292^2	0.5987^2	6.7	-0.6821	-0.2901^2	-0.5389^3	-0.2521^2
6.5	-0.6622	-0.7667^3	-0.1518^3	-0.6591^3	6.8	-0.6624	-0.1018^1	-0.1975^2	-0.8649^2
6.6	-0.6426	-0.8266^2	-0.1702^2	-0.6968^2	6.9	-0.6438	-0.1738^1	-0.3501^2	-0.1449^1
6.7	-0.6241	-0.1567^1	-0.3343^2	-0.1296^1	7.0	-0.6263	-0.2447^1	-0.5103^2	-0.2005^1
6.8	-0.6067	-0.2297^1	-0.5060^2	-0.1867^1	7.1	-0.6097	-0.3147^1	-0.6770^2	-0.2535^1
6.9	-0.5902	-0.3017^1	-0.6840^2	-0.2410^1	7.2	-0.5940	-0.3837^1	-0.8492^2	-0.3041^1
7.0	-0.5747	-0.3726^1	-0.8671^2	-0.2928^1	7.3	-0.5793	-0.4516^1	-0.1025^1	-0.3523^1
7.1	-0.5600	-0.4424^1	-0.1054^1	-0.3422^1	7.4	-0.5652	-0.5185^1	-0.1205^1	-0.3985^1
7.2	-0.5461	-0.5112^1	-0.1245^1	-0.3894^1	7.5	-0.5519	-0.5843^1	-0.1388^1	-0.4427^1
7.3	-0.5330	-0.5789^1	-0.1438^1	-0.4346^1	7.6	-0.5393	-0.6491^1	-0.1572^1	-0.4851^1
7.4	-0.5205	-0.6452^1	-0.1632^1	-0.4778^1	7.7	-0.5272	-0.7128^1	-0.1758^1	-0.5256^1
7.5	-0.5086	-0.7106^1	-0.1827^1	-0.5191^1	7.8	-0.5158	-0.7755^1	-0.1945^1	-0.5646^1
7.6	-0.4974	-0.7748^1	-0.2023^1	-0.5588^1	7.9	-0.5049	-0.8371^1	-0.2133^1	-0.6020^1
7.7	-0.4866	-0.8380^1	-0.2220^1	-0.5969^1	8.0	-0.4945	-0.8977^1	-0.2320^1	-0.6379^1
7.8	-0.4764	-0.9000^1	-0.2415^1	-0.6335^1	8.1	-0.4845	-0.9572^1	-0.2507^1	-0.6724^1
7.9	-0.4667	-0.9609^1	-0.2610^1	-0.6686^1	8.2	-0.4751	-0.1016	-0.2693^1	-0.7057^1
8.0	-0.4574	-0.1021	-0.2804^1	-0.7023^1	8.3	-0.4660	-0.1073	-0.2878^1	-0.7377^1
8.1	-0.4485	-0.1080	-0.2997^1	-0.7349^1	8.4	-0.4573	-0.1130	-0.3062^1	-0.7685^1
8.3	-0.4318	-0.1194	-0.3378^1	-0.7963^1	8.6	-0.4410	-0.1240	-0.3426^1	-0.8270^1
8.5	-0.4165	-0.1305	-0.3751^1	-0.8535^1	8.8	-0.4260	-0.1346	-0.3784^1	-0.8815^1
8.7	-0.4025	-0.1411	-0.4116^1	-0.9068^1	9.0	-0.4122	-0.1449	-0.4133^1	-0.9324^1
8.9	-0.3895	-0.1514	-0.4472^1	-0.9566^1	9.2	-0.3994	-0.1548	-0.4474^1	-0.9801^1
9.1	-0.3775	-0.1613	-0.4818^1	-0.1003	9.4	-0.3875	-0.1644	-0.4806^1	-0.1025
9.3	-0.3664	-0.1709	-0.5154^1	-0.1047	9.6	-0.3765	-0.1736	-0.5129^1	-0.1067
9.5	-0.3560	-0.1801	-0.5481^1	-0.1088	9.8	-0.3662	-0.1826	-0.5444^1	-0.1107
9.7	-0.3464	-0.1891	-0.5797^1	-0.1127	10.0	-0.3566	-0.1912	-0.5749^1	-0.1144
9.9	-0.3373	-0.1977	-0.6104^1	-0.1164	10.2	-0.3476	-0.1996	-0.6044^1	-0.1179
10.1	-0.3289	-0.2061	-0.6401^1	-0.1198	10.4	-0.3391	-0.2077	-0.6331^1	-0.1213
10.4	-0.3172	-0.2181	-0.6829^1	-0.1247	10.7	-0.3274	-0.2193	-0.6746^1	-0.1260
10.7	-0.3065	-0.2295	-0.7236^1	-0.1292	11.0	-0.3166	-0.2305	-0.7141^1	-0.1303
11.0	-0.2967	-0.2404	-0.7625^1	-0.1334	11.3	-0.3068	-0.2411	-0.7518^1	-0.1344
11.3	-0.2877	-0.2509	-0.7995^1	-0.1372	11.6	-0.2977	-0.2512	-0.7878^1	-0.1382
11.6	-0.2794	-0.2608	-0.8347^1	-0.1409	11.9	-0.2893	-0.2609	-0.8222^1	-0.1417
12.1	-0.2669	-0.2764	-0.8899^1	-0.1464	12.4	-0.2764	-0.2765	-0.8771^1	-0.1472
12.6	-0.2558	-0.2910	-0.9410^1	-0.1513	12.9	-0.2652	-0.2907	-0.9271^1	-0.1520
13.1	-0.2460	-0.3046	-0.9884^1	-0.1558	13.4	-0.2552	-0.3039	-0.9735^1	-0.1564
13.6	-0.2371	-0.3173	-0.1032	-0.1599	13.9	-0.2462	-0.3163	-0.1017	-0.1604
14.1	-0.2291	-0.3292	-0.1073	-0.1637	14.4	-0.2381	-0.3280	-0.1057	-0.1641
14.9	-0.2179	-0.3467	-0.1133	-0.1690	15.2	-0.2266	-0.3452	-0.1116	-0.1693
15.7	-0.2082	-0.3627	-0.1187	-0.1737	16.0	-0.2166	-0.3609	-0.1169	-0.1740
16.5	-0.1997	-0.3773	-0.1236	-0.1779	16.8	-0.2079	-0.3753	-0.1218	-0.1781
17.3	-0.1922	-0.3907	-0.1281	-0.1816	17.6	-0.2002	-0.3885	-0.1262	-0.1818
18.1	-0.1855	-0.4030	-0.1321	-0.1850	18.4	-0.1934	-0.4006	-0.1302	-0.1851

$$\hat{\alpha}_3 = 1.50$$

$\hat{\alpha}_4$	$\lambda_1(0,1)$	$\lambda_2(0,1)$	λ_3	λ_4
5.7	-1.0915	0.8480^1	0.2943^3	0.1024
5.8	-1.0422	0.7728^1	0.1952^2	0.8991^1
5.9	-0.9976	0.6974^1	0.3133^2	0.7840^1
6.0	-0.9570	0.6219^1	0.3908^2	0.6770^1
6.1	-0.9197	0.5463^1	0.4329^2	0.5770^1
6.2	-0.8854	0.4708^1	0.4442^2	0.4834^1
6.3	-0.8536	0.3956^1	0.4283^2	0.3954^1
6.4	-0.8242	0.3209^1	0.3885^2	0.3127^1
6.5	-0.7968	0.2465^1	0.3279^2	0.2345^1
6.6	-0.7711	0.1728^1	0.2489^2	0.1606^1
6.7	-0.7471	0.9962^2	0.1538^2	0.9059^2
6.8	-0.7246	0.2719^2	0.4457^3	0.2422^2
6.9	-0.7035	-0.4446^2	-0.7684^3	-0.3881^2
7.0	-0.6836	-0.1153^1	-0.2089^2	-0.9875^2
7.1	-0.6648	-0.1852^1	-0.3500^2	-0.1558^1
7.2	-0.6471	-0.2543^1	-0.4989^2	-0.2102^1
7.3	-0.6304	-0.3224^1	-0.6545^2	-0.2621^1
7.4	-0.6146	-0.3896^1	-0.8157^2	-0.3116^1
7.5	-0.5995	-0.4559^1	-0.9818^2	-0.3591^1
7.6	-0.5853	-0.5212^1	-0.1152^1	-0.4045^1
7.7	-0.5718	-0.5851^1	-0.1324^1	-0.4477^1
7.8	-0.5590	-0.6483^1	-0.1498^1	-0.4893^1
7.9	-0.5468	-0.7105^1	-0.1675^1	-0.5292^1
8.0	-0.5351	-0.7716^1	-0.1853^1	-0.5676^1
8.1	-0.5240	-0.8318^1	-0.2032^1	-0.6044^1
8.2	-0.5134	-0.8910^1	-0.2210^1	-0.6398^1
8.3	-0.5033	-0.9492^1	-0.2389^1	-0.6738^1
8.4	-0.4936	-0.1006	-0.2568^1	-0.7066^1
8.5	-0.4843	-0.1063	-0.2746^1	-0.7382^1
8.6	-0.4754	-0.1118	-0.2923^1	-0.7687^1
8.8	-0.4588	-0.1226	-0.3274^1	-0.8265^1
9.0	-0.4434	-0.1330	-0.3620^1	-0.8804^1
9.2	-0.4292	-0.1431	-0.3959^1	-0.9308^1
9.4	-0.4160	-0.1528	-0.4290^1	-0.9781^1
9.6	-0.4038	-0.1622	-0.4613^1	-0.1022
9.8	-0.3925	-0.1713	-0.4929^1	-0.1064
10.0	-0.3818	-0.1802	-0.5236^1	-0.1104
10.2	-0.3719	-0.1887	-0.5535^1	-0.1141
10.4	-0.3626	-0.1969	-0.5825^1	-0.1176
10.6	-0.3539	-0.2049	-0.6107^1	-0.1209
10.9	-0.3417	-0.2164	-0.6514^1	-0.1256
11.2	-0.3306	-0.2274	-0.6904^1	-0.1299
11.5	-0.3204	-0.2379	-0.7276^1	-0.1340
11.8	-0.3110	-0.2479	-0.7632^1	-0.1377
12.1	-0.3023	-0.2575	-0.7973^1	-0.1413
12.6	-0.2892	-0.2727	-0.8508^1	-0.1466
13.1	-0.2775	-0.2868	-0.9005^1	-0.1515
13.6	-0.2671	-0.2999	-0.9467^1	-0.1559
14.1	-0.2577	-0.3123	-0.9899^1	-0.1599
14.6	-0.2492	-0.3239	-0.1030	-0.1636
15.4	-0.2372	-0.3411	-0.1089	-0.1688
16.2	-0.2268	-0.3567	-0.1143	-0.1735
17.0	-0.2177	-0.3710	-0.1191	-0.1776
17.8	-0.2096	-0.3842	-0.1235	-0.1813
18.6	-0.2025	-0.3963	-0.1276	-0.1847

$$\hat{\alpha}_3 = 1.55$$

$\hat{\alpha}_4$	$\lambda_1(0,1)$	$\lambda_2(0,1)$	λ_3	λ_4
6.0	-1.0683	0.7421	0.6443^3	$.8687^1$
6.1	-1.0232	0.6683^1	0.1921^2	0.7562^1
6.2	-0.9822	0.5945^1	0.2795^2	0.6516^1
6.3	-0.9446	0.5207^1	0.3318^2	0.5540^1
6.4	-0.9100	0.4472^1	0.3536^2	0.4626^1
6.5	-0.8780	0.3740^1	0.3485^2	0.3768^1
6.6	-0.8483	0.3011^1	0.3201^2	0.2959^1
6.7	-0.8206	0.2288^1	0.2710^2	0.2195^1
6.8	-0.7947	0.1570^1	0.2037^2	0.1472^1
6.9	-0.7705	0.8580^2	0.1203^2	0.7874^2
7.0	-0.7478	0.1532^2	0.2299^3	0.1377^2
7.1	-0.7264	-0.5442^2	-0.8663^3	-0.4797^2
7.2	-0.7063	-0.1234^1	-0.2070^2	-0.1067^1
7.3	-0.6873	-0.1915^1	-0.3366^2	-0.1627^1
7.4	-0.6694	-0.2587^1	-0.4742^2	-0.2161^1
7.5	-0.6524	-0.3251^1	-0.6186^2	-0.2670^1
7.6	-0.6363	-0.3906^1	-0.7689^2	-0.3157^1
7.7	-0.6211	-0.4551^1	-0.9242^2	-0.3623^1
7.8	-0.6066	-0.5187^1	-0.1084^1	-0.4069^1
7.9	-0.5929	-0.5814^1	-0.1246^1	-0.4497^1
8.0	-0.5798	-0.6431^1	-0.1412^1	-0.4907^1
8.1	-0.5673	-0.7038^1	-0.1579^1	-0.5300^1
8.2	-0.5554	-0.7636^1	-0.1748^1	-0.5678^1
8.3	-0.5441	-0.8224^1	-0.1918^1	-0.6042^1
8.4	-0.5333	-0.8803^1	-0.2089^1	-0.6391^1
8.5	-0.5229	-0.9373^1	-0.2260^1	-0.6728^1
8.6	-0.5130	-0.9933^1	-0.2431^1	-0.7052^1
8.7	-0.5035	-0.1048	-0.2602^1	-0.7365^1
8.8	-0.4944	-0.1103	-0.2773^1	-0.7667^1
8.9	-0.4857	-0.1156	-0.2942^1	-0.7958^1
9.1	-0.4692	-0.1260	-0.3278^1	-0.8511^1
9.3	-0.4541	-0.1361	-0.3610^1	-0.9028^1
9.5	-0.4400	-0.1458	-0.3935^1	-0.9512^1
9.7	-0.4270	-0.1552	-0.4253^1	-0.9967^1
9.9	-0.4148	-0.1643	-0.4565^1	-0.1039
10.1	-0.4035	-0.1732	-0.4869^1	-0.1080
10.3	-0.3929	-0.1817	-0.5165^1	-0.1118
10.5	-0.3830	-0.1900	-0.5454^1	-0.1154
10.7	-0.3737	-0.1980	-0.5734^1	-0.1188
10.9	-0.3649	-0.2057	-0.6008^1	-0.1220
11.2	-0.3527	-0.2169	-0.6403^1	-0.1266
11.5	-0.3415	-0.2276	-0.6782^1	-0.1308
11.8	-0.3312	-0.2379	-0.7144^1	-0.1347
12.1	-0.3217	-0.2477	-0.7491^1	-0.1384
12.4	-0.3129	-0.2570	-0.7824^1	-0.1419
12.9	-0.2996	-0.2718	-0.8347^1	-0.1471
13.4	-0.2877	-0.2856	-0.8834^1	-0.1519
13.9	-0.2771	-0.2985	-0.9289^1	-0.1562
14.4	-0.2675	-0.3107	-0.9713^1	-0.1601
14.9	-0.2588	-0.3221	-0.1011	-0.1638
15.7	-0.2465	-0.3389	-0.1069	-0.1690
16.5	-0.2359	-0.3543	-0.1122	-0.1736
17.3	-0.2265	-0.3685	-0.1170	-0.1776
18.1	-0.2182	-0.3815	-0.1214	-0.1813
18.9	-0.2109	-0.3935	-0.1254	-0.1846

$\hat{\alpha}_3 = 1.60$

$\hat{\alpha}_4$	$\lambda_1(0,1)$	$\lambda_2(0,1)$	λ_3	λ_4
6.3	-1.0524	0.6456^1	0.7024^3	0.7371^1
6.4	-1.0106	0.5734^1	0.1700^2	0.6343^1
6.5	-0.9725	0.5014^1	0.2344^2	0.5385^1
6.6	-0.9373	0.4296^1	0.2684^2	0.4487^1
6.7	-0.9048	0.3582^1	0.2758^2	0.3644^1
6.8	-0.8747	0.2872^1	0.2597^2	0.2850^1
6.9	-0.8466	0.2167^1	0.2229^2	0.2100^1
7.0	-0.8203	0.1467^1	0.1680^2	0.1390^1
7.1	-0.7958	0.7738^2	0.9694^3	0.7177^2
7.2	-0.7727	0.8695^3	0.1177^3	0.7902^3
7.3	-0.7510	-0.5924^2	-0.8576^3	-0.5279^2
7.4	-0.7305	-0.1265^1	-0.1943^2	-0.1106^1
7.5	-0.7113	-0.1928^1	-0.3121^2	-0.1657^1
7.6	-0.6930	-0.2584^1	-0.4383^2	-0.2182^1
7.7	-0.6757	-0.3232^1	-0.5716^2	-0.2685^1
7.8	-0.6594	-0.3870^1	-0.7111^2	-0.3165^1
7.9	-0.6438	-0.4500^1	-0.8558^2	-0.3624^1
8.0	-0.6291	-0.5121^1	-0.1005^1	-0.4065^1
8.1	-0.6150	-0.5733^1	-0.1158^1	-0.4487^1
8.2	-0.6016	-0.6338^1	-0.1314^1	-0.4893^1
8.3	-0.5889	-0.6931^1	-0.1472^1	-0.5282^1
8.4	-0.5768	-0.7516^1	-0.1632^1	-0.5655^1
8.5	-0.5652	-0.8092^1	-0.1794^1	-0.6015^1
8.6	-0.5541	-0.8659^1	-0.1957^1	-0.6361^1
8.7	-0.5435	-0.9217^1	-0.2120^1	-0.6695^1
8.8	-0.5333	-0.9766^1	-0.2284^1	-0.7016^1
8.9	-0.5236	-0.1031	-0.2448^1	-0.7326^1
9.0	-0.5142	-0.1084	-0.2611^1	-0.7625^1
9.1	-0.5053	-0.1136	-0.2775^1	-0.7914^1
9.2	-0.4967	-0.1188	-0.2937^1	-0.8193^1
9.4	-0.4804	-0.1288	-0.3260^1	-0.8724^1
9.6	-0.4654	-0.1386	-0.3578^1	-0.9222^1
9.8	-0.4515	-0.1480	-0.3890^1	-0.9689^1
10.0	-0.4385	-0.1571	-0.4197^1	-0.1013
10.2	-0.4264	-0.1659	-0.4497^1	-0.1054
10.4	-0.4150	-0.1745	-0.4790^1	-0.1093
10.6	-0.4044	-0.1828	-0.5077^1	-0.1130
10.8	-0.3945	-0.1908	-0.5356^1	-0.1165
11.0	-0.3852	-0.1986	-0.5628^1	-0.1198
11.2	-0.3763	-0.2061	-0.5893^1	-0.1230
11.5	-0.3641	-0.2171	-0.6277^1	-0.1274
11.8	-0.3528	-0.2275	-0.6645^1	-0.1315
12.1	-0.3424	-0.2375	-0.6999^1	-0.1354
12.4	-0.3327	-0.2470	-0.7338^1	-0.1390
12.7	-0.3238	-0.2562	-0.7663^1	-0.1423
13.2	-0.3103	-0.2707	-0.8175^1	-0.1475
13.7	-0.2982	-0.2842	-0.8653^1	-0.1522
14.2	-0.2874	-0.2969	-0.9100^1	-0.1564
14.7	-0.2776	-0.3088	-0.9518^1	-0.1603
15.2	-0.2687	-0.3200	-0.9910^1	-0.1639
16.0	-0.2561	-0.3366	-0.1049	-0.1690
16.8	-0.2452	-0.3518	-0.1101	-0.1736
17.6	-0.2356	-0.3658	-0.1149	-0.1776
18.4	-0.2270	-0.3786	-0.1192	-0.1812
19.2	-0.2195	-0.3905	-0.1232	-0.1845

$\hat{\alpha}_3 = 1.65$

$\hat{\alpha}_4$	$\lambda_1(0,1)$	$\lambda_2(0,1)$	λ_3	λ_4
6.6	-1.0426	0.5584^1	0.5891^3	0.6247^1
6.7	-1.0034	0.4879^1	0.1379^2	0.5300^1
6.8	-0.9675	0.4178^1	0.1860^2	0.4414^1
6.9	-0.9343	0.3480^1	0.2070^2	0.3582^1
7.0	-0.9035	0.2787^1	0.2044^2	0.2798^1
7.1	-0.8748	0.2099^1	0.1809^2	0.2058^1
7.2	-0.8480	0.1417^1	0.1391^2	0.1358^1
7.3	-0.8230	0.7403^2	0.8092^3	0.6949^2
7.4	-0.7995	0.7037^3	0.842563	0.6472^3
7.5	-0.7773	-0.5924^2	-0.7669^3	-0.5340^2
7.6	-0.7565	-0.1248^1	-0.1730^2	-0.1105^1
7.7	-0.7368	-0.1896^1	-0.2790^2	-0.1649^1
7.8	-0.7182	-0.2536^1	-0.3936^2	-0.2169^1
7.9	-0.7005	-0.3168^1	-0.5157^2	-0.2665^1
8.0	-0.6838	-0.3792^1	-0.6443^2	-0.3140^1
8.1	-0.6679	-0.4408^1	-0.7784^2	-0.3595^1
8.2	-0.6528	-0.5015^1	-0.9173^2	-0.4030^1
8.3	-0.6384	-0.5614^1	-0.1060^1	-0.4449^1
8.4	-0.6248	-0.6204^1	-0.1207^1	-0.4850^1
8.5	-0.6116	-0.6787^1	-0.1356^1	-0.5237^1
8.6	-0.5992	-0.7359^1	-0.1508^1	-0.5607^1
8.7	-0.5873	-0.7923^1	-0.1661^1	-0.5964^1
8.8	-0.5759	-0.8479^1	-0.1816^1	-0.6308^1
8.9	-0.5650	-0.9026^1	-0.1971^1	-0.6639^1
9.0	-0.5546	-0.9565^1	-0.2128^1	-0.6958^1
9.1	-0.5446	-0.1009	-0.2284^1	-0.7266^1
9.2	-0.5350	-0.1062	-0.2441^1	-0.7563^1
9.3	-0.5258	-0.1113	-0.2598^1	-0.7851^1
9.4	-0.5169	-0.1164	-0.2755^1	-0.8129^1
9.5	-0.5084	-0.1214	-0.2910^1	-0.8397^1
9.7	-0.4923	-0.1311	-0.3220^1	-0.8909^1
9.9	-0.4774	-0.1405	-0.3526^1	-0.9389^1
10.1	-0.4635	-0.1496	-0.3827^1	-0.9840^1
10.3	-0.4505	-0.1585	-0.4122^1	-0.1027
10.5	-0.4384	-0.1671	-0.4412^1	-0.1067
10.7	-0.4271	-0.1754	-0.4695^1	-0.1105
10.9	-0.4165	-0.1834	-0.4972^1	-0.1140
11.1	-0.4065	-0.1912	-0.5243^1	-0.1174
11.3	-0.3971	-0.1988	-0.5506^1	-0.1207
11.5	-0.3882	-0.2062	-0.5764^1	-0.1237
11.8	-0.3758	-0.2168	-0.6137^1	-0.1281
12.1	-0.3644	-0.2270	-0.6496^1	-0.1321
12.4	-0.3539	-0.2368	-0.6841^1	-0.1359
12.7	-0.3441	-0.2461	-0.7172^1	-0.1394
13.0	-0.3351	-0.2551	-0.7490^1	-0.1427
13.5	-0.3213	-0.2693	-0.7992^1	-0.1478
14.0	-0.3090	-0.2825	-0.8462^1	-0.1524
14.5	-0.2980	-0.2950	-0.8902^1	-0.1566
15.0	-0.2880	-0.3067	-0.9314^1	-0.1604
15.5	-0.2789	-0.3177	-0.9701^1	-0.1639
16.3	-0.2660	-0.3341	-0.1027	-0.1690
17.1	-0.2548	-0.3491	-0.1079	-0.1735
17.9	-0.2449	-0.3629	-0.1127	-0.1775
18.7	-0.2361	-0.3756	-0.1170	-0.1811
19.5	-0.2283	-0.3873	-0.1210	-0.1844

$$\hat{\alpha}_3 = 1.70 \qquad\qquad \hat{\alpha}_3 = 1.75$$

$\hat{\alpha}_4$	$\lambda_1(0,1)$	$\lambda_2(0,1)$	λ_3	λ_4	$\hat{\alpha}_4$	$\lambda_1(0,1)$	$\lambda_2(0,1)$	λ_3	λ_4
6.9	-1.0378	0.4800^1	0.3862^3	0.5286^1	7.2	-1.0375	0.4103^1	0.1488^3	0.4462^1
7.0	-1.0007	0.4114^1	0.1027^2	0.4405^1	7.3	-1.0021	0.3434^1	0.6850^3	0.3637^1
7.1	-0.9666	0.3432^1	0.1391^2	0.3580^1	7.4	-0.9693	0.2771^1	0.9698^3	0.2862^1
7.2	-0.9349	0.2754^1	0.1513^2	0.2802^1	7.5	-0.9389	0.2113^1	0.1034^2	0.2131^1
7.3	-0.9055	0.2082^1	0.1421^2	0.2069^1	7.6	-0.9105	0.1461^1	0.9041^3	0.1441^1
7.4	-0.8780	0.1415^1	0.1142^2	0.1375^1	7.7	-0.8840	0.8145^2	0.6025^3	0.7864^2
7.5	-0.8523	0.7547^2	0.6962^3	0.7179^2	7.8	-0.8604	0.1993^2	0.1688^3	0.1887^2
7.6	-0.8282	0.1002^2	0.1033^3	0.9338^3	7.9	-0.8358	-0.4595^2	-0.4393^3	-0.4254^2
7.7	-0.8055	-0.5470^2	-0.6191^3	-0.5000^2	8.0	-0.8138	-0.1085^1	-0.1147^2	-0.9880^2
7.8	-0.7842	-0.1187^1	-0.1457^2	-0.1066^1	8.1	-0.7931	-0.1704^1	-0.1960^2	-0.1524^1
7.9	-0.7640	-0.1821^1	-0.2396^2	-0.1605^1	8.2	-0.7734	-0.2317^1	-0.2867^2	-0.2037^1
8.0	-0.7449	-0.2446^1	-0.3423^2	-0.2120^1	8.3	-0.7548	-0.2921^1	-0.3856^2	-0.2527^1
8.1	-0.7268	-0.3064^1	-0.4530^2	-0.2612^1	8.4	-0.7372	-0.3519^1	-0.4918^2	-0.2996^1
8.2	-0.7096	-0.3677^1	-0.5710^2	-0.3085^1	8.5	-0.7204	-0.4108^1	-0.6043^2	-0.3445^1
8.3	-0.6934	-0.4277^1	-0.6939^2	-0.3535^1	8.6	-0.7045	-0.4690^1	-0.7225^2	-0.3876^1
8.4	-0.6779	-0.4871^1	-0.8225^2	-0.3967^1	8.7	-0.6893	-0.5264^1	-0.8454^2	-0.4289^1
8.5	-0.6632	-0.5457^1	-0.9556^2	-0.4383^1	8.8	-0.6748	-0.5831^1	-0.9725^2	-0.4686^1
8.6	-0.6491	-0.6035^1	-0.1092^1	-0.4781^1	8.9	-0.6610	-0.6389^1	-0.1103^1	-0.5068^1
8.7	-0.6357	-0.6604^1	-0.1232^1	-0.5165^1	9.0	-0.6478	-0.6940^1	-0.1237^1	-0.5436^1
8.8	-0.6228	-0.7168^1	-0.1376^1	-0.5535^1	9.1	-0.6351	-0.7486^1	-0.1374^1	-0.5791^1
8.9	-0.6106	-0.7720^1	-0.1520^1	-0.5889^1	9.2	-0.6231	-0.8018^1	-0.1512^1	-0.6131^1
9.0	-0.5989	-0.8265^1	-0.1667^1	-0.6231^1	9.3	-0.6115	-0.8546^1	-0.1652^1	-0.6460^1
9.1	-0.5877	-0.8801^1	-0.1815^1	-0.6560^1	9.4	-0.6004	-0.9066^1	-0.1794^1	-0.6777^1
9.2	-0.5769	-0.9330^1	-0.1964^1	-0.6878^1	9.5	-0.5898	-0.9578^1	-0.1936^1	-0.7083^1
9.3	-0.5666	-0.9851^1	-0.2113^1	-0.7185^1	9.6	-0.5796	-0.1008	-0.2080^1	-0.7379^1
9.4	-0.5567	-0.1036	-0.2264^1	-0.7481^1	9.7	-0.5697	-0.1058	-0.2223^1	-0.7665^1
9.5	-0.5472	-0.1087	-0.2414^1	-0.7768^1	9.8	-0.5603	-0.1107	-0.2368^1	-0.7942^1
9.6	-0.5381	-0.1137	-0.2564^1	-0.8045^1	9.9	-0.5512	-0.1155	-0.2512^1	-0.8210^1
9.7	-0.5293	-0.1186	-0.2714^1	-0.8313^1	10.0	-0.5425	-0.1203	-0.2656^1	-0.8469^1
9.8	-0.5208	-0.1234	-0.2864^1	-0.8572^1	10.1	-0.5340	-0.1249	-0.2799^1	-0.8720^1
10.0	-0.5048	-0.1328	-0.3161^1	-0.9066^1	10.3	-0.5181	-0.1341	-0.3085^1	-0.9199^1
10.2	-0.4900	-0.1420	-0.3455^1	-0.9531^1	10.5	-0.5032	-0.1430	-0.3369^1	-0.9650^1
10.4	-0.4761	-0.1508	-0.3745^1	-0.9968^1	10.7	-0.4893	-0.1516	-0.3648^1	-0.1008
10.6	-0.4632	-0.1594	-0.4030^1	-0.1038	10.9	-0.4764	-0.1600	-0.3923^1	-0.1048
10.8	-0.4510	-0.1678	-0.4310^1	-0.1077	11.1	-0.4642	-0.1681	-0.4194^1	-0.1086
11.0	-0.4397	-0.1758	-0.4584^1	-0.1114	11.3	-0.4528	-0.1759	-0.4459^1	-0.1122
11.2	-0.4290	-0.1837	-0.4853^1	-0.1149	11.5	-0.4420	-0.1836	-0.4719^1	-0.1156
11.4	-0.4189	-0.1913	-0.5115^1	-0.1182	11.7	-0.4319	-0.1910	-0.4974^1	-0.1188
11.6	-0.4095	-0.1987	-0.5371^1	-0.1214	11.9	-0.4223	-0.1982	-0.5223^1	-0.1219
11.8	-0.4005	-0.2059	-0.5621^1	-0.1244	12.1	-0.4133	-0.2052	-0.5466^1	-0.1248
12.1	-0.3880	-0.2163	-0.5985^1	-0.1286	12.4	-0.4007	-0.2154	-0.5820^1	-0.1290
12.4	-0.3765	-0.2262	-0.6335^1	-0.1325	12.7	-0.3890	-0.2251	-0.6162^1	-0.1329
12.7	-0.3658	-0.2358	-0.6672^1	-0.1362	13.0	-0.3781	-0.2345	-0.6491^1	-0.1365
13.0	-0.3559	-0.2449	-0.6995^1	-0.1397	13.3	-0.3681	-0.2435	-0.6808^1	-0.1399
13.3	-0.3467	-0.2537	-0.7307^1	-0.1429	13.6	-0.3587	-0.2521	-0.7113^1	-0.1431
13.8	-0.3327	-0.2676	-0.7800^1	-0.1479	14.1	-0.3444	-0.2657	-0.7597^1	-0.1480
14.3	-0.3202	-0.2807	-0.8262^1	-0.1525	14.6	-0.3316	-0.2785	-0.8052^1	-0.1525
14.8	-0.3089	-0.2929	-0.8695^1	-0.1566	15.1	-0.3201	-0.2906	-0.8479^1	-0.1566
15.3	-0.2987	-0.3044	-0.9102^1	-0.1604	15.6	-0.3097	-0.3019	-0.8880^1	-0.1603
15.8	-0.2894	-0.3153	-0.9484^1	-0.1639	16.1	-0.3001	-0.3126	-0.9259^1	-0.1638
16.6	-0.2761	-0.3314	-0.1005	-0.1689	16.9	-0.2866	-0.3286	-0.9819^1	-0.1688
17.4	-0.2646	-0.3462	-0.1056	-0.1734	17.7	-0.2747	-0.3432	-0.1033	-0.1732
18.2	-0.2544	-0.3598	-0.1104	-0.1774	18.5	-0.2642	-0.3567	-0.1080	-0.1772
19.0	-0.2454	-0.3724	-0.1147	-0.1809	19.3	-0.2549	-0.3691	-0.1123	-0.1807
19.8	-0.2373	-0.3840	-0.1186	-0.1842	20.1	-0.2466	-0.3806	-0.1162	-0.1839

	$\hat{\alpha}_3 = 1.80$					$\hat{\alpha}_3 = 1.85$		
$\hat{\alpha}_4$	$\lambda_1(0,1)$	$\lambda_2(0,1)$	λ_3	λ_4	$\hat{\alpha}_4$	$\lambda_1(0,1)$	$\lambda_2(0,1)$	λ_3	λ_4
7.6	-1.0070	0.2835^1	0.3780^3	0.2979^1	7.9	-1.0151	0.2312^1	0.1210^3	0.2416^1
7.7	-0.9753	0.2190^1	0.6135^3	0.2246^1	8.0	-0.9842	0.1684^1	0.3311^3	0.1720^1
7.8	-0.9458	0.1551^1	0.6456^3	0.1555^1	8.1	-0.9555	0.1063^1	0.3515^3	0.1062^1
7.9	-0.9183	0.9177^2	0.4984^3	0.9006^2	8.2	-0.9286	0.4480^2	0.2041^3	0.4384^2
8.0	-0.8925	0.2910^2	0.1925^3	0.2798^2	8.3	-0.9034	-0.1603^2	-0.9175^4	-0.1538^2
8.1	-0.8684	-0.3293^2	-0.2542^3	-0.3104^2	8.4	-0.8798	-0.7621^2	-0.5195^3	-0.7173^2
8.2	-0.8434	-0.1004^1	-0.8891^3	-0.9268^2	8.5	-0.8574	-0.1357^1	-0.1065^2	-0.1254^1
8.3	-0.8221	-0.1610^1	-0.1582^2	-0.1460^1	8.6	-0.8364	-0.1945^1	-0.1714^2	-0.1767^1
8.4	-0.8019	-0.2209^1	-0.2372^2	-0.1969^1	8.7	-0.8164	-0.2527^1	-0.2456^2	-0.2257^1
8.5	-0.7846	-0.2742^1	-0.3157^2	-0.2408^1	8.8	-0.7975	-0.3101^1	-0.3281^2	-0.2726^1
8.6	-0.7664	-0.3327^1	-0.4103^2	-0.2876^1	8.9	-0.7796	-0.3668^1	-0.4179^2	-0.3175^1
8.7	-0.7491	-0.3905^1	-0.5116^2	-0.3325^1	9.0	-0.7625	-0.4228^1	-0.5142^2	-0.3606^1
8.8	-0.7326	-0.4476^1	-0.6191^2	-0.3755^1	9.1	-0.7462	-0.4782^1	-0.6163^2	-0.4020^1
8.9	-0.7169	-0.5039^1	-0.7317^2	-0.4168^1	9.2	-0.7307	-0.5327^1	-0.7233^2	-0.4417^1
9.0	-0.7020	-0.5595^1	-0.8489^2	-0.4565^1	9.3	-0.7160	-0.5866^1	-0.8347^2	-0.4799^1
9.1	-0.6877	-0.6143^1	-0.9701^2	-0.4947^1	9.4	-0.7018	-0.6397^1	-0.9501^2	-0.5167^1
9.2	-0.6741	-0.6684^1	-0.1095^1	-0.5314^1	9.5	-0.6883	-0.6922^1	-0.1069^1	-0.5522^1
9.3	-0.6610	-0.7217^1	-0.1222^1	-0.5668^1	9.6	-0.6754	-0.7439^1	-0.1190^1	-0.5863^1
9.4	-0.6485	-0.7743^1	-0.1353^1	-0.6009^1	9.7	-0.6630	-0.7949^1	-0.1314^1	-0.6193^1
9.5	-0.6366	-0.8261^1	-0.1485^1	-0.6337^1	9.8	-0.6511	-0.8452^1	-0.1441^1	-0.6511^1
9.6	-0.6251	-0.8772^1	-0.1619^1	-0.6655^1	9.9	-0.6397	-0.8948^1	-0.1569^1	-0.6818^1
9.7	-0.6141	-0.9276^1	-0.1754^1	-0.6961^1	10.0	-0.6287	-0.9437^1	-0.1698^1	-0.7115^1
9.8	-0.6035	-0.9773^1	-0.1891^1	-0.7257^1	10.1	-0.6182	-0.9919^1	-0.1828^1	-0.7402^1
9.9	-0.5934	-0.1026	-0.2028^1	-0.7544^1	10.2	-0.6080	-0.1039	-0.1960^1	-0.7680^1
10.0	-0.5836	-0.1074	-0.2166^1	-0.7820^1	10.3	-0.5983	-0.1086	-0.2092^1	-0.7949^1
10.1	-0.5742	-0.1122	-0.2304^1	-0.8088^1	10.4	-0.5889	-0.1132	-0.2224^1	-0.8209^1
10.2	-0.5651	-0.1169	-0.2442^1	-0.8348^1	10.5	-0.5798	-0.1178	-0.2357^1	-0.8462^1
10.3	-0.5564	-0.1215	-0.2580^1	-0.8599^1	10.6	-0.5711	-0.1223	-0.2490^1	-0.8706^1
10.4	-0.5480	-0.1260	-0.2719^1	-0.8843^1	10.7	-0.5627	-0.1267	-0.2623^1	-0.8944^1
10.5	-0.5399	-0.1305	-0.2856^1	-0.9080^1	10.8	-0.5545	-0.1311	-0.2755^1	-0.9174^1
10.7	-0.5244	-0.1393	-0.3130^1	-0.9532^1	11.0	-0.5391	-0.1396	-0.3019^1	-0.9615^1
10.9	-0.5101	-0.1478	-0.3402^1	-0.9958^1	11.2	-0.5246	-0.1479	-0.3280^1	-0.1003
11.1	-0.4966	-0.1561	-0.3669^1	-0.1036	11.4	-0.5111	-0.1559	-0.3539^1	-0.1042
11.3	-0.4840	-0.1641	-0.3933^1	-0.1074	11.6	-0.4983	-0.1638	-0.3794^1	-0.1080
11.5	-0.4721	-0.1719	-0.4193^1	-0.1110	11.8	-0.4864	-0.1714	-0.4046^1	-0.1115
11.7	-0.4609	-0.1794	-0.4447^1	-0.1145	12.0	-0.4751	-0.1788	-0.4293^1	-0.1149
11.9	-0.4504	-0.1868	-0.4697^1	-0.1177	12.2	-0.4644	-0.1860	-0.4535^1	-0.1181
12.1	-0.4405	-0.1940	-0.4942^1	-0.1208	12.4	-0.4544	-0.1929	-0.4773^1	-0.1211
12.3	-0.4311	-0.2009	-0.5181^1	-0.1238	12.6	-0.4449	-0.1997	-0.5007^1	-0.1240
12.5	-0.4222	-0.2077	-0.5416^1	-0.1266	12.8	-0.4358	-0.2064	-0.5235^1	-0.1268
12.8	-0.4097	-0.2175	-0.5757^1	-0.1306	13.1	-0.4228	-0.2163	-0.5579^1	-0.1308
13.1	-0.3981	-0.2269	-0.6087^1	-0.1343	13.4	-0.4110	-0.2255	-0.5901^1	-0.1345
13.4	-0.3874	-0.2359	-0.6405^1	-0.1378	13.7	-0.4001	-0.2343	-0.6212^1	-0.1379
13.7	-0.3774	-0.2446	-0.6712^1	-0.1411	14.0	-0.3900	-0.2429	-0.6513^1	-0.1412
14.0	-0.3680	-0.2530	-0.7008^1	-0.1442	14.4	-0.3789	-0.2537	-0.6878^1	-0.1451
14.5	-0.3538	-0.2662	-0.7477^1	-0.1489	14.8	-0.3663	-0.2638	-0.7256^1	-0.1488
15.0	-0.3410	-0.2786	-0.7920^1	-0.1533	15.3	-0.3532	-0.2761	-0.7692^1	-0.1531
15.5	-0.3295	-0.2904	-0.8336^1	-0.1573	15.8	-0.3413	-0.2876	-0.8102^1	-0.1571
16.0	-0.3190	-0.3014	-0.8728^1	-0.1609	16.3	-0.3306	-0.2985	-0.8490^1	-0.1607
16.5	-0.3094	-0.3119	-0.9098^1	-0.1643	16.8	-0.3207	-0.3088	-0.8856^1	-0.1640
17.3	-0.2957	-0.3274	-0.9647^1	-0.1691	17.6	-0.3067	-0.3242	-0.9401^1	-0.1689
18.1	-0.2837	-0.3417	-0.1015	-0.1735	18.4	-0.2943	-0.3384	-0.9901^1	-0.1732
18.9	-0.2731	-0.3549	-0.1061	-0.1774	19.2	-0.2834	-0.3514	-0.1036	-0.1771
19.7	-0.2636	-0.3671	-0.1103	-0.1809	20.0	-0.2737	-0.3635	-0.1078	-0.1806
20.5	-0.2552	-0.3785	-0.1143	-0.1840	20.8	-0.2649	-0.3748	-0.1117	-0.1837

$\hat{\alpha}_3 = 1.90$

$\hat{\alpha}_4$	$\lambda_1(0,1)$	$\lambda_2(0,1)$	λ_3	λ_4
8.3	-0.9961	0.1249^1	0.1247^3	0.1273^1
8.4	-0.9678	0.6448^2	0.1501^3	0.6431^2
8.5	-0.9414	0.4676^3	0.1668^4	0.4571^3
8.6	-0.9166	-0.5444^2	-0.2575^3	-0.5220^2
8.7	-0.8933	-0.1129^1	-0.6567^3	-0.1063^1
8.8	-0.8713	-0.1707^1	-0.1167^2	-0.1579^1
8.9	-0.8505	-0.2278^1	-0.1777^2	-0.2071^1
9.0	-0.8308	-0.2842^1	-0.2475^2	-0.2543^1
9.1	-0.8121	-0.3399^1	-0.3253^2	-0.2994^1
9.2	-0.7943	-0.3950^1	-0.4100^2	-0.3427^1
9.3	-0.7774	-0.4493^1	-0.5010^2	-0.3843^1
9.4	-0.7613	-0.5030^1	-0.5976^2	-0.4243^1
9.5	-0.7459	-0.5560^1	-0.6989^2	-0.4626^1
9.6	-0.7312	-0.6082^1	-0.8046^2	-0.4996^1
9.7	-0.7172	-0.6598^1	-0.9141^2	-0.5352^1
9.8	-0.7037	-0.7107^1	-0.1027^1	-0.5695^1
9.9	-0.6908	-0.7610^1	-0.1143^1	-0.6026^1
10.0	-0.6785	-0.8105^1	-0.1261^1	-0.6346^1
10.1	-0.6666	-0.8593^1	-0.1381^1	-0.6655^1
10.2	-0.6552	-0.9075^1	-0.1503^1	-0.6953^1
10.3	-0.6443	-0.9551^1	-0.1627^1	-0.7241^1
10.4	-0.6338	-0.1002	-0.1751^1	-0.7521^1
10.5	-0.6236	-0.1048	-0.1877^1	-0.7791^1
10.6	-0.6138	-0.1094	-0.2003^1	-0.8053^1
10.7	-0.6044	-0.1139	-0.2130^1	-0.8307^1
10.8	-0.5953	-0.1183	-0.2258^1	-0.8553^1
10.9	-0.5866	-0.1227	-0.2385^1	-0.8791^1
11.0	-0.5781	-0.1270	-0.2513^1	-0.9023^1
11.1	-0.5699	-0.1312	-0.2640^1	-0.9248^1
11.2	-0.5620	-0.1354	-0.2767^1	-0.9466^1
11.4	-0.5470	-0.1436	-0.3021^1	-0.9885^1
11.6	-0.5329	-0.1516	-0.3272^1	-0.1028
11.8	-0.5196	-0.1593	-0.3521^1	-0.1066
12.0	-0.5072	-0.1669	-0.3766^1	-0.1101
12.2	-0.4954	-0.1742	-0.4008^1	-0.1135
12.4	-0.4843	-0.1813	-0.4245^1	-0.1167
12.6	-0.4739	-0.1883	-0.4479^1	-0.1198
12.8	-0.4639	-0.1950	-0.4708^1	-0.1227
13.0	-0.4545	-0.2016	-0.4933^1	-0.1255
13.2	-0.4456	-0.2080	-0.5154^1	-0.1282
13.5	-0.4326	-0.2176	-0.5487^1	-0.1321
13.8	-0.4210	-0.2265	-0.5798^1	-0.1356
14.1	-0.4101	-0.2351	-0.6100^1	-0.1390
14.4	-0.4000	-0.2434	-0.6392^1	-0.1421
14.7	-0.3905	-0.2513	-0.6674^1	-0.1451
15.2	-0.3763	-0.2637	-0.7114^1	-0.1495
15.7	-0.3632	-0.2757	-0.7539^1	-0.1537
16.2	-0.3513	-0.2869	-0.7940^1	-0.1576
16.7	-0.3404	-0.2976	-0.8319^1	-0.1611
17.2	-0.3305	-0.3077	-0.8678^1	-0.1644
18.0	-0.3163	-0.3227	-0.9213^1	-0.1691
18.8	-0.3038	-0.3366	-0.9705^1	-0.1734
19.6	-0.2927	-0.3494	-0.1016	-0.1772
20.4	-0.2828	-0.3613	-0.1057	-0.1806
21.2	-0.2739	-0.3724	-0.1096	-0.1837

$\hat{\alpha}_3 = 1.95$

$\hat{\alpha}_4$	$\lambda_1(0,1)$	$\lambda_2(0,1)$	λ_3	λ_4
8.7	-0.9827	0.2917^2	0.3576^4	0.2911^2
8.8	-0.9771	0.4111^2	0.5806^5	0.4076^3
8.9	-0.9320	-0.8658^2	-0.3138^3	-0.8313^2
9.0	-0.9088	-0.1434^1	-0.6761^3	-0.1353^1
9.1	-0.8870	-0.1996^1	-0.1145^2	-0.1850^1
9.2	-0.8664	-0.2551^1	-0.1711^2	-0.2326^1
9.3	-0.8468	-0.3100^1	-0.2360^2	-0.2782^1
9.4	-0.8282	-0.3641^1	-0.3087^2	-0.3218^1
9.5	-0.8106	-0.4176^1	-0.3881^2	-0.3637^1
9.6	-0.7937	-0.4704^1	-0.4736^2	-0.4040^1
9.7	-0.7777	-0.5226^1	-0.5645^2	-0.4427^1
9.8	-0.7624	-0.5741^1	-0.6602^2	-0.4799^1
9.9	-0.7477	-0.6249^1	-0.7601^2	-0.5158^1
10.0	-0.7337	-0.6750^1	-0.8637^2	-0.5503^1
10.1	-0.7203	-0.7245^1	-0.9707^2	-0.5837^1
10.2	-0.7074	-0.7733^1	-0.1081^1	-0.6159^1
10.3	-0.6951	-0.8215^1	-0.1193^1	-0.6470^1
10.4	-0.6832	-0.8690^1	-0.1307^1	-0.6770^1
10.5	-0.6718	-0.9159^1	-0.1424^1	-0.7061^1
10.6	-0.6608	-0.9622^1	-0.1541^1	-0.7342^1
10.7	-0.6503	-0.1008	-0.1661^1	-0.7614^1
10.8	-0.6401	-0.1053	-0.1781^1	-0.7878^1
10.9	-0.6303	-0.1097	-0.1902^1	-0.8134^1
11.0	-0.6208	-0.1141	-0.2023^1	-0.8381^1
11.1	-0.6117	-0.1184	-0.2146^1	-0.8622^1
11.2	-0.6029	-0.1227	-0.2268^1	-0.8855^1
11.3	-0.5943	-0.1269	-0.2391^1	-0.9082^1
11.4	-0.5861	-0.1310	-0.2513^1	-0.9302^1
11.5	-0.5781	-0.1351	-0.2636^1	-0.9516^1
11.6	-0.5704	-0.1391	-0.2758^1	-0.9725^1
11.8	-0.5557	-0.1470	-0.3001^1	-0.1012
12.0	-0.5419	-0.1547	-0.3243^1	-0.1050
12.2	-0.5289	-0.1622	-0.3482^1	-0.1086
12.4	-0.5167	-0.1694	-0.3718^1	-0.1120
12.6	-0.5051	-0.1765	-0.3951^1	-0.1153
12.8	-0.4942	-0.1834	-0.4180^1	-0.1184
13.0	-0.4838	-0.1901	-0.4405^1	-0.1213
13.2	-0.4740	-0.1966	-0.4627^1	-0.1241
13.4	-0.4647	-0.2030	-0.4844^1	-0.1268
13.6	-0.4559	-0.2092	-0.5057^1	-0.1294
13.9	-0.4430	-0.2185	-0.5379^1	-0.1332
14.2	-0.4314	-0.2272	-0.5681^1	-0.1366
14.5	-0.4205	-0.2355	-0.5974^1	-0.1398
14.8	-0.4104	-0.2435	-0.6257^1	-0.1429
15.1	-0.4010	-0.2513	-0.6532^1	-0.1458
15.6	-0.3867	-0.2634	-0.6961^1	-0.1501
16.1	-0.3735	-0.2750	-0.7375^1	-0.1542
16.6	-0.3615	-0.2860	-0.7768^1	-0.1580
17.1	-0.3506	-0.2964	-0.8139^1	-0.1614
17.6	-0.3406	-0.3063	-0.8491^1	-0.1646
18.4	-0.3262	-0.3210	-0.9017^1	-0.1693
19.2	-0.3135	-0.3346	-0.9501^1	-0.1735
20.0	-0.3023	-0.3472	-0.9947^1	-0.1772
20.8	-0.2922	-0.3589	-0.1036	-0.1806
21.6	-0.2831	-0.3698	-0.1074	-0.1837

$$\hat{\alpha}_3 = 2.00$$

$\hat{\alpha}_4$	$\lambda_1(0,1)$	$\lambda_2(0,1)$	λ_3	λ_4
9.0	-0.9930	-0.1081^2	-0.4072^5	-0.1076^2
9.1	-0.9740	-0.5675^2	-0.7075^4	-0.5576^2
9.2	-0.9495	-0.1128^1	-0.2724^3	-0.1086^1
9.3	-0.9264	-0.1682^1	-0.5904^3	-0.1592^1
9.4	-0.9047	-0.2229^1	-0.1012^2	-0.2074^1
9.5	-0.8841	-0.2769^1	-0.1527^2	-0.2535^1
9.6	-0.8646	-0.3303^1	-0.2125^2	-0.2977^1
9.7	-0.8460	-0.3830^1	-0.2798^2	-0.3401^1
9.8	-0.8284	-0.4351^1	-0.3537^2	-0.3808^1
9.9	-0.8116	-0.4865^1	-0.4335^2	-0.4199^1
10.0	-0.7955	-0.5372^1	-0.5187^2	-0.4576^1
10.1	-0.7802	-0.5873^1	-0.6086^2	-0.4938^1
10.2	-0.7655	-0.6368^1	-0.7027^2	-0.5287^1
10.3	-0.7515	-0.6856^1	-0.8005^2	-0.5624^1
10.4	-0.7380	-0.7337^1	-0.9016^2	-0.5949^1
10.5	-0.7251	-0.7813^1	-0.1006^1	-0.6263^1
10.6	-0.7127	-0.8282^1	-0.1112^1	-0.6566^1
10.7	-0.7008	-0.8745^1	-0.1221^1	-0.6859^1
10.8	-0.6894	-0.9202^1	-0.1332^1	-0.7143^1
10.9	-0.6783	-0.9652^1	-0.1444^1	-0.7418^1
11.0	-0.6677	-0.1010	-0.1558^1	-0.7684^1
11.1	-0.6575	-0.1054	-0.1672^1	-0.7942^1
11.2	-0.6476	-0.1097	-0.1788^1	-0.8193^1
11.3	-0.6381	-0.1140	-0.1905^1	-0.8436^1
11.4	-0.6289	-0.1182	-0.2022^1	-0.8671^1
11.5	-0.6200	-0.1223	-0.2139^1	-0.8900^1
11.6	-0.6114	-0.1264	-0.2257^1	-0.9122^1
11.7	-0.6030	-0.1305	-0.2375^1	-0.9339^1
11.8	-0.5950	-0.1345	-0.2493^1	-0.9549^1
11.9	-0.5872	-0.1384	-0.2611^1	-0.9753^1
12.1	-0.5723	-0.1461	-0.2845^1	-0.1015
12.3	-0.5584	-0.1536	-0.3079^1	-0.1052
12.5	-0.5452	-0.1609	-0.3310^1	-0.1087
12.7	-0.5328	-0.1680	-0.3539^1	-0.1121
12.9	-0.5210	-0.1750	-0.3765^1	-0.1153
13.1	-0.5099	-0.1817	-0.3988^1	-0.1183
13.3	-0.4994	-0.1883	-0.4207^1	-0.1212
13.5	-0.4894	-0.1947	-0.4423^1	-0.1240
13.7	-0.4799	-0.2010	-0.4635^1	-0.1267
13.9	-0.4709	-0.2071	-0.4843^1	-0.1292
14.2	-0.4578	-0.2162	-0.5159^1	-0.1330
14.5	-0.4459	-0.2247	-0.5455^1	-0.1364
14.8	-0.4349	-0.2329	-0.5742^1	-0.1396
15.1	-0.4245	-0.2408	-0.6021^1	-0.1426
15.4	-0.4148	-0.2485	-0.6290^1	-0.1454
15.9	-0.4002	-0.2604	-0.6713^1	-0.1498
16.4	-0.3867	-0.2719	-0.7123^1	-0.1538
16.9	-0.3744	-0.2827	-0.7511^1	-0.1576
17.4	-0.3632	-0.2930	-0.7879^1	-0.1610
17.9	-0.3529	-0.3028	-0.8228^1	-0.1642
18.7	-0.3381	-0.3174	-0.8751^1	-0.1689
19.5	-0.3250	-0.3309	-0.9233^1	-0.1730
20.3	-0.3134	-0.3434	-0.9678^1	-0.1768
21.1	-0.3030	-0.3550	-0.1009	-0.1802
21.9	-0.2937	-0.3658	-0.1047	-0.1833

Appendix C

Table C–1 for GBD Fits: Method of Moments

For specified α_3 and α_4, Table C–1 gives the β_3 and β_4 values for GBD fits in the region designated in Figure 3.3–5.

The entries of Table C–1 are from "The Extended Generalized Lambda Distribution (EGLD) System for Fitting Distributions to Data with Moments, II: Tables" by E.J. Dudewicz and Z.A. Karian, *American Journal of Mathematical and Management Sciences*, V. 16 3 & 4 (1996), pp. 309–331, copyright ©1996 by American Sciences Press, Inc., 20 Cross Road, Syracuse, New York 13224-2104. Reprinted with permission.

$$0.000 \leq \hat{\alpha}_3 \leq 0.075$$

	$\hat{\alpha}_3 = 0.000$		$\hat{\alpha}_3 = 0.025$		$\hat{\alpha}_3 = 0.050$		$\hat{\alpha}_3 = 0.075$	
$\hat{\alpha}_4$	β_3	β_4	β_3	β_4	β_3	β_4	β_3	β_4
1.05	-0.9615	-0.9615	-0.9625	-0.9616	-0.9644	-0.9626	-0.9673	-0.9647
1.08	-0.9375	-0.9375	-0.9388	-0.9372	-0.9411	-0.9381	-0.9443	-0.9400
1.11	-0.9127	-0.9127	-0.9143	-0.9121	-0.9170	-0.9127	-0.9206	-0.9144
1.14	-0.8871	-0.8871	-0.8891	-0.8862	-0.8921	-0.8866	-0.8962	-0.8881
1.17	-0.8607	-0.8607	-0.8630	-0.8595	-0.8664	-0.8595	-0.8709	-0.8608
1.20	-0.8333	-0.8333	-0.8360	-0.8318	-0.8399	-0.8316	-0.8449	-0.8327
1.23	-0.8051	-0.8051	-0.8082	-0.8033	-0.8125	-0.8027	-0.8179	-0.8036
1.26	-0.7759	-0.7759	-0.7794	-0.7737	-0.7841	-0.7729	-0.7901	-0.7735
1.29	-0.7456	-0.7456	-0.7495	-0.7431	-0.7548	-0.7419	-0.7613	-0.7423
1.32	-0.7143	-0.7143	-0.7187	-0.7113	-0.7244	-0.7099	-0.7314	-0.7100
1.35	-0.6818	-0.6818	-0.6867	-0.6785	-0.6929	-0.6767	-0.7005	-0.6765
1.38	-0.6481	-0.6481	-0.6535	-0.6444	-0.6603	-0.6422	-0.6685	-0.6418
1.41	-0.6132	-0.6132	-0.6190	-0.6090	-0.6264	-0.6065	-0.6353	-0.6057
1.44	-0.5769	-0.5769	-0.5833	-0.5723	-0.5913	-0.5694	-0.6009	-0.5683
1.47	-0.5392	-0.5392	-0.5462	-0.5341	-0.5548	-0.5308	-0.5651	-0.5294
1.50	-0.5000	-0.5000	-0.5076	-0.4943	-0.5169	-0.4906	-0.5279	-0.4889
1.53	-0.4592	-0.4592	-0.4674	-0.4530	-0.4774	-0.4488	-0.4892	-0.4467
1.56	-0.4167	-0.4167	-0.4255	-0.4099	-0.4363	-0.4052	-0.4490	-0.4028
1.59	-0.3723	-0.3723	-0.3819	-0.3649	-0.3935	-0.3598	-0.4071	-0.3570
1.62	-0.3261	-0.3261	-0.3364	-0.3180	-0.3489	-0.3124	-0.3634	-0.3091
1.65	-0.2778	-0.2778	-0.2889	-0.2690	-0.3023	-0.2628	-0.3178	-0.2592
1.68	-0.2273	-0.2273	-0.2392	-0.2178	-0.2536	-0.2110	-0.2702	-0.2069
1.71	-0.1744	-0.1744	-0.1873	-0.1642	-0.2027	-0.1567	-0.2204	-0.1522
1.74	-0.1190	-0.1190	-0.1329	-0.1080	-0.1494	-0.0999	-0.1683	-0.0948
1.77	-0.0610	-0.0610	-0.0758	-0.0491	-0.0935	-0.0402	-0.1137	-0.0346
1.80	-0.0000	-0.0000	-0.0160	0.0128	-0.0349	0.0224	-0.0565	0.0286
1.83	0.0641	0.0641	0.0470	0.0779	0.0267	0.0883	0.0036	0.0951
1.86	0.1316	0.1316	0.1132	0.1465	0.0915	0.1577	0.0667	0.1651
1.89	0.2027	0.2027	0.1829	0.2188	0.1597	0.2309	0.1332	0.2390
1.92	0.2778	0.2778	0.2565	0.2951	0.2316	0.3082	0.2033	0.3170
1.95	0.3571	0.3571	0.3343	0.3757	0.3076	0.3899	0.2772	0.3995
1.98	0.4412	0.4412	0.4167	0.4612	0.3880	0.4765	0.3554	0.4869
2.01	0.5303	0.5303	0.5039	0.5519	0.4731	0.5684	0.4382	0.5797
2.04	0.6250	0.6250	0.5966	0.6482	0.5635	0.6661	0.5260	0.6783
2.07	0.7258	0.7258	0.6952	0.7509	0.6596	0.7701	0.6193	0.7833
2.10	0.8333	0.8333	0.8004	0.8604	0.7620	0.8812	0.7186	0.8955
2.13	0.9483	0.9483	0.9127	0.9775	0.8713	0.9999	0.8245	1.0154
2.16	1.0714	1.0714	1.0330	1.1030	0.9883	1.1273	0.9378	1.1440
2.19	1.2037	1.2037	1.1621	1.2379	1.1137	1.2641	1.0591	1.2822
2.22	1.3462	1.3462	1.3011	1.3832	1.2487	1.4116	1.1895	1.4311
2.25	1.5000	1.5000	1.4511	1.5402	1.3941	1.5710	1.3300	1.5921
2.28	1.6667	1.6667	1.6134	1.7103	1.5515	1.7438	1.4817	1.7667
2.31	1.8478	1.8478	1.7898	1.8954	1.7222	1.9318	1.6462	1.9565
2.34	2.0455	2.0455	1.9820	2.0974	1.9082	2.1370	1.8250	2.1638
2.37	2.2619	2.2619	2.1924	2.3187	2.1114	2.3620	2.0202	2.3910
2.40	2.5000	2.5000	2.4236	2.5624	2.3344	2.6097	2.2341	2.6411
2.43	2.7632	2.7632	2.6788	2.8318	2.5804	2.8837	2.4696	2.9178
2.46	3.0556	3.0556	2.9622	3.1314	2.8530	3.1884	2.7301	3.2254
2.49	3.3824	3.3824	3.2784	3.4665	3.1568	3.5293	3.0199	3.5696
2.52	3.7500	3.7500	3.6338	3.8438	3.4976	3.9132	3.3443	3.9570
2.55	4.1667	4.1667	4.0360	4.2717	3.8826	4.3488	3.7098	4.3966
2.58	4.6429	4.6429	4.4949	4.7612	4.3209	4.8473	4.1248	4.8993
2.61	5.1923	5.1923	5.0235	5.3266	4.8245	5.4231	4.6003	5.4797
2.64	5.8333	5.8333	5.6391	5.9869	5.4094	6.0958	5.1505	6.1574
2.67	6.5909	6.5909	6.3649	6.7682	6.0969	6.8918	5.7947	6.9587

$$0.100 \leq \hat{\alpha}_3 \leq 0.175$$

	$\hat{\alpha}_3 = 0.100$		$\hat{\alpha}_3 = 0.125$		$\hat{\alpha}_3 = 0.150$		$\hat{\alpha}_3 = 0.175$	
$\hat{\alpha}_4$	β_3	β_4	β_3	β_4	β_3	β_4	β_3	β_4
1.05	-0.9710	-0.9679	-0.9755	-0.9722	-0.9808	-0.9777	-0.9867	-0.9842
1.08	-0.9484	-0.9430	-0.9534	-0.9472	-0.9592	-0.9525	-0.9656	-0.9590
1.11	-0.9252	-0.9173	-0.9306	-0.9214	-0.9369	-0.9267	-0.9439	-0.9331
1.14	-0.9012	-0.8908	-0.9071	-0.8947	-0.9139	-0.8999	-0.9214	-0.9064
1.17	-0.8765	-0.8634	-0.8829	-0.8672	-0.8902	-0.8724	-0.8983	-0.8788
1.20	-0.8509	-0.8351	-0.8579	-0.8388	-0.8658	-0.8439	-0.8745	-0.8503
1.23	-0.8245	-0.8058	-0.8320	-0.8094	-0.8405	-0.8144	-0.8498	-0.8208
1.26	-0.7972	-0.7755	-0.8053	-0.7790	-0.8144	-0.7839	-0.8244	-0.7903
1.29	-0.7689	-0.7441	-0.7777	-0.7475	-0.7874	-0.7524	-0.7981	-0.7588
1.32	-0.7397	-0.7116	-0.7491	-0.7149	-0.7595	-0.7197	-0.7709	-0.7261
1.35	-0.7094	-0.6780	-0.7195	-0.6811	-0.7306	-0.6858	-0.7428	-0.6923
1.38	-0.6781	-0.6430	-0.6888	-0.6460	-0.7007	-0.6507	-0.7137	-0.6572
1.41	-0.6456	-0.6068	-0.6571	-0.6096	-0.6698	-0.6143	-0.6835	-0.6208
1.44	-0.6118	-0.5691	-0.6242	-0.5718	-0.6377	-0.5764	-0.6523	-0.5830
1.47	-0.5768	-0.5300	-0.5900	-0.5325	-0.6044	-0.5371	-0.6199	-0.5437
1.50	-0.5405	-0.4892	-0.5545	-0.4917	-0.5698	-0.4962	-0.5864	-0.5028
1.53	-0.5027	-0.4468	-0.5176	-0.4491	-0.5340	-0.4536	-0.5515	-0.4603
1.56	-0.4634	-0.4026	-0.4793	-0.4048	-0.4967	-0.4092	-0.5153	-0.4159
1.59	-0.4224	-0.3565	-0.4394	-0.3585	-0.4579	-0.3629	-0.4776	-0.3698
1.62	-0.3798	-0.3084	-0.3978	-0.3102	-0.4175	-0.3146	-0.4385	-0.3216
1.65	-0.3353	-0.2582	-0.3545	-0.2598	-0.3754	-0.2642	-0.3977	-0.2712
1.68	-0.2888	-0.2056	-0.3093	-0.2071	-0.3315	-0.2114	-0.3552	-0.2186
1.71	-0.2403	-0.1505	-0.2621	-0.1519	-0.2857	-0.1562	-0.3109	-0.1635
1.74	-0.1895	-0.0928	-0.2128	-0.0940	-0.2379	-0.0983	-0.2646	-0.1058
1.77	-0.1364	-0.0323	-0.1612	-0.0333	-0.1879	-0.0376	-0.2163	-0.0453
1.80	-0.0807	0.0313	-0.1071	0.0305	-0.1355	0.0261	-0.1657	0.0182
1.83	-0.0222	0.0982	-0.0504	0.0975	-0.0807	0.0931	-0.1128	0.0850
1.86	0.0392	0.1686	0.0091	0.1682	-0.0231	0.1637	-0.0573	0.1553
1.89	0.1038	0.2429	0.0717	0.2426	0.0373	0.2381	0.0009	0.2294
1.92	0.1718	0.3214	0.1375	0.3212	0.1009	0.3166	0.0621	0.3076
1.95	0.2435	0.4043	0.2070	0.4044	0.1678	0.3997	0.1265	0.3902
1.98	0.3193	0.4922	0.2802	0.4925	0.2384	0.4876	0.1943	0.4777
2.01	0.3995	0.5855	0.3576	0.5859	0.3129	0.5809	0.2658	0.5704
2.04	0.4845	0.6847	0.4396	0.6853	0.3917	0.6800	0.3414	0.6689
2.07	0.5747	0.7903	0.5266	0.7910	0.4752	0.7855	0.4213	0.7737
2.10	0.6707	0.9031	0.6189	0.9039	0.5639	0.8980	0.5061	0.8854
2.13	0.7730	1.0236	0.7173	1.0246	0.6581	1.0182	0.5961	1.0048
2.16	0.8822	1.1529	0.8222	1.1539	0.7585	1.1471	0.6918	1.1325
2.19	0.9991	1.2918	0.9344	1.2929	0.8657	1.2854	0.7939	1.2696
2.22	1.1246	1.4415	1.0546	1.4425	0.9804	1.4344	0.9029	1.4171
2.25	1.2595	1.6032	1.1837	1.6042	1.1034	1.5951	1.0197	1.5763
2.28	1.4051	1.7785	1.3228	1.7794	1.2358	1.7692	1.1451	1.7484
2.31	1.5627	1.9692	1.4731	1.9697	1.3785	1.9583	1.2801	1.9352
2.34	1.7338	2.1773	1.6360	2.1774	1.5329	2.1644	1.4258	2.1386
2.37	1.9203	2.4053	1.8132	2.4049	1.7005	2.3899	1.5836	2.3609
2.40	2.1243	2.6562	2.0067	2.6550	1.8830	2.6377	1.7551	2.6048
2.43	2.3484	2.9337	2.2187	2.9313	2.0827	2.9111	1.9422	2.8737
2.46	2.5958	3.2420	2.4523	3.2382	2.3020	3.2143	2.1471	3.1713
2.49	2.8704	3.5868	2.7109	3.5809	2.5440	3.5525	2.3725	3.5027
2.52	3.1769	3.9746	2.9986	3.9660	2.8126	3.9320	2.6217	3.8738
2.55	3.5213	4.4142	3.3209	4.4019	3.1122	4.3606	2.8988	4.2921
2.58	3.9112	4.9164	3.6844	4.8991	3.4489	4.8486	3.2088	4.7671
2.61	4.3561	5.4957	4.0976	5.4715	3.8299	5.4090	3.5579	5.3109
2.64	4.8690	6.1710	4.5716	6.1374	4.2648	6.0590	3.9543	5.9395
2.67	5.4667	6.9681	5.1212	6.9213	4.7660	6.8216	4.4084	6.6739

$$0.200 \leq \hat{\alpha}_3 \leq 0.275$$

	$\hat{\alpha}_3 = 0.200$		$\hat{\alpha}_3 = 0.225$		$\hat{\alpha}_3 = 0.250$		$\hat{\alpha}_3 = 0.275$	
$\hat{\alpha}_4$	β_3	β_4	β_3	β_4	β_3	β_4	β_3	β_4
1.10	-0.9587	-0.9495	-0.9667	-0.9583	-0.9753	-0.9683	-0.9843	-0.9794
1.13	-0.9370	-0.9231	-0.9457	-0.9320	-0.9548	-0.9420	-0.9645	-0.9533
1.16	-0.9147	-0.8958	-0.9240	-0.9048	-0.9338	-0.9150	-0.9441	-0.9264
1.19	-0.8917	-0.8676	-0.9016	-0.8767	-0.9121	-0.8870	-0.9230	-0.8986
1.22	-0.8680	-0.8385	-0.8785	-0.8477	-0.8897	-0.8582	-0.9013	-0.8700
1.25	-0.8435	-0.8085	-0.8547	-0.8177	-0.8665	-0.8284	-0.8789	-0.8404
1.28	-0.8182	-0.7773	-0.8301	-0.7867	-0.8427	-0.7976	-0.8558	-0.8098
1.31	-0.7920	-0.7451	-0.8047	-0.7546	-0.8181	-0.7657	-0.8320	-0.7782
1.34	-0.7650	-0.7118	-0.7785	-0.7214	-0.7926	-0.7327	-0.8074	-0.7454
1.37	-0.7370	-0.6772	-0.7513	-0.6870	-0.7664	-0.6985	-0.7819	-0.7115
1.40	-0.7081	-0.6413	-0.7233	-0.6513	-0.7392	-0.6630	-0.7557	-0.6764
1.43	-0.6781	-0.6041	-0.6942	-0.6143	-0.7111	-0.6263	-0.7285	-0.6400
1.46	-0.6471	-0.5654	-0.6642	-0.5758	-0.6820	-0.5881	-0.7004	-0.6022
1.49	-0.6149	-0.5252	-0.6330	-0.5359	-0.6519	-0.5484	-0.6714	-0.5629
1.52	-0.5815	-0.4834	-0.6007	-0.4943	-0.6207	-0.5072	-0.6413	-0.5221
1.55	-0.5469	-0.4399	-0.5672	-0.4510	-0.5883	-0.4643	-0.6101	-0.4797
1.58	-0.5109	-0.3945	-0.5324	-0.4060	-0.5547	-0.4197	-0.5777	-0.4355
1.61	-0.4735	-0.3472	-0.4963	-0.3590	-0.5199	-0.3731	-0.5442	-0.3895
1.64	-0.4346	-0.2978	-0.4587	-0.3100	-0.4837	-0.3246	-0.5094	-0.3415
1.67	-0.3940	-0.2462	-0.4196	-0.2588	-0.4461	-0.2739	-0.4732	-0.2914
1.70	-0.3518	-0.1923	-0.3789	-0.2052	-0.4069	-0.2209	-0.4356	-0.2391
1.73	-0.3078	-0.1358	-0.3365	-0.1492	-0.3662	-0.1655	-0.3965	-0.1845
1.76	-0.2619	-0.0766	-0.2923	-0.0905	-0.3237	-0.1074	-0.3558	-0.1272
1.79	-0.2139	-0.0145	-0.2461	-0.0289	-0.2794	-0.0466	-0.3134	-0.0673
1.82	-0.1637	0.0508	-0.1979	0.0357	-0.2331	0.0172	-0.2691	-0.0044
1.85	-0.1111	0.1194	-0.1475	0.1036	-0.1848	0.0843	-0.2229	0.0616
1.88	-0.0561	0.1916	-0.0946	0.1751	-0.1343	0.1549	-0.1746	0.1311
1.91	0.0017	0.2678	-0.0393	0.2505	-0.0813	0.2292	-0.1241	0.2041
1.94	0.0624	0.3483	0.0188	0.3300	-0.0259	0.3076	-0.0713	0.2811
1.97	0.1263	0.4333	0.0798	0.4141	0.0323	0.3903	-0.0159	0.3624
2.00	0.1935	0.5234	0.1441	0.5030	0.0935	0.4779	0.0423	0.4483
2.03	0.2645	0.6190	0.2117	0.5974	0.1579	0.5707	0.1034	0.5393
2.06	0.3394	0.7206	0.2831	0.6975	0.2257	0.6691	0.1676	0.6357
2.09	0.4187	0.8288	0.3585	0.8041	0.2972	0.7738	0.2354	0.7381
2.12	0.5027	0.9441	0.4383	0.9177	0.3728	0.8852	0.3068	0.8470
2.15	0.5919	1.0675	0.5228	1.0391	0.4528	1.0041	0.3823	0.9631
2.18	0.6867	1.1996	0.6127	1.1690	0.5376	1.1313	0.4622	1.0872
2.21	0.7878	1.3416	0.7082	1.3084	0.6277	1.2676	0.5469	1.2199
2.24	0.8958	1.4944	0.8102	1.4583	0.7236	1.4140	0.6369	1.3624
2.27	1.0115	1.6594	0.9191	1.6199	0.8258	1.5717	0.7326	1.5156
2.30	1.1356	1.8381	1.0357	1.7948	0.9351	1.7420	0.8348	1.6808
2.33	1.2692	2.0322	1.1609	1.9844	1.0522	1.9265	0.9439	1.8594
2.36	1.4134	2.2437	1.2958	2.1909	1.1780	2.1270	1.0609	2.0532
2.39	1.5695	2.4753	1.4415	2.4164	1.3134	2.3456	1.1865	2.2641
2.42	1.7391	2.7296	1.5993	2.6637	1.4598	2.5848	1.3218	2.4943
2.45	1.9241	3.0103	1.7708	2.9361	1.6184	2.8477	1.4680	2.7468
2.48	2.1267	3.3216	1,9581	3.2375	1.7909	3.1378	1.6265	3.0247
2.51	2.3495	3.6687	2.1633	3.5727	1.9793	3.4597	1.7989	3.3321
2.54	2.5958	4.0581	2.3893	3.9478	2.1858	3.8187	1.9872	3.6738
2.57	2.8695	4.4979	2.6393	4.3699	2.4134	4.2214	2.1938	4.0558
2.60	3.1756	4.9983	2.9176	4.8487	2.6654	4.6763	2.4214	4.4855
2.63	3.5204	5.5726	3.2293	5.3958	2.9462	5.1940	2.6735	4.9723
2.66	3.9117	6.2383	3.5808	6.0270	3.2609	5.7881	2.9545	5.5279
2.69	4.3598	7.0186	3.9807	6.7626	3.6164	6.4765	3.2696	6.1680
2.72	4.8784	7.9452	4.4396	7.6306	4.0211	7.2832	3.6258	6.9127

$$0.300 \leq \hat{\alpha}_3 \leq 0.375$$

	$\hat{\alpha}_3 = 0.300$		$\hat{\alpha}_3 = 0.325$		$\hat{\alpha}_3 = 0.350$		$\hat{\alpha}_3 = 0.375$	
$\hat{\alpha}_4$	β_3	β_4	β_3	β_4	β_3	β_4	β_3	β_4
1.15	-0.9614	-0.9479	-0.9722	-0.9615	-0.9832	-0.9762	-0.9944	-0.9919
1.18	-0.9412	-0.9207	-0.9527	-0.9346	-0.9644	-0.9495	-0.9763	-0.9656
1.21	-0.9204	-0.8926	-0.9326	-0.9067	-0.9450	-0.9220	-0.9576	-0.9384
1.24	-0.8990	-0.8636	-0.9119	-0.8780	-0.9250	-0.8936	-0.9383	-0.9104
1.27	-0.8769	-0.8336	-0.8905	-0.8484	-0.9044	-0.8644	-0.9185	-0.8815
1.30	-0.8541	-0.8026	-0.8685	-0.8177	-0.8832	-0.8341	-0.8981	-0.8517
1.33	-0.8305	-0.7706	-0.8458	-0.7861	-0.8613	-0.8029	-0.8770	-0.8209
1.36	-0.8063	-0.7374	-0.8224	-0.7533	-0.8388	-0.7705	-0.8553	-0.7891
1.39	-0.7812	-0.7031	-0.7982	-0.7194	-0.8155	-0.7371	-0.8330	-0.7562
1.42	-0.7553	-0.6675	-0.7733	-0.6842	-0.7915	-0.7025	-0.8099	-0.7221
1.45	-0.7285	-0.6306	-0.7475	-0.6478	-0.7667	-0.6666	-0.7861	-0.6868
1.48	-0.7008	-0.5922	-0.7208	-0.6100	-0.7411	-0.6294	-0.7615	-0.6503
1.51	-0.6721	-0.5525	-0.6933	-0.5708	-0.7146	-0.5908	-0.7361	-0.6124
1.54	-0.6425	-0.5111	-0.6647	-0.5301	-0.6873	-0.5508	-0.7099	-0.5731
1.57	-0.6118	-0.4681	-0.6352	-0.4877	-0.6590	-0.5091	-0.6828	-0.5322
1.60	-0.5799	-0.4233	-0.6047	-0.4437	-0.6297	-0.4658	-0.6547	-0.4898
1.63	-0.5469	-0.3767	-0.5730	-0.3978	-0.5993	-0.4208	-0.6257	-0.4456
1.66	-0.5126	-0.3281	-0.5402	-0.3499	-0.5679	-0.3739	-0.5957	-0.3997
1.69	-0.4770	-0.2773	-0.5061	-0.3001	-0.5353	-0.3249	-0.5646	-0.3518
1.72	-0.4400	-0.2243	-0.4707	-0.2480	-0.5015	-0.2739	-0.5323	-0.3019
1.75	-0.4016	-0.1688	-0.4339	-0.1935	-0.4664	-0.2206	-0.4989	-0.2498
1.78	-0.3616	-0.1108	-0.3957	-0.1366	-0.4300	-0.1649	-0.4642	-0.1954
1.81	-0.3199	-0.0500	-0.3559	-0.0770	-0.3921	-0.1066	-0.4281	-0.1385
1.84	-0.2764	0.0138	-0.3145	-0.0145	-0.3527	-0.0455	-0.3907	-0.0789
1.87	-0.2310	0.0808	-0.2713	0.0511	-0.3116	0.0186	-0.3517	-0.0165
1.90	-0.1837	0.1513	-0.2263	0.1200	-0.2688	0.0858	-0.3111	0.0489
1.93	-0.1341	0.2254	-0.1792	0.1925	-0.2242	0.1564	-0.2689	0.1176
1.96	-0.0823	0.3036	-0.1300	0.2688	-0.1776	0.2308	-0.2248	0.1808
1.99	-0.0280	0.3861	-0.0786	0.3493	-0.1290	0.3091	-0.1788	0.2659
2.02	0.0290	0.4733	-0.0247	0.4343	-0.0780	0.3917	-0.1308	0.3460
2.05	0.0888	0.5656	0.0318	0.5242	-0.0247	0.4790	-0.0806	0.4305
2.08	0.1517	0.6635	0.0912	0.6194	0.0312	0.5713	-0.0280	0.5198
2.11	0.2179	0.7675	0.1535	0.7204	0.0898	0.6692	0.0271	0.6144
2.14	0.2877	0.8781	0.2192	0.8278	0.1515	0.7731	0.0849	0.7146
2.17	0.3614	0.9961	0.2884	0.9421	0.2164	0.8835	0.1457	0.8210
2.20	0.4394	1.1221	0.3615	1.0640	0.2847	1.0012	0.2095	0.9342
2.23	0.5220	1.2570	0.4387	1.1944	0.3569	1.1267	0.2768	1.0548
2.26	0.6097	1.4017	0.5206	1.3340	0.4331	1.2610	0.3478	1.1836
2.29	0.7030	1.5574	0.6074	1.4840	0.5139	1.4050	0.4228	1.3214
2.32	0.8023	1.7253	0.6997	1.6454	0.5995	1.5597	0.5021	1.4692
2.35	0.9084	1.9069	0.7981	1.8197	0.6905	1.7263	0.5862	1.6280
2.38	1.0220	2.1039	0.9030	2.0083	0.7874	1.9062	0.6755	1.7991
2.41	1.1438	2.3182	1.0154	2.2130	0.8908	2.1012	0.7705	1.9841
2.44	1.2750	2.5524	1.1359	2.4361	1.0013	2.3130	0.8719	2.1845
2.47	1.4165	2.8091	1.2655	2.6801	1.1199	2.5439	0.9802	2.4024
2.50	1.5697	3.0917	1.4053	2.9478	1.2473	2.7966	1.0962	2.6400
2.53	1.7362	3.4043	1.5566	3.2430	1.3847	3.0742	1.2209	2.9003
2.56	1.9177	3.7519	1.7210	3.5699	1.5333	3.3805	1.3552	3.1864
2.59	2.1165	4.1404	1.9001	3.9339	1.6946	3.7202	1.5003	3.5022
2.62	2.3351	4.5775	2.0962	4.3414	1.8702	4.0987	1.6576	3.8527
2.65	2.5769	5.0725	2.3118	4.8007	2.0624	4.5231	1.8288	4.2436
2.68	2.8458	5.6377	2.5501	5.3219	2.2735	5.0020	2.0158	4.6822
2.71	3.1467	6.2886	2.8149	5.9182	2.5066	5.5462	2.2211	5.1776
2.74	3.4859	7.0458	3.1111	6.6066	2.7654	6.1699	2.4474	5.7410
2.77	3.8712	7.9370	3.4448	7.4096	3.0545	6.8911	2.6984	6.3874

$$0.400 \leq \hat{\alpha}_3 \leq 0.475$$

	$\hat{\alpha}_3 = 0.400$		$\hat{\alpha}_3 = 0.425$		$\hat{\alpha}_3 = 0.450$		$\hat{\alpha}_3 = 0.475$	
$\hat{\alpha}_4$	β_3	β_4	β_3	β_4	β_3	β_4	β_3	β_4
1.25	-0.9455	-0.9188	-0.9592	-0.9378	-0.9729	-0.9577	-0.9865	-0.9784
1.28	-0.9262	-0.8901	-0.9407	-0.9095	-0.9552	-0.9299	-0.9695	-0.9512
1.31	-0.9064	-0.8604	-0.9217	-0.8804	-0.9369	-0.9013	-0.9520	-0.9232
1.34	-0.8860	-0.8298	-0.9021	-0.8503	-0.9181	-0.8718	-0.9340	-0.8943
1.37	-0.8649	-0.7982	-0.8819	-0.8193	-0.8988	-0.8414	-0.9155	-0.8644
1.40	-0.8432	-0.7655	-0.8611	-0.7872	-0.8788	-0.8099	-0.8965	-0.8337
1.43	-0.8208	-0.7316	-0.8396	-0.7540	-0.8583	-0.7774	-0.8768	-0.8019
1.46	-0.7978	-0.6966	-0.8175	-0.7197	-0.8372	-0.7439	-0.8566	-0.7691
1.49	-0.7740	-0.6603	-0.7947	-0.6841	-0.8154	-0.7091	-0.8358	-0.7351
1.52	-0.7494	-0.6227	-0.7712	-0.6473	-0.7929	-0.6731	-0.8143	-0.7000
1.55	-0.7240	-0.5837	-0.7470	-0.6091	-0.7697	-0.6358	-0.7922	-0.6637
1.58	-0.6978	-0.5432	-0.7219	-0.5695	-0.7458	-0.5972	-0.7694	-0.6260
1.61	-0.6707	-0.5011	-0.6960	-0.5284	-0.7211	-0.5571	-0.7459	-0.5870
1.64	-0.6426	-0.4574	-0.6693	-0.4857	-0.6956	-0.5155	-0.7216	-0.5464
1.67	-0.6136	-0.4118	-0.6416	-0.4413	-0.6693	-0.4722	-0.6966	-0.5044
1.70	-0.5836	-0.3644	-0.6130	-0.3951	-0.6421	-0.4273	-0.6707	-0.4607
1.73	-0.5525	-0.3150	-0.5835	-0.3470	-0.6140	-0.3805	-0.6440	-0.4153
1.76	-0.5203	-0.2635	-0.5528	-0.2968	-0.5849	-0.3317	-0.6164	-0.3680
1.79	-0.4869	-0.2097	-0.5211	-0.2445	-0.5548	-0.2809	-0.5878	-0.3187
1.82	-0.4522	-0.1534	-0.4882	-0.1898	-0.5236	-0.2279	-0.5583	-0.2674
1.85	-0.4162	-0.0946	-0.4540	-0.1327	-0.4912	-0.1726	-0.5277	-0.2139
1.88	-0.3788	-0.0330	-0.4186	-0.0730	-0.4577	-0.1147	-0.4960	-0.1580
1.91	-0.3398	0.0316	-0.3818	-0.0104	-0.4229	-0.0542	-0.4631	-0.0996
1.94	-0.2993	0.0993	-0.3435	0.0552	-0.3868	0.0091	-0.4291	-0.0384
1.97	-0.2571	0.1705	-0.3037	0.1240	-0.3492	0.0756	-0.3937	0.0256
2.00	-0.2131	0.2454	-0.2622	0.1963	-0.3102	0.1453	-0.3570	0.0927
2.03	-0.1672	0.3242	-0.2190	0.2724	-0.2696	0.2186	-0.3188	0.1631
2.06	-0.1192	0.4074	-0.1739	0.3525	-0.2273	0.2956	-0.2791	0.2371
2.09	-0.0691	0.4952	-0.1269	0.4370	-0.1832	0.3768	-0.2378	0.3149
2.12	-0.0166	0.5880	-0.0777	0.5263	-0.1371	0.4624	-0.1947	0.3968
2.15	0.0384	0.6863	-0.0263	0.6206	-0.0891	0.5528	-0.1498	0.4832
2.18	0.0961	0.7907	0.0276	0.7206	-0.0388	0.6484	-0.1029	0.5745
2.21	0.1567	0.9015	0.0841	0.8267	0.0138	0.7496	-0.0540	0.6710
2.24	0.2205	1.0195	0.1434	0.9394	0.0689	0.8570	-0.0027	0.7732
2.27	0.2876	1.1454	0.2057	1.0594	0.1267	0.9712	0.0509	0.8816
2.30	0.3584	1.2799	0.2713	1.1874	0.1875	1.0927	0.1071	0.9967
2.33	0.4332	1.4239	0.3404	1.3241	0.2513	1.2223	0.1661	1.1193
2.36	0.5124	1.5785	0.4134	1.4707	0.3186	1.3609	0.2281	1.2501
2.39	0.5963	1.7449	0.4905	1.6279	0.3895	1.5092	0.2933	1.3898
2.42	0.6853	1.9244	0.5722	1.7972	0.4645	1.6685	0.3620	1.5393
2.45	0.7801	2.1186	0.6589	1.9798	0.5437	1.8399	0.4345	1.6999
2.48	0.8812	2.3294	0.7511	2.1774	0.6278	2.0247	0.5112	1.8725
2.51	0.9892	2.5588	0.8492	2.3918	0.7170	2.2247	0.5923	2.0588
2.54	1.1049	2.8094	0.9540	2.6253	0.8119	2.4417	0.6784	2.2602
2.57	1.2291	3.0843	1.0662	2.8803	0.9132	2.6780	0.7698	2.4786
2.60	1.3630	3.3870	1.1865	3.1600	1.0214	2.9360	0.8672	2.7163
2.63	1.5077	3.7218	1.3159	3.4681	1.1373	3.2189	0.9712	2.9759
2.66	1.6645	4.0941	1.4556	3.8089	1.2618	3.5304	1.0824	3.2603
2.69	1.8351	4.5103	1.6068	4.1877	1.3960	3.8749	1.2017	3.5733
2.72	2.0216	4.9784	1.7710	4.6113	1.5410	4.2577	1.3299	3.9192
2.75	2.2261	5.5085	1.9502	5.0876	1.6982	4.6855	1.4683	4.3034
2.78	2.4518	6.1135	2.1464	5.6270	1.8694	5.1664	1.6180	4.7323
2.81	2.7019	6.8097	2.3623	6.2424	2.0564	5.7106	1.7807	5.2140
2.84	2.9810	7.6190	2.6012	6.9506	2.2618	6.3312	1.9580	5.7587
2.87	3.2944	8.5704	2.8670	7.7736	2.4884	7.0448	2.1521	6.3789

$$0.500 \leq \hat{\alpha}_3 \leq 0.575$$

	$\hat{\alpha}_3 = 0.500$		$\hat{\alpha}_3 = 0.525$		$\hat{\alpha}_3 = 0.550$		$\hat{\alpha}_3 = 0.575$	
$\hat{\alpha}_4$	β_3	β_4	β_3	β_4	β_3	β_4	β_3	β_4
1.35	-0.9439	-0.9079	-0.9597	-0.9322	-0.9751	-0.9571	-0.9902	-0.9827
1.38	-0.9261	-0.8785	-0.9426	-0.9034	-0.9588	-0.9291	-0.9747	-0.9553
1.41	-0.9077	-0.8481	-0.9250	-0.8737	-0.9421	-0.9001	-0.9587	-0.9271
1.44	-0.8887	-0.8167	-0.9070	-0.8432	-0.9248	-0.8703	-0.9423	-0.8981
1.47	-0.8692	-0.7843	-0.8884	-0.8116	-0.9071	-0.8396	-0.9254	-0.8682
1.50	-0.8492	-0.7508	-0.8692	-0.7790	-0.8889	-0.8078	-0.9081	-0.8373
1.53	-0.8285	-0.7162	-0.8495	-0.7453	-0.8702	-0.7751	-0.8903	-0.8055
1.56	-0.8072	-0.6804	-0.8293	-0.7104	-0.8509	-0.7412	-0.8719	-0.7726
1.59	-0.7853	-0.6433	-0.8084	-0.6744	-0.8310	-0.7062	-0.8530	-0.7387
1.62	-0.7626	-0.6049	-0.7869	-0.6371	-0.8106	-0.6700	-0.8336	-0.7036
1.65	-0.7393	-0.5650	-0.7647	-0.5984	-0.7895	-0.6325	-0.8136	-0.6673
1.68	-0.7152	-0.5237	-0.7419	-0.5583	-0.7678	-0.5937	-0.7931	-0.6298
1.71	-0.6904	-0.4808	-0.7183	-0.5167	-0.7455	-0.5535	-0.7719	-0.5908
1.74	-0.6648	-0.4361	-0.6940	-0.4735	-0.7224	-0.5117	-0.7501	-0.5505
1.77	-0.6383	-0.3897	-0.6689	-0.4286	-0.6987	-0.4684	-0.7276	-0.5087
1.80	-0.6109	-0.3414	-0.6430	-0.3820	-0.6742	-0.4233	-0.7044	-0.4653
1.83	-0.5827	-0.2911	-0.6162	-0.3334	-0.6489	-0.3765	-0.6805	-0.4202
1.86	-0.5534	-0.2387	-0.5886	-0.2829	-0.6228	-0.3278	-0.6559	-0.3734
1.89	-0.5231	-0.1840	-0.5600	-0.2301	-0.5958	-0.2771	-0.6304	-0.3246
1.92	-0.4917	-0.1268	-0.5304	-0.1751	-0.5679	-0.2242	-0.6042	-0.2739
1.95	-0.4592	-0.0671	-0.4998	-0.1177	-0.5391	-0.1691	-0.5771	-0.2210
1.98	-0.4255	-0.0046	-0.4681	-0.0577	-0.5093	-0.1115	-0.5490	-0.1659
2.01	-0.3906	0.0609	-0.4353	0.0051	-0.4785	-0.0514	-0.5201	-0.1083
2.04	-0.3543	0.1296	-0.4012	0.0709	-0.4465	0.0115	-0.4901	-0.0482
2.07	-0.3166	0.2016	-0.3659	0.1398	-0.4134	0.0774	-0.4591	0.0146
2.10	-0.2774	0.2773	-0.3292	0.2121	-0.3791	0.1464	-0.4270	0.0804
2.13	-0.2365	0.3570	-0.2910	0.2881	-0.3434	0.2188	-0.3937	0.1493
2.16	-0.1940	0.4409	-0.2514	0.3681	-0.3064	0.2948	-0.3591	0.2216
2.19	-0.1497	0.5295	-0.2101	0.4523	-0.2680	0.3748	-0.3233	0.2975
2.22	-0.1035	0.6230	-0.1671	0.5410	-0.2280	0.4590	-0.2861	0.3772
2.25	-0.0552	0.7219	-0.1223	0.6348	-0.1864	0.5477	-0.2475	0.4612
2.28	-0.0047	0.8267	-0.0755	0.7339	-0.1430	0.6414	-0.2073	0.5496
2.31	0.0481	0.9378	-0.0267	0.8389	-0.0978	0.7404	-0.1654	0.6429
2.34	0.1034	1.0560	0.0244	0.9502	-0.0506	0.8452	-0.1218	0.7415
2.37	0.1615	1.1818	0.0779	1.0686	-0.0013	0.9564	-0.0764	0.8457
2.40	0.2224	1.3160	0.1339	1.1945	0.0502	1.0744	-0.0290	0.9563
2.43	0.2865	1.4595	0.1927	1.3287	0.1041	1.1999	0.0206	1.0735
2.46	0.3541	1.6132	0.2545	1.4722	0.1607	1.3337	0.0724	1.1982
2.49	0.4253	1.7781	0.3195	1.6257	0.2200	1.4765	0.1266	1.3310
2.52	0.5005	1.9557	0.3879	1.7905	0.2823	1.6293	0.1835	1.4726
2.55	0.5801	2.1472	0.4601	1.9677	0.3479	1.7931	0.2431	1.6241
2.58	0.6644	2.3544	0.5364	2.1588	0.4170	1.9692	0.3058	1.7863
2.61	0.7540	2.5793	0.6172	2.3653	0.4899	2.1589	0.3717	1.9605
2.64	0.8493	2.8241	0.7028	2.5893	0.5670	2.3639	0.4412	2.1480
2.67	0.9510	3.0916	0.7937	2.8330	0.6486	2.5859	0.5146	2.3504
2.70	1.0596	3.3848	0.8906	3.0989	0.7351	2.8271	0.5920	2.5693
2.73	1.1760	3.7077	0.9939	3.3902	0.8271	3.0900	0.6741	2.8069
2.76	1.3011	4.0647	1.1044	3.7105	0.9250	3.3777	0.7611	3.0656
2.79	1.4359	4.4615	1.2229	4.0643	1.0295	3.6937	0.8536	3.3482
2.82	1.5816	4.9049	1.3503	4.4570	1.1413	4.0421	0.9522	3.6581
2.85	1.7397	5.4032	1.4877	4.8950	1.2613	4.4282	1.0573	3.9993
2.88	1.9117	5.9670	1.6363	5.3865	1.3903	4.8581	1.1699	4.3766
2.91	2.0999	6.6098	1.7977	5.9416	1.5295	5.3395	1.2906	4.7958
2.94	2.3066	7.3486	1.9736	6.5730	1.6802	5.8819	1.4205	5.2641
2.97	2.5348	8.2059	2.1661	7.2969	1.8439	6.4972	1.5607	5.7904

C. Table C–1 for GBD Fits: Method of Moments

$$0.600 \leq \hat{\alpha}_3 \leq 0.675$$

	$\hat{\alpha}_3 = 0.600$		$\hat{\alpha}_3 = 0.625$		$\hat{\alpha}_3 = 0.650$		$\hat{\alpha}_3 = 0.675$	
$\hat{\alpha}_4$	β_3	β_4	β_3	β_4	β_3	β_4	β_3	β_4
1.50	-0.9268	-0.8674	-0.9449	-0.8978	-0.9624	-0.9287	-0.9793	-0.9598
1.53	-0.9098	-0.8365	-0.9288	-0.8679	-0.9471	-0.8996	-0.9648	-0.9316
1.56	-0.8924	-0.8046	-0.9122	-0.8370	-0.9314	-0.8697	-0.9499	-0.9026
1.59	-0.8745	-0.7717	-0.8952	-0.8051	-0.9153	-0.8389	-0.9346	-0.8728
1.62	-0.8560	-0.7377	-0.8777	-0.7723	-0.8987	-0.8071	-0.9189	-0.8421
1.65	-0.8371	-0.7026	-0.8598	-0.7383	-0.8817	-0.7743	-0.9028	-0.8104
1.68	-0.8176	-0.6663	-0.8413	-0.7032	-0.8642	-0.7404	-0.8862	-0.7778
1.71	-0.7975	-0.6287	-0.8223	-0.6670	-0.8462	-0.7055	-0.8692	-0.7441
1.74	-0.7768	-0.5898	-0.8027	-0.6295	-0.8277	-0.6693	-0.8517	-0.7093
1.77	-0.7556	-0.5495	-0.7826	-0.5906	-0.8087	-0.6319	-0.8338	-0.6733
1.80	-0.7337	-0.5077	-0.7619	-0.5504	-0.7891	-0.5933	-0.8153	-0.6361
1.83	-0.7111	-0.4643	-0.7406	-0.5087	-0.7690	-0.5532	-0.7963	-0.5977
1.86	-0.6879	-0.4193	-0.7187	-0.4655	-0.7483	-0.5117	-0.7768	-0.5579
1.89	-0.6639	-0.3725	-0.6961	-0.4206	-0.7270	-0.4687	-0.7568	-0.5166
1.92	-0.6391	-0.3238	-0.6728	-0.3739	-0.7051	-0.4240	-0.7361	-0.4739
1.95	-0.6136	-0.2732	-0.6488	-0.3254	-0.6825	-0.3776	-0.7149	-0.4295
1.98	-0.5873	-0.2204	-0.6240	-0.2750	-0.6593	-0.3294	-0.6930	-0.3835
2.01	-0.5601	-0.1654	-0.5985	-0.2225	-0.6353	-0.2793	-0.6705	-0.3357
2.04	-0.5320	-0.1080	-0.5721	-0.1677	-0.6106	-0.2271	-0.6473	-0.2860
2.07	-0.5029	-0.0481	-0.5449	-0.1106	-0.5850	-0.1728	-0.6234	-0.2343
2.10	-0.4729	0.0145	-0.5168	-0.0511	-0.5587	-0.1161	-0.5987	-0.1804
2.13	-0.4417	0.0800	-0.4877	0.0112	-0.5315	-0.0570	-0.5733	-0.1243
2.16	-0.4095	0.1487	-0.4576	0.0763	-0.5035	0.0047	-0.5471	-0.0658
2.19	-0.3761	0.2206	-0.4265	0.1444	-0.4744	0.0692	-0.5200	-0.0048
2.22	-0.3415	0.2961	-0.3943	0.2158	-0.4444	0.1367	-0.4921	0.0590
2.25	-0.3056	0.3754	-0.3609	0.2907	-0.4134	0.2074	-0.4632	0.1257
2.28	-0.2683	0.4588	-0.3263	0.3694	-0.3813	0.2815	-0.4333	0.1955
2.31	-0.2296	0.5467	-0.2904	0.4520	-0.3480	0.3593	-0.4024	0.2686
2.34	-0.1893	0.6393	-0.2531	0.5390	-0.3135	0.4410	-0.3705	0.3453
2.37	-0.1474	0.7371	-0.2144	0.6307	-0.2777	0.5269	-0.3374	0.4259
2.40	-0.1037	0.8405	-0.1741	0.7275	-0.2405	0.6174	-0.3030	0.5105
2.43	-0.0581	0.9500	-0.1322	0.8297	-0.2019	0.7128	-0.2675	0.5996
2.46	-0.0106	1.0662	-0.0886	0.9379	-0.1618	0.8136	-0.2305	0.6934
2.49	0.0390	1.1895	-0.0431	1.0525	-0.1201	0.9201	-0.1922	0.7924
2.52	0.0909	1.3208	0.0043	1.1742	-0.0766	1.0329	-0.1524	0.8969
2.55	0.1453	1.4608	0.0539	1.3036	-0.0314	1.1525	-0.1109	1.0075
2.58	0.2022	1.6103	0.1057	1.4413	0.0159	1.2795	-0.0678	1.1247
2.61	0.2619	1.7703	0.1600	1.5884	0.0652	1.4147	-0.0229	1.2491
2.64	0.3247	1.9419	0.2168	1.7456	0.1167	1.5587	0.0240	1.3812
2.67	0.3908	2.1265	0.2764	1.9140	0.1707	1.7126	0.0729	1.5219
2.70	0.4603	2.3254	0.3390	2.0948	0.2272	1.8772	0.1240	1.6720
2.73	0.5338	2.5403	0.4049	2.2895	0.2864	2.0538	0.1775	1.8324
2.76	0.6114	2.7733	0.4743	2.4996	0.3487	2.2436	0.2334	2.0041
2.79	0.6935	3.0265	0.5475	2.7270	0.4141	2.4481	0.2921	2.1885
2.82	0.7806	3.3027	0.6248	2.9737	0.4830	2.6691	0.3537	2.3868
2.85	0.8733	3.6050	0.7067	3.2423	0.5557	2.9084	0.4185	2.6007
2.88	0.9719	3.9372	0.7935	3.5358	0.6325	3.1685	0.4866	2.8319
2.91	1.0772	4.3038	0.8858	3.8576	0.7137	3.4521	0.5585	3.0827
2.94	1.1898	4.7102	0.9840	4.2119	0.7998	3.7624	0.6343	3.3556
2.97	1.3107	5.1630	1.0889	4.6037	0.8913	4.1031	0.7145	3.6534
3.00	1.4407	5.6704	1.2010	5.0390	0.9886	4.4789	0.7995	3.9796
3.03	1.5810	6.2425	1.3213	5.5253	1.0924	4.8952	0.8898	4.3384
3.06	1.7330	6.8920	1.4507	6.0716	1.2035	5.3586	0.9857	4.7346
3.09	1.8981	7.6352	1.5902	6.6895	1.3225	5.8775	1.0880	5.1742
3.12	2.0784	8.4931	1.7414	7.3935	1.4504	6.4620	1.1974	5.6645

$$0.700 \leq \hat{\alpha}_3 \leq 0.775$$

	$\hat{\alpha}_3 = 0.700$		$\hat{\alpha}_3 = 0.725$		$\hat{\alpha}_3 = 0.750$		$\hat{\alpha}_3 = 0.775$	
$\hat{\alpha}_4$	β_3	β_4	β_3	β_4	β_3	β_4	β_3	β_4
1.63	-0.9333	-0.8672	-0.9522	-0.9027	-0.9703	-0.9382	-0.9876	-0.9736
1.66	-0.9179	-0.8363	-0.9377	-0.8729	-0.9566	-0.9095	-0.9746	-0.9459
1.69	-0.9021	-0.8045	-0.9227	-0.8423	-0.9425	-0.8799	-0.9613	-0.9174
1.72	-0.8859	-0.7717	-0.9074	-0.8107	-0.9280	-0.8495	-0.9476	-0.8882
1.75	-0.8692	-0.7378	-0.8917	-0.7781	-0.9131	-0.8182	-0.9336	-0.8580
1.78	-0.8521	-0.7029	-0.8755	-0.7445	-0.8979	-0.7859	-0.9192	-0.8270
1.81	-0.8345	-0.6668	-0.8589	-0.7099	-0.8822	-0.7527	-0.9045	-0.7951
1.84	-0.8165	-0.6295	-0.8419	-0.6741	-0.8662	-0.7184	-0.8893	-0.7622
1.87	-0.7979	-0.5908	-0.8244	-0.6371	-0.8497	-0.6829	-0.8738	-0.7283
1.90	-0.7788	-0.5509	-0.8064	-0.5988	-0.8328	-0.6463	-0.8579	-0.6933
1.93	-0.7592	-0.5095	-0.7879	-0.5593	-0.8154	-0.6085	-0.8415	-0.6571
1.96	-0.7390	-0.4666	-0.7690	-0.5183	-0.7975	-0.5694	-0.8247	-0.6198
1.99	-0.7183	-0.4221	-0.7495	-0.4759	-0.7792	-0.5289	-0.8075	-0.5812
2.02	-0.6969	-0.3760	-0.7294	-0.4319	-0.7604	-0.4870	-0.7898	-0.5413
2.05	-0.6749	-0.3280	-0.7088	-0.3863	-0.7410	-0.4436	-0.7717	-0.5000
2.08	-0.6523	-0.2782	-0.6876	-0.3389	-0.7212	-0.3986	-0.7530	-0.4572
2.11	-0.6290	-0.2264	-0.6658	-0.2897	-0.7007	-0.3519	-0.7339	-0.4129
2.14	-0.6050	-0.1724	-0.6433	-0.2386	-0.6797	-0.3034	-0.7142	-0.3669
2.17	-0.5802	-0.1162	-0.6202	-0.1854	-0.6581	-0.2530	-0.6940	-0.3192
2.20	-0.5547	-0.0577	-0.5964	-0.1300	-0.6358	-0.2007	-0.6732	-0.2697
2.23	-0.5284	0.0035	-0.5718	-0.0723	-0.6129	-0.1462	-0.6518	-0.2182
2.26	-0.5012	0.0673	-0.5465	-0.0121	-0.5893	-0.0894	-0.6298	-0.1647
2.29	-0.4731	0.1340	-0.5204	0.0507	-0.5650	-0.0303	-0.6072	-0.1090
2.32	-0.4441	0.2038	-0.4934	0.1163	-0.5400	0.0314	-0.5839	-0.0510
2.35	-0.4142	0.2770	-0.4656	0.1850	-0.5142	0.0958	-0.5599	0.0094
2.38	-0.3831	0.3536	-0.4369	0.2568	-0.4875	0.1630	-0.5351	0.0724
2.41	-0.3511	0.4341	-0.4072	0.3320	-0.4600	0.2334	-0.5097	0.1383
2.44	-0.3178	0.5186	-0.3765	0.4108	-0.4317	0.3070	-0.4834	0.2070
2.47	-0.2834	0.6075	-0.3448	0.4936	-0.4023	0.3841	-0.4563	0.2789
2.50	-0.2477	0.7012	-0.3119	0.5807	-0.3720	0.4651	-0.4283	0.3542
2.53	-0.2107	0.7999	-0.2779	0.6722	-0.3407	0.5500	-0.3994	0.4332
2.56	-0.1722	0.9041	-0.2426	0.7687	-0.3083	0.6393	-0.3696	0.5159
2.59	-0.1322	1.0143	-0.2060	0.8704	-0.2747	0.7333	-0.3387	0.6028
2.62	-0.0907	1.1310	-0.1680	0.9779	-0.2399	0.8324	-0.3068	0.6942
2.65	-0.0474	1.2548	-0.1286	1.0915	-0.2039	0.9369	-0.2738	0.7904
2.68	-0.0024	1.3862	-0.0876	1.2119	-0.1665	1.0472	-0.2396	0.8918
2.71	0.0446	1.5261	-0.0449	1.3396	-0.1276	1.1640	-0.2042	0.9987
2.74	0.0937	1.6752	-0.0005	1.4753	-0.0873	1.2877	-0.1674	1.1117
2.77	0.1449	1.8344	0.0459	1.6197	-0.0453	1.4190	-0.1293	1.2313
2.80	0.1985	2.0047	0.0942	1.7737	-0.0016	1.5585	-0.0897	1.3580
2.83	0.2546	2.1874	0.1446	1.9382	0.0439	1.7071	-0.0486	1.4924
2.86	0.3134	2.3837	0.1973	2.1144	0.0913	1.8655	-0.0058	1.6353
2.89	0.3752	2.5952	0.2525	2.3033	0.1408	2.0349	0.0388	1.7875
2.92	0.4401	2.8237	0.3103	2.5065	0.1925	2.2162	0.0852	1.9499
2.95	0.5084	3.0713	0.3710	2.7256	0.2465	2.4108	0.1336	2.1234
2.98	0.5803	3.3402	0.4346	2.9623	0.3032	2.6201	0.1841	2.3093
3.01	0.6563	3.6333	0.5016	3.2190	0.3625	2.8459	0.2369	2.5088
3.04	0.7367	3.9539	0.5722	3.4979	0.4248	3.0900	0.2922	2.7234
3.07	0.8219	4.3060	0.6466	3.8023	0.4902	3.3547	0.3500	2.9550
3.10	0.9122	4.6941	0.7253	4.1353	0.5591	3.6426	0.4107	3.2054
3.13	1.0084	5.1239	0.8086	4.5014	0.6317	3.9567	0.4744	3.4769
3.16	1.1109	5.6022	0.8969	4.9052	0.7084	4.3008	0.5414	3.7724
3.19	1.2203	6.1375	0.9907	5.3529	0.7894	4.6790	0.6120	4.0948
3.22	1.3376	6.7400	1.0906	5.8517	0.8753	5.0966	0.6864	4.4481
3.25	1.4636	7.4230	1.1972	6.4104	0.9665	5.5599	0.7649	4.8366

$$0.800 \leq \hat{\alpha}_3 \leq 0.875$$

| | $\hat{\alpha}_3 = 0.800$ | | $\hat{\alpha}_3 = 0.825$ | | $\hat{\alpha}_3 = 0.850$ | | $\hat{\alpha}_3 = 0.875$ | |
$\hat{\alpha}_4$	β_3	β_4	β_3	β_4	β_3	β_4	β_3	β_4
1.80	-0.9303	-0.8474	-0.9502	-0.8886	-0.9690	-0.9292	-0.9868	-0.9692
1.83	-0.9162	-0.8162	-0.9369	-0.8586	-0.9565	-0.9004	-0.9751	-0.9416
1.86	-0.9017	-0.7840	-0.9232	-0.8278	-0.9436	-0.8709	-0.9630	-0.9133
1.89	-0.8868	-0.7509	-0.9092	-0.7960	-0.9305	-0.8405	-0.9506	-0.8842
1.92	-0.8716	-0.7167	-0.8949	-0.7633	-0.9170	-0.8092	-0.9379	-0.8543
1.95	-0.8559	-0.6814	-0.8802	-0.7296	-0.9031	-0.7770	-0.9249	-0.8235
1.98	-0.8399	-0.6450	-0.8651	-0.6949	-0.8890	-0.7438	-0.9116	-0.7918
2.01	-0.8234	-0.6074	-0.8496	-0.6590	-0.8744	-0.7097	-0.8979	-0.7593
2.04	-0.8065	-0.5685	-0.8337	-0.6220	-0.8595	-0.6744	-0.8839	-0.7257
2.07	-0.7892	-0.5283	-0.8175	-0.5838	-0.8443	-0.6380	-0.8696	-0.6911
2.10	-0.7714	-0.4867	-0.8008	-0.5443	-0.8286	-0.6005	-0.8549	-0.6554
2.13	-0.7531	-0.4436	-0.7837	-0.5034	-0.8126	-0.5617	-0.8399	-0.6186
2.16	-0.7343	-0.3990	-0.7661	-0.4611	-0.7961	-0.5216	-0.8245	-0.5805
2.19	-0.7151	-0.3527	-0.7481	-0.4173	-0.7792	-0.4802	-0.8087	-0.5413
2.22	-0.6953	-0.3047	-0.7296	-0.3719	-0.7619	-0.4373	-0.7925	-0.5007
2.25	-0.6749	-0.2549	-0.7106	-0.3249	-0.7442	-0.3928	-0.7759	-0.4587
2.28	-0.6540	-0.2031	-0.6911	-0.2761	-0.7260	-0.3468	-0.7588	-0.4153
2.31	-0.6325	-0.1493	-0.6710	-0.2254	-0.7073	-0.2991	-0.7414	-0.3703
2.34	-0.6104	-0.0933	-0.6504	-0.1728	-0.6881	-0.2496	-0.7235	-0.3237
2.37	-0.5877	-0.0350	-0.6292	-0.1181	-0.6683	-0.1982	-0.7051	-0.2754
2.40	-0.5642	0.0258	-0.6075	-0.0611	-0.6481	-0.1448	-0.6862	-0.2253
2.43	-0.5401	0.0891	-0.5851	-0.0018	-0.6273	-0.0893	-0.6669	-0.1733
2.46	-0.5153	0.1553	-0.5621	0.0600	-0.6059	-0.0315	-0.6470	-0.1192
2.49	-0.4897	0.2244	-0.5383	0.1244	-0.5839	0.0286	-0.6266	-0.0630
2.52	-0.4633	0.2966	-0.5139	0.1916	-0.5613	0.0913	-0.6056	-0.0046
2.55	-0.4361	0.3722	-0.4888	0.2619	-0.5381	0.1566	-0.5841	0.0562
2.58	-0.4080	0.4514	-0.4629	0.3353	-0.5141	0.2248	-0.5620	0.1196
2.61	-0.3790	0.5345	-0.4362	0.4122	-0.4895	0.2961	-0.5392	0.1857
2.64	-0.3490	0.6217	-0.4086	0.4928	-0.4641	0.3706	-0.5157	0.2547
2.67	-0.3180	0.7134	-0.3801	0.5772	-0.4379	0.4485	-0.4916	0.3268
2.70	-0.2860	0.8098	-0.3508	0.6659	-0.4109	0.5302	-0.4668	0.4021
2.73	-0.2528	0.9114	-0.3204	0.7591	-0.3831	0.6158	-0.4412	0.4809
2.76	-0.2185	1.0185	-0.2891	0.8571	-0.3544	0.7057	-0.4149	0.5634
2.79	-0.1830	1.1317	-0.2567	0.9604	-0.3247	0.8001	-0.3877	0.6500
2.82	-0.1461	1.2514	-0.2231	1.0693	-0.2941	0.8995	-0.3597	0.7408
2.85	-0.1078	1.3781	-0.1884	1.1844	-0.2625	1.0041	-0.3308	0.8362
2.88	-0.0681	1.5126	-0.1523	1.3060	-0.2297	1.1145	-0.3010	0.9366
2.91	-0.0268	1.6554	-0.1150	1.4348	-0.1959	1.2310	-0.2701	1.0423
2.94	0.0161	1.8074	-0.0763	1.5715	-0.1608	1.3542	-0.2383	1.1537
2.97	0.0608	1.9695	-0.0360	1.7167	-0.1244	1.4847	-0.2053	1.2714
3.00	0.1073	2.1427	0.0058	1.8712	-0.0868	1.6231	-0.1713	1.3957
3.03	0.1558	2.3280	0.0492	2.0359	-0.0477	1.7701	-0.1360	1.5274
3.06	0.2065	2.5268	0.0945	2.2118	-0.0070	1.9265	-0.0994	1.6670
3.09	0.2595	2.7405	0.1417	2.4001	0.0352	2.0931	-0.0615	1.8152
3.12	0.3148	2.9709	0.1909	2.6021	0.0790	2.2711	-0.0221	1.9729
3.15	0.3728	3.2198	0.2422	2.8192	0.1247	2.4616	0.0187	2.1409
3.18	0.4337	3.4894	0.2959	3.0531	0.1723	2.6658	0.0612	2.3202
3.21	0.4975	3.7825	0.3520	3.3058	0.2220	2.8853	0.1053	2.5120
3.24	0.5646	4.1020	0.4108	3.5796	0.2738	3.1217	0.1513	2.7176
3.27	0.6353	4.4517	0.4725	3.8771	0.3280	3.3770	0.1992	2.9385
3.30	0.7098	4.8356	0.5373	4.2014	0.3847	3.6536	0.2491	3.1763
3.33	0.7885	5.2591	0.6054	4.5561	0.4441	3.9539	0.3013	3.4330
3.36	0.8718	5.7282	0.6771	4.9456	0.5064	4.2812	0.3558	3.7108
3.39	0.9601	6.2504	0.7527	5.3750	0.5718	4.6390	0.4128	4.0125
3.42	1.0539	6.8350	0.8327	5.8506	0.6406	5.0317	0.4725	4.3409

$$0.900 \leq \hat{\alpha}_3 \leq 0.950$$

	$\hat{\alpha}_3 = 0.900$		$\hat{\alpha}_3 = 0.925$		$\hat{\alpha}_3 = 0.950$	
$\hat{\alpha}_4$	β_3	β_4	β_3	β_4	β_3	β_4
1.94	-0.9496	-0.8790	-0.9687	-0.9233	-0.9867	-0.9667
1.98	-0.9329	-0.8389	-0.9531	-0.8849	-0.9721	-0.9300
2.02	-0.9158	-0.7972	-0.9370	-0.8452	-0.9569	-0.8920
2.06	-0.8981	-0.7539	-0.9204	-0.8039	-0.9414	-0.8527
2.10	-0.8798	-0.7089	-0.9032	-0.7610	-0.9253	-0.8118
2.14	-0.8609	-0.6620	-0.8855	-0.7165	-0.9087	-0.7695
2.18	-0.8414	-0.6132	-0.8672	-0.6701	-0.8916	-0.7255
2.22	-0.8213	-0.5622	-0.8484	-0.6219	-0.8739	-0.6797
2.26	-0.8005	-0.5090	-0.8289	-0.5715	-0.8557	-0.6321
2.30	-0.7790	-0.4534	-0.8088	-0.5190	-0.8369	-0.5825
2.34	-0.7568	-0.3952	-0.7881	-0.4642	-0.8175	-0.5307
2.38	-0.7338	-0.3343	-0.7667	-0.4068	-0.7975	-0.4767
2.42	-0.7101	-0.2704	-0.7446	-0.3468	-0.7769	-0.4203
2.46	-0.6856	-0.2033	-0.7217	-0.2839	-0.7556	-0.3613
2.50	-0.6602	-0.1328	-0.6981	-0.2180	-0.7336	-0.2995
2.54	-0.6339	-0.0587	-0.6736	-0.1487	-0.7109	-0.2347
2.58	-0.6066	0.0195	-0.6484	-0.0759	-0.6874	-0.1667
2.62	-0.5784	0.1019	-0.6222	0.0008	-0.6631	-0.0954
2.66	-0.5491	0.1890	-0.5951	0.0816	-0.6380	-0.0203
2.70	-0.5187	0.2811	-0.5671	0.1668	-0.6120	0.0588
2.74	-0.4872	0.3787	-0.5380	0.2569	-0.5852	0.1421
2.78	-0.4544	0.4823	-0.5078	0.3523	-0.5573	0.2301
2.82	-0.4204	0.5924	-0.4765	0.4534	-0.5285	0.3231
2.86	-0.3849	0.7096	-0.4440	0.5607	-0.4986	0.4216
2.90	-0.3480	0.8347	-0.4101	0.6749	-0.4676	0.5260
2.94	-0.3095	0.9683	-0.3750	0.7965	-0.4354	0.6370
2.98	-0.2693	1.1114	-0.3384	0.9263	-0.4019	0.7550
3.02	-0.2274	1.2650	-0.3002	1.0650	-0.3671	0.8807
3.06	-0.1836	1.4303	-0.2605	1.2137	-0.3309	1.0150
3.10	-0.1377	1.6085	-0.2190	1.3735	-0.2933	1.1587
3.14	-0.0896	1.8013	-0.1756	1.5454	-0.2540	1.3127
3.18	-0.0392	2.0103	-0.1303	1.7309	-0.2130	1.4782
3.22	0.0137	2.2376	-0.0828	1.9317	-0.1702	1.6564
3.26	0.0694	2.4856	-0.0330	2.1496	-0.1255	1.8489
3.30	0.1280	2.7573	0.0193	2.3868	-0.0787	2.0574
3.34	0.1898	3.0559	0.0742	2.6459	-0.0297	2.2837
3.38	0.2552	3.3856	0.1320	2.9299	0.0217	2.5302
3.42	0.3243	3.7512	0.1929	3.2425	0.0757	2.7998
3.46	0.3976	4.1587	0.2572	3.5879	0.1325	3.0954
3.50	0.4755	4.6154	0.3252	3.9715	0.1924	3.4211
3.54	0.5585	5.1304	0.3973	4.3997	0.2555	3.7814
3.58	0.6470	5.7152	0.4738	4.8802	0.3222	4.1819
3.62	0.7417	6.3843	0.5552	5.4231	0.3927	4.6294
3.66	0.8433	7.1567	0.6420	6.0407	0.4676	5.1324
3.70	0.9527	8.0570	0.7348	6.7489	0.5471	5.7014
3.74	1.0707	9.1188	0.8341	7.5684	0.6319	6.3498
3.78	1.1986	10.3876	0.9410	8.5264	0.7223	7.0947
3.82	1.3376	11.9282	1.0561	9.6600	0.8191	7.9584
3.86	1.4895	13.8345	1.1807	11.0200	0.9230	8.9707
3.90	1.6561	16.2486	1.3160	12.6791	1.0348	10.1718
3.94	1.8399	19.3964	1.4634	14.7440	1.1556	11.6178
3.98	2.0439	23.6582	1.6250	17.3781	1.2865	13.3891
4.02	2.2718	29.7282	1.8028	20.8449	1.4289	15.6048
4.06	2.5283	39.0200	1.9996	25.5976	1.5846	18.4495
4.10	2.8197	54.9194	2.2190	32.4863	1.7555	22.2248

$$0.975 \leq \hat{\alpha}_3 \leq 1.025$$

	$\hat{\alpha}_3 = 0.975$		$\hat{\alpha}_3 = 1.000$		$\hat{\alpha}_3 = 1.025$	
$\hat{\alpha}_4$	β_3	β_4	β_3	β_4	β_3	β_4
2.10	-0.9460	-0.8613	-0.9655	-0.9095	-0.9837	-0.9564
2.14	-0.9305	-0.8210	-0.9509	-0.8711	-0.9701	-0.9198
2.18	-0.9144	-0.7792	-0.9359	-0.8313	-0.9560	-0.8820
2.22	-0.8979	-0.7358	-0.9204	-0.7901	-0.9415	-0.8428
2.26	-0.8808	-0.6907	-0.9044	-0.7473	-0.9266	-0.8022
2.30	-0.8633	-0.6437	-0.8880	-0.7029	-0.9112	-0.7601
2.34	-0.8451	-0.5949	-0.8711	-0.6567	-0.8954	-0.7164
2.38	-0.8265	-0.5439	-0.8536	-0.6087	-0.8790	-0.6710
2.42	-0.8072	-0.4908	-0.8356	-0.5586	-0.8622	-0.6238
2.46	-0.7873	-0.4354	-0.8170	-0.5065	-0.8449	-0.5747
2.50	-0.7668	-0.3774	-0.7979	-0.4521	-0.8270	-0.5235
2.54	-0.7457	-0.3168	-0.7782	-0.3953	-0.8087	-0.4702
2.58	-0.7238	-0.2533	-0.7579	-0.3359	-0.7897	-0.4146
2.62	-0.7013	-0.1868	-0.7369	-0.2737	-0.7702	-0.3565
2.66	-0.6780	-0.1169	-0.7153	-0.2086	-0.7501	-0.2957
2.70	-0.6539	-0.0435	-0.6930	-0.1404	-0.7294	-0.2322
2.74	-0.6291	0.0337	-0.6699	-0.0687	-0.7080	-0.1656
2.78	-0.6034	0.1150	-0.6461	0.0066	-0.6859	-0.0957
2.82	-0.5767	0.2008	-0.6215	0.0858	-0.6631	-0.0224
2.86	-0.5492	0.2914	-0.5961	0.1693	-0.6396	0.0547
2.90	-0.5207	0.3871	-0.5698	0.2573	-0.6154	0.1358
2.94	-0.4911	0.4886	-0.5426	0.3503	-0.5903	0.2212
2.98	-0.4604	0.5962	-0.5145	0.4487	-0.5644	0.3114
3.02	-0.4286	0.7105	-0.4853	0.5529	-0.5376	0.4066
3.06	-0.3956	0.8321	-0.4551	0.6634	-0.5098	0.5073
3.10	-0.3613	0.9618	-0.4237	0.7808	-0.4811	0.6141
3.14	-0.3256	1.1003	-0.3912	0.9059	-0.4514	0.7273
3.18	-0.2884	1.2486	-0.3574	1.0392	-0.4205	0.8477
3.22	-0.2497	1.4076	-0.3223	1.1816	-0.3885	0.9758
3.26	-0.2094	1.5785	-0.2857	1.3341	-0.3554	1.1124
3.30	-0.1673	1.7626	-0.2477	1.4977	-0.3209	1.2584
3.34	-0.1233	1.9616	-0.2081	1.6736	-0.2851	1.4148
3.38	-0.0774	2.1771	-0.1668	1.8632	-0.2478	1.5825
3.42	-0.0292	2.4113	-0.1237	2.0681	-0.2090	1.7629
3.46	0.0212	2.6666	-0.0786	2.2901	-0.1685	1.9574
3.50	0.0742	2.9457	-0.0315	2.5315	-0.1264	2.1676
3.54	0.1298	3.2522	0.0179	2.7946	-0.0823	2.3954
3.58	0.1884	3.5900	0.0697	3.0825	-0.0363	2.6431
3.62	0.2501	3.9639	0.1240	3.3987	0.0119	2.9133
3.66	0.3152	4.3800	0.1811	3.7474	0.0623	3.2089
3.70	0.3841	4.8453	0.2413	4.1337	0.1152	3.5337
3.74	0.4571	5.3689	0.3047	4.5637	0.1708	3.8920
3.78	0.5345	5.9618	0.3717	5.0450	0.2293	4.2890
3.82	0.6169	6.6384	0.4426	5.5868	0.2908	4.7311
3.86	0.7047	7.4167	0.5177	6.2010	0.3558	5.2261
3.90	0.7986	8.3208	0.5975	6.9025	0.4244	5.7836
3.94	0.8992	9.3825	0.6825	7.7103	0.4971	6.4160
3.98	1.0073	10.6451	0.7732	8.6499	0.5741	7.1385
4.02	1.1238	12.1695	0.8702	9.7548	0.6560	7.9713
4.06	1.2499	14.0430	0.9742	11.0712	0.7433	8.9405
4.10	1.3867	16.3966	1.0862	12.6638	0.8365	10.0815
4.14	1.5360	19.4345	1.2070	14.6263	0.9362	11.4423
4.18	1.6994	23.4950	1.3379	17.0996	1.0433	13.0908
4.22	1.8794	29.1798	1.4802	20.3055	1.1587	15.1258
4.26	2.0787	37.6727	1.6357	24.6143	1.2833	17.6959

$$1.050 \leq \hat{\alpha}_3 \leq 1.100$$

	$\hat{\alpha}_3 = 1.050$		$\hat{\alpha}_3 = 1.075$		$\hat{\alpha}_3 = 1.100$	
$\hat{\alpha}_4$	β_3	β_4	β_3	β_4	β_3	β_4
2.24	-0.9544	-0.8747	-0.9733	-0.9252	-0.9909	-0.9741
2.28	-0.9402	-0.8355	-0.9600	-0.8879	-0.9786	-0.9386
2.32	-0.9256	-0.7948	-0.9464	-0.8493	-0.9658	-0.9019
2.36	-0.9105	-0.7527	-0.9323	-0.8094	-0.9526	-0.8641
2.40	-0.8950	-0.7090	-0.9178	-0.7680	-0.9391	-0.8249
2.44	-0.8791	-0.6635	-0.9029	-0.7251	-0.9252	-0.7843
2.48	-0.8627	-0.6163	-0.8876	-0.6806	-0.9109	-0.7423
2.52	-0.8458	-0.5672	-0.8718	-0.6343	-0.8961	-0.6987
2.56	-0.8284	-0.5161	-0.8556	-0.5863	-0.8810	-0.6534
2.60	-0.8105	-0.4628	-0.8388	-0.5363	-0.8654	-0.6064
2.64	-0.7920	-0.4073	-0.8217	-0.4842	-0.8493	-0.5576
2.68	-0.7730	-0.3492	-0.8040	-0.4300	-0.8328	-0.5068
2.72	-0.7534	-0.2886	-0.7857	-0.3733	-0.8159	-0.4538
2.76	-0.7332	-0.2252	-0.7670	-0.3143	-0.7984	-0.3987
2.80	-0.7125	-0.1587	-0.7477	-0.2525	-0.7805	-0.3411
2.84	-0.6910	-0.0891	-0.7278	-0.1879	-0.7620	-0.2811
2.88	-0.6689	-0.0161	-0.7073	-0.1202	-0.7430	-0.2183
2.92	-0.6461	0.0607	-0.6862	-0.0493	-0.7234	-0.1526
2.96	-0.6226	0.1414	-0.6644	0.0251	-0.7032	-0.0838
3.00	-0.5983	0.2264	-0.6420	0.1033	-0.6825	-0.0117
3.04	-0.5732	0.3160	-0.6188	0.1856	-0.6611	0.0639
3.08	-0.5473	0.4106	-0.5950	0.2721	-0.6390	0.1434
3.12	-0.5204	0.5107	-0.5703	0.3634	-0.6163	0.2269
3.16	-0.4927	0.6166	-0.5448	0.4598	-0.5929	0.3149
3.20	-0.4640	0.7288	-0.5185	0.5617	-0.5687	0.4077
3.24	-0.4343	0.8481	-0.4913	0.6696	-0.5438	0.5057
3.28	-0.4035	0.9749	-0.4632	0.7840	-0.5180	0.6092
3.32	-0.3715	1.1101	-0.4341	0.9054	-0.4913	0.7188
3.36	-0.3384	1.2544	-0.4039	1.0346	-0.4638	0.8350
3.40	-0.3040	1.4087	-0.3727	1.1723	-0.4354	0.9583
3.44	-0.2683	1.5741	-0.3403	1.3193	-0.4059	1.0895
3.48	-0.2311	1.7518	-0.3067	1.4764	-0.3754	1.2293
3.52	-0.1924	1.9432	-0.2718	1.6449	-0.3438	1.3785
3.56	-0.1521	2.1498	-0.2355	1.8259	-0.3110	1.5381
3.60	-0.1100	2.3734	-0.1978	2.0208	-0.2770	1.7091
3.64	-0.0662	2.6161	-0.1585	2.2311	-0.2417	1.8928
3.68	-0.0203	2.8804	-0.1176	2.4588	-0.2050	2.0905
3.72	0.0276	3.1692	-0.0750	2.7059	-0.1669	2.3039
3.76	0.0778	3.4858	-0.0305	2.9749	-0.1272	2.5348
3.80	0.1304	3.8344	0.0160	3.2689	-0.0858	2.7853
3.84	0.1857	4.2199	0.0647	3.5911	-0.0427	3.0581
3.88	0.2437	4.6480	0.1156	3.9459	0.0023	3.3560
3.92	0.3049	5.1261	0.1691	4.3381	0.0494	3.6825
3.96	0.3694	5.6631	0.2251	4.7736	0.0986	4.0418
4.00	0.4375	6.2700	0.2841	5.2600	0.1501	4.4389
4.04	0.5095	6.9610	0.3462	5.8061	0.2041	4.8797
4.08	0.5859	7.7540	0.4117	6.4233	0.2609	5.3718
4.12	0.6670	8.6726	0.4809	7.1258	0.3205	5.9241
4.16	0.7534	9.7479	0.5541	7.9320	0.3833	6.5480
4.20	0.8455	11.0222	0.6317	8.8656	0.4496	7.2577
4.24	0.9441	12.5544	0.7142	9.9584	0.5196	8.0717
4.28	1.0499	14.4284	0.8021	11.2531	0.5936	9.0138
4.32	1.1637	16.7689	0.8958	12.8094	0.6722	10.1157
4.36	1.2866	19.7687	0.9962	14.7126	0.7557	11.4201
4.40	1.4197	23.7429	1.1039	17.0892	0.8446	12.9867

$$1.125 \leq \hat{\alpha}_3 \leq 1.175$$

	$\hat{\alpha}_3 = 1.125$		$\hat{\alpha}_3 = 1.150$		$\hat{\alpha}_3 = 1.175$	
$\hat{\alpha}_4$	β_3	β_4	β_3	β_4	β_3	β_4
2.44	-0.9460	-0.8413	-0.9653	-0.8961	-0.9833	-0.9490
2.48	-0.9326	-0.8015	-0.9528	-0.8585	-0.9716	-0.9133
2.52	-0.9188	-0.7603	-0.9400	-0.8196	-0.9596	-0.8764
2.56	-0.9046	-0.7177	-0.9267	-0.7793	-0.9473	-0.8384
2.60	-0.8901	-0.6735	-0.9132	-0.7376	-0.9346	-0.7990
2.64	-0.8751	-0.6276	-0.8992	-0.6944	-0.9216	-0.7583
2.68	-0.8598	-0.5799	-0.8848	-0.6496	-0.9082	-0.7161
2.72	-0.8439	-0.5303	-0.8701	-0.6031	-0.8945	-0.6725
2.76	-0.8277	-0.4788	-0.8550	-0.5548	-0.8803	-0.6272
2.80	-0.8110	-0.4251	-0.8394	-0.5046	-0.8658	-0.5802
2.84	-0.7938	-0.3691	-0.8234	-0.4524	-0.8509	-0.5313
2.88	-0.7761	-0.3107	-0.8070	-0.3980	-0.8356	-0.4806
2.92	-0.7579	-0.2498	-0.7901	-0.3414	-0.8199	-0.4278
2.96	-0.7393	-0.1861	-0.7727	-0.2823	-0.8038	-0.3728
3.00	-0.7200	-0.1195	-0.7549	-0.2206	-0.7872	-0.3156
3.04	-0.7002	-0.0497	-0.7365	-0.1561	-0.7702	-0.2558
3.08	-0.6798	0.0234	-0.7176	-0.0887	-0.7527	-0.1935
3.12	-0.6589	0.1001	-0.6982	-0.0181	-0.7347	-0.1284
3.16	-0.6372	0.1807	-0.6782	0.0559	-0.7162	-0.0603
3.20	-0.6150	0.2654	-0.6577	0.1335	-0.6972	0.0110
3.24	-0.5920	0.3546	-0.6365	0.2150	-0.6776	0.0857
3.28	-0.5683	0.4486	-0.6147	0.3007	-0.6575	0.1641
3.32	-0.5439	0.5479	-0.5923	0.3909	-0.6368	0.2463
3.36	-0.5187	0.6528	-0.5692	0.4860	-0.6155	0.3328
3.40	-0.4927	0.7638	-0.5453	0.5863	-0.5936	0.4238
3.44	-0.4658	0.8815	-0.5207	0.6924	-0.5711	0.5197
3.48	-0.4380	1.0065	-0.4953	0.8045	-0.5478	0.6209
3.52	-0.4093	1.1393	-0.4691	0.9234	-0.5238	0.7277
3.56	-0.3796	1.2808	-0.4420	1.0496	-0.4991	0.8408
3.60	-0.3488	1.4318	-0.4141	1.1837	-0.4736	0.9605
3.64	-0.3169	1.5933	-0.3852	1.3266	-0.4473	1.0876
3.68	-0.2839	1.7663	-0.3552	1.4789	-0.4201	1.2226
3.72	-0.2496	1.9520	-0.3243	1.6418	-0.3920	1.3663
3.76	-0.2140	2.1519	-0.2922	1.8162	-0.3630	1.5196
3.80	-0.1770	2.3676	-0.2589	2.0034	-0.3329	1.6833
3.84	-0.1385	2.6009	-0.2244	2.2048	-0.3019	1.8586
3.88	-0.0985	2.8539	-0.1886	2.4219	-0.2697	2.0466
3.92	-0.0568	3.1293	-0.1515	2.6567	-0.2363	2.2488
3.96	-0.0133	3.4299	-0.1128	2.9113	-0.2017	2.4667
4.00	0.0321	3.7592	-0.0726	3.1882	-0.1658	2.7022
4.04	0.0795	4.1215	-0.0307	3.4903	-0.1286	2.9573
4.08	0.1291	4.5216	0.0130	3.8210	-0.0898	3.2346
4.12	0.1810	4.9655	0.0586	4.1845	-0.0495	3.5369
4.16	0.2354	5.4607	0.1062	4.5857	-0.0075	3.8676
4.20	0.2926	6.0161	0.1560	5.0306	0.0362	4.2307
4.24	0.3527	6.6430	0.2082	5.5263	0.0819	4.6312
4.28	0.4160	7.3556	0.2628	6.0818	0.1295	5.0748
4.32	0.4827	8.1721	0.3202	6.7080	0.1793	5.5685
4.36	0.5532	9.1160	0.3805	7.4191	0.2315	6.1210
4.40	0.6278	10.2185	0.4440	8.2327	0.2862	6.7432
4.44	0.7070	11.5221	0.5110	9.1718	0.3435	7.4484
4.48	0.7911	13.0850	0.5817	10.2670	0.4038	8.2540
4.52	0.8806	14.9906	0.6565	11.5591	0.4673	9.1822
4.56	0.9762	17.3618	0.7358	13.1049	0.5341	10.2622
4.60	1.0785	20.3874	0.8201	14.9845	0.6048	11.5333

$$1.200 \le \hat{\alpha}_3 \le 1.250$$

	$\hat{\alpha}_3 = 1.200$		$\hat{\alpha}_3 = 1.225$		$\hat{\alpha}_3 = 1.250$	
$\hat{\alpha}_4$	β_3	β_4	β_3	β_4	β_3	β_4
2.60	-0.9546	-0.8579	-0.9732	-0.9144	-0.9904	-0.9686
2.64	-0.9425	-0.8194	-0.9618	-0.8780	-0.9798	-0.9342
2.68	-0.9300	-0.7797	-0.9502	-0.8405	-0.9690	-0.8987
2.72	-0.9171	-0.7386	-0.9382	-0.8017	-0.9578	-0.8621
2.76	-0.9040	-0.6960	-0.9259	-0.7617	-0.9463	-0.8243
2.80	-0.8904	-0.6519	-0.9133	-0.7202	-0.9346	-0.7853
2.84	-0.8766	-0.6062	-0.9004	-0.6773	-0.9225	-0.7450
2.88	-0.8623	-0.5588	-0.8871	-0.6329	-0.9101	-0.7033
2.92	-0.8477	-0.5095	-0.8735	-0.5868	-0.8974	-0.6601
2.96	-0.8326	-0.4583	-0.8595	-0.5390	-0.8844	-0.6153
3.00	-0.8172	-0.4050	-0.8451	-0.4893	-0.8710	-0.5690
3.04	-0.8014	-0.3496	-0.8304	-0.4377	-0.8573	-0.5209
3.08	-0.7851	-0.2918	-0.8153	-0.3841	-0.8433	-0.4709
3.12	-0.7684	-0.2316	-0.7998	-0.3283	-0.8288	-0.4191
3.16	-0.7513	-0.1687	-0.7839	-0.2701	-0.8140	-0.3651
3.20	-0.7337	-0.1030	-0.7675	-0.2095	-0.7989	-0.3090
3.24	-0.7156	-0.0344	-0.7508	-0.1462	-0.7833	-0.2505
3.28	-0.6970	0.0375	-0.7336	-0.0801	-0.7674	-0.1896
3.32	-0.6779	0.1127	-0.7159	-0.0111	-0.7510	-0.1261
3.36	-0.6583	0.1916	-0.6977	0.0612	-0.7342	-0.0598
3.40	-0.6381	0.2745	-0.6791	0.1368	-0.7170	0.0096
3.44	-0.6173	0.3615	-0.6600	0.2161	-0.6993	0.0821
3.48	-0.5960	0.4531	-0.6403	0.2994	-0.6811	0.1580
3.52	-0.5740	0.5496	-0.6200	0.3868	-0.6624	0.2375
3.56	-0.5513	0.6513	-0.5992	0.4787	-0.6433	0.3209
3.60	-0.5280	0.7588	-0.5778	0.5756	-0.6236	0.4085
3.64	-0.5040	0.8724	-0.5558	0.6776	-0.6034	0.5007
3.68	-0.4792	0.9927	-0.5331	0.7854	-0.5826	0.5976
3.72	-0.4536	1.1203	-0.5098	0.8993	-0.5612	0.6998
3.76	-0.4272	1.2558	-0.4858	1.0198	-0.5392	0.8075
3.80	-0.4000	1.4000	-0.4610	1.1476	-0.5166	0.9214
3.84	-0.3719	1.5537	-0.4354	1.2832	-0.4932	1.0419
3.88	-0.3428	1.7178	-0.4090	1.4275	-0.4692	1.1695
3.92	-0.3127	1.8934	-0.3818	1.5812	-0.4445	1.3049
3.96	-0.2816	2.0816	-0.3537	1.7452	-0.4190	1.4488
4.00	-0.2495	2.2839	-0.3247	1.9205	-0.3927	1.6020
4.04	-0.2161	2.5018	-0.2947	2.1084	-0.3656	1.7654
4.08	-0.1815	2.7371	-0.2637	2.3101	-0.3376	1.9400
4.12	-0.1456	2.9918	-0.2315	2.5272	-0.3087	2.1269
4.16	-0.1084	3.2684	-0.1983	2.7614	-0.2788	2.3274
4.20	-0.0697	3.5697	-0.1638	3.0148	-0.2479	2.5430
4.24	-0.0294	3.8990	-0.1280	3.2897	-0.2160	2.7754
4.28	0.0125	4.2602	-0.0909	3.5888	-0.1829	3.0266
4.32	0.0562	4.6581	-0.0523	3.9154	-0.1486	3.2988
4.36	0.1017	5.0983	-0.0123	4.2732	-0.1131	3.5946
4.40	0.1493	5.5876	0.0295	4.6668	-0.0762	3.9172
4.44	0.1990	6.1344	0.0729	5.1017	-0.0379	4.2703
4.48	0.2509	6.7491	0.1182	5.5843	0.0019	4.6580
4.52	0.3054	7.4446	0.1654	6.1227	0.0433	5.0857
4.56	0.3625	8.2375	0.2148	6.7268	0.0864	5.5596
4.60	0.4226	9.1489	0.2665	7.4090	0.1312	6.0872
4.64	0.4857	10.2066	0.3205	8.1848	0.1781	6.6781
4.68	0.5522	11.4478	0.3772	9.0744	0.2270	7.3437
4.72	0.6224	12.9230	0.4367	10.1036	0.2781	8.0987
4.76	0.6966	14.7034	0.4993	11.3072	0.3315	8.9618

C. Table C-1 for GBD Fits: Method of Moments

$$1.275 \leq \hat{\alpha}_3 \leq 1.325$$

$\hat{\alpha}_4$	$\hat{\alpha}_3 = 1.275$		$\hat{\alpha}_3 = 1.300$		$\hat{\alpha}_3 = 1.325$	
	β_3	β_4	β_3	β_4	β_3	β_4
2.80	-0.9543	-0.8474	-0.9726	-0.9068	-0.9895	-0.9635
2.84	-0.9431	-0.8094	-0.9621	-0.8709	-0.9797	-0.9297
2.88	-0.9315	-0.7702	-0.9514	-0.8340	-0.9697	-0.8948
2.92	-0.9197	-0.7297	-0.9403	-0.7958	-0.9594	-0.8588
2.96	-0.9075	-0.6877	-0.9290	-0.7564	-0.9489	-0.8218
3.00	-0.8951	-0.6443	-0.9174	-0.7157	-0.9381	-0.7835
3.04	-0.8823	-0.5994	-0.9055	-0.6737	-0.9270	-0.7440
3.08	-0.8692	-0.5528	-0.8933	-0.6301	-0.9156	-0.7033
3.12	-0.8558	-0.5045	-0.8808	-0.5850	-0.9040	-0.6611
3.16	-0.8420	-0.4544	-0.8680	-0.5383	-0.8921	-0.6175
3.20	-0.8279	-0.4023	-0.8549	-0.4899	-0.8799	-0.5723
3.24	-0.8135	-0.3481	-0.8414	-0.4396	-0.8673	-0.5255
3.28	-0.7987	-0.2918	-0.8277	-0.3874	-0.8545	-0.4770
3.32	-0.7835	-0.2332	-0.8135	-0.3332	-0.8414	-0.4267
3.36	-0.7679	-0.1721	-0.7991	-0.2768	-0.8280	-0.3745
3.40	-0.7519	-0.1084	-0.7843	-0.2181	-0.8142	-0.3203
3.44	-0.7356	-0.0419	-0.7691	-0.1569	-0.8001	-0.2639
3.48	-0.7188	0.0275	-0.7535	-0.0932	-0.7857	-0.2052
3.52	-0.7015	0.1001	-0.7376	-0.0267	-0.7709	-0.1441
3.56	-0.6838	0.1761	-0.7212	0.0427	-0.7557	-0.0804
3.60	-0.6657	0.2557	-0.7045	0.1153	-0.7402	-0.0140
3.64	-0.6471	0.3392	-0.6873	0.1912	-0.7243	0.0553
3.68	-0.6279	0.4268	-0.6696	0.2707	-0.7080	0.1276
3.72	-0.6083	0.5188	-0.6515	0.3540	-0.6913	0.2033
3.76	-0.5881	0.6157	-0.6330	0.4414	-0.6742	0.2826
3.80	-0.5674	0.7177	-0.6139	0.5332	-0.6567	0.3655
3.84	-0.5460	0.8252	-0.5944	0.6298	-0.6387	0.4526
3.88	-0.5241	0.9388	-0.5743	0.7314	-0.6202	0.5439
3.92	-0.5016	1.0589	-0.5536	0.8385	-0.6013	0.6399
3.96	-0.4784	1.1860	-0.5324	0.9515	-0.5818	0.7409
4.00	-0.4545	1.3208	-0.5106	1.0709	-0.5619	0.8473
4.04	-0.4298	1.4640	-0.4882	1.1972	-0.5414	0.9595
4.08	-0.4045	1.6164	-0.4651	1.3311	-0.5203	1.0779
4.12	-0.3783	1.7787	-0.4414	1.4732	-0.4987	1.2032
4.16	-0.3514	1.9520	-0.4169	1.6242	-0.4764	1.3358
4.20	-0.3236	2.1373	-0.3918	1.7850	-0.4536	1.4764
4.24	-0.2948	2.3360	-0.3658	1.9566	-0.4300	1.6258
4.28	-0.2652	2.5495	-0.3391	2.1399	-0.4058	1.7847
4.32	-0.2345	2.7793	-0.3115	2.3362	-0.3809	1.9540
4.36	-0.2028	3.0275	-0.2830	2.5469	-0.3552	2.1348
4.40	-0.1699	3.2960	-0.2536	2.7735	-0.3287	2.3282
4.44	-0.1360	3.5876	-0.2233	3.0178	-0.3015	2.5356
4.48	-0.1007	3.9051	-0.1919	3.2819	-0.2733	2.7583
4.52	-0.0642	4.2520	-0.1594	3.5683	-0.2443	2.9981
4.56	-0.0263	4.6325	-0.1258	3.8797	-0.2143	3.2571
4.60	0.0131	5.0514	-0.0910	4.2193	-0.1833	3.5374
4.64	0.0540	5.5147	-0.0549	4.5912	-0.1513	3.8417
4.68	0.0966	6.0295	-0.0175	4.9999	-0.1181	4.1732
4.72	0.1410	6.6046	0.0214	5.4510	-0.0838	4.5354
4.76	0.1872	7.2509	0.0617	5.9512	-0.0483	4.9327
4.80	0.2354	7.9819	0.1037	6.5086	-0.0114	5.3703
4.84	0.2858	8.8150	0.1474	7.1332	0.0268	5.8543
4.88	0.3385	9.7723	0.1929	7.8377	0.0665	6.3923
4.92	0.3937	10.8830	0.2403	8.6377	0.1077	6.9935
4.96	0.4516	12.1859	0.2899	9.5536	0.1506	7.6694

$$1.350 \le \hat{\alpha}_3 \le 1.400$$

$\hat{\alpha}_4$	$\hat{\alpha}_3 = 1.350$		$\hat{\alpha}_3 = 1.375$		$\hat{\alpha}_3 = 1.400$	
	β_3	β_4	β_3	β_4	β_3	β_4
3.00	-0.9572	-0.8480	-0.9750	-0.9093	-0.9913	-0.9679
3.04	-0.9469	-0.8109	-0.9653	-0.8744	-0.9824	-0.9349
3.08	-0.9363	-0.7726	-0.9555	-0.8384	-0.9732	-0.9009
3.12	-0.9255	-0.7330	-0.9454	-0.8012	-0.9638	-0.8660
3.16	-0.9144	-0.6922	-0.9351	-0.7630	-0.9542	-0.8300
3.20	-0.9030	-0.6500	-0.9245	-0.7234	-0.9443	-0.7929
3.24	-0.8914	-0.6064	-0.9136	-0.6826	-0.9342	-0.7547
3.28	-0.8794	-0.5612	-0.9025	-0.6405	-0.9239	-0.7152
3.32	-0.8672	-0.5145	-0.8911	-0.5969	-0.9133	-0.6745
3.36	-0.8547	-0.4660	-0.8795	-0.5518	-0.9024	-0.6325
3.40	-0.8419	-0.4157	-0.8676	-0.5051	-0.8913	-0.5890
3.44	-0.8288	-0.3636	-0.8554	-0.4567	-0.8800	-0.5440
3.48	-0.8154	-0.3094	-0.8429	-0.4066	-0.8684	-0.4975
3.52	-0.8017	-0.2531	-0.8302	-0.3546	-0.8565	-0.4493
3.56	-0.7876	-0.1945	-0.8171	-0.3006	-0.8444	-0.3993
3.60	-0.7732	-0.1336	-0.8038	-0.2445	-0.8320	-0.3475
3.64	-0.7585	-0.0701	-0.7901	-0.1861	-0.8193	-0.2937
3.68	-0.7434	-0.0040	-0.7761	-0.1255	-0.8063	-0.2379
3.72	-0.7280	0.0651	-0.7618	-0.0623	-0.7930	-0.1799
3.76	-0.7122	0.1371	-0.7472	0.0035	-0.7795	-0.1196
3.80	-0.6960	0.2125	-0.7322	0.0722	-0.7656	-0.0568
3.84	-0.6794	0.2913	-0.7169	0.1438	-0.7514	0.0085
3.88	-0.6624	0.3738	-0.7011	0.2186	-0.7369	0.0767
3.92	-0.6449	0.4602	-0.6851	0.2969	-0.7220	0.1477
3.96	-0.6271	0.5510	-0.6686	0.3787	-0.7068	0.2219
4.00	-0.6087	0.6462	-0.6517	0.4645	-0.6912	0.2995
4.04	-0.5899	0.7464	-0.6344	0.5544	-0.6753	0.3805
4.08	-0.5707	0.8518	-0.6167	0.6488	-0.6590	0.4654
4.12	-0.5509	0.9630	-0.5986	0.7479	-0.6423	0.5544
4.16	-0.5306	1.0802	-0.5800	0.8522	-0.6252	0.6476
4.20	-0.5097	1.2041	-0.5609	0.9620	-0.6076	0.7456
4.24	-0.4883	1.3351	-0.5413	1.0778	-0.5897	0.8485
4.28	-0.4663	1.4740	-0.5212	1.2001	-0.5713	0.9568
4.32	-0.4436	1.6214	-0.5006	1.3293	-0.5524	1.0710
4.36	-0.4204	1.7780	-0.4794	1.4661	-0.5331	1.1913
4.40	-0.3964	1.9447	-0.4577	1.6111	-0.5133	1.3185
4.44	-0.3718	2.1225	-0.4353	1.7651	-0.4929	1.4529
4.48	-0.3464	2.3125	-0.4124	1.9288	-0.4720	1.5953
4.52	-0.3203	2.5160	-0.3887	2.1033	-0.4506	1.7464
4.56	-0.2934	2.7343	-0.3645	2.2895	-0.4286	1.9069
4.60	-0.2656	2.9691	-0.3395	2.4886	-0.4060	2.0776
4.64	-0.2370	3.2221	-0.3137	2.7020	-0.3827	2.2597
4.68	-0.2075	3.4957	-0.2872	2.9312	-0.3588	2.4541
4.72	-0.1769	3.7922	-0.2599	3.1779	-0.3343	2.6621
4.76	-0.1454	4.1146	-0.2318	3.4441	-0.3090	2.8852
4.80	-0.1128	4.4662	-0.2027	3.7321	-0.2829	3.1250
4.84	-0.0791	4.8511	-0.1728	4.0447	-0.2561	3.3834
4.88	-0.0442	5.2740	-0.1418	4.3850	-0.2285	3.6625
4.92	-0.0080	5.7407	-0.1098	4.7567	-0.2000	3.9647
4.96	0.0295	6.2581	-0.0767	5.1642	-0.1706	4.2931
5.00	0.0684	6.8345	-0.0425	5.6126	-0.1403	4.6510
5.04	0.1088	7.4804	-0.0071	6.1084	-0.1090	5.0423
5.08	0.1508	8.2087	0.0296	6.6591	-0.0767	5.4720
5.12	0.1945	9.0357	0.0676	7.2742	-0.0432	5.9457
5.16	0.2399	9.9823	0.1071	7.9650	-0.0086	6.4702

　　　　　　　C. Table C–1 for GBD Fits: Method of Moments

$$1.425 \leq \hat{\alpha}_3 \leq 1.475$$

$\hat{\alpha}_4$	$\hat{\alpha}_3 = 1.425$		$\hat{\alpha}_3 = 1.450$		$\hat{\alpha}_3 = 1.475$	
	β_3	β_4	β_3	β_4	β_3	β_4
3.20	-0.9626	-0.8588	-0.9796	-0.9214	-0.9951	-0.9810
3.24	-0.9532	-0.8229	-0.9708	-0.8876	-0.9870	-0.9491
3.28	-0.9436	-0.7859	-0.9619	-0.8528	-0.9787	-0.9163
3.32	-0.9337	-0.7478	-0.9527	-0.8171	-0.9702	-0.8827
3.36	-0.9237	-0.7085	-0.9433	-0.7802	-0.9614	-0.8481
3.40	-0.9133	-0.6679	-0.9337	-0.7422	-0.9525	-0.8124
3.44	-0.9028	-0.6259	-0.9238	-0.7031	-0.9433	-0.7758
3.48	-0.8920	-0.5826	-0.9138	-0.6626	-0.9340	-0.7380
3.52	-0.8809	-0.5378	-0.9035	-0.6209	-0.9244	-0.6990
3.56	-0.8696	-0.4915	-0.8930	-0.5778	-0.9146	-0.6589
3.60	-0.8581	-0.4435	-0.8823	-0.5333	-0.9046	-0.6174
3.64	-0.8463	-0.3939	-0.8713	-0.4873	-0.8944	-0.5746
3.68	-0.8342	-0.3424	-0.8601	-0.4396	-0.8839	-0.5304
3.72	-0.8219	-0.2889	-0.8486	-0.3902	-0.8733	-0.4846
3.76	-0.8093	-0.2335	-0.8369	-0.3391	-0.8624	-0.4374
3.80	-0.7964	-0.1759	-0.8249	-0.2861	-0.8512	-0.3884
3.84	-0.7832	-0.1160	-0.8127	-0.2311	-0.8399	-0.3377
3.88	-0.7698	-0.0538	-0.8002	-0.1740	-0.8282	-0.2852
3.92	-0.7560	0.0110	-0.7874	-0.1147	-0.8164	-0.2307
3.96	-0.7419	0.0786	-0.7744	-0.0530	-0.8043	-0.1742
4.00	-0.7275	0.1490	-0.7610	0.0112	-0.7919	-0.1155
4.04	-0.7128	0.2224	-0.7474	0.0780	-0.7793	-0.0545
4.08	-0.6978	0.2991	-0.7335	0.1476	-0.7664	0.0090
4.12	-0.6824	0.3793	-0.7192	0.2202	-0.7532	0.0750
4.16	-0.6666	0.4631	-0.7047	0.2959	-0.7397	0.1437
4.20	-0.6505	0.5509	-0.6898	0.3750	-0.7260	0.2153
4.24	-0.6340	0.6430	-0.6746	0.4577	-0.7119	0.2900
4.28	-0.6171	0.7395	-0.6590	0.5443	-0.6975	0.3679
4.32	-0.5998	0.8410	-0.6431	0.6349	-0.6829	0.4494
4.36	-0.5821	0.9476	-0.6268	0.7300	-0.6679	0.5346
4.40	-0.5639	1.0598	-0.6102	0.8297	-0.6525	0.6237
4.44	-0.5453	1.1781	-0.5931	0.9345	-0.6369	0.7171
4.48	-0.5263	1.3030	-0.5757	1.0447	-0.6208	0.8150
4.52	-0.5067	1.4348	-0.5578	1.1607	-0.6044	0.9177
4.56	-0.4867	1.5744	-0.5395	1.2830	-0.5876	1.0257
4.60	-0.4661	1.7222	-0.5207	1.4121	-0.5705	1.1393
4.64	-0.4450	1.8791	-0.5015	1.5486	-0.5529	1.2590
4.68	-0.4234	2.0459	-0.4818	1.6930	-0.5349	1.3851
4.72	-0.4012	2.2235	-0.4616	1.8461	-0.5165	1.5183
4.76	-0.3783	2.4128	-0.4409	2.0086	-0.4976	1.6591
4.80	-0.3549	2.6152	-0.4197	2.1814	-0.4783	1.8082
4.84	-0.3307	2.8319	-0.3979	2.3655	-0.4585	1.9663
4.88	-0.3059	3.0644	-0.3755	2.5619	-0.4382	2.1341
4.92	-0.2804	3.3146	-0.3525	2.7719	-0.4174	2.3127
4.96	-0.2542	3.5842	-0.3289	2.9969	-0.3960	2.5029
5.00	-0.2271	3.8757	-0.3046	3.2385	-0.3741	2.7060
5.04	-0.1992	4.1918	-0.2796	3.4985	-0.3516	2.9232
5.08	-0.1705	4.5354	-0.2539	3.7789	-0.3285	3.1561
5.12	-0.1409	4.9104	-0.2275	4.0824	-0.3048	3.4062
5.16	-0.1103	5.3209	-0.2003	4.4116	-0.2804	3.6755
5.20	-0.0787	5.7721	-0.1722	4.7699	-0.2553	3.9662
5.24	-0.0461	6.2702	-0.1433	5.1611	-0.2296	4.2809
5.28	-0.0124	6.8227	-0.1136	5.5900	-0.2030	4.6225
5.32	0.0225	7.4386	-0.0828	6.0618	-0.1757	4.9945
5.36	0.0585	8.1292	-0.0511	6.5834	-0.1476	5.4012

$$1.500 \leq \hat{\alpha}_3 \leq 1.550$$

	$\hat{\alpha}_3 = 1.500$		$\hat{\alpha}_3 = 1.525$		$\hat{\alpha}_3 = 1.550$	
$\hat{\alpha}_4$	β_3	β_4	β_3	β_4	β_3	β_4
3.45	-0.9591	-0.8357	-0.9759	-0.9013	-0.9912	-0.9636
3.50	-0.9482	-0.7910	-0.9657	-0.8593	-0.9818	-0.9240
3.55	-0.9369	-0.7446	-0.9551	-0.8158	-0.9720	-0.8831
3.60	-0.9253	-0.6964	-0.9443	-0.7707	-0.9619	-0.8408
3.65	-0.9133	-0.6463	-0.9332	-0.7239	-0.9516	-0.7970
3.70	-0.9011	-0.5943	-0.9218	-0.6754	-0.9410	-0.7517
3.75	-0.8885	-0.5401	-0.9101	-0.6250	-0.9301	-0.7046
3.80	-0.8756	-0.4837	-0.8981	-0.5726	-0.9189	-0.6558
3.85	-0.8623	-0.4248	-0.8857	-0.5180	-0.9074	-0.6051
3.90	-0.8487	-0.3634	-0.8731	-0.4612	-0.8956	-0.5524
3.95	-0.8347	-0.2993	-0.8601	-0.4020	-0.8835	-0.4976
4.00	-0.8204	-0.2322	-0.8468	-0.3402	-0.8711	-0.4405
4.05	-0.8057	-0.1620	-0.8331	-0.2757	-0.8584	-0.3810
4.10	-0.7906	-0.0885	-0.8191	-0.2083	-0.8453	-0.3189
4.15	-0.7751	-0.0114	-0.8047	-0.1377	-0.8320	-0.2541
4.20	-0.7593	0.0696	-0.7900	-0.0638	-0.8183	-0.1864
4.25	-0.7430	0.1547	-0.7748	0.0137	-0.8042	-0.1156
4.30	-0.7262	0.2443	-0.7593	0.0950	-0.7899	-0.0414
4.35	-0.7090	0.3387	-0.7434	0.1805	-0.7751	0.0363
4.40	-0.6914	0.4382	-0.7271	0.2704	-0.7600	0.1179
4.45	-0.6733	0.5434	-0.7104	0.3651	-0.7445	0.2036
4.50	-0.6547	0.6547	-0.6932	0.4650	-0.7286	0.2936
4.55	-0.6355	0.7725	-0.6755	0.5704	-0.7123	0.3885
4.60	-0.6159	0.8976	-0.6574	0.6819	-0.6956	0.4885
4.65	-0.5957	1.0304	-0.6388	0.8000	-0.6784	0.5940
4.70	-0.5749	1.1719	-0.6198	0.9252	-0.6608	0.7055
4.75	-0.5535	1.3227	-0.6001	1.0582	-0.6428	0.8235
4.80	-0.5315	1.4839	-0.5800	1.1996	-0.6242	0.9486
4.85	-0.5089	1.6564	-0.5593	1.3504	-0.6052	1.0813
4.90	-0.4855	1.8415	-0.5380	1.5114	-0.5857	1.2225
4.95	-0.4615	2.0405	-0.5160	1.6836	-0.5656	1.3729
5.00	-0.4368	2.2549	-0.4935	1.8682	-0.5450	1.5333
5.05	-0.4112	2.4867	-0.4703	2.0666	-0.5238	1.7048
5.10	-0.3849	2.7378	-0.4463	2.2802	-0.5020	1.8885
5.15	-0.3577	3.0107	-0.4217	2.5109	-0.4796	2.0858
5.20	-0.3296	3.3083	-0.3963	2.7606	-0.4565	2.2980
5.25	-0.3006	3.6339	-0.3701	3.0318	-0.4328	2.5270
5.30	-0.2706	3.9915	-0.3431	3.3271	-0.4083	2.7746
5.35	-0.2395	4.3859	-0.3152	3.6498	-0.3831	3.0432
5.40	-0.2074	4.8228	-0.2864	4.0039	-0.3572	3.3354
5.45	-0.1741	5.3092	-0.2566	4.3939	-0.3304	3.6543
5.50	-0.1396	5.8539	-0.2259	4.8253	-0.3027	4.0038
5.55	-0.1038	6.4674	-0.1940	5.3049	-0.2742	4.3881
5.60	-0.0666	7.1634	-0.1610	5.8409	-0.2447	4.8126
5.65	-0.0280	7.9590	-0.1268	6.4436	-0.2143	5.2837
5.70	0.0122	8.8767	-0.0914	7.1259	-0.1828	5.8093
5.75	0.0541	9.9459	-0.0546	7.9039	-0.1502	6.3991
5.80	0.0977	11.2067	-0.0164	8.7990	-0.1164	7.0651
5.85	0.1431	12.7140	0.0233	9.8386	-0.0814	7.8227
5.90	0.1906	14.5462	0.0646	11.0602	-0.0451	8.6917
5.95	0.2403	16.8185	0.1077	12.5145	-0.0074	9.6977
6.00	0.2923	19.7077	0.1525	14.2736	0.0317	10.8751
6.05	0.3467	23.4994	0.1993	16.4422	0.0724	12.2707
6.10	0.4039	28.6870	0.2482	19.1790	0.1147	13.9498
6.15	0.4640	36.2027	0.2994	22.7367	0.1588	16.0065

$$1.575 \leq \hat{\alpha}_3 \leq 1.625$$

	$\hat{\alpha}_3 = 1.575$		$\hat{\alpha}_3 = 1.600$		$\hat{\alpha}_3 = 1.625$	
$\hat{\alpha}_4$	β_3	β_4	β_3	β_4	β_3	β_4
3.68	-0.9627	-0.8407	-0.9787	-0.9074	-0.9934	-0.9706
3.73	-0.9526	-0.7973	-0.9693	-0.8667	-0.9847	-0.9323
3.78	-0.9423	-0.7523	-0.9597	-0.8246	-0.9758	-0.8928
3.83	-0.9317	-0.7057	-0.9499	-0.7810	-0.9666	-0.8519
3.88	-0.9208	-0.6574	-0.9397	-0.7359	-0.9572	-0.8097
3.93	-0.9097	-0.6073	-0.9294	-0.6892	-0.9475	-0.7661
3.98	-0.8982	-0.5552	-0.9187	-0.6407	-0.9376	-0.7208
4.03	-0.8865	-0.5010	-0.9078	-0.5904	-0.9274	-0.6740
4.08	-0.8745	-0.4446	-0.8966	-0.5382	-0.9170	-0.6254
4.13	-0.8622	-0.3859	-0.8851	-0.4839	-0.9063	-0.5751
4.18	-0.8496	-0.3247	-0.8734	-0.4274	-0.8954	-0.5227
4.23	-0.8367	-0.2608	-0.8614	-0.3685	-0.8842	-0.4684
4.28	-0.8234	-0.1941	-0.8490	-0.3072	-0.8727	-0.4118
4.33	-0.8098	-0.1244	-0.8364	-0.2433	-0.8610	-0.3529
4.38	-0.7959	-0.0515	-0.8235	-0.1765	-0.8489	-0.2916
4.43	-0.7817	0.0249	-0.8103	-0.1067	-0.8366	-0.2277
4.48	-0.7671	0.1050	-0.7967	-0.0338	-0.8240	-0.1610
4.53	-0.7522	0.1890	-0.7828	0.0426	-0.8112	-0.0913
4.58	-0.7368	0.2773	-0.7686	0.1226	-0.7980	-0.0184
4.63	-0.7211	0.3702	-0.7541	0.2066	-0.7845	0.0578
4.68	-0.7051	0.4680	-0.7392	0.2948	-0.7706	0.1377
4.73	-0.6886	0.5712	-0.7239	0.3875	-0.7565	0.2214
4.78	-0.6717	0.6800	-0.7083	0.4850	-0.7420	0.3093
4.83	-0.6543	0.7951	-0.6923	0.5878	-0.7272	0.4016
4.88	-0.6365	0.9170	-0.6759	0.6963	-0.7120	0.4988
4.93	-0.6183	1.0461	-0.6591	0.8109	-0.6965	0.6011
4.98	-0.5996	1.1833	-0.6419	0.9322	-0.6806	0.7090
5.03	-0.5804	1.3292	-0.6242	1.0607	-0.6643	0.8229
5.08	-0.5606	1.4846	-0.6061	1.1970	-0.6476	0.9434
5.13	-0.5404	1.6506	-0.5875	1.3419	-0.6305	1.0709
5.18	-0.5196	1.8280	-0.5684	1.4961	-0.6130	1.2061
5.23	-0.4982	2.0182	-0.5489	1.6607	-0.5951	1.3497
5.28	-0.4762	2.2224	-0.5288	1.8365	-0.5766	1.5025
5.33	-0.4536	2.4423	-0.5082	2.0247	-0.5578	1.6653
5.38	-0.4303	2.6797	-0.4870	2.2267	-0.5384	1.8390
5.43	-0.4064	2.9365	-0.4652	2.4439	-0.5185	2.0249
5.48	-0.3817	3.2153	-0.4428	2.6781	-0.4981	2.2242
5.53	-0.3563	3.5187	-0.4198	2.9313	-0.4772	2.4383
5.58	-0.3301	3.8503	-0.3961	3.2057	-0.4557	2.6688
5.63	-0.3032	4.2138	-0.3718	3.5040	-0.4336	2.9177
5.68	-0.2753	4.6140	-0.3467	3.8295	-0.4109	3.1871
5.73	-0.2466	5.0564	-0.3209	4.1857	-0.3875	3.4797
5.78	-0.2169	5.5480	-0.2942	4.5773	-0.3635	3.7982
5.83	-0.1862	6.0971	-0.2668	5.0094	-0.3388	4.1464
5.88	-0.1545	6.7142	-0.2385	5.4885	-0.3133	4.5284
5.93	-0.1217	7.4121	-0.2092	6.0224	-0.2871	4.9491
5.98	-0.0877	8.2075	-0.1790	6.6209	-0.2601	5.4146
6.03	-0.0525	9.1217	-0.1478	7.2960	-0.2322	5.9321
6.08	-0.0160	10.1828	-0.1156	8.0629	-0.2035	6.5107
6.13	0.0219	11.4282	-0.0822	8.9413	-0.1738	7.1614
6.18	0.0612	12.9093	-0.0476	9.9567	-0.1432	7.8983
6.23	0.1021	14.6987	-0.0118	11.1429	-0.1115	8.7392
6.28	0.1447	16.9014	0.0254	12.5461	-0.0788	9.7071
6.33	0.1890	19.6767	0.0639	14.2302	-0.0450	10.8325
6.38	0.2352	23.2772	0.1039	16.2874	-0.0099	12.1562

$$1.650 \leq \hat{\alpha}_3 \leq 1.700$$

	$\hat{\alpha}_3 = 1.650$		$\hat{\alpha}_3 = 1.675$		$\hat{\alpha}_3 = 1.700$	
$\hat{\alpha}_4$	β_3	β_4	β_3	β_4	β_3	β_4
3.95	-0.9606	-0.8216	-0.9763	-0.8908	-0.9906	-0.9562
4.00	-0.9513	-0.7788	-0.9676	-0.8507	-0.9826	-0.9185
4.05	-0.9417	-0.7344	-0.9587	-0.8092	-0.9743	-0.8796
4.10	-0.9319	-0.6885	-0.9496	-0.7663	-0.9658	-0.8394
4.15	-0.9219	-0.6410	-0.9402	-0.7220	-0.9571	-0.7980
4.20	-0.9116	-0.5917	-0.9306	-0.6761	-0.9482	-0.7551
4.25	-0.9010	-0.5405	-0.9208	-0.6286	-0.9391	-0.7109
4.30	-0.8903	-0.4874	-0.9108	-0.5793	-0.9297	-0.6650
4.35	-0.8792	-0.4322	-0.9005	-0.5282	-0.9201	-0.6176
4.40	-0.8679	-0.3748	-0.8899	-0.4751	-0.9103	-0.5684
4.45	-0.8563	-0.3150	-0.8792	-0.4200	-0.9003	-0.5174
4.50	-0.8445	-0.2527	-0.8681	-0.3627	-0.8900	-0.4645
4.55	-0.8324	-0.1878	-0.8569	-0.3031	-0.8795	-0.4096
4.60	-0.8200	-0.1201	-0.8454	-0.2410	-0.8687	-0.3525
4.65	-0.8074	-0.0493	-0.8336	-0.1763	-0.8578	-0.2931
4.70	-0.7944	0.0246	-0.8215	-0.1088	-0.8466	-0.2312
4.75	-0.7812	0.1020	-0.8092	-0.0384	-0.8351	-0.1668
4.80	-0.7676	0.1830	-0.7966	0.0352	-0.8234	-0.0997
4.85	-0.7537	0.2680	-0.7837	0.1122	-0.8114	-0.0296
4.90	-0.7396	0.3572	-0.7706	0.1928	-0.7992	0.0435
4.95	-0.7250	0.4509	-0.7571	0.2772	-0.7867	0.1200
5.00	-0.7102	0.5494	-0.7434	0.3658	-0.7739	0.2000
5.05	-0.6950	0.6532	-0.7293	0.4588	-0.7608	0.2838
5.10	-0.6794	0.7626	-0.7149	0.5566	-0.7475	0.3717
5.15	-0.6635	0.8781	-0.7002	0.6595	-0.7339	0.4638
5.20	-0.6472	1.0002	-0.6851	0.7679	-0.7199	0.5607
5.25	-0.6305	1.1295	-0.6697	0.8823	-0.7057	0.6625
5.30	-0.6134	1.2666	-0.6540	1.0031	-0.6911	0.7697
5.35	-0.5959	1.4121	-0.6378	1.1309	-0.6762	0.8828
5.40	-0.5780	1.5669	-0.6213	1.2662	-0.6610	1.0021
5.45	-0.5596	1.7317	-0.6044	1.4098	-0.6454	1.1282
5.50	-0.5407	1.9078	-0.5871	1.5624	-0.6295	1.2617
5.55	-0.5214	2.0960	-0.5694	1.7249	-0.6132	1.4031
5.60	-0.5016	2.2977	-0.5513	1.8981	-0.5965	1.5533
5.65	-0.4812	2.5144	-0.5327	2.0831	-0.5795	1.7130
5.70	-0.4604	2.7476	-0.5136	2.2812	-0.5620	1.8831
5.75	-0.4389	2.9994	-0.4940	2.4936	-0.5441	2.0646
5.80	-0.4169	3.2718	-0.4740	2.7221	-0.5258	2.2587
5.85	-0.3943	3.5675	-0.4534	2.9683	-0.5070	2.4666
5.90	-0.3710	3.8895	-0.4323	3.2343	-0.4878	2.6899
5.95	-0.3471	4.2413	-0.4107	3.5227	-0.4681	2.9302
6.00	-0.3225	4.6270	-0.3884	3.8361	-0.4478	3.1894
6.05	-0.2972	5.0517	-0.3656	4.1778	-0.4271	3.4699
6.10	-0.2711	5.5214	-0.3421	4.5518	-0.4058	3.7743
6.15	-0.2443	6.0435	-0.3179	4.9628	-0.3840	4.1055
6.20	-0.2166	6.6268	-0.2931	5.4161	-0.3616	4.4673
6.25	-0.1881	7.2825	-0.2676	5.9187	-0.3386	4.8639
6.30	-0.1586	8.0247	-0.2412	6.4786	-0.3149	5.3004
6.35	-0.1283	8.8710	-0.2142	7.1060	-0.2906	5.7830
6.40	-0.0969	9.8446	-0.1862	7.8136	-0.2656	6.3190
6.45	-0.0644	10.9757	-0.1574	8.6173	-0.2398	6.9178
6.50	-0.0309	12.3050	-0.1277	9.5375	-0.2134	7.5906
6.60	0.0398	15.8052	-0.0654	11.8436	-0.1580	9.2192
6.71	0.1236	21.8563	0.0080	15.4729	-0.0931	11.6291
6.83	0.2234	34.5155	0.0947	21.8805	-0.0169	15.4823

$$1.725 \leq \hat{\alpha}_3 \leq 1.775$$

$\hat{\alpha}_4$	$\hat{\alpha}_3 = 1.725$		$\hat{\alpha}_3 = 1.750$		$\hat{\alpha}_3 = 1.775$	
	β_3	β_4	β_3	β_4	β_3	β_4
4.22	-0.9610	-0.8129	-0.9761	-0.8838	-0.9900	-0.9506
4.27	-0.9524	-0.7710	-0.9681	-0.8446	-0.9826	-0.9138
4.32	-0.9436	-0.7277	-0.9600	-0.8041	-0.9750	-0.8758
4.37	-0.9346	-0.6829	-0.9516	-0.7623	-0.9672	-0.8367
4.42	-0.9254	-0.6366	-0.9430	-0.7191	-0.9592	-0.7964
4.47	-0.9159	-0.5886	-0.9342	-0.6745	-0.9510	-0.7547
4.52	-0.9063	-0.5389	-0.9252	-0.6283	-0.9426	-0.7118
4.57	-0.8964	-0.4873	-0.9160	-0.5805	-0.9340	-0.6673
4.62	-0.8863	-0.4338	-0.9065	-0.5310	-0.9252	-0.6214
4.67	-0.8760	-0.3783	-0.8969	-0.4797	-0.9163	-0.5739
4.72	-0.8654	-0.3206	-0.8871	-0.4265	-0.9071	-0.5246
4.77	-0.8547	-0.2605	-0.8771	-0.3713	-0.8977	-0.4736
4.82	-0.8437	-0.1981	-0.8668	-0.3139	-0.8882	-0.4208
4.87	-0.8324	-0.1330	-0.8563	-0.2543	-0.8784	-0.3659
4.92	-0.8209	-0.0652	-0.8456	-0.1922	-0.8684	-0.3090
4.97	-0.8092	0.0056	-0.8347	-0.1277	-0.8583	-0.2498
5.02	-0.7972	0.0795	-0.8236	-0.0604	-0.8479	-0.1883
5.07	-0.7850	0.1567	-0.8122	0.0098	-0.8373	-0.1243
5.12	-0.7725	0.2375	-0.8006	0.0830	-0.8264	-0.0576
5.17	-0.7597	0.3221	-0.7887	0.1595	-0.8154	0.0119
5.22	-0.7467	0.4107	-0.7766	0.2394	-0.8041	0.0843
5.27	-0.7334	0.5037	-0.7642	0.3231	-0.7926	0.1600
5.32	-0.7198	0.6014	-0.7516	0.4107	-0.7809	0.2390
5.37	-0.7059	0.7041	-0.7387	0.5026	-0.7689	0.3217
5.42	-0.6917	0.8122	-0.7255	0.5990	-0.7567	0.4082
5.47	-0.6772	0.9261	-0.7121	0.7003	-0.7442	0.4988
5.52	-0.6623	1.0463	-0.6984	0.8069	-0.7315	0.5938
5.57	-0.6472	1.1733	-0.6844	0.9191	-0.7185	0.6936
5.62	-0.6317	1.3077	-0.6700	1.0374	-0.7053	0.7985
5.67	-0.6158	1.4501	-0.6554	1.1622	-0.6918	0.9088
5.72	-0.5996	1.6012	-0.6405	1.2943	-0.6779	1.0250
5.77	-0.5830	1.7618	-0.6252	1.4340	-0.6638	1.1476
5.82	-0.5661	1.9329	-0.6096	1.5821	-0.6495	1.2771
5.87	-0.5487	2.1153	-0.5937	1.7394	-0.6347	1.4140
5.92	-0.5309	2.3102	-0.5774	1.9067	-0.6197	1.5590
5.97	-0.5127	2.5190	-0.5607	2.0849	-0.6044	1.7128
6.02	-0.4941	2.7431	-0.5436	2.2751	-0.5887	1.8761
6.07	-0.4750	2.9841	-0.5262	2.4786	-0.5727	2.0500
6.12	-0.4555	3.2440	-0.5083	2.6966	-0.5563	2.2352
6.17	-0.4354	3.5251	-0.4900	2.9308	-0.5396	2.4331
6.22	-0.4149	3.8298	-0.4713	3.1829	-0.5225	2.6449
6.27	-0.3938	4.1613	-0.4521	3.4550	-0.5050	2.8720
6.32	-0.3722	4.5231	-0.4325	3.7495	-0.4871	3.1161
6.37	-0.3500	4.9193	-0.4124	4.0692	-0.4687	3.3791
6.42	-0.3273	5.3551	-0.3918	4.4173	-0.4500	3.6632
6.47	-0.3039	5.8364	-0.3706	4.7978	-0.4308	3.9709
6.52	-0.2798	6.3706	-0.3490	5.2151	-0.4111	4.3054
6.57	-0.2551	6.9666	-0.3267	5.6747	-0.3909	4.6699
6.62	-0.2297	7.6354	-0.3039	6.1832	-0.3703	5.0688
6.67	-0.2036	8.3911	-0.2804	6.7486	-0.3491	5.5069
6.72	-0.1767	9.2511	-0.2563	7.3808	-0.3274	5.9901
6.82	-0.1205	11.3823	-0.2061	8.8976	-0.2823	7.1219
6.92	-0.0607	14.3149	-0.1531	10.8766	-0.2348	8.5435
7.04	0.0161	19.6946	-0.0851	14.2160	-0.1743	10.8112
7.16	0.0993	29.3837	-0.0121	19.4357	-0.1096	14.0621

$$1.800 \leq \hat{\alpha}_3 \leq 1.850$$

	$\hat{\alpha}_3 = 1.800$		$\hat{\alpha}_3 = 1.825$		$\hat{\alpha}_3 = 1.850$	
$\hat{\alpha}_4$	β_3	β_4	β_3	β_4	β_3	β_4
4.48	-0.9649	-0.8221	-0.9793	-0.8933	-0.9924	-0.9604
4.54	-0.9555	-0.7734	-0.9705	-0.8478	-0.9843	-0.9176
4.60	-0.9458	-0.7229	-0.9615	-0.8006	-0.9759	-0.8734
4.66	-0.9358	-0.6704	-0.9522	-0.7516	-0.9672	-0.8276
4.72	-0.9256	-0.6158	-0.9426	-0.7008	-0.9584	-0.7802
4.78	-0.9151	-0.5589	-0.9329	-0.6479	-0.9492	-0.7310
4.84	-0.9043	-0.4997	-0.9228	-0.5930	-0.9399	-0.6799
4.90	-0.8933	-0.4378	-0.9125	-0.5358	-0.9303	-0.6268
4.96	-0.8819	-0.3732	-0.9019	-0.4761	-0.9204	-0.5715
5.02	-0.8703	-0.3057	-0.8911	-0.4139	-0.9103	-0.5140
5.08	-0.8585	-0.2351	-0.8800	-0.3490	-0.9000	-0.4541
5.14	-0.8463	-0.1611	-0.8687	-0.2811	-0.8893	-0.3917
5.20	-0.8338	-0.0835	-0.8570	-0.2101	-0.8785	-0.3264
5.26	-0.8210	-0.0020	-0.8451	-0.1358	-0.8674	-0.2583
5.32	-0.8079	0.0835	-0.8329	-0.0579	-0.8560	-0.1871
5.38	-0.7946	0.1736	-0.8205	0.0239	-0.8444	-0.1125
5.44	-0.7808	0.2684	-0.8077	0.1098	-0.8325	-0.0344
5.50	-0.7668	0.3685	-0.7946	0.2001	-0.8203	0.0476
5.56	-0.7524	0.4742	-0.7812	0.2953	-0.8078	0.1337
5.62	-0.7377	0.5859	-0.7676	0.3956	-0.7951	0.2242
5.68	-0.7226	0.7043	-0.7535	0.5015	-0.7821	0.3195
5.74	-0.7072	0.8298	-0.7392	0.6134	-0.7687	0.4199
5.80	-0.6913	0.9632	-0.7245	0.7319	-0.7551	0.5258
5.86	-0.6751	1.1051	-0.7095	0.8576	-0.7411	0.6378
5.92	-0.6585	1.2565	-0.6941	0.9910	-0.7269	0.7562
5.98	-0.6415	1.4181	-0.6784	1.1329	-0.7123	0.8818
6.04	-0.6241	1.5911	-0.6623	1.2841	-0.6973	1.0150
6.10	-0.6062	1.7766	-0.6458	1.4455	-0.6821	1.1566
6.16	-0.5878	1.9761	-0.6288	1.6181	-0.6664	1.3073
6.22	-0.5690	2.1910	-0.6115	1.8031	-0.6504	1.4682
6.28	-0.5497	2.4231	-0.5938	2.0019	-0.6340	1.6401
6.34	-0.5299	2.6746	-0.5756	2.2159	-0.6172	1.8242
6.40	-0.5096	2.9479	-0.5569	2.4468	-0.6001	2.0218
6.46	-0.4887	3.2458	-0.5377	2.6969	-0.5825	2.2344
6.52	-0.4672	3.5717	-0.5181	2.9682	-0.5644	2.4637
6.58	-0.4452	3.9295	-0.4980	3.2638	-0.5459	2.7117
6.64	-0.4225	4.3241	-0.4773	3.5867	-0.5270	2.9806
6.70	-0.3992	4.7613	-0.4560	3.9410	-0.5076	3.2730
6.76	-0.3752	5.2479	-0.4342	4.3311	-0.4876	3.5923
6.82	-0.3505	5.7928	-0.4118	4.7626	-0.4672	3.9420
6.88	-0.3251	6.4067	-0.3888	5.2423	-0.4462	4.3266
6.94	-0.2989	7.1032	-0.3651	5.7785	-0.4247	4.7514
7.00	-0.2719	7.8998	-0.3407	6.3815	-0.4026	5.2228
7.06	-0.2440	8.8190	-0.3156	7.0642	-0.3798	5.7489
7.12	-0.2153	9.8910	-0.2898	7.8431	-0.3565	6.3392
7.18	-0.1856	11.1562	-0.2632	8.7396	-0.3325	7.0061
7.24	-0.1550	12.6711	-0.2357	9.7818	-0.3077	7.7651
7.30	-0.1233	14.5162	-0.2074	11.0076	-0.2823	8.6362
7.36	-0.0905	16.8105	-0.1783	12.4692	-0.2561	9.6455
7.42	-0.0566	19.7384	-0.1481	14.2404	-0.2292	10.8281
7.48	-0.0215	23.6004	-0.1170	16.4295	-0.2014	12.2320
7.54	0.0150	28.9225	-0.0848	19.2020	-0.1727	13.9243
7.60	0.0528	36.7165	-0.0515	22.8234	-0.1431	16.0026
7.66	0.0920	49.2080	-0.0170	27.7491	-0.1126	18.6139
7.72	0.1328	72.4385	0.0187	34.8307	-0.0811	21.9904

$$1.875 \leq \hat{\alpha}_3 \leq 1.925$$

	$\hat{\alpha}_3 = 1.875$		$\hat{\alpha}_3 = 1.900$		$\hat{\alpha}_3 = 1.925$	
$\hat{\alpha}_4$	β_3	β_4	β_3	β_4	β_3	β_4
4.75	-0.9686	-0.8318	-0.9822	-0.9032	-0.9946	-0.9704
4.81	-0.9600	-0.7850	-0.9742	-0.8594	-0.9872	-0.9293
4.87	-0.9512	-0.7365	-0.9660	-0.8140	-0.9796	-0.8867
4.93	-0.9421	-0.6861	-0.9575	-0.7670	-0.9717	-0.8428
4.99	-0.9328	-0.6338	-0.9488	-0.7183	-0.9636	-0.7973
5.05	-0.9232	-0.5794	-0.9399	-0.6678	-0.9553	-0.7502
5.11	-0.9135	-0.5229	-0.9308	-0.6153	-0.9468	-0.7014
5.17	-0.9034	-0.4640	-0.9214	-0.5608	-0.9381	-0.6508
5.23	-0.8932	-0.4026	-0.9119	-0.5041	-0.9291	-0.5982
5.29	-0.8827	-0.3386	-0.9021	-0.4451	-0.9200	-0.5436
5.35	-0.8720	-0.2718	-0.8920	-0.3836	-0.9106	-0.4869
5.41	-0.8610	-0.2020	-0.8818	-0.3195	-0.9010	-0.4278
5.47	-0.8497	-0.1290	-0.8713	-0.2526	-0.8912	-0.3663
5.53	-0.8382	-0.0525	-0.8606	-0.1828	-0.8812	-0.3021
5.59	-0.8265	0.0276	-0.8496	-0.1097	-0.8710	-0.2352
5.65	-0.8145	0.1117	-0.8384	-0.0332	-0.8605	-0.1654
5.71	-0.8022	0.2000	-0.8269	0.0469	-0.8498	-0.0924
5.77	-0.7897	0.2928	-0.8152	0.1309	-0.8389	-0.0160
5.83	-0.7768	0.3906	-0.8033	0.2191	-0.8277	0.0640
5.89	-0.7637	0.4936	-0.7911	0.3119	-0.8163	0.1479
5.95	-0.7503	0.6023	-0.7786	0.4094	-0.8047	0.2359
6.01	-0.7366	0.7172	-0.7658	0.5122	-0.7928	0.3283
6.07	-0.7226	0.8389	-0.7528	0.6206	-0.7807	0.4255
6.13	-0.7082	0.9677	-0.7395	0.7351	-0.7683	0.5279
6.19	-0.6936	1.1045	-0.7259	0.8562	-0.7556	0.6358
6.25	-0.6786	1.2500	-0.7119	0.9845	-0.7427	0.7497
6.31	-0.6632	1.4049	-0.6977	1.1205	-0.7295	0.8701
6.37	-0.6475	1.5702	-0.6832	1.2651	-0.7160	0.9976
6.43	-0.6315	1.7469	-0.6683	1.4190	-0.7022	1.1327
6.49	-0.6151	1.9363	-0.6531	1.5830	-0.6881	1.2761
6.55	-0.5983	2.1395	-0.6376	1.7583	-0.6737	1.4287
6.61	-0.5810	2.3582	-0.6217	1.9458	-0.6590	1.5912
6.67	-0.5634	2.5942	-0.6055	2.1470	-0.6440	1.7647
6.73	-0.5454	2.8495	-0.5889	2.3634	-0.6287	1.9502
6.79	-0.5269	3.1264	-0.5719	2.5965	-0.6130	2.1489
6.85	-0.5079	3.4278	-0.5545	2.8484	-0.5969	2.3624
6.91	-0.4885	3.7569	-0.5366	3.1214	-0.5805	2.5923
6.97	-0.4686	4.1176	-0.5184	3.4180	-0.5637	2.8404
7.03	-0.4482	4.5145	-0.4997	3.7416	-0.5466	3.1088
7.09	-0.4272	4.9532	-0.4805	4.0956	-0.5290	3.4003
7.15	-0.4057	5.4405	-0.4609	4.4846	-0.5110	3.7176
7.21	-0.3836	5.9845	-0.4408	4.9138	-0.4926	4.0644
7.27	-0.3609	6.5957	-0.4202	5.3896	-0.4738	4.4447
7.33	-0.3376	7.2869	-0.3990	5.9198	-0.4545	4.8635
7.39	-0.3137	8.0745	-0.3773	6.5139	-0.4347	5.3269
7.45	-0.2890	8.9797	-0.3550	7.1840	-0.4144	5.8421
7.51	-0.2637	10.0304	-0.3321	7.9454	-0.3937	6.4181
7.57	-0.2376	11.2638	-0.3086	8.8175	-0.3723	7.0660
7.63	-0.2108	12.7314	-0.2845	9.8259	-0.3505	7.7998
7.69	-0.1831	14.5055	-0.2596	11.0044	-0.3280	8.6375
7.75	-0.1546	16.6919	-0.2341	12.3995	-0.3050	9.6023
7.83	-0.1151	20.5380	-0.1988	14.7116	-0.2733	11.1407
7.97	-0.0420	32.1944	-0.1337	20.7854	-0.2150	14.8576
8.11	0.-370	63.9400	-0.0639	32.4901	-0.1527	20.9276
8.25	0.1227	475.8773	0.0114	64.1704	-0.0859	32.5601

$$1.950 \leq \hat{\alpha}_3 \leq 2.000$$

	$\hat{\alpha}_3 = 1.950$		$\hat{\alpha}_3 = 1.975$		$\hat{\alpha}_3 = 2.000$	
$\hat{\alpha}_4$	β_3	β_4	β_3	β_4	β_3	β_4
5.08	-0.9655	-0.8047	-0.9788	-0.8785	-0.9910	-0.9477
5.15	-0.9562	-0.7505	-0.9701	-0.8278	-0.9829	-0.9002
5.22	-0.9465	-0.6940	-0.9611	-0.7750	-0.9745	-0.8509
5.29	-0.9365	-0.6350	-0.9518	-0.7201	-0.9659	-0.7996
5.36	-0.9263	-0.5735	-0.9423	-0.6630	-0.9570	-0.7463
5.43	-0.9158	-0.5092	-0.9325	-0.6033	-0.9478	-0.6908
5.50	-0.9050	-0.4418	-0.9224	-0.5410	-0.9384	-0.6330
5.57	-0.8940	-0.3713	-0.9121	-0.4759	-0.9288	-0.5727
5.64	-0.8826	-0.2973	-0.9014	-0.4078	-0.9189	-0.5097
5.71	-0.8710	-0.2196	-0.8906	-0.3364	-0.9087	-0.4439
5.78	-0.8591	-0.1379	-0.8794	-0.2615	-0.8983	-0.3750
5.85	-0.8468	-0.0520	-0.8680	-0.1829	-0.8876	-0.3029
5.92	-0.8343	0.0386	-0.8563	-0.1003	-0.8766	-0.2273
5.99	-0.8215	0.1342	-0.8443	-0.0134	-0.8654	-0.1479
6.06	-0.8083	0.2353	-0.8320	0.0782	-0.8540	-0.0644
6.13	-0.7949	0.3422	-0.8195	0.1749	-0.8422	0.0234
6.20	-0.7811	0.4556	-0.8066	0.2771	-0.8302	0.1159
6.27	-0.7670	0.5759	-0.7934	0.3851	-0.8179	0.2135
6.34	-0.7525	0.7038	-0.7799	0.4997	-0.8053	0.3166
6.41	-0.7377	0.8401	-0.7661	0.6212	-0.7924	0.4256
6.48	-0.7225	0.9856	-0.7520	0.7504	-0.7792	0.5411
6.55	-0.7069	1.1411	-0.7375	0.8880	-0.7657	0.6637
6.62	-0.6910	1.3077	-0.7227	1.0348	-0.7520	0.7940
6.69	-0.6747	1.4866	-0.7076	1.1917	-0.7378	0.9327
6.76	-0.6579	1.6792	-0.6920	1.3598	-0.7234	1.0806
6.83	-0.6408	1.8870	-0.6761	1.5402	-0.7086	1.2386
6.90	-0.6232	2.1118	-0.6598	1.7343	-0.6935	1.4078
6.97	-0.6052	2.3557	-0.6432	1.9437	-0.6780	1.5894
7.04	-0.5867	2.6213	-0.6261	2.1702	-0.6622	1.7847
7.11	-0.5677	2.9113	-0.6086	2.4158	-0.6460	1.9952
7.18	-0.5483	3.2292	-0.5906	2.6830	-0.6294	2.2228
7.25	-0.5283	3.5790	-0.5722	2.9747	-0.6124	2.4695
7.32	-0.5078	3.9658	-0.5533	3.2943	-0.5950	2.7378
7.39	-0.4867	4.3954	-0.5340	3.6459	-0.5771	3.0306
7.46	-0.4651	4.8751	-0.5141	4.0342	-0.5589	3.3511
7.53	-0.4428	5.4141	-0.4938	4.4654	-0.5401	3.7034
7.60	-0.4199	6.0235	-0.4728	4.9465	-0.5209	4.0924
7.67	-0.3964	6.7180	-0.4513	5.4865	-0.5012	4.5238
7.74	-0.3722	7.5161	-0.4293	6.0968	-0.4810	5.0048
7.81	-0.3472	8.4423	-0.4066	6.7915	-0.4603	5.5443
7.88	-0.3215	9.5294	-0.3833	7.5889	-0.4390	6.1533
7.95	-0.2951	10.8224	-0.3593	8.5133	-0.4171	6.8457
8.02	-0.2678	12.3848	-0.3346	9.5967	-0.3947	7.6396
8.09	-0.2396	14.3089	-0.3091	10.8833	-0.3716	8.5584
8.16	-0.2106	16.7347	-0.2830	12.4349	-0.3479	9.6336
8.23	-0.1806	19.8851	-0.2560	14.3415	-0.3235	10.9079
8.30	-0.1496	24.1372	-0.2282	16.7387	-0.2984	12.4412
8.37	-0.1175	30.1849	-0.1995	19.8408	-0.2725	14.3201
8.44	-0.0843	39.4587	-0.1699	24.0086	-0.2459	16.6745
8.51	-0.0499	55.4564	-0.1393	29.8991	-0.2185	19.7083
8.58	-0.0143	89.5900	-0.1076	38.8481	-0.1902	23.7616
8.65	0.0226	212.7215	-0.0749	54.0532	-0.1610	29.4467
8.72			-0.0411	85.5496	-0.1309	37.9880
8.79			-0.0061	189.4380	-0.0998	52.2426
8.86					-0.0676	80.7819

Appendix D

Tables for GLD Fits: Method of Percentiles

For specified ρ_3 and ρ_4, Tables D–1 through D–5 give the λ_3 and λ_4 values for GLD fits in the regions designated by "T1," "T2," "T3," "T4," and "T5," respectively, in Figure 4.3–1.

The entries for Tables D–1, D–2, and D–5 are from "Fitting the Generalized Lambda Distribution (GLD) System by a Method of Percentiles, II: Tables" by E.J. Dudewicz and Z.A. Karian, *American Journal of Mathematical and Management Sciences*, V. 19, 1 & 2 (1999), pp. 1–73, copyright ©1999 by American Sciences Press, Inc., 20 Cross Road, Syracuse, New York 13224-2104. Reprinted with permission.

Table D–1: $0.005 \leq \hat{\rho}_4 \leq 0.008$ and $\lambda_3,\ \lambda_4 > 0$

	$\hat{\rho}_4 = 0.005$		$\hat{\rho}_4 = 0.006$		$\hat{\rho}_4 = 0.007$		$\hat{\rho}_4 = 0.008$	
$\hat{\rho}_3$	λ_3	λ_4	λ_3	λ_4	λ_3	λ_4	λ_3	λ_4
.02	28.9513	66.0812	27.9517	65.0816	27.1063	64.2362	26.3739	63.5039
.04	28.8455	59.3966	27.8459	58.3970	27.0005	57.5516	26.2681	56.8192
.06	28.7427	55.4454	27.7431	54.4458	26.8977	53.6004	26.1653	52.8681
.08	28.6432	52.6155	27.6436	51.6159	26.7982	50.7705	26.0658	50.0381
.10	28.5472	50.4016	27.5476	49.4020	26.7022	48.5566	25.9698	47.8242
.12	28.4549	48.5788	27.4553	47.5792	26.6099	46.7339	25.8776	46.0015
.14	28.3666	47.0274	27.3670	46.0278	26.5216	45.1824	25.7892	44.4501
.16	28.2824	45.6758	27.2828	44.6762	26.4374	43.8309	25.7050	43.0985
.18	28.2025	44.4780	27.2029	43.4784	26.3575	42.6330	25.6251	41.9007
.20	28.1271	43.4026	27.1275	42.4030	26.2821	41.5576	25.5497	40.8252
.22	28.0562	42.4272	27.0566	41.4276	26.2113	40.5822	25.4789	39.8498
.24	27.9902	41.5353	26.9906	40.5357	26.1452	39.6903	25.4128	38.9579
.26	27.9290	40.7144	26.9294	39.7148	26.0840	38.8694	25.3517	38.1371
.28	27.8729	39.9549	26.8733	38.9553	26.0279	38.1099	25.2955	37.3775
.30	27.8219	39.2490	26.8223	38.2495	25.9769	37.4041	25.2445	36.6717
.32	27.7761	38.5907	26.7765	37.5911	25.9311	36.7457	25.1987	36.0133
.34	27.7355	37.9748	26.7359	36.9752	25.8905	36.1298	25.1582	35.3974
.36	27.7003	37.3971	26.7007	36.3975	25.8554	35.5521	25.1230	34.8197
.38	27.6706	36.8541	26.6710	35.8545	25.8256	35.0091	25.0932	34.2768
.40	27.6462	36.3429	26.6466	35.3433	25.8012	34.4979	25.0688	33.7656
.42	27.6273	35.8609	26.6277	34.8613	25.7823	34.0159	25.0499	33.2835
.44	27.6138	35.4059	26.6142	34.4063	25.7688	33.5609	25.0364	32.8285
.46	27.6057	34.9759	26.6061	33.9763	25.7607	33.1309	25.0283	32.3985
.48	27.6030	34.5693	26.6034	33.5697	25.7580	32.7243	25.0256	31.9919
.50	27.6057	34.1845	26.6061	33.1849	25.7607	32.3395	25.0283	31.6071
.52	27.6136	33.8202	26.6140	32.8206	25.7686	31.9752	25.0362	31.2428
.54	27.6268	33.4751	26.6272	32.4755	25.7818	31.6301	25.0494	30.8978
.56	27.6450	33.1482	26.6454	32.1486	25.8000	31.3032	25.0677	30.5709
.58	27.6683	32.8384	26.6687	31.8388	25.8233	30.9934	25.0909	30.2611
.60	27.6965	32.5449	26.6969	31.5453	25.8515	30.6999	25.1191	29.9675
.62	27.7295	32.2666	26.7299	31.2670	25.8845	30.4216	25.1521	29.6893
.64	27.7671	32.0029	26.7675	31.0033	25.9221	30.1579	25.1897	29.4256
.66	27.8092	31.7530	26.8096	30.7534	25.9642	29.9080	25.2319	29.1756
.68	27.8557	31.5161	26.8561	30.5165	26.0107	29.6712	25.2784	28.9388
.70	27.9064	31.2917	26.9068	30.2921	26.0614	29.4467	25.3291	28.7144
.72	27.9612	31.0791	26.9616	30.0795	26.1162	29.2341	25.3838	28.5017
.74	28.0198	30.8777	27.0202	29.8781	26.1748	29.0327	25.4425	28.3003
.76	28.0822	30.6869	27.0826	29.6873	26.2372	28.8420	25.5048	28.1096
.78	28.1481	30.5063	27.1485	29.5067	26.3031	28.6613	25.5708	27.9290
.80	28.2174	30.3353	27.2178	29.3357	26.3724	28.4903	25.6400	27.7580
.82	28.2899	30.1735	27.2903	29.1739	26.4449	28.3285	25.7126	27.5961
.84	28.3655	30.0203	27.3659	29.0207	26.5205	28.1753	25.7881	27.4429
.86	28.4439	29.8754	27.4443	28.8758	26.5989	28.0304	25.8665	27.2980
.88	28.5250	29.7383	27.5254	28.7387	26.6800	27.8933	25.9477	27.1610
.90	28.6087	29.6087	27.6091	28.6091	26.7637	27.7637	26.0313	27.0313
.92	28.6948	29.4862	27.6952	28.4866	26.8498	27.6412	26.1174	26.9088
.94	28.7831	29.3703	27.7835	28.3707	26.9381	27.5253	26.2057	26.7930
.96	28.8734	29.2609	27.8738	28.2613	27.0284	27.4159	26.2960	26.6835
.98	28.9657	29.1574	27.9661	28.1578	27.1207	27.3125	26.3883	26.5801
1.00	29.0598	29.0598	28.0602	28.0602	27.2148	27.2148	26.4824	26.4824

Table D–1: $0.009 \leq \hat{\rho}_4 \leq 0.012$ and $\lambda_3,\ \lambda_4 > 0$

	$\hat{\rho}_4 = 0.009$		$\hat{\rho}_4 = 0.010$		$\hat{\rho}_4 = 0.011$		$\hat{\rho}_4 = 0.012$	
$\hat{\rho}_3$	λ_3	λ_4	λ_3	λ_4	λ_3	λ_4	λ_3	λ_4
.02	25.7279	62.8579	25.1500	62.2801	24.6273	61.7574	24.1500	61.2802
.04	25.6221	56.1732	25.0442	55.5954	24.5215	55.0727	24.0442	54.5955
.06	25.5193	52.2221	24.9414	51.6442	24.4187	51.1215	23.9414	50.6443
.08	25.4198	49.3921	24.8419	48.8142	24.3192	48.2915	23.8419	47.8143
.10	25.3238	47.1782	24.7459	46.6003	24.2232	46.0776	23.7459	45.6004
.12	25.2316	45.3555	24.6537	44.7776	24.1309	44.2549	23.6537	43.7777
.14	25.1432	43.8041	24.5654	43.2262	24.0426	42.7035	23.5654	42.2262
.16	25.0590	42.4525	24.4811	41.8746	23.9584	41.3519	23.4811	40.8747
.18	24.9791	41.2547	24.4012	40.6768	23.8785	40.1541	23.4012	39.6768
.20	24.9037	40.1792	24.3258	39.6014	23.8030	39.0786	23.3258	38.6014
.22	24.8329	39.2038	24.2550	38.6259	23.7322	38.1032	23.2550	37.6260
.24	24.7668	38.3119	24.1889	37.7340	23.6662	37.2113	23.1889	36.7341
.26	24.7057	37.4911	24.1278	36.9132	23.6050	36.3904	23.1278	35.9132
.28	24.6495	36.7315	24.0716	36.1537	23.5489	35.6309	23.0716	35.1537
.30	24.5985	36.0257	24.0206	35.4478	23.4979	34.9251	23.0206	34.4478
.32	24.5527	35.3673	23.9748	34.7895	23.4520	34.2667	22.9748	33.7895
.34	24.5122	34.7514	23.9343	34.1735	23.4115	33.6508	22.9343	33.1735
.36	24.4770	34.1737	23.8991	33.5958	23.3763	33.0731	22.8991	32.5959
.38	24.4472	33.6308	23.8693	33.0529	23.3465	32.5301	22.8693	32.0529
.40	24.4228	33.1195	23.8449	32.5417	23.3222	32.0189	22.8449	31.5417
.42	24.4039	32.6375	23.8260	32.0597	23.3032	31.5369	22.8260	31.0597
.44	24.3904	32.1825	23.8125	31.6046	23.2897	31.0819	22.8125	30.6047
.46	24.3823	31.7525	23.8044	31.1747	23.2817	30.6519	22.8044	30.1747
.48	24.3796	31.3459	23.8017	30.7680	23.2790	30.2453	22.8017	29.7681
.50	24.3823	30.9611	23.8044	30.3832	23.2816	29.8605	22.8044	29.3833
.52	24.3902	30.5968	23.8123	30.0189	23.2896	29.4962	22.8123	29.0190
.54	24.4034	30.2518	23.8255	29.6739	23.3027	29.1511	22.8255	28.6739
.56	24.4216	29.9248	23.8438	29.3470	23.3210	28.8242	22.8438	28.3470
.58	24.4449	29.6151	23.8670	29.0372	23.3443	28.5144	22.8670	28.0372
.60	24.4731	29.3215	23.8952	28.7436	23.3725	28.2209	22.8952	27.7436
.62	24.5061	29.0433	23.9282	28.4654	23.4055	27.9426	22.9282	27.4654
.64	24.5437	28.7795	23.9658	28.2017	23.4431	27.6789	22.9658	27.2017
.66	24.5859	28.5296	24.0080	27.9517	23.4852	27.4290	23.0080	26.9518
.68	24.6324	28.2928	24.0545	27.7149	23.5317	27.1921	23.0545	26.7149
.70	24.6831	28.0684	24.1052	27.4905	23.5824	26.9677	23.1052	26.4905
.72	24.7378	27.8557	24.1599	27.2779	23.6372	26.7551	23.1599	26.2779
.74	24.7965	27.6543	24.2186	27.0764	23.6958	26.5537	23.2186	26.0765
.76	24.8588	27.4636	24.2809	26.8857	23.7582	26.3629	23.2809	25.8857
.78	24.9247	27.2829	24.3469	26.7051	23.8241	26.1823	23.3469	25.7051
.80	24.9940	27.1119	24.4162	26.5341	23.8934	26.0113	23.4162	25.5341
.82	25.0665	26.9501	24.4887	26.3722	23.9659	25.8495	23.4887	25.3722
.84	25.1421	26.7969	24.5642	26.2190	24.0415	25.6963	23.5642	25.2191
.86	25.2205	26.6520	24.6426	26.0741	24.1199	25.5514	23.6426	25.0741
.88	25.3016	26.5149	24.7238	25.9371	24.2010	25.4143	23.7238	24.9371
.90	25.3853	26.3853	24.8074	25.8074	24.2847	25.2847	23.8074	24.8075
.92	25.4714	26.2628	24.8935	25.6849	24.3707	25.1621	23.8935	24.6849
.94	25.5597	26.1469	24.9818	25.5691	24.4590	25.0463	23.9818	24.5691
.96	25.6500	26.0375	25.0722	25.4596	24.5494	24.9368	24.0722	24.4596
.98	25.7423	25.9341	25.1644	25.3562	24.6417	24.8334	24.1644	24.3562
1.00	25.8364	25.8364	25.2585	25.2585	24.7357	24.7357	24.2585	24.2585

Table D–1: $0.0135 \leq \hat{\rho}_4 \leq 0.0180$ and $\lambda_3,\ \lambda_4 > 0$

	$\hat{\rho}_4 = 0.0135$		$\hat{\rho}_4 = 0.0150$		$\hat{\rho}_4 = 0.0165$		$\hat{\rho}_4 = 0.0180$	
$\hat{\rho}_3$	λ_3	λ_4	λ_3	λ_4	λ_3	λ_4	λ_3	λ_4
.02	23.5038	60.6342	22.9260	60.0566	22.4033	59.5341	21.9261	59.0572
.04	23.3980	53.9493	22.8202	53.3717	22.2975	52.8491	21.8203	52.3721
.06	23.2952	49.9981	22.7175	49.4204	22.1948	48.8978	21.7175	48.4207
.08	23.1957	47.1681	22.6180	46.5904	22.0953	46.0678	21.6180	45.5907
.10	23.0997	44.9542	22.5220	44.3765	21.9993	43.8539	21.5220	43.3767
.12	23.0074	43.1314	22.4297	42.5538	21.9070	42.0311	21.4298	41.5540
.14	22.9191	41.5800	22.3414	41.0023	21.8187	40.4797	21.3415	40.0025
.16	22.8349	40.2284	22.2572	39.6507	21.7345	39.1281	21.2572	38.6509
.18	22.7550	39.0306	22.1773	38.4529	21.6546	37.9303	21.1773	37.4531
.20	22.6796	37.9552	22.1018	37.3775	21.5791	36.8548	21.1019	36.3777
.22	22.6087	36.9798	22.0310	36.4021	21.5083	35.8794	21.0311	35.4022
.24	22.5427	36.0879	21.9650	35.5102	21.4423	34.9875	20.9650	34.5103
.26	22.4815	35.2670	21.9038	34.6893	21.3811	34.1666	20.9039	33.6895
.28	22.4254	34.5075	21.8477	33.9298	21.3250	33.4071	20.8477	32.9300
.30	22.3744	33.8016	21.7966	33.2239	21.2739	32.7013	20.7967	32.2241
.32	22.3286	33.1433	21.7508	32.5656	21.2281	32.0429	20.7509	31.5657
.34	22.2880	32.5273	21.7103	31.9496	21.1876	31.4270	20.7104	30.9498
.36	22.2528	31.9496	21.6751	31.3719	21.1524	30.8493	20.6752	30.3721
.38	22.2230	31.4067	21.6453	30.8290	21.1226	30.3063	20.6454	29.8291
.40	22.1987	30.8955	21.6209	30.3178	21.0982	29.7951	20.6210	29.3179
.42	22.1797	30.4135	21.6020	29.8358	21.0793	29.3131	20.6021	28.8359
.44	22.1663	29.9584	21.5885	29.3807	21.0658	28.8581	20.5886	28.3809
.46	22.1582	29.5285	21.5804	28.9507	21.0577	28.4281	20.5805	27.9509
.48	22.1555	29.1218	21.5778	28.5441	21.0551	28.0214	20.5778	27.5443
.50	22.1582	28.7370	21.5804	28.1593	21.0577	27.6366	20.5805	27.1595
.52	22.1661	28.3727	21.5884	27.7950	21.0657	27.2723	20.5884	26.7951
.54	22.1792	28.0277	21.6015	27.4499	21.0788	26.9273	20.6016	26.4501
.56	22.1975	27.7007	21.6198	27.1230	21.0971	26.6004	20.6198	26.1232
.58	22.2208	27.3910	21.6431	26.8133	21.1204	26.2906	20.6431	25.8134
.60	22.2490	27.0974	21.6712	26.5197	21.1485	25.9970	20.6713	25.5198
.62	22.2820	26.8191	21.7042	26.2414	21.1815	25.7187	20.7043	25.2416
.64	22.3196	26.5554	21.7418	25.9777	21.2191	25.4550	20.7419	24.9778
.66	22.3617	26.3055	21.7840	25.7278	21.2613	25.2051	20.7840	24.7279
.68	22.4082	26.0687	21.8305	25.4909	21.3078	24.9683	20.8305	24.4911
.70	22.4589	25.8442	21.8812	25.2665	21.3585	24.7438	20.8813	24.2666
.72	22.5137	25.6316	21.9359	25.0539	21.4132	24.5312	20.9360	24.0540
.74	22.5723	25.4302	21.9946	24.8525	21.4719	24.3298	20.9947	23.8526
.76	22.6347	25.2395	22.0569	24.6617	21.5342	24.1390	21.0570	23.6618
.78	22.7006	25.0588	22.1229	24.4811	21.6002	23.9584	21.1229	23.4812
.80	22.7699	24.8878	22.1922	24.3101	21.6695	23.7874	21.1922	23.3102
.82	22.8424	24.7260	22.2647	24.1482	21.7420	23.6255	21.2647	23.1483
.84	22.9180	24.5728	22.3402	23.9951	21.8175	23.4724	21.3403	22.9952
.86	22.9964	24.4279	22.4186	23.8502	21.8959	23.3275	21.4187	22.8502
.88	23.0775	24.2908	22.4998	23.7131	21.9771	23.1904	21.4998	22.7132
.90	23.1612	24.1612	22.5835	23.5835	22.0607	23.0608	21.5835	22.5836
.92	23.2472	24.0386	22.6695	23.4609	22.1468	22.9382	21.6696	22.4610
.94	23.3355	23.9228	22.7578	23.3451	22.2351	22.8224	21.7579	22.3452
.96	23.4259	23.8134	22.8482	23.2356	22.3255	22.7129	21.8482	22.2357
.98	23.5182	23.7099	22.9404	23.1322	22.4177	22.6095	21.9405	22.1323
1.00	23.6122	23.6122	23.0345	23.0345	22.5118	22.5118	22.0346	22.0346

Table D–1: $0.0195 \leq \hat{\rho}_4 \leq 0.0240$ and $\lambda_3, \lambda_4 > 0$

	$\hat{\rho}_4 = 0.0195$		$\hat{\rho}_4 = 0.0210$		$\hat{\rho}_4 = 0.0225$		$\hat{\rho}_4 = 0.0240$	
$\hat{\rho}_3$	λ_3	λ_4	λ_3	λ_4	λ_3	λ_4	λ_3	λ_4
.02	21.4871	58.6186	21.0807	58.2125	20.7022	57.8346	20.3483	57.4812
.04	21.3813	51.9332	20.9749	51.5270	20.5965	51.1489	20.2425	50.7952
.06	21.2785	47.9819	20.8721	47.5756	20.4937	47.1974	20.1397	46.8436
.08	21.1790	45.1518	20.7726	44.7455	20.3942	44.3672	20.0402	44.0134
.10	21.0830	42.9378	20.6766	42.5315	20.2982	42.1532	19.9442	41.7994
.12	20.9908	41.1151	20.5843	40.7087	20.2059	40.3304	19.8519	39.9766
.14	20.9024	39.5636	20.4960	39.1573	20.1176	38.7790	19.7636	38.4251
.16	20.8182	38.2120	20.4118	37.8056	20.0334	37.4273	19.6794	37.0735
.18	20.7383	37.0142	20.3319	36.6078	19.9535	36.2295	19.5995	35.8757
.20	20.6629	35.9387	20.2564	35.5324	19.8780	35.1540	19.5241	34.8002
.22	20.5921	34.9633	20.1856	34.5569	19.8072	34.1786	19.4532	33.8247
.24	20.5260	34.0714	20.1196	33.6650	19.7412	33.2867	19.3872	32.9328
.26	20.4649	33.2505	20.0584	32.8441	19.6800	32.4658	19.3260	32.1120
.28	20.4087	32.4910	20.0023	32.0846	19.6239	31.7063	19.2699	31.3524
.30	20.3577	31.7851	19.9513	31.3788	19.5728	31.0004	19.2189	30.6466
.32	20.3119	31.1268	19.9054	30.7204	19.5270	30.3421	19.1731	29.9882
.34	20.2714	30.5108	19.8649	30.1044	19.4865	29.7261	19.1325	29.3722
.36	20.2362	29.9331	19.8297	29.5267	19.4513	29.1484	19.0973	28.7945
.38	20.2064	29.3902	19.7999	28.9838	19.4215	28.6054	19.0675	28.2516
.40	20.1820	28.8790	19.7756	28.4726	19.3971	28.0942	19.0432	27.7403
.42	20.1631	28.3969	19.7566	27.9905	19.3782	27.6122	19.0242	27.2583
.44	20.1496	27.9419	19.7431	27.5355	19.3647	27.1572	19.0107	26.8033
.46	20.1415	27.5119	19.7350	27.1055	19.3566	26.7272	19.0027	26.3733
.48	20.1388	27.1053	19.7324	26.6989	19.3539	26.3205	19.0000	25.9666
.50	20.1415	26.7205	19.7350	26.3141	19.3566	25.9357	19.0026	25.5818
.52	20.1494	26.3562	19.7430	25.9498	19.3645	25.5714	19.0106	25.2175
.54	20.1626	26.0111	19.7561	25.6047	19.3777	25.2263	19.0237	24.8724
.56	20.1808	25.6842	19.7744	25.2778	19.3959	24.8994	19.0420	24.5455
.58	20.2041	25.3744	19.7976	24.9680	19.4192	24.5896	19.0652	24.2357
.60	20.2323	25.0808	19.8258	24.6744	19.4474	24.2960	19.0934	23.9421
.62	20.2653	24.8026	19.8588	24.3961	19.4804	24.0178	19.1264	23.6638
.64	20.3029	24.5389	19.8964	24.1324	19.5180	23.7541	19.1640	23.4001
.66	20.3450	24.2889	19.9386	23.8825	19.5601	23.5041	19.2061	23.1502
.68	20.3915	24.0521	19.9851	23.6456	19.6066	23.2673	19.2526	22.9133
.70	20.4422	23.8276	20.0358	23.4212	19.6573	23.0428	19.3034	22.6889
.72	20.4970	23.6150	20.0905	23.2086	19.7121	22.8302	19.3581	22.4762
.74	20.5556	23.4136	20.1492	23.0072	19.7707	22.6288	19.4167	22.2748
.76	20.6180	23.2228	20.2115	22.8164	19.8331	22.4380	19.4791	22.0840
.78	20.6839	23.0422	20.2774	22.6358	19.8990	22.2574	19.5450	21.9034
.80	20.7532	22.8712	20.3467	22.4647	19.9683	22.0864	19.6143	21.7324
.82	20.8257	22.7093	20.4192	22.3029	20.0408	21.9245	19.6868	21.5705
.84	20.9013	22.5562	20.4948	22.1497	20.1164	21.7713	19.7624	21.4173
.86	20.9797	22.4112	20.5732	22.0048	20.1948	21.6264	19.8408	21.2724
.88	21.0608	22.2742	20.6544	21.8677	20.2759	21.4893	19.9219	21.1353
.90	21.1445	22.1445	20.7380	21.7381	20.3596	21.3597	20.0056	21.0057
.92	21.2306	22.0220	20.8241	21.6155	20.4457	21.2371	20.0917	20.8832
.94	21.3189	21.9062	20.9124	21.4997	20.5340	21.1213	20.1800	20.7673
.96	21.4092	21.7967	21.0028	21.3902	20.6243	21.0118	20.2704	20.6578
.98	21.5015	21.6933	21.0950	21.2868	20.7166	20.9084	20.3626	20.5544
1.00	21.5956	21.5956	21.1891	21.1891	20.8107	20.8107	20.4567	20.4567

Table D-1: $0.026 \leq \hat{\rho}_4 \leq 0.032$ and $\lambda_3, \lambda_4 > 0$

	$\hat{\rho}_4 = 0.026$		$\hat{\rho}_4 = 0.028$		$\hat{\rho}_4 = 0.030$		$\hat{\rho}_4 = 0.032$	
$\hat{\rho}_3$	λ_3	λ_4	λ_3	λ_4	λ_3	λ_4	λ_3	λ_4
.02	19.9093	57.0431	19.5028	56.6378	19.1244	56.2606	18.7704	55.9081
.04	19.8035	50.3567	19.3970	49.9508	19.0185	49.5730	18.6646	49.2199
.06	19.7007	46.4050	19.2942	45.9989	18.9157	45.6209	18.5618	45.2676
.08	19.6012	43.5747	19.1947	43.1686	18.8162	42.7905	18.4623	42.4371
.10	19.5052	41.3606	19.0987	40.9545	18.7202	40.5763	18.3663	40.2228
.12	19.4129	39.5378	19.0065	39.1316	18.6280	38.7534	18.2741	38.3999
.14	19.3246	37.9863	18.9182	37.5801	18.5397	37.2019	18.1858	36.8483
.16	19.2404	36.6347	18.8339	36.2285	18.4555	35.8502	18.1016	35.4966
.18	19.1605	35.4368	18.7540	35.0306	18.3756	34.6524	18.0217	34.2987
.20	19.0851	34.3614	18.6786	33.9551	18.3001	33.5769	17.9462	33.2232
.22	19.0142	33.3859	18.6078	32.9796	18.2293	32.6014	17.8754	32.2478
.24	18.9482	32.4940	18.5417	32.0877	18.1632	31.7095	17.8093	31.3558
.26	18.8870	31.6731	18.4806	31.2668	18.1021	30.8886	17.7482	30.5349
.28	18.8309	30.9136	18.4244	30.5073	18.0459	30.1290	17.6920	29.7754
.30	18.7799	30.2077	18.3734	29.8014	17.9949	29.4231	17.6410	29.0695
.32	18.7340	29.5493	18.3276	29.1430	17.9491	28.7647	17.5952	28.4111
.34	18.6935	28.9334	18.2871	28.5271	17.9086	28.1488	17.5547	27.7951
.36	18.6583	28.3557	18.2519	27.9494	17.8734	27.5711	17.5195	27.2174
.38	18.6285	27.8127	18.2221	27.4064	17.8436	27.0281	17.4897	26.6744
.40	18.6042	27.3015	18.1977	26.8952	17.8192	26.5168	17.4653	26.1631
.42	18.5852	26.8194	18.1788	26.4131	17.8002	26.0348	17.4463	25.6811
.44	18.5717	26.3644	18.1652	25.9581	17.7867	25.5797	17.4328	25.2260
.46	18.5636	25.9344	18.1572	25.5281	17.7787	25.1497	17.4247	24.7960
.48	18.5609	25.5277	18.1545	25.1214	17.7760	24.7431	17.4220	24.3893
.50	18.5636	25.1429	18.1571	24.7366	17.7786	24.3582	17.4247	24.0045
.52	18.5715	24.7786	18.1651	24.3723	17.7865	23.9939	17.4326	23.6402
.54	18.5847	24.4335	18.1782	24.0272	17.7997	23.6488	17.4458	23.2950
.56	18.6029	24.1066	18.1965	23.7002	17.8179	23.3219	17.4640	22.9681
.58	18.6262	23.7968	18.2197	23.3904	17.8412	23.0120	17.4873	22.6583
.60	18.6544	23.5032	18.2479	23.0968	17.8694	22.7184	17.5155	22.3646
.62	18.6874	23.2249	18.2809	22.8185	17.9024	22.4401	17.5484	22.0864
.64	18.7250	22.9612	18.3185	22.5548	17.9400	22.1764	17.5860	21.8226
.66	18.7671	22.7112	18.3606	22.3048	17.9821	21.9264	17.6282	21.5726
.68	18.8136	22.4744	18.4071	22.0680	18.0286	21.6895	17.6747	21.3357
.70	18.8643	22.2499	18.4578	21.8435	18.0793	21.4651	17.7254	21.1113
.72	18.9191	22.0373	18.5126	21.6309	18.1341	21.2524	17.7801	20.8986
.74	18.9777	21.8358	18.5712	21.4294	18.1927	21.0510	17.8388	20.6971
.76	19.0401	21.6451	18.6336	21.2387	18.2551	20.8602	17.9011	20.5064
.78	19.1060	21.4644	18.6995	21.0580	18.3210	20.6795	17.9670	20.3257
.80	19.1753	21.2934	18.7688	20.8870	18.3903	20.5085	18.0363	20.1546
.82	19.2478	21.1315	18.8413	20.7251	18.4628	20.3466	18.1088	19.9928
.84	19.3234	20.9783	18.9169	20.5719	18.5383	20.1934	18.1844	19.8396
.86	19.4018	20.8334	18.9953	20.4270	18.6168	20.0485	18.2628	19.6946
.88	19.4829	20.6963	19.0764	20.2899	18.6979	19.9114	18.3440	19.5575
.90	19.5666	20.5667	19.1601	20.1603	18.7816	19.7818	18.4277	19.4279
.92	19.6527	20.4441	19.2462	20.0377	18.8677	19.6592	18.5137	19.3053
.94	19.7410	20.3283	19.3345	19.9218	18.9560	19.5433	18.6020	19.1894
.96	19.8313	20.2188	19.4249	19.8124	19.0463	19.4339	18.6924	19.0799
.98	19.9236	20.1154	19.5172	19.7089	19.1386	19.3304	18.7847	18.9765
1.00	20.0177	20.0177	19.6112	19.6112	19.2327	19.2327	18.8788	18.8788

Table D–1: $0.034 \le \hat{\rho}_4 \le 0.040$ and $\lambda_3,\ \lambda_4 > 0$

	$\hat{\rho}_4 = 0.034$		$\hat{\rho}_4 = 0.036$		$\hat{\rho}_4 = 0.038$		$\hat{\rho}_4 = 0.040$	
$\hat{\rho}_3$	λ_3	λ_4	λ_3	λ_4	λ_3	λ_4	λ_3	λ_4
.02	18.4379	55.5773	18.1245	55.2658	17.8279	54.9714	17.5466	54.6926
.04	18.3321	48.8884	18.0187	48.5759	17.7222	48.2806	17.4408	48.0006
.06	18.2294	44.9358	17.9159	44.6231	17.6194	44.3274	17.3381	44.0471
.08	18.1299	42.1051	17.8164	41.7923	17.5199	41.4964	17.2386	41.2159
.10	18.0339	39.8908	17.7204	39.5779	17.4239	39.2820	17.1426	39.0013
.12	17.9416	38.0678	17.6281	37.7548	17.3316	37.4589	17.0503	37.1782
.14	17.8533	36.5162	17.5398	36.2032	17.2433	35.9072	16.9620	35.6264
.16	17.7691	35.1645	17.4556	34.8515	17.1591	34.5554	16.8778	34.2746
.18	17.6892	33.9666	17.3757	33.6535	17.0792	33.3575	16.7979	33.0766
.20	17.6137	32.8911	17.3003	32.5780	17.0038	32.2819	16.7225	32.0011
.22	17.5429	31.9156	17.2295	31.6025	16.9330	31.3064	16.6517	31.0255
.24	17.4769	31.0237	17.1634	30.7105	16.8669	30.4144	16.5856	30.1335
.26	17.4157	30.2027	17.1023	29.8896	16.8057	29.5935	16.5244	29.3126
.28	17.3596	29.4432	17.0461	29.1300	16.7496	28.8339	16.4683	28.5530
.30	17.3085	28.7373	16.9951	28.4241	16.6986	28.1280	16.4173	27.8471
.32	17.2627	28.0789	16.9492	27.7657	16.6527	27.4695	16.3714	27.1886
.34	17.2222	27.4629	16.9087	27.1497	16.6122	26.8536	16.3309	26.5726
.36	17.1870	26.8852	16.8735	26.5720	16.5770	26.2758	16.2957	25.9949
.38	17.1572	26.3422	16.8437	26.0290	16.5472	25.7328	16.2659	25.4518
.40	17.1328	25.8309	16.8193	25.5177	16.5228	25.2215	16.2415	24.9406
.42	17.1139	25.3489	16.8004	25.0357	16.5038	24.7394	16.2225	24.4585
.44	17.1003	24.8938	16.7869	24.5806	16.4903	24.2843	16.2090	24.0034
.46	17.0923	24.4638	16.7788	24.1505	16.4822	23.8543	16.2009	23.5733
.48	17.0896	24.0571	16.7761	23.7438	16.4795	23.4476	16.1982	23.1666
.50	17.0922	23.6722	16.7787	23.3590	16.4821	23.0627	16.2008	22.7817
.52	17.1001	23.3079	16.7866	22.9946	16.4901	22.6983	16.2087	22.4173
.54	17.1133	22.9628	16.7997	22.6495	16.5032	22.3532	16.2218	22.0721
.56	17.1315	22.6358	16.8180	22.3225	16.5214	22.0262	16.2401	21.7451
.58	17.1548	22.3259	16.8413	22.0127	16.5447	21.7163	16.2633	21.4352
.60	17.1830	22.0323	16.8694	21.7190	16.5729	21.4227	16.2915	21.1416
.62	17.2159	21.7540	16.9024	21.4407	16.6058	21.1443	16.3245	20.8632
.64	17.2535	21.4902	16.9400	21.1769	16.6434	20.8805	16.3621	20.5994
.66	17.2956	21.2403	16.9821	20.9269	16.6856	20.6305	16.4042	20.3494
.68	17.3421	21.0034	17.0286	20.6900	16.7320	20.3936	16.4507	20.1124
.70	17.3928	20.7789	17.0793	20.4655	16.7827	20.1691	16.5014	19.8879
.72	17.4476	20.5662	17.1341	20.2528	16.8375	19.9564	16.5561	19.6752
.74	17.5062	20.3647	17.1927	20.0514	16.8961	19.7549	16.6148	19.4737
.76	17.5686	20.1739	17.2551	19.8605	16.9585	19.5641	16.6771	19.2829
.78	17.6345	19.9933	17.3210	19.6799	17.0244	19.3834	16.7430	19.1022
.80	17.7038	19.8222	17.3903	19.5088	17.0937	19.2123	16.8123	18.9311
.82	17.7763	19.6603	17.4628	19.3469	17.1662	19.0504	16.8849	18.7692
.84	17.8519	19.5071	17.5384	19.1937	17.2418	18.8972	16.9604	18.6159
.86	17.9303	19.3622	17.6168	19.0487	17.3202	18.7522	17.0389	18.4709
.88	18.0115	19.2251	17.6979	18.9116	17.4014	18.6151	17.1200	18.3338
.90	18.0952	19.0954	17.7816	18.7819	17.4851	18.4854	17.2037	18.2041
.92	18.1812	18.9728	17.8677	18.6593	17.5712	18.3628	17.2898	18.0815
.94	18.2695	18.8569	17.9560	18.5435	17.6595	18.2469	17.3781	17.9656
.96	18.3599	18.7475	18.0464	18.4340	17.7499	18.1374	17.4685	17.8561
.98	18.4522	18.6440	18.1387	18.3305	17.8422	18.0340	17.5608	17.7527
1.00	18.5463	18.5463	18.2328	18.2328	17.9363	17.9363	17.6549	17.6549

Table D–1: $0.0425 \leq \hat{\rho}_4 \leq 0.0500$ and $\lambda_3,\ \lambda_4 > 0$

$\hat{\rho}_3$	$\hat{\rho}_4 = 0.0425$		$\hat{\rho}_4 = 0.0450$		$\hat{\rho}_4 = 0.0475$		$\hat{\rho}_4 = 0.0500$	
	λ_3	λ_4	λ_3	λ_4	λ_3	λ_4	λ_3	λ_4
.02	17.2142	54.3636	16.9007	54.0541	16.6042	53.7621	16.3229	53.4858
.04	17.1084	47.6700	16.7949	47.3587	16.4984	47.0647	16.2171	46.7861
.06	17.0056	43.7160	16.6921	43.4040	16.3957	43.1093	16.1144	42.8300
.08	16.9061	40.8846	16.5927	40.5724	16.2962	40.2773	16.0149	39.9976
.10	16.8101	38.6698	16.4966	38.3574	16.2002	38.0622	15.9189	37.7823
.12	16.7179	36.8465	16.4044	36.5341	16.1079	36.2387	15.8267	35.9586
.14	16.6295	35.2948	16.3160	34.9821	16.0196	34.6867	15.7384	34.4066
.16	16.5453	33.9429	16.2318	33.6302	15.9354	33.3347	15.6542	33.0545
.18	16.4654	32.7449	16.1520	32.4322	15.8555	32.1366	15.5743	31.8564
.20	16.3900	31.6693	16.0765	31.3565	15.7801	31.0609	15.4988	30.7806
.22	16.3192	30.6937	16.0057	30.3810	15.7093	30.0853	15.4280	29.8050
.24	16.2531	29.8017	15.9396	29.4889	15.6432	29.1933	15.3619	28.9129
.26	16.1920	28.9807	15.8784	28.6679	15.5820	28.3722	15.3008	28.0919
.28	16.1358	28.2211	15.8223	27.9082	15.5259	27.6126	15.2446	27.3322
.30	16.0848	27.5152	15.7712	27.2023	15.4748	26.9066	15.1936	26.6262
.32	16.0389	26.8567	15.7254	26.5438	15.4290	26.2481	15.1477	25.9676
.34	15.9984	26.2407	15.6849	25.9278	15.3884	25.6321	15.1072	25.3516
.36	15.9632	25.6629	15.6496	25.3500	15.3532	25.0542	15.0719	24.7737
.38	15.9334	25.1199	15.6198	24.8069	15.3234	24.5111	15.0421	24.2306
.40	15.9090	24.6086	15.5954	24.2956	15.2990	23.9998	15.0177	23.7193
.42	15.8900	24.1265	15.5764	23.8135	15.2800	23.5177	14.9987	23.2371
.44	15.8765	23.6713	15.5629	23.3583	15.2665	23.0625	14.9852	22.7819
.46	15.8684	23.2413	15.5548	22.9282	15.2583	22.6324	14.9770	22.3517
.48	15.8657	22.8345	15.5521	22.5214	15.2556	22.2256	14.9743	21.9449
.50	15.8683	22.4496	15.5547	22.1365	15.2582	21.8406	14.9769	21.5599
.52	15.8762	22.0852	15.5626	21.7721	15.2661	21.4761	14.9848	21.1954
.54	15.8893	21.7400	15.5757	21.4269	15.2792	21.1309	14.9979	20.8502
.56	15.9075	21.4130	15.5939	21.0998	15.2974	20.8038	15.0161	20.5230
.58	15.9308	21.1031	15.6172	20.7899	15.3207	20.4939	15.0393	20.2131
.60	15.9590	20.8094	15.6453	20.4962	15.3488	20.2001	15.0674	19.9193
.62	15.9919	20.5310	15.6783	20.2178	15.3817	19.9217	15.1004	19.6408
.64	16.0295	20.2672	15.7159	19.9539	15.4193	19.6578	15.1379	19.3769
.66	16.0716	20.0171	15.7580	19.7038	15.4614	19.4077	15.1800	19.1268
.68	16.1181	19.7802	15.8044	19.4669	15.5079	19.1707	15.2265	18.8897
.70	16.1688	19.5556	15.8552	19.2423	15.5586	18.9461	15.2772	18.6651
.72	16.2235	19.3429	15.9099	19.0295	15.6133	18.7333	15.3319	18.4523
.74	16.2822	19.1414	15.9685	18.8280	15.6720	18.5318	15.3906	18.2507
.76	16.3445	18.9505	16.0309	18.6371	15.7344	18.3409	15.4529	18.0598
.78	16.4105	18.7698	16.0968	18.4564	15.8003	18.1601	15.5189	17.8790
.80	16.4798	18.5987	16.1661	18.2853	15.8696	17.9889	15.5882	17.7078
.82	16.5523	18.4367	16.2386	18.1233	15.9421	17.8270	15.6607	17.5458
.84	16.6279	18.2835	16.3142	17.9700	16.0177	17.6737	15.7363	17.3925
.86	16.7063	18.1385	16.3927	17.8250	16.0961	17.5286	15.8147	17.2474
.88	16.7875	18.0014	16.4738	17.6878	16.1773	17.3915	15.8959	17.1102
.90	16.8712	17.8717	16.5575	17.5581	16.2610	17.2617	15.9796	16.9805
.92	16.9573	17.7490	16.6436	17.4355	16.3471	17.1391	16.0658	16.8578
.94	17.0456	17.6331	16.7320	17.3196	16.4355	17.0232	16.1541	16.7419
.96	17.1360	17.5236	16.8224	17.2101	16.5259	16.9136	16.2445	16.6323
.98	17.2283	17.4202	16.9147	17.1066	16.6182	16.8101	16.3369	16.5288
1.00	17.3224	17.3224	17.0089	17.0089	16.7123	16.7123	16.4310	16.4310

Table D–1: $0.053 \leq \hat{\rho}_4 \leq 0.062$ and $\lambda_3,\ \lambda_4 > 0$

	$\hat{\rho}_4 = 0.053$		$\hat{\rho}_4 = 0.056$		$\hat{\rho}_4 = 0.059$		$\hat{\rho}_4 = 0.062$	
$\hat{\rho}_3$	λ_3	λ_4	λ_3	λ_4	λ_3	λ_4	λ_3	λ_4
.02	16.0034	53.1731	15.7015	52.8790	15.4154	52.6016	15.1434	52.3395
.04	15.8976	46.4703	15.5957	46.1726	15.3096	45.8912	15.0377	45.6246
.06	15.7949	42.5132	15.4930	42.2143	15.2069	41.9316	14.9350	41.6635
.08	15.6954	39.6803	15.3935	39.3808	15.1074	39.0975	14.8355	38.8287
.10	15.5994	37.4646	15.2975	37.1649	15.0114	36.8811	14.7395	36.6119
.12	15.5072	35.6408	15.2053	35.3408	14.9192	35.0568	14.6473	34.7873
.14	15.4188	34.0886	15.1170	33.7885	14.8309·	33.5043	14.5590	33.2346
.16	15.3347	32.7364	15.0328	32.4362	14.7467	32.1520	14.4748	31.8821
.18	15.2548	31.5382	14.9529	31.2379	14.6668	30.9535	14.3949	30.6836
.20	15.1793	30.4624	14.8774	30.1621	14.5913	29.8776	14.3195	29.6076
.22	15.1085	29.4867	14.8066	29.1863	14.5205	28.9018	14.2486	28.6317
.24	15.0424	28.5946	14.7406	28.2941	14.4545	28.0096	14.1826	27.7394
.26	14.9813	27.7735	14.6794	27.4730	14.3933	27.1884	14.1214	26.9182
.28	14.9251	27.0138	14.6232	26.7132	14.3371	26.4286	14.0652	26.1583
.30	14.8740	26.3078	14.5722	26.0072	14.2860	25.7225	14.0141	25.4522
.32	14.8282	25.6492	14.5263	25.3486	14.2402	25.0639	13.9683	24.7935
.34	14.7876	25.0331	14.4857	24.7324	14.1996	24.4477	13.9277	24.1773
.36	14.7524	24.4553	14.4505	24.1545	14.1643	23.8697	13.8924	23.5993
.38	14.7225	23.9121	14.4206	23.6114	14.1345	23.3265	13.8625	23.0560
.40	14.6981	23.4007	14.3962	23.0999	14.1100	22.8150	13.8381	22.5445
.42	14.6791	22.9185	14.3772	22.6177	14.0910	22.3327	13.8190	22.0622
.44	14.6656	22.4633	14.3636	22.1624	14.0774	21.8774	13.8054	21.6068
.46	14.6574	22.0331	14.3555	21.7322	14.0692	21.4471	13.7972	21.1764
.48	14.6547	21.6262	14.3527	21.3253	14.0665	21.0402	13.7944	20.7694
.50	14.6573	21.2412	14.3553	20.9402	14.0690	20.6551	13.7970	20.3843
.52	14.6651	20.8766	14.3631	20.5756	14.0769	20.2904	13.8048	20.0196
.54	14.6782	20.5313	14.3762	20.2302	14.0899	19.9450	13.8178	19.6741
.56	14.6964	20.2042	14.3944	19.9030	14.1081	19.6177	13.8360	19.3468
.58	14.7196	19.8941	14.4176	19.5930	14.1313·	19.3076	13.8592	19.0366
.60	14.7478	19.6003	14.4457	19.2991	14.1594	19.0137	13.8873	18.7426
.62	14.7807	19.3218	14.4786	19.0205	14.1923	18.7351	13.9201	18.4639
.64	14.8183	19.0579	14.5162	18.7565	14.2298	18.4710	13.9577	18.1998
.66	14.8603	18.8077	14.5583	18.5063	14.2719	18.2207	13.9997	17.9495
.68	14.9068	18.5706	14.6047	18.2692	14.3184	17.9835	14.0462	17.7122
.70	14.9575	18.3459	14.6554	18.0445	14.3690	17.7588	14.0968	17.4874
.72	15.0122	18.1331	14.7101	17.8316	14.4238	17.5458	14.1516	17.2744
.74	15.0709	17.9315	14.7688	17.6299	14.4824	17.3441	14.2102	17.0726
.76	15.1332	17.7405	14.8311	17.4389	14.5447	17.1530	14.2726	16.8815
.78	15.1991	17.5596	14.8970	17.2580	14.6107	16.9721	14.3385	16.7005
.80	15.2685	17.3885	14.9664	17.0867	14.6800	16.8008	14.4078	16.5292
.82	15.3410	17.2264	15.0389	16.9247	14.7525	16.6387	14.4804	16.3670
.84	15.4166	17.0731	15.1145	16.7713	14.8281	16.4853	14.5560	16.2135
·.86	15.4950	16.9280	15.1930	16.6262	14.9066	16.3401	14.6345	16.0683
.88	15.5762	16.7907	15.2742	16.4889	14.9878	16.2028	14.7157	15.9310
.90	15.6600	16.6610	15.3579	16.3591	15.0716	16.0730	14.7995	15.8011
.92	15.7461	16.5383	15.4440	16.2364	15.1577	15.9503	14.8856	15.6783
.94	15.8345	16.4223	15.5324	16.1204	15.2461	15.8342	14.9740	15.5623
.96	15.9249	16.3127	15.6229	16.0108	15.3366	15.7246	15.0645	15.4526
.98	16.0173	16.2092	15.7153	15.9073	15.4290	15.6210	15.1570	15.3490
1.00	16.1114	16.1114	15.8095	15.8095	15.5232	15.5232	15.2512	15.2512

Table D–1: $0.0655 \le \hat{\rho}_4 \le 0.0760$ and $\lambda_3,\ \lambda_4 > 0$

	$\hat{\rho}_4 = 0.0655$		$\hat{\rho}_4 = 0.0690$		$\hat{\rho}_4 = 0.0725$		$\hat{\rho}_4 = 0.0760$	
$\hat{\rho}_3$	λ_3	λ_4	λ_3	λ_4	λ_3	λ_4	λ_3	λ_4
.02	14.8424	52.0513	14.5570	51.7804	14.2858	51.5255	14.0274	51.2852
.04	14.7366	45.3305	14.4513	45.0530	14.1801	44.7905	13.9217	44.5420
.06	14.6339	41.3675	14.3486	41.0878	14.0774	40.8229	13.8190	40.5716
.08	14.5344	38.5317	14.2491	38.2509	13.9779	37.9849	13.7196	37.7322
.10	14.4385	36.3144	14.1532	36.0330	13.8820	35.7662	13.6236	35.5128
.12	14.3463	34.4895	14.0609	34.2077	13.7898	33.9404	13.5314	33.6865
.14	14.2580	32.9365	13.9726	32.6544	13.7015	32.3869	13.4432	32.1326
.16	14.1738	31.5838	13.8885	31.3015	13.6173	31.0337	13.3590	30.7792
.18	14.0939	30.3851	13.8086	30.1027	13.5374	29.8347	13.2791	29.5800
.20	14.0184	29.3090	13.7331	29.0264	13.4620	28.7584	13.2037	28.5035
.22	13.9476	28.3330	13.6623	28.0504	13.3911	27.7822	13.1328	27.5271
.24	13.8815	27.4407	13.5962	27.1579	13.3250	26.8896	13.0667	26.6345
.26	13.8204	26.6194	13.5350	26.3365	13.2639	26.0682	13.0055	25.8129
.28	13.7642	25.8595	13.4788	25.5766	13.2076	25.3081	12.9493	25.0528
.30	13.7131	25.1533	13.4277	24.8703	13.1565	24.6017	12.8982	24.3464
.32	13.6672	24.4945	13.3819	24.2115	13.1106	23.9429	12.8523	23.6874
.34	13.6266	23.8782	13.3412	23.5951	13.0700	23.3265	12.8116	23.0709
.36	13.5913	23.3002	13.3059	23.0170	13.0347	22.7483	12.7763	22.4926
.38	13.5614	22.7569	13.2760	22.4736	13.0047	22.2048	12.7463	21.9491
.40	13.5369	22.2453	13.2515	21.9620	12.9802	21.6930	12.7218	21.4372
.42	13.5179	21.7629	13.2324	21.4795	12.9611	21.2105	12.7026	20.9546
.44	13.5043	21.3074	13.2188	21.0240	12.9474	20.7549	12.6889	20.4989
.46	13.4961	20.8770	13.2106	20.5935	12.9392	20.3243	12.6806	20.0682
.48	13.4932	20.4699	13.2077	20.1863	12.9363	19.9170	12.6777	19.6609
.50	13.4957	20.0847	13.2102	19.8010	12.9388	19.5316	12.6802	19.2753
.52	13.5035	19.7199	13.2180	19.4361	12.9465	19.1666	12.6879	18.9102
.54	13.5166	19.3744	13.2310	19.0905	12.9595	18.8209	12.7008	18.5644
.56	13.5347	19.0469	13.2491	18.7630	12.9775	18.4933	12.7188	18.2367
.58	13.5578	18.7367	13.2722	18.4526	13.0006	18.1829	12.7419	17.9261
.60	13.5859	18.4426	13.3002	18.1584	13.0287	17.8886	12.7699	17.6317
.62	13.6188	18.1639	13.3331	17.8796	13.0615	17.6096	12.8027	17.3527
.64	13.6563	17.8996	13.3706	17.6153	13.0990	17.3452	12.8402	17.0882
.66	13.6984	17.6492	13.4126	17.3648	13.1410	17.0946	12.8822	16.8374
.68	13.7448	17.4119	13.4590	17.1274	13.1874	16.8571	12.9286	16.5998
.70	13.7954	17.1870	13.5097	16.9024	13.2380	16.6320	12.9792	16.3746
.72	13.8502	16.9739	13.5644	16.6892	13.2927	16.4188	13.0339	16.1612
.74	13.9088	16.7721	13.6230	16.4873	13.3514	16.2167	13.0925	15.9591
.76	13.9711	16.5809	13.6854	16.2960	13.4137	16.0254	13.1549	15.7677
.78	14.0371	16.3998	13.7513	16.1149	13.4797	15.8442	13.2208	15.5864
.80	14.1064	16.2284	13.8207	15.9434	13.5490	15.6726	13.2901	15.4147
.82	14.1790	16.0662	13.8932	15.7811	13.6216	15.5102	13.3627	15.2522
.84	14.2546	15.9127	13.9689	15.6275	13.6972	15.3566	13.4384	15.0985
.86	14.3331	15.7674	14.0474	15.4822	13.7758	15.2112	13.5170	14.9530
.88	14.4143	15.6300	14.1286	15.3448	13.8570	15.0737	13.5983	14.8155
.90	14.4981	15.5001	14.2124	15.2148	13.9409	14.9437	13.6821	14.6854
.92	14.5843	15.3773	14.2987	15.0919	14.0271	14.8207	13.7684	14.5624
.94	14.6727	15.2612	14.3871	14.9758	14.1156	14.7045	13.8569	14.4461
.96	14.7633	15.1515	14.4777	14.8661	14.2062	14.5947	13.9476	14.3363
.98	14.8557	15.0479	14.5702	14.7624	14.2987	14.4910	14.0401	14.2325
1.00	14.9500	14.9500	14.6645	14.6645	14.3931	14.3931	14.1345	14.1345

Table D–1: $0.080 \le \hat{\rho}_4 \le 0.092$ and $\lambda_3,\ \lambda_4 > 0$

$\hat{\rho}_3$	$\hat{\rho}_4 = 0.080$ λ_3	λ_4	$\hat{\rho}_4 = 0.084$ λ_3	λ_4	$\hat{\rho}_4 = 0.088$ λ_3	λ_4	$\hat{\rho}_4 = 0.092$ λ_3	λ_4
.02	13.7464	51.0273	13.4790	50.7860	13.2242	50.5601	12.9808	50.3487
.04	13.6407	44.2734	13.3734	44.0201	13.1185	43.7808	12.8751	43.5546
.06	13.5380	40.2995	13.2707	40.0423	13.0159	39.7987	12.7725	39.5677
.08	13.4385	37.4585	13.1713	37.1993	12.9165	36.9536	12.6731	36.7202
.10	13.3426	35.2380	13.0754	34.9778	12.8206	34.7308	12.5772	34.4960
.12	13.2504	33.4111	12.9832	33.1501	12.7284	32.9023	12.4850	32.6667
.14	13.1621	31.8568	12.8949	31.5953	12.6402	31.3469	12.3968	31.1106
.16	13.0780	30.5030	12.8107	30.2412	12.5560	29.9924	12.3126	29.7557
.18	12.9981	29.3036	12.7309	29.0415	12.4761	28.7924	12.2328	28.5553
.20	12.9227	28.2268	12.6554	27.9645	12.4007	27.7152	12.1573	27.4778
.22	12.8518	27.2504	12.5846	26.9878	12.3298	26.7383	12.0865	26.5008
.24	12.7857	26.3576	12.5185	26.0949	12.2637	25.8452	12.0203	25.6074
.26	12.7245	25.5359	12.4573	25.2731	12.2025	25.0232	11.9591	24.7853
.28	12.6683	24.7756	12.4010	24.5127	12.1462	24.2627	11.9028	24.0246
.30	12.6172	24.0691	12.3499	23.8060	12.0951	23.5559	11.8516	23.3177
.32	12.5712	23.4100	12.3039	23.1469	12.0491	22.8966	11.8056	22.6582
.34	12.5305	22.7934	12.2632	22.5301	12.0083	22.2798	11.7648	22.0412
.36	12.4952	22.2150	12.2278	21.9516	11.9729	21.7011	11.7294	21.4625
.38	12.4652	21.6714	12.1978	21.4078	11.9429	21.1572	11.6993	20.9184
.40	12.4406	21.1594	12.1732	20.8958	11.9182	20.6450	11.6746	20.4060
.42	12.4214	20.6767	12.1539	20.4129	11.8989	20.1620	11.6553	19.9228
.44	12.4077	20.2208	12.1402	19.9569	11.8851	19.7059	11.6414	19.4665
.46	12.3993	19.7900	12.1318	19.5260	11.8767	19.2748	11.6329	19.0353
.48	12.3964	19.3826	12.1288	19.1184	11.8736	18.8670	11.6298	18.6273
.50	12.3988	18.9969	12.1311	18.7325	11.8759	18.4810	11.6320	18.2411
.52	12.4065	18.6317	12.1388	18.3672	11.8835	18.1154	11.6395	17.8754
.54	12.4193	18.2857	12.1516	18.0210	11.8963	17.7691	11.6523	17.5289
.56	12.4374	17.9578	12.1696	17.6930	11.9142	17.4409	11.6701	17.2005
.58	12.4604	17.6471	12.1926	17.3821	11.9372	17.1299	11.6930	16.8892
.60	12.4884	17.3526	12.2205	17.0874	11.9650	16.8350	11.7209	16.5942
.62	12.5211	17.0734	12.2532	16.8081	11.9977	16.5555	11.7535	16.3144
.64	12.5586	16.8087	12.2906	16.5432	12.0351	16.2905	11.7909	16.0492
.66	12.6005	16.5579	12.3326	16.2922	12.0770	16.0392	11.8327	15.7978
.68	12.6469	16.3201	12.3789	16.0543	12.1233	15.8011	11.8790	15.5595
.70	12.6975	16.0948	12.4295	15.8288	12.1739	15.5755	11.9296	15.3337
.72	12.7522	15.8813	12.4842	15.6151	12.2286	15.3617	11.9842	15.1197
.74	12.8108	15.6790	12.5428	15.4127	12.2872	15.1591	12.0428	14.9169
.76	12.8732	15.4874	12.6051	15.2210	12.3495	14.9672	12.1052	14.7249
.78	12.9391	15.3060	12.6711	15.0395	12.4155	14.7855	12.1711	14.5430
.80	13.0085	15.1342	12.7404	14.8676	12.4848	14.6135	12.2405	14.3708
.82	13.0811	14.9716	12.8131	14.7049	12.5575	14.4506	12.3132	14.2078
.84	13.1567	14.8178	12.8888	14.5509	12.6332	14.2965	12.3889	14.0535
.86	13.2353	14.6722	12.9674	14.4052	12.7119	14.1507	12.4676	13.9076
.88	13.3167	14.5346	13.0487	14.2674	12.7933	14.0128	12.5490	13.7696
.90	13.4006	14.4044	13.1327	14.1372	12.8772	13.8824	12.6331	13.6391
.92	13.4869	14.2813	13.2190	14.0140	12.9637	13.7592	12.7195	13.5157
.94	13.5754	14.1650	13.3077	13.8976	13.0523	13.6427	12.8083	13.3991
.96	13.6661	14.0550	13.3984	13.7876	13.1431	13.5326	12.8991	13.2889
.98	13.7587	13.9512	13.4911	13.6837	13.2358	13.4286	12.9919	13.1848
1.00	13.8532	13.8532	13.5855	13.5855	13.3304	13.3304	13.0866	13.0866

Table D–1: $0.0965 \leq \hat{\rho}_4 \leq 0.1100$ and $\lambda_3, \lambda_4 > 0$

$\hat{\rho}_3$	$\hat{\rho}_4 = 0.0965$		$\hat{\rho}_4 = 0.1010$		$\hat{\rho}_4 = 0.1055$		$\hat{\rho}_4 = 0.1100$	
	λ_3	λ_4	λ_3	λ_4	λ_3	λ_4	λ_3	λ_4
.02	12.7193	50.1273	12.4698	49.9225	12.2312	49.7336	12.0027	49.5598
.04	12.6136	43.3146	12.3642	43.0890	12.1257	42.8769	11.8972	42.6776
.06	12.5110	39.3217	12.2616	39.0894	12.0231	38.8699	11.7947	38.6623
.08	12.4117	36.4713	12.1622	36.2357	11.9238	36.0126	11.6954	35.8010
.10	12.3158	34.2454	12.0664	34.0080	11.8280	33.7827	11.5996	33.5688
.12	12.2237	32.4149	11.9743	32.1762	11.7359	31.9496	11.5075	31.7342
.14	12.1354	30.8581	11.8861	30.6186	11.6477	30.3910	11.4193	30.1746
.16	12.0513	29.5026	11.8019	29.2624	11.5635	29.0341	11.3352	28.8170
.18	11.9714	28.3018	11.7221	28.0611	11.4837	27.8324	11.2554	27.6146
.20	11.8960	27.2240	11.6466	26.9830	11.4082	26.7537	11.1799	26.5355
.22	11.8251	26.2466	11.5758	26.0053	11.3374	25.7757	11.1091	25.5571
.24	11.7590	25.3531	11.5097	25.1114	11.2712	24.8816	11.0429	24.6627
.26	11.6977	24.5307	11.4484	24.2889	11.2099	24.0588	10.9816	23.8396
.28	11.6414	23.7698	11.3920	23.5278	11.1536	23.2974	10.9252	23.0780
.30	11.5902	23.0627	11.3408	22.8205	11.1023	22.5899	10.8739	22.3702
.32	11.5442	22.4031	11.2947	22.1606	11.0562	21.9299	10.8277	21.7099
.34	11.5034	21.7859	11.2538	21.5433	11.0153	21.3123	10.7867	21.0921
.36	11.4679	21.2070	11.2183	20.9641	10.9796	20.7329	10.7511	20.5125
.38	11.4377	20.6627	11.1881	20.4197	10.9494	20.1882	10.7207	19.9676
.40	11.4129	20.1502	11.1632	19.9069	10.9245	19.6753	10.6957	19.4543
.42	11.3936	19.6668	11.1438	19.4233	10.9050	19.1914	10.6762	18.9702
.44	11.3796	19.2103	11.1298	18.9666	10.8909	18.7344	10.6620	18.5130
.46	11.3711	18.7788	11.1212	18.5349	10.8822	18.3024	10.6532	18.0807
.48	11.3679	18.3706	11.1179	18.1264	10.8788	17.8937	10.6497	17.6717
.50	11.3701	17.9842	11.1200	17.7398	10.8808	17.5068	10.6517	17.2844
.52	11.3775	17.6183	11.1274	17.3736	10.8881	17.1403	10.6588	16.9176
.54	11.3902	17.2715	11.1400	17.0265	10.9006	16.7930	10.6713	16.5700
.56	11.4080	16.9429	11.1577	16.6976	10.9183	16.4638	10.6888	16.2405
.58	11.4308	16.6314	11.1805	16.3858	10.9409	16.1517	10.7114	15.9280
.60	11.4586	16.3360	11.2082	16.0902	10.9686	15.8558	10.7390	15.6318
.62	11.4912	16.0560	11.2407	15.8100	11.0010	15.5752	10.7713	15.3509
.64	11.5285	15.7906	11.2779	15.5442	11.0382	15.3092	10.8084	15.0845
.66	11.5703	15.5389	11.3197	15.2923	11.0799	15.0569	10.8501	14.8319
.68	11.6165	15.3004	11.3659	15.0535	11.1261	14.8178	10.8962	14.5925
.70	11.6671	15.0743	11.4164	14.8272	11.1766	14.5912	10.9467	14.3655
.72	11.7217	14.8601	11.4710	14.6127	11.2311	14.3764	11.0012	14.1504
.74	11.7803	14.6571	11.5296	14.4094	11.2897	14.1729	11.0597	13.9466
.76	11.8426	14.4648	11.5919	14.2169	11.3520	13.9801	11.1221	13.7536
.78	11.9086	14.2827	11.6579	14.0346	11.4180	13.7975	11.1881	13.5707
.80	11.9780	14.1103	11.7273	13.8620	11.4875	13.6247	11.2575	13.3976
.82	12.0507	13.9472	11.8000	13.6986	11.5602	13.4610	11.3303	13.2337
.84	12.1265	13.7927	11.8759	13.5440	11.6361	13.3062	11.4062	13.0786
.86	12.2052	13.6466	11.9546	13.3977	11.7149	13.1597	11.4851	12.9318
.88	12.2867	13.5084	12.0362	13.2593	11.7965	13.0211	11.5668	12.7931
.90	12.3708	13.3778	12.1203	13.1285	11.8808	12.8901	11.6511	12.6618
.92	12.4573	13.2543	12.2069	13.0048	11.9674	12.7663	11.7379	12.5378
.94	12.5461	13.1375	12.2958	12.8879	12.0564	12.6492	11.8270	12.4206
.96	12.6370	13.0272	12.3869	12.7775	12.1475	12.5386	11.9182	12.3098
.98	12.7299	12.9230	12.4798	12.6732	12.2406	12.4342	12.0114	12.2052
1.00	12.8246	12.8246	12.5747	12.5746	12.3355	12.3355	12.1065	12.1065

Table D–1: $0.115 \leq \hat{\rho}_4 \leq 0.130$ and $\lambda_3,\ \lambda_4 > 0$

	$\hat{\rho}_4 = 0.115$		$\hat{\rho}_4 = 0.120$		$\hat{\rho}_4 = 0.125$		$\hat{\rho}_4 = 0.130$	
$\hat{\rho}_3$	λ_3	λ_4	λ_3	λ_4	λ_3	λ_4	λ_3	λ_4
.02	11.7596	49.3841	11.5270	49.2261	11.3040	49.0857	11.0898	48.9624
.04	11.6541	42.4703	11.4215	42.2771	11.1986	42.0976	10.9844	41.9311
.06	11.5516	38.4449	11.3191	38.2406	11.0962	38.0487	10.8821	37.8686
.08	11.4524	35.5787	11.2199	35.3690	10.9970	35.1711	10.7830	34.9845
.10	11.3566	33.3436	11.1242	33.1307	10.9013	32.9294	10.6873	32.7389
.12	11.2646	31.5072	11.0321	31.2923	10.8093	31.0886	10.5954	30.8957
.14	11.1764	29.9463	10.9440	29.7299	10.7212	29.5247	10.5073	29.3301
.16	11.0923	28.5877	10.8599	28.3703	10.6371	28.1640	10.4232	27.9682
.18	11.0125	27.3846	10.7801	27.1664	10.5573	26.9593	10.3434	26.7625
.20	10.9371	26.3050	10.7047	26.0862	10.4819	25.8783	10.2680	25.6808
.22	10.8662	25.3261	10.6338	25.1068	10.4110	24.8984	10.1971	24.7002
.24	10.8000	24.4313	10.5676	24.2116	10.3448	24.0027	10.1309	23.8040
.26	10.7387	23.6078	10.5062	23.3877	10.2834	23.1784	10.0694	22.9793
.28	10.6822	22.8460	10.4498	22.6255	10.2269	22.4158	10.0129	22.2162
.30	10.6309	22.1379	10.3983	21.9171	10.1754	21.7070	9.9613	21.5071
.32	10.5846	21.4773	10.3520	21.2562	10.1290	21.0458	9.9149	20.8455
.34	10.5436	20.8592	10.3109	20.6378	10.0879	20.4270	9.8736	20.2263
.36	10.5079	20.2793	10.2751	20.0575	10.0519	19.8465	9.8376	19.6454
.38	10.4774	19.7341	10.2446	19.5120	10.0213	19.3005	9.8069	19.0990
.40	10.4524	19.2205	10.2194	18.9981	9.9960	18.7863	9.7815	18.5844
.42	10.4327	18.7360	10.1996	18.5133	9.9761	18.3011	9.7614	18.0987
.44	10.4184	18.2785	10.1852	18.0553	9.9615	17.8427	9.7467	17.6400
.46	10.4095	17.8459	10.1762	17.6223	9.9524	17.4093	9.7374	17.2061
.48	10.4059	17.4365	10.1725	17.2126	9.9486	16.9991	9.7335	16.7954
.50	10.4077	17.0489	10.1742	16.8245	9.9501	16.6106	9.7348	16.4064
.52	10.4148	16.6817	10.1811	16.4569	9.9569	16.2425	9.7415	16.0378
.54	10.4271	16.3336	10.1933	16.1084	9.9689	15.8936	9.7534	15.6884
.56	10.4445	16.0037	10.2106	15.7781	9.9861	15.5627	9.7704	15.3570
.58	10.4670	15.6909	10.2330	15.4648	10.0084	15.2490	9.7925	15.0427
.60	10.4945	15.3942	10.2603	15.1677	10.0356	14.9513	9.8196	14.7445
.62	10.5268	15.1129	10.2925	14.8859	10.0677	14.6691	9.8516	14.4617
.64	10.5638	14.8461	10.3294	14.6187	10.1045	14.4013	9.8883	14.1934
.66	10.6054	14.5931	10.3710	14.3652	10.1460	14.1474	9.9296	13.9390
.68	10.6515	14.3533	10.4170	14.1250	10.1919	13.9067	9.9755	13.6977
.70	10.7018	14.1259	10.4673	13.8972	10.2422	13.6784	10.0257	13.4689
.72	10.7564	13.9105	10.5218	13.6813	10.2966	13.4620	10.0801	13.2520
.74	10.8149	13.7063	10.5803	13.4767	10.3551	13.2570	10.1385	13.0465
.76	10.8772	13.5128	10.6426	13.2828	10.4174	13.0627	10.2009	12.8517
.78	10.9432	13.3296	10.7086	13.0992	10.4834	12.8787	10.2669	12.6672
.80	11.0127	13.1562	10.7781	12.9254	10.5530	12.7044	10.3365	12.4926
.82	11.0855	12.9920	10.8510	12.7608	10.6259	12.5395	10.4095	12.3272
.84	11.1615	12.8366	10.9270	12.6051	10.7020	12.3834	10.4857	12.1707
.86	11.2405	12.6895	11.0061	12.4578	10.7811	12.2357	10.5649	12.0226
.88	11.3222	12.5505	11.0879	12.3184	10.8631	12.0961	10.6469	11.8826
.90	11.4066	12.4190	11.1725	12.1867	10.9477	11.9640	10.7317	11.7502
.92	11.4935	12.2947	11.2595	12.0622	11.0349	11.8392	10.8190	11.6251
.94	11.5827	12.1773	11.3488	11.9445	11.1244	11.7212	10.9087	11.5069
.96	11.6741	12.0664	11.4403	11.8333	11.2160	11.6098	11.0005	11.3952
.98	11.7675	11.9616	11.5338	11.7283	11.3098	11.5046	11.0944	11.2897
1.00	11.8627	11.8627	11.6292	11.6292	11.4053	11.4053	11.1902	11.1902

Table D–1: $0.136 \leq \hat{\rho}_4 \leq 0.154$ and $\lambda_3,\ \lambda_4 > 0$

| | $\hat{\rho}_4 = 0.136$ | | $\hat{\rho}_4 = 0.142$ | | $\hat{\rho}_4 = 0.148$ | | $\hat{\rho}_4 = 0.154$ | |
$\hat{\rho}_3$	λ_3	λ_4	λ_3	λ_4	λ_3	λ_4	λ_3	λ_4
.02	10.8436	48.8373	10.6082	48.7371	10.3827	48.6621	10.1664	48.6129
.04	10.7383	41.7481	10.5030	41.5827	10.2776	41.4347	10.0613	41.3036
.06	10.6360	37.6675	10.4007	37.4820	10.1754	37.3115	9.9592	37.1556
.08	10.5369	34.7746	10.3017	34.5794	10.0765	34.3982	9.8604	34.2304
.10	10.4414	32.5240	10.2062	32.3232	9.9810	32.1358	9.7650	31.9612
.12	10.3494	30.6775	10.1143	30.4731	9.8892	30.2817	9.6732	30.1027
.14	10.2614	29.1097	10.0263	28.9027	9.8012	28.7085	9.5853	28.5264
.16	10.1774	27.7461	9.9423	27.5372	9.7173	27.3411	9.5014	27.1568
.18	10.0976	26.5391	9.8625	26.3289	9.6375	26.1312	9.4217	25.9453
.20	10.0222	25.4564	9.7871	25.2452	9.5621	25.0462	9.3463	24.8590
.22	9.9512	24.4751	9.7162	24.2629	9.4911	24.0630	9.2753	23.8747
.24	9.8849	23.5782	9.6499	23.3652	9.4248	23.1645	9.2089	22.9752
.26	9.8235	22.7528	9.5884	22.5393	9.3632	22.3378	9.1473	22.1478
.28	9.7669	21.9893	9.5317	21.7751	9.3065	21.5729	9.0905	21.3822
.30	9.7152	21.2796	9.4800	21.0648	9.2547	20.8621	9.0386	20.6706
.32	9.6687	20.6175	9.4334	20.4022	9.2079	20.1988	8.9917	20.0067
.34	9.6273	19.9978	9.3918	19.7820	9.1663	19.5780	8.9499	19.3853
.36	9.5912	19.4164	9.3556	19.2000	9.1299	18.9954	8.9133	18.8020
.38	9.5603	18.8695	9.3245	18.6526	9.0987	18.4474	8.8819	18.2533
.40	9.5347	18.3543	9.2988	18.1368	9.0727	17.9310	8.8558	17.7362
.42	9.5145	17.8681	9.2784	17.6500	9.0521	17.4435	8.8349	17.2480
.44	9.4997	17.4088	9.2633	17.1900	9.0368	16.9828	8.8194	16.7866
.46	9.4902	16.9743	9.2536	16.7549	9.0269	16.5470	8.8092	16.3500
.48	9.4860	16.5630	9.2492	16.3429	9.0223	16.1343	8.8044	15.9364
.50	9.4872	16.1734	9.2502	15.9526	9.0230	15.7432	8.8048	15.5445
.52	9.4936	15.8041	9.2564	15.5826	9.0290	15.3724	8.8105	15.1728
.54	9.5053	15.4540	9.2679	15.2317	9.0402	15.0207	8.8215	14.8202
.56	9.5222	15.1219	9.2845	14.8988	9.0566	14.6870	8.8376	14.4856
.58	9.5441	14.8069	9.3062	14.5830	9.0781	14.3703	8.8588	14.1680
.60	9.5710	14.5080	9.3329	14.2834	9.1046	14.0698	8.8851	13.8666
.62	9.6028	14.2245	9.3645	13.9991	9.1360	13.7846	8.9163	13.5804
.64	9.6393	13.9555	9.4009	13.7293	9.1722	13.5140	8.9523	13.3089
.66	9.6806	13.7004	9.4420	13.4733	9.2131	13.2572	8.9930	13.0511
.68	9.7263	13.4584	9.4876	13.2306	9.2586	13.0135	9.0384	12.8065
.70	9.7764	13.2289	9.5377	13.0003	9.3085	12.7825	9.0882	12.5745
.72	9.8307	13.0113	9.5919	12.7820	9.3627	12.5633	9.1423	12.3545
.74	9.8892	12.8051	9.6503	12.5751	9.4210	12.3556	9.2006	12.1459
.76	9.9515	12.6097	9.7126	12.3790	9.4833	12.1587	9.2628	11.9481
.78	10.0176	12.4246	9.7787	12.1932	9.5494	11.9722	9.3290	11.7608
.80	10.0872	12.2493	9.8484	12.0173	9.6192	11.7956	9.3988	11.5834
.82	10.1603	12.0834	9.9215	11.8507	9.6924	11.6283	9.4721	11.4154
.84	10.2365	11.9264	9.9979	11.6931	9.7689	11.4700	9.5487	11.2564
.86	10.3159	11.7778	10.0774	11.5440	9.8485	11.3203	9.6285	11.1060
.88	10.3981	11.6373	10.1598	11.4030	9.9311	11.1787	9.7113	10.9637
.90	10.4830	11.5045	10.2449	11.2697	10.0165	11.0448	9.7968	10.8293
.92	10.5705	11.3790	10.3326	11.1437	10.1044	10.9183	9.8851	10.7022
.94	10.6604	11.2604	10.4227	11.0246	10.1948	10.7988	9.9757	10.5822
.96	10.7525	11.1483	10.5151	10.9122	10.2875	10.6860	10.0688	10.4688
.98	10.8467	11.0426	10.6095	10.8061	10.3822	10.5794	10.1639	10.3619
1.00	10.9427	10.9427	10.7059	10.7059	10.4789	10.4789	10.2609	10.2609

Table D–1: $0.161 \leq \hat{\rho}_4 \leq 0.182$ and $\lambda_3, \lambda_4 > 0$

	$\hat{\rho}_4 = 0.161$		$\hat{\rho}_4 = 0.168$		$\hat{\rho}_4 = 0.175$		$\hat{\rho}_4 = 0.182$	
$\hat{\rho}_3$	λ_3	λ_4	λ_3	λ_4	λ_3	λ_4	λ_3	λ_4
.02	9.9246	48.5893	9.6935	48.6037	9.4721	48.6584	9.2597	48.7564
.04	9.8197	41.1721	9.5887	41.0635	9.3675	40.9779	9.1552	40.9158
.06	9.7177	36.9917	9.4869	36.8468	9.2658	36.7206	9.0537	36.6130
.08	9.6190	34.0512	9.3882	33.8891	9.1672	33.7438	8.9553	33.6150
.10	9.5237	31.7730	9.2930	31.6011	9.0722	31.4449	8.8603	31.3040
.12	9.4320	29.9089	9.2014	29.7306	8.9807	29.5674	8.7689	29.4187
.14	9.3442	28.3287	9.1137	28.1461	8.8930	27.9781	8.6813	27.8241
.16	9.2603	26.9562	9.0298	26.7705	8.8092	26.5989	8.5976	26.4410
.18	9.1806	25.7425	8.9502	25.5544	8.7295	25.3801	8.5180	25.2194
.20	9.1052	24.6545	8.8748	24.4645	8.6541	24.2882	8.4426	24.1251
.22	9.0342	23.6689	8.8038	23.4773	8.5831	23.2993	8.3716	23.1344
.24	8.9678	22.7682	8.7373	22.5754	8.5166	22.3960	8.3050	22.2295
.26	8.9061	21.9397	8.6755	21.7457	8.4547	21.5651	8.2430	21.3973
.28	8.8492	21.1732	8.6185	20.9782	8.3976	20.7965	8.1858	20.6275
.30	8.7971	20.4608	8.5663	20.2648	8.3453	20.0821	8.1333	19.9119
.32	8.7501	19.7961	8.5191	19.5992	8.2979	19.4154	8.0857	19.2442
.34	8.7081	19.1738	8.4770	18.9760	8.2555	18.7913	8.0431	18.6190
.36	8.6713	18.5897	8.4399	18.3910	8.2182	18.2053	8.0055	18.0319
.38	8.6397	18.0402	8.4080	17.8405	8.1860	17.6538	7.9730	17.4793
.40	8.6133	17.5221	8.3813	17.3216	8.1590	17.1338	7.9456	16.9582
.42	8.5921	17.0330	8.3599	16.8315	8.1372	16.6426	7.9235	16.4659
.44	8.5763	16.5707	8.3437	16.3681	8.1207	16.1781	7.9066	16.0001
.46	8.5658	16.1331	8.3328	15.9294	8.1094	15.7382	7.8949	15.5590
.48	8.5606	15.7185	8.3273	15.5137	8.1035	15.3213	7.8885	15.1407
.50	8.5607	15.3255	8.3270	15.1195	8.1028	14.9258	7.8873	14.7439
.52	8.5660	14.9527	8.3320	14.7455	8.1073	14.5505	7.8914	14.3671
.54	8.5767	14.5989	8.3422	14.3905	8.1171	14.1942	7.9008	14.0093
.56	8.5924	14.2632	8.3576	14.0535	8.1322	13.8557	7.9154	13.6694
.58	8.6133	13.9444	8.3782	13.7334	8.1523	13.5342	7.9351	13.3463
.60	8.6393	13.6418	8.4038	13.4294	8.1776	13.2288	7.9599	13.0393
.62	8.6702	13.3544	8.4344	13.1407	8.2078	12.9386	7.9898	12.7475
.64	8.7060	13.0816	8.4699	12.8666	8.2430	12.6630	8.0246	12.4703
.66	8.7465	12.8226	8.5101	12.6063	8.2830	12.4012	8.0643	12.2069
.68	8.7917	12.5768	8.5551	12.3592	8.3277	12.1526	8.1087	11.9567
.70	8.8413	12.3437	8.6045	12.1246	8.3769	11.9167	8.1578	11.7192
.72	8.8953	12.1225	8.6584	11.9022	8.4307	11.6928	8.2113	11.4937
.74	8.9535	11.9127	8.7165	11.6911	8.4887	11.4804	8.2693	11.2798
.76	9.0158	11.7139	8.7788	11.4911	8.5509	11.2790	8.3314	11.0769
.78	9.0819	11.5255	8.8449	11.3015	8.6171	11.0881	8.3976	10.8846
.80	9.1518	11.3470	8.9149	11.1219	8.6871	10.9072	8.4677	10.7023
.82	9.2252	11.1780	8.9884	10.9519	8.7608	10.7360	8.5416	10.5297
.84	9.3020	11.0181	9.0654	10.7909	8.8380	10.5739	8.6190	10.3664
.86	9.3820	10.8668	9.1457	10.6386	8.9185	10.4205	8.6998	10.2118
.88	9.4650	10.7237	9.2290	10.4946	9.0021	10.2755	8.7837	10.0657
.90	9.5509	10.5885	9.3152	10.3585	9.0887	10.1384	8.8707	9.9275
.92	9.6395	10.4607	9.4041	10.2299	9.1781	10.0089	8.9605	9.7970
.94	9.7305	10.3400	9.4956	10.1085	9.2700	9.8866	9.0530	9.6738
.96	9.8239	10.2261	9.5895	9.9938	9.3644	9.7712	9.1479	9.5575
.98	9.9195	10.1185	9.6855	9.8856	9.4610	9.6623	9.2452	9.4479
1.00	10.0171	10.0171	9.7836	9.7836	9.5597	9.5597	9.3445	9.3445

Table D–1: $0.190 \le \hat{\rho}_4 \le 0.214$ and $\lambda_3, \ \lambda_4 > 0$

	$\hat{\rho}_4 = 0.190$		$\hat{\rho}_4 = 0.198$		$\hat{\rho}_4 = 0.206$		$\hat{\rho}_4 = 0.214$	
$\hat{\rho}_3$	λ_3	λ_4	λ_3	λ_4	λ_3	λ_4	λ_3	λ_4
.02	9.0272	48.9259	8.8047	49.1635	8.5916	49.4781	8.3870	49.8814
.04	8.9229	40.8741	8.7006	40.8647	8.4877	40.8892	8.2834	40.9492
.06	8.8215	36.5128	8.5994	36.4371	8.3867	36.3863	8.1826	36.3610
.08	8.7233	33.4878	8.5014	33.3817	8.2889	33.2968	8.0850	33.2332
.10	8.6285	31.1613	8.4067	31.0380	8.1943	30.9339	7.9906	30.8488
.12	8.5372	29.2662	8.3156	29.1319	8.1034	29.0156	7.8998	28.9171
.14	8.4497	27.6648	8.2282	27.5230	8.0161	27.3983	7.8127	27.2906
.16	8.3661	26.2769	8.1447	26.1297	7.9326	25.9990	7.7293	25.8847
.18	8.2865	25.0516	8.0651	24.9003	7.8532	24.7651	7.6499	24.6457
.20	8.2111	23.9545	7.9898	23.8000	7.7778	23.6613	7.5746	23.5380
.22	8.1400	22.9614	7.9187	22.8043	7.7067	22.6628	7.5034	22.5364
.24	8.0734	22.0545	7.8520	21.8953	7.6399	21.7514	7.4366	21.6225
.26	8.0113	21.2206	7.7898	21.0596	7.5776	20.9136	7.3741	20.7824
.28	7.9539	20.4493	7.7322	20.2865	7.5198	20.1388	7.3162	20.0057
.30	7.9012	19.7323	7.6793	19.5681	7.4667	19.4186	7.2628	19.2837
.32	7.8534	19.0633	7.6312	18.8975	7.4183	18.7465	7.2141	18.6099
.34	7.8105	18.4367	7.5880	18.2696	7.3748	18.1170	7.1702	17.9787
.36	7.7726	17.8483	7.5497	17.6798	7.3361	17.5257	7.1311	17.3857
.38	7.7397	17.2944	7.5164	17.1244	7.3023	16.9687	7.0969	16.8271
.40	7.7119	16.7719	7.4881	16.6004	7.2736	16.4431	7.0676	16.2997
.42	7.6893	16.2781	7.4650	16.1051	7.2499	15.9461	7.0434	15.8010
.44	7.6719	15.8109	7.4471	15.6362	7.2314	15.4755	7.0242	15.3285
.46	7.6597	15.3682	7.4343	15.1918	7.2180	15.0293	7.0101	14.8803
.48	7.6527	14.9483	7.4267	14.7701	7.2098	14.6057	7.0012	14.4547
.50	7.6510	14.5497	7.4244	14.3697	7.2068	14.2033	6.9975	14.0502
.52	7.6545	14.1712	7.4273	13.9893	7.2090	13.8208	6.9990	13.6655
.54	7.6633	13.8116	7.4355	13.6276	7.2165	13.4570	7.0057	13.2994
.56	7.6773	13.4697	7.4489	13.2837	7.2292	13.1109	7.0176	12.9508
.58	7.6965	13.1447	7.4674	12.9566	7.2471	12.7815	7.0348	12.6190
.60	7.7208	12.8357	7.4912	12.6455	7.2702	12.4680	7.0571	12.3029
.62	7.7502	12.5420	7.5200	12.3495	7.2984	12.1697	7.0847	12.0020
.64	7.7845	12.2627	7.5539	12.0681	7.3317	11.8858	7.1174	11.7155
.66	7.8238	11.9973	7.5927	11.8004	7.3700	11.6157	7.1551	11.4428
.68	7.8679	11.7451	7.6364	11.5460	7.4133	11.3589	7.1979	11.1833
.70	7.9167	11.5056	7.6848	11.3043	7.4614	11.1147	7.2456	10.9364
.72	7.9700	11.2781	7.7380	11.0747	7.5142	10.8827	7.2981	10.7017
.74	8.0279	11.0623	7.7956	10.8567	7.5717	10.6623	7.3553	10.4788
.76	8.0899	10.8575	7.8576	10.6498	7.6336	10.4531	7.4172	10.2670
.78	8.1562	10.6634	7.9239	10.4536	7.6999	10.2547	7.4834	10.0661
.80	8.2264	10.4794	7.9942	10.2676	7.7703	10.0666	7.5540	9.8755
.82	8.3004	10.3051	8.0684	10.0915	7.8448	9.8883	7.6287	9.6949
.84	8.3781	10.1401	8.1464	9.9247	7.9230	9.7195	7.7073	9.5239
.86	8.4592	9.9840	8.2279	9.7669	8.0050	9.5598	7.7897	9.3621
.88	8.5436	9.8364	8.3128	9.6177	8.0904	9.4088	7.8757	9.2091
.90	8.6311	9.6969	8.4009	9.4767	8.1791	9.2661	7.9651	9.0646
.92	8.7215	9.5651	8.4919	9.3435	8.2709	9.1314	8.0576	8.9281
.94	8.8147	9.4408	8.5858	9.2178	8.3655	9.0042	8.1532	8.7993
.96	8.9103	9.3234	8.6822	9.0993	8.4629	8.8843	8.2515	8.6779
.98	9.0083	9.2128	8.7811	8.9875	8.5627	8.7713	8.3525	8.5635
1.00	9.1085	9.1085	8.8822	8.8822	8.6649	8.6649	8.4558	8.4558

Table D–1: $0.223 \leq \hat{\rho}_4 \leq 0.250$ and $\lambda_3, \lambda_4 > 0$

	$\hat{\rho}_4 = 0.223$		$\hat{\rho}_4 = 0.232$		$\hat{\rho}_4 = 0.241$		$\hat{\rho}_4 = 0.250$	
$\hat{\rho}_3$	λ_3	λ_4	λ_3	λ_4	λ_3	λ_4	λ_3	λ_4
.02	8.1664	50.4604	7.9552	51.2028	7.7528	52.1529	7.5585	53.3831
.04	8.0631	41.0621	7.8522	41.2271	7.6501	41.4492	7.4561	41.7349
.06	7.9626	36.3640	7.7520	36.4018	7.5502	36.4761	7.3566	36.5894
.08	7.8652	33.1877	7.6549	33.1703	7.4534	33.1818	7.2601	33.2235
.10	7.7711	30.7763	7.5610	30.7284	7.3598	30.7057	7.1668	30.7087
.12	7.6805	28.8275	7.4706	28.7606	7.2696	28.7164	7.0768	28.6955
.14	7.5935	27.1895	7.3838	27.1096	7.1830	27.0510	6.9904	27.0140
.16	7.5103	25.7752	7.3007	25.6861	7.1000	25.6172	6.9076	25.5687
.18	7.4309	24.5301	7.2215	24.4340	7.0209	24.3575	6.8286	24.3005
.20	7.3556	23.4176	7.1462	23.3162	6.9456	23.2337	6.7534	23.1701
.22	7.2844	22.4121	7.0750	22.3064	6.8744	22.2192	6.6821	22.1504
.24	7.2175	21.4950	7.0079	21.3857	6.8073	21.2946	6.6149	21.2216
.26	7.1549	20.6522	6.9452	20.5400	6.7443	20.4456	6.5518	20.3689
.28	7.0967	19.8730	6.8868	19.7581	6.6857	19.6608	6.4929	19.5811
.30	7.0431	19.1488	6.8328	19.0315	6.6314	18.9317	6.4383	18.8491
.32	6.9940	18.4729	6.7834	18.3534	6.5816	18.2511	6.3880	18.1660
.34	6.9496	17.8397	6.7386	17.7181	6.5363	17.6135	6.3422	17.5259
.36	6.9100	17.2447	6.6984	17.1209	6.4955	17.0141	6.3009	16.9241
.38	6.8752	16.6841	6.6630	16.5582	6.4595	16.4491	6.2641	16.3567
.40	6.8453	16.1547	6.6324	16.0266	6.4282	15.9152	6.2320	15.8204
.42	6.8204	15.6538	6.6067	15.5235	6.4016	15.4097	6.2046	15.3123
.44	6.8004	15.1791	6.5859	15.0464	6.3800	14.9301	6.1821	14.8302
.46	6.7856	14.7286	6.5702	14.5934	6.3633	14.4745	6.1643	14.3718
.48	6.7758	14.3006	6.5595	14.1628	6.3516	14.0411	6.1516	13.9356
.50	6.7712	13.8935	6.5539	13.7530	6.3449	13.6284	6.1438	13.5198
.52	6.7717	13.5061	6.5534	13.3627	6.3434	13.2351	6.1411	13.1233
.54	6.7775	13.1372	6.5582	12.9908	6.3471	12.8600	6.1435	12.7449
.56	6.7885	12.7858	6.5682	12.6362	6.3559	12.5021	6.1512	12.3835
.58	6.8048	12.4509	6.5834	12.2981	6.3701	12.1606	6.1641	12.0382
.60	6.8263	12.1318	6.6039	11.9757	6.3895	11.8345	6.1823	11.7083
.62	6.8530	11.8277	6.6297	11.6681	6.4142	11.5232	6.2058	11.3931
.64	6.8849	11.5379	6.6607	11.3749	6.4442	11.2262	6.2348	11.0919
.66	6.9219	11.2619	6.6970	11.0953	6.4795	10.9427	6.2691	10.8042
.68	6.9641	10.9991	6.7384	10.8289	6.5202	10.6724	6.3087	10.5296
.70	7.0112	10.7490	6.7850	10.5751	6.5660	10.4146	6.3538	10.2675
.72	7.0633	10.5111	6.8366	10.3336	6.6171	10.1691	6.4042	10.0175
.74	7.1203	10.2848	6.8931	10.1037	6.6732	9.9352	6.4599	9.7793
.76	7.1819	10.0699	6.9546	9.8852	6.7344	9.7128	6.5207	9.5526
.78	7.2482	9.8658	7.0208	9.6777	6.8005	9.5015	6.5867	9.3369
.80	7.3189	9.6723	7.0916	9.4808	6.8714	9.3007	6.6577	9.1320
.82	7.3938	9.4888	7.1668	9.2940	6.9469	9.1103	6.7335	8.9375
.84	7.4729	9.3150	7.2463	9.1170	7.0269	8.9298	6.8140	8.7530
.86	7.5558	9.1505	7.3299	8.9496	7.1112	8.7590	6.8990	8.5784
.88	7.6425	8.9950	7.4174	8.7912	7.1996	8.5974	6.9883	8.4132
.90	7.7328	8.8480	7.5086	8.6416	7.2918	8.4448	7.0817	8.2572
.92	7.8264	8.7093	7.6033	8.5004	7.3878	8.3007	7.1790	8.1099
.94	7.9230	8.5785	7.7012	8.3672	7.4871	8.1649	7.2800	7.9711
.96	8.0226	8.4551	7.8023	8.2417	7.5897	8.0370	7.3843	7.8405
.98	8.1249	8.3390	7.9061	8.1236	7.6953	7.9167	7.4918	7.7176
1.00	8.2297	8.2297	8.0125	8.0125	7.8036	7.8036	7.6022	7.6022

Table D–1: $0.26 \leq \hat{\rho}_4 \leq 0.29$ and $\lambda_3,\ \lambda_4 > 0$

	$\hat{\rho}_4 = 0.26$		$\hat{\rho}_4 = 0.27$		$\hat{\rho}_4 = 0.28$		$\hat{\rho}_4 = 0.29$	
$\hat{\rho}_3$	λ_3	λ_4	λ_3	λ_4	λ_3	λ_4	λ_3	λ_4
.02	7.3514	55.2232	7.1532	57.8791	6.9630	62.1808	6.7805	71.9138
.04	7.2495	42.1368	7.0517	42.6421	6.8621	43.2708	6.6801	44.0506
.06	7.1504	36.7643	6.9530	36.9955	6.7638	37.2891	6.5823	37.6529
.08	7.0543	33.3069	6.8573	33.4318	6.6685	33.6009	6.4875	33.8181
.10	6.9613	30.7433	6.7647	30.8122	6.5763	30.9172	6.3957	31.0604
.12	6.8717	28.7001	6.6754	28.7350	6.4873	28.8012	6.3071	28.9004
.14	6.7855	26.9986	6.5894	27.0108	6.4017	27.0514	6.2217	27.1215
.16	6.7029	25.5390	6.5070	25.5352	6.3195	25.5578	6.1398	25.6077
.18	6.6239	24.2602	6.4282	24.2445	6.2409	24.2539	6.0613	24.2890
.20	6.5488	23.1217	6.3532	23.0970	6.1659	23.0964	5.9864	23.1202
.22	6.4775	22.0956	6.2819	22.0638	6.0946	22.0552	5.9151	22.0702
.24	6.4102	21.1616	6.2144	21.1240	6.0271	21.1090	5.8475	21.1170
.26	6.3469	20.3046	6.1510	20.2622	5.9634	20.2419	5.7836	20.2440
.28	6.2877	19.5129	6.0915	19.4664	5.9036	19.4416	5.7236	19.4389
.30	6.2327	18.7776	6.0361	18.7275	5.8478	18.6988	5.6674	18.6917
.32	6.1820	18.0914	5.9848	18.0379	5.7960	18.0057	5.6151	17.9949
.34	6.1356	17.4484	5.9378	17.3918	5.7484	17.3563	5.5668	17.3420
.36	6.0935	16.8438	5.8951	16.7843	5.7049	16.7456	5.5225	16.7281
.38	6.0560	16.2736	5.8567	16.2112	5.6656	16.1695	5.4824	16.1487
.40	6.0230	15.7345	5.8227	15.6691	5.6307	15.6243	5.4464	15.6004
.42	5.9946	15.2235	5.7933	15.1551	5.6001	15.1072	5.4147	15.0802
.44	5.9709	14.7383	5.7684	14.6667	5.5740	14.6157	5.3873	14.5854
.46	5.9520	14.2768	5.7482	14.2020	5.5525	14.1475	5.3643	14.1138
.48	5.9379	13.8372	5.7328	13.7589	5.5356	13.7009	5.3459	13.6637
.50	5.9288	13.4180	5.7222	13.3360	5.5235	13.2743	5.3321	13.2332
.52	5.9247	13.0177	5.7166	12.9319	5.5162	12.8662	5.3231	12.8211
.54	5.9257	12.6353	5.7160	12.5454	5.5139	12.4755	5.3190	12.4260
.56	5.9318	12.2698	5.7205	12.1755	5.5166	12.1011	5.3198	12.0470
.58	5.9432	11.9202	5.7303	11.8213	5.5246	11.7421	5.3258	11.6831
.60	5.9600	11.5857	5.7453	11.4821	5.5379	11.3978	5.3371	11.3335
.62	5.9821	11.2657	5.7658	11.1570	5.5566	11.0674	5.3538	10.9976
.64	6.0096	10.9597	5.7918	10.8457	5.5808	10.7505	5.3761	10.6748
.66	6.0426	10.6669	5.8233	10.5475	5.6107	10.4464	5.4040	10.3645
.68	6.0811	10.3871	5.8605	10.2620	5.6462	10.1549	5.4378	10.0665
.70	6.1251	10.1197	5.9033	9.9888	5.6876	9.8754	5.4776	9.7803
.72	6.1747	9.8644	5.9518	9.7277	5.7349	9.6078	5.5234	9.5057
.74	6.2296	9.6208	6.0059	9.4782	5.7880	9.3518	5.5753	9.2425
.76	6.2901	9.3887	6.0658	9.2401	5.8471	9.1071	5.6334	8.9905
.78	6.3558	9.1677	6.1312	9.0131	5.9121	8.8735	5.6978	8.7495
.80	6.4268	8.9576	6.2022	8.7971	5.9830	8.6509	5.7684	8.5195
.82	6.5029	8.7579	6.2786	8.5917	6.0596	8.4389	5.8452	8.3002
.84	6.5840	8.5686	6.3603	8.3967	6.1419	8.2375	5.9280	8.0915
.86	6.6699	8.3892	6.4471	8.2118	6.2297	8.0465	6.0168	7.8934
.88	6.7604	8.2194	6.5388	8.0369	6.3228	7.8656	6.1115	7.7057
.90	6.8552	8.0591	6.6353	7.8716	6.4211	7.6946	6.2117	7.5282
.92	6.9543	7.9078	6.7362	7.7157	6.5242	7.5334	6.3172	7.3607
.94	7.0572	7.7652	6.8414	7.5688	6.6318	7.3815	6.4278	7.2031
.96	7.1638	7.6311	6.9505	7.4307	6.7438	7.2387	6.5430	7.0550
.98	7.2737	7.5050	7.0633	7.3010	6.8598	7.1048	6.6627	6.9162
1.00	7.3868	7.3868	7.1793	7.1793	6.9794	6.9794	6.7863	6.7863

Table D–1: $0.301 \leq \hat{\rho}_4 \leq 0.334$ and λ_3, $\lambda_4 > 0$

$\hat{\rho}_3$	$\hat{\rho}_4 = 0.301$		$\hat{\rho}_4 = 0.312$		$\hat{\rho}_4 = 0.323$		$\hat{\rho}_4 = 0.334$	
	λ_3	λ_4	λ_3	λ_4	λ_3	λ_4	λ_3	λ_4
.04	6.4882	45.1317	6.3045	46.5288	6.1283	48.3901	5.9594	51.0040
.06	6.3910	38.1458	6.2079	38.7527	6.0324	39.4976	5.8641	40.4142
.08	6.2967	34.1179	6.1141	34.4888	5.9392	34.9405	5.7716	35.4857
.10	6.2054	31.2653	6.0233	31.5239	5.8489	31.8419	5.6818	32.2257
.12	6.1172	29.0494	5.9356	29.2433	5.7617	29.4854	5.5951	29.7800
.14	6.0322	27.2343	5.8510	27.3864	5.6775	27.5803	5.5113	27.8189
.16	5.9506	25.6952	5.7696	25.8184	5.5965	25.9793	5.4307	26.1800
.18	5.8723	24.3581	5.6916	24.4604	5.5187	24.5975	5.3532	24.7710
.20	5.7975	23.1754	5.6169	23.2620	5.4442	23.3812	5.2789	23.5346
.22	5.7262	22.1146	5.5457	22.1891	5.3730	22.2946	5.2078	22.4325
.24	5.6585	21.1528	5.4779	21.2176	5.3052	21.3123	5.1399	21.4381
.26	5.5945	20.2728	5.4137	20.3297	5.2408	20.4157	5.0754	20.5318
.28	5.5341	19.4617	5.3530	19.5122	5.1798	19.5910	5.0141	19.6991
.30	5.4774	18.7095	5.2959	18.7544	5.1223	18.8272	4.9562	18.9288
.32	5.4246	18.0081	5.2425	18.0483	5.0683	18.1159	4.9016	18.2118
.34	5.3756	17.3513	5.1927	17.3871	5.0178	17.4502	4.8505	17.5412
.36	5.3304	16.7335	5.1467	16.7654	4.9710	16.8244	4.8027	16.9111
.38	5.2893	16.1506	5.1045	16.1788	4.9278	16.2339	4.7585	16.3167
.40	5.2521	15.5988	5.0662	15.6234	4.8882	15.6749	4.7178	15.7540
.42	5.2191	15.0750	5.0318	15.0960	4.8525	15.1440	4.6807	15.2196
.44	5.1902	14.5766	5.0015	14.5940	4.8206	14.6384	4.6472	14.7106
.46	5.1657	14.1013	4.9752	14.1150	4.7926	14.1558	4.6175	14.2245
.48	5.1455	13.6471	4.9532	13.6570	4.7687	13.6940	4.5917	13.7591
.50	5.1298	13.2125	4.9355	13.2182	4.7490	13.2513	4.5698	13.3127
.52	5.1187	12.7959	4.9223	12.7972	4.7335	12.8260	4.5520	12.8834
.54	5.1124	12.3961	4.9137	12.3926	4.7225	12.4168	4.5385	12.4699
.56	5.1110	12.0119	4.9099	12.0034	4.7161	12.0226	4.5294	12.0710
.58	5.1147	11.6425	4.9110	11.6284	4.7145	11.6423	4.5249	11.6856
.60	5.1236	11.2870	4.9173	11.2670	4.7179	11.2750	4.5253	11.3126
.62	5.1378	10.9448	4.9289	10.9183	4.7265	10.9199	4.5307	10.9514
.64	5.1577	10.6152	4.9460	10.5818	4.7407	10.5764	4.5415	10.6011
.66	5.1833	10.2978	4.9689	10.2570	4.7605	10.2441	4.5580	10.2613
.68	5.2148	9.9921	4.9978	9.9434	4.7864	9.9223	4.5805	9.9313
.70	5.2524	9.6980	5.0329	9.6408	4.8186	9.6110	4.6093	9.6110
.72	5.2963	9.4151	5.0745	9.3490	4.8574	9.3098	4.6448	9.3001
.74	5.3466	9.1433	5.1227	9.0679	4.9031	9.0186	4.6874	8.9984
.76	5.4034	8.8824	5.1778	8.7972	4.9560	8.7374	4.7375	8.7058
.78	5.4668	8.6324	5.2399	8.5372	5.0164	8.4662	4.7955	8.4225
.80	5.5369	8.3932	5.3092	8.2876	5.0844	8.2051	4.8618	8.1487
.82	5.6136	8.1647	5.3857	8.0487	5.1604	7.9543	4.9366	7.8846
.84	5.6971	7.9469	5.4695	7.8204	5.2443	7.7140	5.0203	7.6306
.86	5.7870	7.7399	5.5606	7.6030	5.3364	7.4845	5.1131	7.3871
.88	5.8834	7.5434	5.6588	7.3963	5.4365	7.2659	5.2150	7.1544
.90	5.9860	7.3576	5.7640	7.2006	5.5445	7.0585	5.3261	6.9330
.92	6.0945	7.1822	5.8760	7.0158	5.6603	6.8624	5.4461	6.7234
.94	6.2087	7.0170	5.9943	6.8418	5.7834	6.6778	5.5748	6.5259
.96	6.3282	6.8620	6.1186	6.6785	5.9135	6.5046	5.7116	6.3406
.98	6.4525	6.7169	6.2485	6.5258	6.0500	6.3428	5.8560	6.1677
1.00	6.5813	6.5813	6.3835	6.3835	6.1923	6.1923	6.0072	6.0072

Table D-1: $0.346 \le \hat{\rho}_4 \le 0.382$ and λ_3, $\lambda_4 > 0$

$\hat{\rho}_3$	$\hat{\rho}_4 = 0.346$		$\hat{\rho}_4 = 0.358$		$\hat{\rho}_4 = 0.370$		$\hat{\rho}_4 = 0.382$	
	λ_3	λ_4	λ_3	λ_4	λ_3	λ_4	λ_3	λ_4
.04	5.7828	55.5853	5.6137	66.5241				
.06	5.6882	41.6700	5.5199	43.2973	5.3587	45.4912	5.2042	48.6694
.08	5.5964	36.2068	5.4288	37.0896	5.2684	38.1788	5.1147	39.5434
.10	5.5073	32.7298	5.3404	33.3372	5.1807	34.0678	5.0278	34.9489
.12	5.4211	30.1677	5.2549	30.6332	5.0958	31.1878	4.9435	31.8466
.14	5.3379	28.1348	5.1722	28.5147	5.0137	28.9659	4.8620	29.4982
.16	5.2577	26.4478	5.0924	26.7708	4.9344	27.1547	4.7832	27.6064
.18	5.1805	25.0046	5.0156	25.2878	4.8580	25.6249	4.7072	26.0212
.20	5.1064	23.7430	4.9418	23.9969	4.7844	24.3000	4.6340	24.6565
.22	5.0354	22.6217	4.8709	22.8536	4.7138	23.1311	4.5636	23.4578
.24	4.9676	21.6122	4.8031	21.8270	4.6460	22.0848	4.4959	22.3888
.26	4.9029	20.6940	4.7383	20.8953	4.5812	21.1377	4.4310	21.4239
.28	4.8414	19.8517	4.6765	20.0421	4.5192	20.2722	4.3689	20.5443
.30	4.7830	19.0734	4.6178	19.2549	4.4601	19.4750	4.3096	19.7359
.32	4.7279	18.3498	4.5622	18.5240	4.4040	18.7360	4.2530	18.9878
.34	4.6760	17.6735	4.5096	17.8415	4.3508	18.0468	4.1992	18.2912
.36	4.6274	17.0385	4.4601	17.2012	4.3004	17.4009	4.1481	17.6392
.38	4.5820	16.4397	4.4137	16.5979	4.2530	16.7928	4.0998	17.0261
.40	4.5400	15.8730	4.3704	16.0271	4.2086	16.2179	4.0542	16.4470
.42	4.5014	15.3348	4.3304	15.4853	4.1672	15.6725	4.0114	15.8981
.44	4.4663	14.8222	4.2936	14.9692	4.1287	15.1533	3.9715	15.3758
.46	4.4347	14.3325	4.2601	14.4763	4.0934	14.6574	3.9344	14.8773
.48	4.4067	13.8635	4.2300	14.0041	4.0612	14.1824	3.9002	14.4000
.50	4.3825	13.4133	4.2034	13.5506	4.0323	13.7261	3.8690	13.9416
.52	4.3621	12.9801	4.1804	13.1139	4.0067	13.2867	3.8409	13.5001
.54	4.3457	12.5624	4.1612	12.6926	3.9846	12.8624	3.8159	13.0738
.56	4.3336	12.1588	4.1458	12.2851	3.9661	12.4518	3.7943	12.6611
.58	4.3258	11.7683	4.1346	11.8902	3.9514	12.0535	3.7761	12.2605
.60	4.3226	11.3897	4.1277	11.5067	3.9407	11.6662	3.7616	11.8706
.62	4.3243	11.0222	4.1255	11.1337	3.9343	11.2888	3.7510	11.4903
.64	4.3312	10.6649	4.1281	10.7703	3.9325	10.9204	3.7445	11.1183
.66	4.3436	10.3173	4.1360	10.4156	3.9355	10.5599	3.7426	10.7538
.68	4.3618	9.9787	4.1496	10.0691	3.9440	10.2067	3.7456	10.3955
.70	4.3864	9.6488	4.1693	9.7301	3.9583	9.8598	3.7540	10.0427
.72	4.4177	9.3271	4.1956	9.3982	3.9789	9.5188	3.7683	9.6944
.74	4.4562	9.0136	4.2292	9.0731	4.0066	9.1831	3.7894	9.3500
.76	4.5025	8.7082	4.2706	8.7547	4.0422	8.8523	3.8179	9.0086
.78	4.5570	8.4109	4.3207	8.4428	4.0864	8.5262	3.8549	8.6698
.80	4.6205	8.1219	4.3801	8.1377	4.1404	8.2047	3.9017	8.3331
.82	4.6933	7.8416	4.4497	7.8397	4.2051	7.8881	3.9595	7.9983
.84	4.7760	7.5704	4.5302	7.5493	4.2818	7.5767	4.0300	7.6655
.86	4.8690	7.3091	4.6225	7.2673	4.3717	7.2714	4.1150	7.3351
.88	4.9726	7.0581	4.7272	6.9946	4.4761	6.9733	4.2167	7.0081
.90	5.0871	6.8185	4.8448	6.7325	4.5960	6.6839	4.3370	6.6861
.92	5.2122	6.5907	4.9755	6.4822	4.7323	6.4052	4.4780	6.3717
.94	5.3479	6.3757	5.1193	6.2450	4.8854	6.1392	4.6411	6.0681
.96	5.4936	6.1740	5.2758	6.0220	5.0550	5.8883	4.8268	5.7791
.98	5.6486	5.9859	5.4440	5.8144	5.2402	5.6544	5.0343	5.5090
1.00	5.8118	5.8118	5.6227	5.6227	5.4393	5.4393	5.2613	5.2613

Table D–1: $0.394 \leq \hat{\rho}_4 \leq 0.430$ and λ_3, $\lambda_4 > 0$

$\hat{\rho}_3$	$\hat{\rho}_4 = 0.394$		$\hat{\rho}_4 = 0.406$		$\hat{\rho}_4 = 0.418$		$\hat{\rho}_4 = 0.430$	
	λ_3	λ_4	λ_3	λ_4	λ_3	λ_4	λ_3	λ_4
.06	5.0560	54.0416	4.9138	70.4214				
.08	4.9673	41.2989	4.8259	43.6607	4.6901	47.1054	4.5596	53.1079
.10	4.8812	36.0202	4.7405	37.3425	4.6056	39.0148	4.4760	41.2144
.12	4.7976	32.6299	4.6577	33.5666	4.5235	34.6979	4.3947	36.0876
.14	4.7167	30.1244	4.5775	30.8616	4.4440	31.7329	4.3159	32.7713
.16	4.6385	28.1347	4.4998	28.7512	4.3669	29.4712	4.2394	30.3151
.18	4.5629	26.4834	4.4248	27.0200	4.2924	27.6419	4.1655	28.3636
.20	4.4901	25.0715	4.3523	25.5519	4.2204	26.1060	4.0939	26.7445
.22	4.4199	23.8380	4.2824	24.2772	4.1508	24.7821	4.0247	25.3613
.24	4.3524	22.7426	4.2151	23.1507	4.0837	23.6189	3.9579	24.1541
.26	4.2875	21.7571	4.1503	22.1414	4.0191	22.5814	3.8934	23.0833
.28	4.2253	20.8615	4.0881	21.2270	3.9568	21.6452	3.8313	22.1212
.30	4.1658	20.0402	4.0283	20.3910	3.8970	20.7920	3.7714	21.2479
.32	4.1088	19.2818	3.9711	19.6208	3.8395	20.0083	3.7137	20.4483
.34	4.0544	18.5769	3.9163	18.9066	3.7843	19.2833	3.6582	19.7108
.36	4.0027	17.9182	3.8639	18.2404	3.7314	18.6086	3.6049	19.0264
.38	3.9535	17.2997	3.8139	17.6159	3.6808	17.9775	3.5537	18.3877
.40	3.9069	16.7163	3.7664	17.0279	3.6324	17.3844	3.5046	17.7888
.42	3.8629	16.1638	3.7213	16.4718	3.5862	16.8244	3.4575	17.2246
.44	3.8215	15.6387	3.6785	15.9439	3.5423	16.2938	3.4126	16.6911
.46	3.7827	15.1379	3.6382	15.4411	3.5006	15.7892	3.3696	16.1847
.48	3.7467	14.6587	3.6004	14.9605	3.4612	15.3076	3.3287	15.7024
.50	3.7133	14.1988	3.5650	14.4998	3.4240	14.8465	3.2899	15.2414
.52	3.6827	13.7561	3.5322	14.0566	3.3890	14.4036	3.2531	14.7994
.54	3.6551	13.3287	3.5019	13.6291	3.3564	13.9769	3.2183	14.3744
.56	3.6304	12.9150	3.4743	13.2155	3.3261	13.5646	3.1856	13.9644
.58	3.6088	12.5133	3.4495	12.8142	3.2983	13.1650	3.1550	13.5678
.60	3.5905	12.1224	3.4276	12.4237	3.2730	12.7765	3.1267	13.1829
.62	3.5757	11.7407	3.4088	12.0425	3.2503	12.3977	3.1005	12.8083
.64	3.5646	11.3672	3.3932	11.6694	3.2305	12.0272	3.0768	12.4425
.66	3.5576	11.0005	3.3811	11.3030	3.2136	11.6636	3.0556	12.0841
.68	3.5550	10.6395	3.3729	10.9421	3.2000	11.3056	3.0370	11.7319
.70	3.5572	10.2832	3.3688	10.5853	3.1899	10.9518	3.0213	11.3843
.72	3.5648	9.9304	3.3695	10.2315	3.1838	10.6007	3.0087	11.0400
.74	3.5784	9.5801	3.3754	9.8792	3.1820	10.2510	2.9996	10.6974
.76	3.5990	9.2313	3.3874	9.5271	3.1852	9.9010	2.9945	10.3550
.78	3.6275	8.8829	3.4064	9.1739	3.1942	9.5490	2.9938	10.0108
.80	3.6653	8.5339	3.4336	8.8178	3.2101	9.1929	2.9985	9.6627
.82	3.7138	8.1836	3.4708	8.4574	3.2344	8.8306	3.0096	9.3083
.84	3.7752	7.8311	3.5200	8.0908	3.2689	8.4593	3.0285	8.9442
.86	3.8519	7.4762	3.5840	7.7162	3.3164	8.0756	3.0574	8.5666
.88	3.9469	7.1189	3.6668	7.3319	3.3809	7.6757	3.0995	8.1700
.90	4.0637	6.7602	3.7735	6.9368	3.4683	7.2547	3.1599	7.7470
.92	4.2064	6.4023	3.9108	6.5305	3.5878	6.8069	3.2470	7.2863
.94	4.3787	6.0492	4.0872	6.1153	3.7536	6.3264	3.3766	6.7703
.96	4.5834	5.7072	4.3115	5.6983	3.9875	5.8108	3.5817	6.1696
.98	4.8210	5.3841	4.5899	5.2931	4.3172	5.2721	3.9399	5.4389
1.00	5.0881	5.0881	4.9194	4.9194	4.7547	4.7548	4.5939	4.5939

Table D–1: $0.442 \leq \hat{\rho}_4 \leq 0.478$ and $\lambda_3,\ \lambda_4 > 0$

$\hat{\rho}_3$	$\hat{\rho}_4 = 0.442$		$\hat{\rho}_4 = 0.454$		$\hat{\rho}_4 = 0.466$		$\hat{\rho}_4 = 0.478$	
	λ_3	λ_4	λ_3	λ_4	λ_3	λ_4	λ_3	λ_4
.08	4.4343	79.8432						
.10	4.3515	44.3122	4.2317	49.3212	4.1164	61.9369		
.12	4.2710	37.8406	4.1520	40.1475	4.0376	43.4154	3.9274	48.8048
.14	4.1928	34.0258	4.0746	35.5738	3.9610	37.5463	3.8517	40.1963
.16	4.1171	31.3119	3.9996	32.5038	3.8867	33.9555	3.7781	35.7737
.18	4.0437	29.2044	3.9268	30.1916	3.8145	31.3642	3.7066	32.7818
.20	3.9726	27.4819	3.8563	28.3375	3.7446	29.3378	3.6372	30.5218
.22	3.9039	26.0260	3.7880	26.7907	3.6767	27.6755	3.5699	28.7079
.24	3.8374	24.7655	3.7218	25.4647	3.6110	26.2674	3.5047	27.1948
.26	3.7731	23.6545	3.6579	24.3049	3.5474	25.0473	3.4414	25.8985
.28	3.7111	22.6617	3.5960	23.2749	3.4858	23.9716	3.3801	24.7659
.30	3.6512	21.7644	3.5362	22.3488	3.4261	23.0104	3.3206	23.7613
.32	3.5934	20.9460	3.4784	21.5080	3.3683	22.1424	3.2630	22.8597
.34	3.5377	20.1939	3.4226	20.7383	3.3125	21.3514	3.2071	22.0426
.36	3.4841	19.4980	3.3687	20.0287	3.2584	20.6252	3.1530	21.2961
.38	3.4324	18.8504	3.3166	19.3706	3.2061	19.9544	3.1006	20.6095
.40	3.3827	18.2448	3.2665	18.7571	3.1556	19.3311	3.0498	19.9742
.42	3.3349	17.6759	3.2181	18.1823	3.1067	18.7493	3.0006	19.3835
.44	3.2890	17.1391	3.1714	17.6417	3.0595	18.2038	2.9529	18.8318
.46	3.2450	16.6308	3.1265	17.1311	3.0138	17.6903	2.9067	18.3144
.48	3.2029	16.1477	3.0833	16.6473	2.9697	17.2053	2.8619	17.8275
.50	3.1626	15.6872	3.0418	16.1872	2.9272	16.7457	2.8185	17.3677
.52	3.1241	15.2467	3.0019	15.7484	2.8861	16.3087	2.7765	16.9323
.54	3.0874	14.8240	2.9636	15.3287	2.8465	15.8921	2.7358	16.5188
.56	3.0526	14.4173	2.9269	14.9260	2.8083	15.4937	2.6964	16.1250
.58	3.0196	14.0248	2.8919	14.5385	2.7716	15.1119	2.6583	15.7492
.60	2.9885	13.6449	2.8585	14.1646	2.7362	14.7449	2.6213	15.3896
.62	2.9594	13.2760	2.8267	13.8029	2.7022	14.3913	2.5856	15.0447
.64	2.9322	12.9168	2.7965	13.4519	2.6696	14.0497	2.5511	14.7133
.66	2.9070	12.5659	2.7681	13.1103	2.6384	13.7188	2.5177	14.3940
.68	2.8840	12.2220	2.7413	12.7768	2.6086	13.3975	2.4854	14.0858
.70	2.8633	11.8837	2.7164	12.4503	2.5802	13.0845	2.4543	13.7876
.72	2.8451	11.5497	2.6933	12.1294	2.5532	12.7789	2.4243	13.4984
.74	2.8295	11.2185	2.6722	11.8130	2.5277	12.4794	2.3954	13.2173
.76	2.8169	10.8886	2.6533	11.4995	2.5037	12.1851	2.3676	12.9434
.78	2.8076	10.5583	2.6367	11.1877	2.4814	11.8947	2.3410	12.6758
.80	2.8021	10.2256	2.6227	10.8759	2.4608	11.6070	2.3155	12.4136
.82	2.8011	9.8881	2.6116	10.5622	2.4420	11.3208	2.2913	12.1560
.84	2.8055	9.5430	2.6040	10.2444	2.4253	11.0344	2.2682	11.9019
.86	2.8166	9.1865	2.6004	9.9197	2.4109	10.7462	2.2465	11.6505
.88	2.8364	8.8137	2.6019	9.5845	2.3992	10.4541	2.2261	11.4005
.90	2.8681	8.4171	2.6098	9.2338	2.3907	10.1552	2.2073	11.1507
.92	2.9167	7.9856	2.6263	8.8605	2.3860	9.8460	2.1901	10.8996
.94	2.9923	7.4999	2.6553	8.4527	2.3862	9.5215	2.1748	10.6453
.96	3.1170	6.9211	2.7042	7.9896	2.3932	9.1740	2.1618	10.3856
.98	3.3615	6.1418	2.7910	7.4240	2.4100	8.7907	2.1514	10.1172
1.00	3.3635	5.2636	2.9874	6.5859	2.4430	8.3468	2.1444	9.8357

Table D–1: $0.490 \leq \hat{\rho}_4 \leq 0.526$ and $\lambda_3,\ \lambda_4 > 0$

$\hat{\rho}_3$	$\hat{\rho}_4 = 0.490$		$\hat{\rho}_4 = 0.502$		$\hat{\rho}_4 = 0.514$		$\hat{\rho}_4 = 0.526$	
	λ_3	λ_4	λ_3	λ_4	λ_3	λ_4	λ_3	λ_4
.12	3.8214	64.2211						
.14	3.7464	44.1155	3.6450	51.3837				
.16	3.6736	38.1536	3.5729	41.5147	3.4759	47.0716	3.3824	63.7499
.18	3.6028	34.5411	3.5029	36.8151	3.4066	39.9625	3.3138	44.9480
.20	3.5341	31.9479	3.4348	33.7114	3.3392	35.9819	3.2471	39.1087
.22	3.4673	29.9280	3.3686	31.3973	3.2737	33.2157	3.1823	35.5637
.24	3.4025	28.2765	3.3044	29.5562	3.2100	31.1011	3.1191	33.0233
.26	3.3397	26.8819	3.2420	28.0307	3.1481	29.3943	3.0578	31.0502
.28	3.2787	25.6769	3.1814	26.7312	3.0879	27.9671	2.9981	29.4427
.30	3.2195	24.6177	3.1225	25.6014	3.0294	26.7438	2.9400	28.0907
.32	3.1621	23.6740	3.0653	24.6041	2.9725	25.6761	2.8835	26.9280
.34	3.1063	22.8242	3.0098	23.7128	2.9172	24.7310	2.8285	25.9108
.36	3.0522	22.0523	2.9558	22.9087	2.8634	23.8850	2.7749	25.0094
.38	2.9997	21.3460	2.9033	22.1771	2.8111	23.1211	2.7228	24.2024
.40	2.9488	20.6956	2.8523	21.5074	2.7602	22.4260	2.6720	23.4739
.42	2.8993	20.0935	2.8028	20.8906	2.7106	21.7898	2.6225	22.8117
.44	2.8513	19.5336	2.7546	20.3198	2.6623	21.2044	2.5743	22.2064
.46	2.8047	19.0108	2.7077	19.7894	2.6153	20.6633	2.5272	21.6505
.48	2.7594	18.5208	2.6620	19.2946	2.5694	20.1614	2.4813	21.1380
.50	2.7154	18.0601	2.6176	18.8317	2.5248	19.6943	2.4365	20.6639
.52	2.6727	17.6256	2.5744	18.3972	2.4812	19.2582	2.3927	20.2241
.54	2.6312	17.2149	2.5323	17.9885	2.4386	18.8503	2.3500	19.8152
.56	2.5908	16.8256	2.4912	17.6030	2.3971	18.4679	2.3081	19.4343
.58	2.5516	16.4557	2.4512	17.2389	2.3566	18.1087	2.2673	19.0789
.60	2.5135	16.1037	2.4122	16.8942	2.3169	17.7708	2.2273	18.7470
.62	2.4764	15.7679	2.3741	16.5674	2.2782	17.4526	2.1881	18.4367
.64	2.4404	15.4470	2.3370	16.2571	2.2403	17.1527	2.1497	18.1466
.66	2.4053	15.1398	2.3007	15.9621	2.2032	16.8697	2.1121	17.8752
.68	2.3712	14.8452	2.2653	15.6814	2.1668	16.6026	2.0751	17.6215
.70	2.3380	14.5623	2.2306	15.4139	2.1312	16.3505	2.0389	17.3843
.72	2.3058	14.2902	2.1968	15.1589	2.0963	16.1124	2.0033	17.1629
.74	2.2744	14.0280	2.1637	14.9156	2.0620	15.8877	1.9684	16.9565
.76	2.2439	13.7749	2.1313	14.6833	2.0284	15.6757	1.9340	16.7644
.78	2.2142	13.5304	2.0995	14.4614	1.9953	15.4758	1.9002	16.5860
.80	2.1854	13.2937	2.0684	14.2494	1.9628	15.2877	1.8668	16.4209
.82	2.1573	13.0642	2.0379	14.0469	1.9308	15.1108	1.8340	16.2687
.84	2.1301	12.8414	2.0080	13.8534	1.8993	14.9448	1.8017	16.1290
.86	2.1036	12.6246	1.9787	13.6685	1.8683	14.7895	1.7697	16.0016
.88	2.0779	12.4133	1.9498	13.4920	1.8377	14.6446	1.7382	15.8861
.90	2.0530	12.2069	1.9215	13.3237	1.8075	14.5100	1.7071	15.7825
.92	2.0289	12.0049	1.8936	13.1631	1.7776	14.3855	1.6763	15.6907
.94	2.0055	11.8067	1.8661	13.0103	1.7481	14.2709	1.6459	15.6104
.96	1.9830	11.6118	1.8390	12.8651	1.7189	14.1664	1.6158	15.5416
.98	1.9613	11.4193	1.8123	12.7274	1.6900	14.0720	1.5859	15.4844
1.00	1.9405	11.2286	1.7859	12.5972	1.6613	13.9875	1.5564	15.4387

Table D–1: $0.538 \leq \hat{\rho}_4 \leq 0.574$ and $\lambda_3,\ \lambda_4 > 0$

$\hat{\rho}_3$	$\hat{\rho}_4 = 0.538$ λ_3	λ_4	$\hat{\rho}_4 = 0.550$ λ_3	λ_4	$\hat{\rho}_4 = 0.562$ λ_3	λ_4	$\hat{\rho}_4 = 0.574$ λ_3	λ_4
.18	3.2243	56.8961						
.20	3.1584	44.0215	3.0726	55.4584				
.22	3.0942	38.8200	3.0091	44.0299	2.9270	57.3016		
.24	3.0316	35.5298	2.9472	39.0740	2.8658	45.0216	2.7871	66.3575
.26	2.9708	33.1312	2.8870	35.8939	2.8062	39.9409	2.7282	47.4182
.28	2.9116	31.2523	2.8284	33.5636	2.7481	36.7209	2.6707	41.6362
.30	2.8540	29.7140	2.7713	31.7340	2.6915	34.3782	2.6147	38.1597
.32	2.7979	28.4169	2.7156	30.2357	2.6364	32.5487	2.5600	35.6918
.34	2.7432	27.2999	2.6614	28.9730	2.5826	31.0571	2.5067	33.7943
.36	2.6900	26.3226	2.6085	27.8870	2.5301	29.8056	2.4547	32.2646
.38	2.6381	25.4570	2.5569	26.9387	2.4790	28.7337	2.4040	30.9928
.40	2.5876	24.6829	2.5067	26.1006	2.4290	27.8016	2.3544	29.9124
.42	2.5382	23.9853	2.4576	25.3533	2.3802	26.9817	2.3060	28.9800
.44	2.4901	23.3526	2.4097	24.6820	2.3326	26.2541	2.2587	28.1658
.46	2.4432	22.7759	2.3629	24.0754	2.2860	25.6038	2.2124	27.4486
.48	2.3973	22.2479	2.3171	23.5248	2.2405	25.0195	2.1672	26.8124
.50	2.3525	21.7629	2.2724	23.0231	2.1960	24.4923	2.1229	26.2455
.52	2.3087	21.3161	2.2287	22.5647	2.1524	24.0151	2.0796	25.7384
.54	2.2658	20.9036	2.1859	22.1449	2.1098	23.5824	2.0372	25.2838
.56	2.2239	20.5221	2.1440	21.7599	2.0680	23.1894	1.9956	24.8758
.58	2.1829	20.1690	2.1029	21.4065	2.0271	22.8323	1.9549	24.5097
.60	2.1427	19.8417	2.0627	21.0820	1.9869	22.5080	1.9149	24.1814
.62	2.1033	19.5384	2.0233	20.7842	1.9476	22.2137	1.8758	23.8877
.64	2.0646	19.2573	1.9846	20.5111	1.9089	21.9472	1.8373	23.6259
.66	2.0267	18.9969	1.9466	20.2610	1.8710	21.7066	1.7995	23.3937
.68	1.9895	18.7561	1.9093	20.0326	1.8338	21.4902	1.7625	23.1892
.70	1.9530	18.5336	1.8726	19.8246	1.7972	21.2967	1.7260	23.0107
.72	1.9170	18.3286	1.8366	19.6360	1.7612	21.1249	1.6902	22.8568
.74	1.8817	18.1402	1.8011	19.4659	1.7258	20.9737	1.6550	22.7265
.76	1.8469	17.9678	1.7662	19.3134	1.6910	20.8423	1.6204	22.6186
.78	1.8127	17.8107	1.7319	19.1780	1.6567	20.7300	1.5863	22.5323
.80	1.7790	17.6684	1.6981	19.0590	1.6230	20.6360	1.5528	22.4669
.82	1.7458	17.5405	1.6648	18.9560	1.5897	20.5598	1.5198	22.4219
.84	1.7130	17.4266	1.6319	18.8685	1.5570	20.5011	1.4873	22.3968
.86	1.6807	17.3264	1.5995	18.7963	1.5248	20.4593	1.4553	22.3910
.88	1.6488	17.2396	1.5676	18.7389	1.4930	20.4342	1.4237	22.4044
.90	1.6174	17.1660	1.5361	18.6961	1.4616	20.4254	1.3927	22.4367
.92	1.5862	17.1055	1.5050	18.6678	1.4307	20.4329	1.3621	22.4877
.94	1.5555	17.0579	1.4743	18.6537	1.4002	20.4563	1.3319	22.5574
.96	1.5251	17.0230	1.4440	18.6538	1.3702	20.4957	1.3022	22.6456
.98	1.4951	17.0009	1.4140	18.6678	1.3405	20.5508	1.2729	22.7525
1.00	1.4654	16.9915	1.3844	18.6959	1.3112	20.6218	1.2440	22.8781

Table D–1: $0.586 \leq \hat{\rho}_4 \leq 0.622$ and λ_3, $\lambda_4 > 0$

	$\hat{\rho}_4 = 0.586$		$\hat{\rho}_4 = 0.598$		$\hat{\rho}_4 = 0.610$		$\hat{\rho}_4 = 0.622$	
$\hat{\rho}_3$	λ_3	λ_4	λ_3	λ_4	λ_3	λ_4	λ_3	λ_4
.28	2.5959	52.8597						
.30	2.5405	44.7330						
.32	2.4864	40.5450	2.4152	51.3449				
.34	2.4336	37.7435	2.3629	44.8007				
.36	2.3821	35.6598	2.3119	41.1053	2.2441	55.7224		
.38	2.3318	34.0174	2.2621	38.5612	2.1948	47.7863		
.40	2.2826	32.6749	2.2134	36.6469	2.1466	43.7192		
.42	2.2346	31.5503	2.1658	35.1318	2.0995	41.0277	2.0353	60.2810
.44	2.1877	30.5915	2.1193	33.8936	2.0534	39.0527	1.9897	51.3813
.46	2.1418	29.7640	2.0738	32.8599	2.0083	37.5208	1.9451	47.1513
.48	2.0969	29.0430	2.0293	31.9837	1.9642	36.2915	1.9014	44.4470
.50	2.0529	28.4108	1.9857	31.2338	1.9211	35.2839	1.8587	42.5175
.52	2.0099	27.8539	1.9431	30.5873	1.8788	34.4465	1.8168	41.0617
.54	1.9678	27.3622	1.9013	30.0279	1.8374	33.7450	1.7758	39.9293
.56	1.9265	26.9273	1.8603	29.5432	1.7968	33.1555	1.7356	39.0347
.58	1.8860	26.5431	1.8202	29.1237	1.7570	32.6604	1.6962	38.3246
.60	1.8464	26.2042	1.7808	28.7618	1.7180	32.2471	1.6576	37.7641
.62	1.8074	25.9065	1.7422	28.4516	1.6797	31.9055	1.6197	37.3292
.64	1.7692	25.6465	1.7043	28.1884	1.6422	31.6282	1.5825	37.0031
.66	1.7317	25.4214	1.6671	27.9684	1.6053	31.4094	1.5460	36.7740
.68	1.6949	25.2288	1.6306	27.7884	1.5691	31.2445	1.5102	36.6336
.70	1.6587	25.0667	1.5947	27.6460	1.5336	31.1301	1.4750	36.5762
.72	1.6232	24.9334	1.5595	27.5390	1.4987	31.0636	1.4404	36.5985
.74	1.5882	24.8276	1.5248	27.4659	1.4644	31.0431	1.4065	36.6989
.76	1.5538	24.7482	1.4908	27.4254	1.4307	31.0673	1.3731	36.8779
.78	1.5201	24.6941	1.4573	27.4166	1.3975	31.1355	1.3403	37.1375
.80	1.4868	24.6647	1.4244	27.4386	1.3650	31.2473	1.3081	37.4817
.82	1.4541	24.6594	1.3920	27.4910	1.3329	31.4032	1.2764	37.9170
.84	1.4219	24.6776	1.3601	27.5736	1.3014	31.6040	1.2452	38.4527
.86	1.3902	24.7192	1.3288	27.6864	1.2704	31.8510	1.2146	39.1019
.88	1.3590	24.7837	1.2979	27.8295	1.2399	32.1462	1.1844	39.8834
.90	1.3283	24.8712	1.2676	28.0034	1.2099	32.4923	1.1547	40.8236
.92	1.2980	24.9818	1.2376	28.2088	1.1803	32.8928	1.1255	41.9624
.94	1.2682	25.1154	1.2082	28.4466	1.1513	33.3524	1.0967	43.3598
.96	1.2388	25.2724	1.1792	28.7179	1.1226	33.8767	1.0684	45.1157
.98	1.2099	25.4531	1.1507	29.0242	1.0944	34.4737	1.0405	47.4101
1.00	1.1814	25.6581	1.1225	29.3676	1.0666	35.1528	1.0131	50.6210

Table D–2: $0.0965 \leq \hat{\rho}_4 \leq 0.2500$ and λ_3, $\lambda_4 < 0$

	$\hat{\rho}_4 = 0.0965$		$\hat{\rho}_4 = 0.1010$		$\hat{\rho}_4 = 0.1055$		$\hat{\rho}_4 = 0.1100$	
$\hat{\rho}_3$	λ_3	λ_4	λ_3	λ_4	λ_3	λ_4	λ_3	λ_4
.02	14.9985	.007375	14.5980	.007678	14.2241	.007968	13.8734	.008246
.04	19.4018	.009919	18.4847	.010926	17.6989	.011868	17.0114	.012758

	$\hat{\rho}_4 = 0.115$		$\hat{\rho}_4 = 0.120$		$\hat{\rho}_4 = 0.125$		$\hat{\rho}_4 = 0.130$	
$\hat{\rho}_3$	λ_3	λ_4	λ_3	λ_4	λ_3	λ_4	λ_3	λ_4
.02	13.5076	.008540	13.1639	.008819	12.8396	.009083	12.5326	.009331
.04	16.3361	.013694	15.7342	.014583	15.1912	.054308	14.6963	.016241
.06	24.1625	.009595	21.8283	.012275	20.2013	.014576	18.9505	.016633

	$\hat{\rho}_4 = 0.136$		$\hat{\rho}_4 = 0.142$		$\hat{\rho}_4 = 0.148$		$\hat{\rho}_4 = 0.154$	
$\hat{\rho}_3$	λ_3	λ_4	λ_3	λ_4	λ_3	λ_4	λ_3	λ_4
.02	12.1846	.009607	11.8561	.009859	11.5451	.010086	11.2497	.010285
.04	14.1549	.017169	13.6610	.018052	13.2067	.018891	12.7860	.019690
.06	17.7511	.001888	16.7680	.020937	15.9343	.022854	15.2103	.024654
.08			28.4134	.008739	23.2586	.015052	20.6849	.019749

	$\hat{\rho}_4 = 0.161$		$\hat{\rho}_4 = 0.168$		$\hat{\rho}_4 = 0.175$		$\hat{\rho}_4 = 0.182$	
$\hat{\rho}_3$	λ_3	λ_4	λ_3	λ_4	λ_3	λ_4	λ_3	λ_4
.02	10.9227	.010480	10.6126	.010632	10.3177	.010738	10.0365	.010794
.04	12.3312	.020570	11.9095	.021395	11.5163	.022164	11.1478	.022874
.06	14.4700	.026630	13.8166	.028490	13.2313	.030247	12.7009	.031908
.08	18.7108	.024323	17.2687	.028317	16.1306	.031920	15.1896	.035233
.10					25.5898	.015803	21.2151	.025055

	$\hat{\rho}_4 = 0.190$		$\hat{\rho}_4 = 0.198$		$\hat{\rho}_4 = 0.206$		$\hat{\rho}_4 = 0.214$	
$\hat{\rho}_3$	λ_3	λ_4	λ_3	λ_4	λ_3	λ_4	λ_3	λ_4
.02	9.73016	.010791	9.43839	.010709	9.15975	.010541	8.89307	.010278
.04	10.7530	.023611	10.3826	.024262	10.0336	.024820	9.70358	.025278
.06	12.1496	.033696	11.6464	.035368	11.1834	.036925	10.7542	.038363
.08	14.2809	.038738	13.5007	.041997	12.8164	.045043	12.2064	.047898
.10	18.5972	.033008	16.8527	.039657	15.5401	.045513	14.4858	.050812
.12					23.8252	.024628	19.6285	.038262

	$\hat{\rho}_4 = 0.223$		$\hat{\rho}_4 = 0.232$		$\hat{\rho}_4 = 0.241$		$\hat{\rho}_4 = 0.250$	
$\hat{\rho}_3$	λ_3	λ_4	λ_3	λ_4	λ_3	λ_4	λ_3	λ_4
.02	8.60607	.009858	8.33168	.009294	8.06884	.008571	7.81665	.007675
.04	9.35238	.025663	9.02009	.025896	8.70467	.025960	8.40445	.025838
.06	10.3058	.039833	9.88868	.041134	9.49843	.042254	9.13161	.043174
.08	11.5903	.050895	11.0338	.053670	10.5260	.056222	10.0585	.058543
.10	13.5025	.056267	12.6691	.061285	11.9448	.065925	11.3035	.070222
.12	17.1282	.049722	15.4479	.059261	14.1772	.067624	13.1528	.075148
.14			25.7910	.025339	19.3559	.049624	16.6760	.065567

Table D–2: $0.260 \le \hat{\rho}_4 \le 0.382$ and λ_3, $\lambda_4 < 0$

	$\hat{\rho}_4 = 0.260$		$\hat{\rho}_4 = 0.270$		$\hat{\rho}_4 = 0.280$		$\hat{\rho}_4 = 0.290$	
$\hat{\rho}_3$	λ_3	λ_4	λ_3	λ_4	λ_3	λ_4	λ_3	λ_4
.02	7.54804	.006456	7.29078	.004982	7.04415	.003231	6.80757	.001182
.04	8.08698	.025462	7.78492	.024805	7.49694	.023839	7.22196	.022533
.06	8.74814	.043940	8.38704	.044405	8.04579	.044535	7.72234	.044292
.08	9.57880	.060834	9.13454	.062795	8.72054	.064394	8.33276	.065592
.10	10.6667	.074609	10.0934	.078590	9.57138	.082147	9.09181	.085253
.12	12.2052	.082738	11.3984	.089645	10.6946	.095933	10.0694	.101628
.14	14.7797	.079769	13.4016	.091871	12.3147	.102544	11.4147	.112106
.16	23.7047	.038842	17.7305	.072665	15.1968	.094167	13.5468	.111299
.18							19.0640	.077452

	$\hat{\rho}_4 = 0.301$		$\hat{\rho}_4 = 0.312$		$\hat{\rho}_4 = 0.323$		$\hat{\rho}_4 = 0.334$	
$\hat{\rho}_3$	λ_3	λ_4	λ_3	λ_4	λ_3	λ_4	λ_3	λ_4
.04	6.93341	.020667	6.65856	.018312	6.39667	.015432	6.14717	.011993
.06	7.38512	.043547	7.06565	.042247	6.76250	.040333	6.47458	.037750
.08	7.93267	.066390	7.55704	.066579	7.20317	.066085	6.86896	.064829
.10	8.60525	.088099	8.15503	.090282	7.73594	.091721	7.34393	.092322
.12	9.45282	.107200	8.89575	.112007	8.38716	.115984	7.91886	.119042
.14	10.5737	.121534	9.84557	.129907	9.20235	.137241	8.62528	.143499
.16	12.2057	.127238	11.1483	.141044	10.2720	.153145	9.52160	.163727
.18	15.2413	.113826	13.2200	.139106	11.8174	.159456	10.7345	.176652
.20			19.1473	.092960	14.6904	.144014	12.6286	.175609
.22							17.9296	.125390

	$\hat{\rho}_4 = 0.346$		$\hat{\rho}_4 = 0.358$		$\hat{\rho}_4 = 0.370$		$\hat{\rho}_4 = 0.382$	
$\hat{\rho}_3$	λ_3	λ_4	λ_3	λ_4	λ_3	λ_4	λ_3	λ_4
.04	5.88870	.007574	5.64421	.002441				
.06	6.17689	.034107	5.89558	.029549	5.63019	.024036	5.38050	.017556
.08	6.52489	.062495	6.20069	.059057	5.89523	.054424	5.60781	.048525
.10	6.94351	.091902	6.56849	.090222	6.21652	.087137	5.88596	.082509
.12	7.44679	.121201	7.00939	.121972	6.60215	.121170	6.22173	.118599
.14	8.05593	.149002	7.53770	.152975	7.06182	.155225	6.62186	.155525
.16	8.80771	.173598	8.17658	.181671	7.60992	.187810	7.09504	.191808
.18	9.77462	.192656	8.96861	.206155	8.27148	.217252	7.65554	.225873
.20	11.1288	.202121	10.0031	.223662	9.09496	.241468	8.32982	.255994
.22	13.4798	.190940	11.5146	.228712	10.1879	.257200	9.17124	.279957
.24			14.6527	.200377	11.8673	.256937	10.3043	.293986
.26					18.1400	.171123	12.1431	.287898

Table D–2: $0.394 \leq \hat{\rho}_4 \leq 0.478$ and $\lambda_3, \ \lambda_4 < 0$

$\hat{\rho}_3$	$\hat{\rho}_4 = 0.394$ λ_3	λ_4	$\hat{\rho}_4 = 0.406$ λ_3	λ_4	$\hat{\rho}_4 = 0.418$ λ_3	λ_4	$\hat{\rho}_4 = 0.430$ λ_3	λ_4
.06	5.14648	.010131	4.92823	.001824				
.08	5.33809	.041326	5.08603	.032838	4.85183	.023135	4.63580	.012363
.10	5.57574	.076223	5.28529	.068202	5.01456	.058438	4.76385	.047009
.12	5.86568	.114061	5.53240	.107377	5.22100	.098411	4.93132	.087108
.14	6.21303	.153620	5.83187	.149242	5.47599	.142127	5.14401	.132052
.16	6.62284	.193401	6.18666	.192274	5.78168	.188080	5.40450	.180453
.18	7.10256	.231829	6.59992	.234832	6.13864	.234516	5.71217	.230448
.20	7.66600	.267329	7.07776	.275346	6.54810	.279762	6.06521	.280169
.22	8.34012	.298201	7.63302	.312336	7.01474	.322364	6.46318	.328039
.24	9.18074	.322169	8.29169	.344205	7.54983	.361017	6.90924	.372811
.26	10.3230	.335029	9.10812	.368726	8.17661	.394307	7.41250	.413412
.28	12.2680	.323744	10.2158	.381544	8.94332	.420203	7.99185	.448685
.30			12.1391	.367957	9.96654	.434642	8.68558	.476942
.32							9.58155	.494863
.34							10.9695	.492530

$\hat{\rho}_3$	$\hat{\rho}_4 = 0.442$ λ_3	λ_4	$\hat{\rho}_4 = 0.454$ λ_3	λ_4	$\hat{\rho}_4 = 0.466$ λ_3	λ_4	$\hat{\rho}_4 = 0.478$ λ_3	λ_4
.08	4.43824	.000738						
.10	4.53373	.034109	4.32479	.020049	4.13740	.005237		
.12	4.66379	.073542	4.41934	.057960	4.19908	.040807	4.00382	.022699
.14	4.83556	.118887	4.55121	.102672	4.29242	.837115	4.06108	.062638
.16	5.05303	.169048	4.72655	.153615	4.42575	.134118	4.15282	.110907
.18	5.31586	.222142	4.94674	.209097	4.60353	.190882	4.28702	.167313
.20	5.62059	.276049	5.20817	.266777	4.82387	.251655	4.46571	.229992
.22	5.96339	.328927	5.50486	.324421	5.07996	.313740	4.68340	.295932
.24	6.34245	.379408	5.83149	.380354	5.36386	.374949	4.93044	.362224
.26	6.75945	.426506	6.18525	.433487	5.66938	.433878	5.19766	.426864
.28	7.22091	.469421	6.56646	.483123	5.99307	.489751	5.47855	.488778
.30	7.74053	.507297	6.97905	.528742	6.33418	.542156	5.76940	.547503
.32	8.34498	.538890	7.43162	.569804	6.69440	.590839	6.06870	.602891
.34	9.09082	.561863	7.94021	.605526	7.07815	.635550	6.37657	.654918
.36	10.1322	.570281	8.53574	.634516	7.49348	.675916	6.69455	.703581
.38			9.28746	.653763	7.95452	.711277	7.02565	.748823
.40					8.48787	.740354	7.37490	.790473
.42					9.15249	.760307	7.75048	.828159
.44					10.1394	.762243	8.16661	.861145
.46							8.65100	.887904
.48							9.27138	.904692
.50							10.3282	.895569

Table D–2: $0.490 \leq \hat{\rho}_4 \leq 0.526$ and λ_3, $\lambda_4 < 0$

$\hat{\rho}_3$	$\hat{\rho}_4 = 0.490$		$\hat{\rho}_4 = 0.502$		$\hat{\rho}_4 = 0.514$		$\hat{\rho}_4 = 0.526$	
	λ_3	λ_4	λ_3	λ_4	λ_3	λ_4	λ_3	λ_4
.12	3.83361	.004340						
.14	3.85879	.040387	3.68590	.018038				
.16	3.91090	.084877	3.70284	.057466	3.52951	.030338	3.38879	.004896
.18	4.00037	.138753	3.74867	.106479	3.53682	.072733	3.36622	.040089
.20	4.13450	.201343	3.83474	.166034	3.57456	.125918	3.36202	.084587
.22	4.31234	.269941	3.96750	.234895	3.65533	.190993	3.38885	.141080
.24	4.52468	.340901	4.14266	.309397	3.78501	.266158	3.46121	.211152
.26	4.75964	.411239	4.34743	.385286	3.95590	.346644	3.58550	.292636
.28	5.00749	.479237	4.56831	.459600	4.15174	.427520	3.75098	.379423
.30	5.26187	.544167	4.79551	.530952	4.35773	.505874	3.93761	.465623
.32	5.51912	.605858	5.02312	.598958	4.56439	.580663	4.12927	.548289
.34	5.77736	.664390	5.24781	.663709	4.76644	.651831	4.31680	.626716
.36	6.03580	.719936	5.46768	.725482	4.96105	.719710	4.49571	.701175
.38	6.29436	.772682	5.68162	.784604	5.14661	.784744	4.66378	.772231
.40	6.55346	.822790	5.88888	.841400	5.32207	.847385	4.81986	.840484
.42	6.81396	.870387	6.08885	.896180	5.48662	.908061	4.96325	.906494
.44	7.07720	.915544	6.28093	.949233	5.63951	.967171	5.09339	.970762
.46	7.34517	.958265	6.46445	1.00084	5.77993	1.02509	5.20978	1.03374
.48	7.62083	.998462	6.63856	1.05126	5.90697	1.08218	5.31188	1.09586
.50	7.90886	1.03590	6.80217	1.10079	6.01962	1.13880	5.39915	1.15749
.52	8.21722	1.07010	6.95386	1.14971	6.11670	1.19533	5.47099	1.21903
.54	8.56118	1.10001	7.09179	1.19838	6.19692	1.25214	5.52684	1.28084
.56	8.97677	1.12315	7.21358	1.24717	6.25888	1.30965	5.56612	1.34328
.58	9.60174	1.13013	7.31614	1.29658	6.30112	1.36830	5.58837	1.40673
.60			7.39559	1.34720	6.32221	1.42855	5.59323	1.47156
.62			7.44719	1.39977	6.32085	1.49091	5.58050	1.53812
.64			7.46549	1.45521	6.29605	1.55591	5.55025	1.60679
.66			7.44484	1.51460	6.24723	1.62409	5.50278	1.67791
.68			7.38038	1.57914	6.17441	1.69599	5.43873	1.75182
.70			7.26943	1.65003	6.07832	1.77213	5.35903	1.82882
.72			7.11278	1.72832	5.96044	1.85298	5.26495	1.90917
.74			6.91505	1.81476	5.82295	1.93892	5.15802	1.99309
.76			6.68388	1.90979	5.66859	2.03028	5.03997	2.08075
.78			6.42828	2.01356	5.50051	2.12729	4.91270	2.17222
.80			6.15717	2.12606	5.32202	2.23005	4.77817	2.26756
.82			5.87838	2.24719	5.13642	2.33861	4.63835	2.36672
.84	7.03552	2.13737	5.59829	2.37678	4.94681	2.45288	4.49510	2.46959
.86	6.38593	2.33102	5.32185	2.51463	4.75599	2.57273	4.35016	2.57600
.88	5.87348	2.52119	5.05272	2.66049	4.56637	2.69790	4.20510	2.68570
.90	5.43394	2.71637	4.79349	2.81407	4.37996	2.82808	4.06127	2.79842
.92	5.04513	2.91944	4.54588	2.97505	4.19834	2.96291	3.91982	2.91379
.94	4.69622	3.13158	4.31093	3.14307	4.02272	3.10194	3.78166	3.03143
.96	4.38057	3.35402	4.08912	3.31771	3.85396	3.24468	3.64755	3.15090
.98	4.09369	3.58721	3.88054	3.49852	3.69264	3.39060	3.51801	3.27174
1.00	3.83170	3.83170	3.68500	3.68500	3.53910	3.53910	3.39345	3.39345

Table D–2: $0.538 \leq \hat{\rho}_4 \leq 0.574$ and $\lambda_3, \; \lambda_4 < 0$

	$\hat{\rho}_4 = 0.538$		$\hat{\rho}_4 = 0.550$		$\hat{\rho}_4 = 0.562$		$\hat{\rho}_4 = 0.574$	
$\hat{\rho}_3$	λ_3	λ_4	λ_3	λ_4	λ_3	λ_4	λ_3	λ_4
.18	3.23328	.010462						
.20	3.19873	.045969	3.07799	.012406				
.22	3.18073	.091087	3.03099	.046730	2.92725	.010178		
.24	3.19250	.149191	2.99706	.090281	2.86847	.041649	2.78571	.003807
.26	3.25123	.223004	2.98855	.147025	2.81803	.080997	2.71751	.031020
.28	3.36657	.310720	3.02506	.221630	2.78508	.132063	2.65306	.063947
.30	3.52595	.404569	3.12429	.314715	2.78758	.201153	2.59712	.105215
.32	3.70446	.496937	3.27660	.417098	2.85390	.294908	2.55962	.159906
.34	3.88375	.584543	3.44978	.517498	2.99341	.407609	2.56418	.238368
.36	4.05484	.667015	3.62048	.611397	3.16634	.520313	2.65642	.353873
.38	4.21400	.744991	3.77918	.698752	3.33530	.623650	2.83262	.488804
.40	4.35978	.819296	3.92295	.780709	3.48781	.717768	3.01472	.610728
.42	4.49160	.890698	4.05123	.858453	3.62186	.804761	3.17262	.716684
.44	4.60923	.959861	4.16416	.932976	3.73812	.886475	3.30561	.811446
.46	4.71254	1.02735	4.26205	1.00508	3.83767	.964291	3.41688	.898608
.48	4.80140	1.09367	4.34525	1.07540	3.92156	1.03923	3.50930	.980461
.50	4.87570	1.15925	4.41408	1.14448	3.99066	1.11206	3.58499	1.05851
.52	4.93533	1.22448	4.46882	1.21277	4.04573	1.18339	3.64558	1.13378
.54	4.98019	1.28973	4.50976	1.28066	4.08745	1.25369	3.69236	1.20704
.56	5.01025	1.35534	4.53721	1.34849	4.11642	1.32337	3.72640	1.27882
.58	5.02554	1.42163	4.55152	1.41656	4.13325	1.39274	3.74861	1.34957
.60	5.02619	1.48891	4.55308	1.48515	4.13856	1.46207	3.75985	1.41962
.62	5.01247	1.55745	4.54239	1.55449	4.13297	1.53161	3.76091	1.48922
.64	4.98478	1.62754	4.52002	1.62480	4.11716	1.60153	3.75256	1.55859
.66	4.94371	1.69940	4.48663	1.69625	4.09182	1.67198	3.73555	1.62789
.68	4.89000	1.77326	4.44298	1.76901	4.05771	1.74310	3.71066	1.69722
.70	4.82455	1.84932	4.38990	1.84318	4.01560	1.81496	3.67862	1.76668
.72	4.74843	1.92771	4.32831	1.91887	3.96631	1.88762	3.64019	1.83631
.74	4.66282	2.00854	4.25919	1.99610	3.91067	1.96110	3.59612	1.90612
.76	4.56901	2.09190	4.18353	2.07491	3.84951	2.03539	3.54713	1.97611
.78	4.46834	2.17778	4.10238	2.15527	3.78367	2.11045	3.49392	2.04623
.80	4.36216	2.26617	4.01674	2.23711	3.71396	2.18620	3.43719	2.11643
.82	4.25181	2.35697	3.92762	2.32033	3.64116	2.26256	3.37758	2.18661
.84	4.13857	2.45005	3.83595	2.40479	3.56601	2.33939	3.31570	2.25668
.86	4.02364	2.54524	3.74263	2.49034	3.48920	2.41655	3.25211	2.32651
.88	3.90811	2.64232	3.64845	2.57677	3.41136	2.49388	3.18733	2.39597
.90	3.79295	2.74101	3.55415	2.66385	3.33307	2.57118	3.12183	2.46491
.92	3.67898	2.84102	3.46035	2.75135	3.25481	2.64826	3.05603	2.53318
.94	3.56692	2.94202	3.36762	2.83899	3.17705	2.72491	2.99029	2.60063
.96	3.45734	3.04366	3.27640	2.92650	3.10015	2.80092	2.92493	2.66709
.98	3.35070	3.14557	3.18708	3.01359	3.02443	2.87608	2.86023	2.73240
1.00	3.24736	3.24736	3.09997	3.09997	2.95016	2.95016	2.79641	2.79641

Table D–2: $0.586 \leq \hat{\rho}_4 \leq 0.622$ and $\lambda_3, \lambda_4 < 0$

$\hat{\rho}_3$	$\hat{\rho}_4 = 0.586$		$\hat{\rho}_4 = 0.598$		$\hat{\rho}_4 = 0.610$		$\hat{\rho}_4 = 0.622$	
	λ_3	λ_4	λ_3	λ_4	λ_3	λ_4	λ_3	λ_4
.28	2.58357	.016265						
.30	2.51056	.041966						
.32	2.44008	.072213	2.39377	.018533				
.34	2.37504	.109053	2.31856	.039593				
.36	2.32155	.156827	2.24445	.062899				
.38	2.29739	.227101	2.17225	.089033				
.40	2.37404	.356359	2.10308	.119078				
.42	2.60480	.536847	2.03907	.155386				
.44	2.80566	.676842	1.98626	.204788				
.46	2.95769	.788904	2.00065	.321329				
.48	3.07646	.886201	2.49982	.689537				
.50	3.17066	.974588	2.67832	.824975				
.52	3.24526	1.05708	2.79756	.930794				
.54	3.30339	1.13545	2.88479	1.02317	2.17967	.723125		
.56	3.34725	1.21082	2.94991	1.10763	2.43072	.911825		
.58	3.37845	1.28396	2.99796	1.18682	2.54082	1.02242		
.60	3.39833	1.35542	3.03202	1.26225	2.61264	1.11488		
.62	3.40797	1.42559	3.05425	1.33486	2.66181	1.19788		
.64	3.40837	1.49478	3.06622	1.40526	2.69484	1.27483	2.10969	.988449
.66	3.40039	1.56320	3.06923	1.47390	2.71536	1.34748	2.23420	1.11870
.68	3.38488	1.63101	3.06435	1.54108	2.72570	1.41685	2.29480	1.21200
.70	3.36261	1.69831	3.05252	1.60699	2.72753	1.48358	2.33032	1.29212
.72	3.33432	1.76518	3.03457	1.67179	2.72215	1.54809	2.35036	1.36463
.74	3.30072	1.83165	3.01125	1.73558	2.71060	1.61066	2.35934	1.43187
.76	3.26249	1.89773	2.98325	1.79839	2.69375	1.67149	2.35983	1.49510
.78	3.22025	1.96340	2.95120	1.86026	2.67235	1.73070	2.35354	1.55506
.80	3.17463	2.02861	2.91568	1.92118	2.64707	1.78837	2.34170	1.61219
.82	3.12618	2.09330	2.87724	1.98112	2.61848	1.84454	2.32527	1.66681
.84	3.07544	2.15740	2.83635	2.04004	2.58709	1.89923	2.30502	1.71912
.86	3.02289	2.22082	2.79348	2.09789	2.55335	1.95243	2.28156	1.76925
.88	2.96897	2.28345	2.74901	2.15460	2.51769	2.00413	2.25542	1.81728
.90	2.91409	2.34519	2.70333	2.21010	2.48046	2.05430	2.22706	1.86328
.92	2.85861	2.40592	2.65675	2.26432	2.44198	2.10290	2.19684	1.90728
.94	2.80285	2.46553	2.60957	2.31718	2.40254	2.14989	2.16511	1.94929
.96	2.74709	2.52391	2.56204	2.36860	2.36238	2.19523	2.13214	1.98933
.98	2.69157	2.58093	2.51439	2.41850	2.32174	2.23888	2.09818	2.02738
1.00	2.63649	2.63649	2.46680	2.46680	2.28080	2.28080	2.06344	2.06344

Table D–3: $0.502 \leq \hat{\rho}_4 \leq 0.526$ and $\lambda_3, \ \lambda_4 > 0$

$\hat{\rho}_3$	$\hat{\rho}_4 = 0.502$		$\hat{\rho}_4 = 0.514$		$\hat{\rho}_4 = 0.526$	
	λ_3	λ_4	λ_3	λ_4	λ_3	λ_4
.38	.00013	.01106				
.40	.00039	.01130	.00019	.07795		
.42	.00067	.01153	.00181	.07966		
.44	.00097	.01176	.00353	.08134	.00154	.14887
.46	.00127	.01197	.00535	.08296	.00431	.15202
.48	.00160	.01217	.00728	.08451	.00725	.15511
.50	.00193	.01236	.00930	.08600	.01037	.15811
.52	.00228	.01252	.01143	.08741	.01366	.16102
.54	.00264	.01268	.01366	.08872	.01714	.16383
.56	.00301	.01281	.01599	.08994	.02082	.16650
.58	.00339	.01293	.01842	.09106	.02469	.16904
.60	.00378	.01303	.02095	.09205	.02877	.17142
.62	.00418	.01311	.02357	.09292	.03306	.17362
.64	.00458	.01316	.02628	.09364	.03756	.17561
.66	.00498	.01320	.02907	.09422	.04229	.17738
.68	.00539	.01321	.03194	.09465	.04725	.17890
.70	.00579	.01320	.03487	.09490	.05244	.18014
.72	.00620	.01317	.03786	.09499	.05786	.18107
.74	.00660	.01312	.04090	.09489	.06351	.18166
.76	.00699	.01305	.04398	.09461	.06939	.18187
.78	.00739	.01296	.04707	.09414	.07549	.18167
.80	.00777	.01285	.05017	.09348	.08180	.18103
.82	.00814	.01272	.05325	.09263	.08829	.17991
.84	.00850	.01258	.05631	.09159	.09494	.17829
.86	.00886	.01241	.05932	.09037	.10171	.17614
.88	.00919	.01223	.06227	.08898	.10854	.17345
.90	.00952	.01204	.06513	.08742	.11538	.17022
.92	.00983	.01184	.06790	.08572	.12216	.16646
.94	.01013	.01162	.07056	.08388	.12879	.16219
.96	.01041	.01140	.07309	.08193	.13519	.15749
.98	.01068	.01117	.07548	.07987	.14129	.15240
1.00	.01093	.01093	.07773	.07773	.14701	.14701

Table D-3: $0.538 \le \hat{\rho}_4 \le 0.622$ and $\lambda_3, \lambda_4 > 0$

	$\hat{\rho}_4 = 0.538$		$\hat{\rho}_4 = 0.550$		$\hat{\rho}_4 = 0.562$		$\hat{\rho}_4 = 0.574$	
$\hat{\rho}_3$	λ_3	λ_4	λ_3	λ_4	λ_3	λ_4	λ_3	λ_4
.48	.00301	.22336						
.50	.00668	.22795						
.52	.01057	.23246	.00381	.30117				
.54	.01468	.23689	.00808	.30721				
.56	.01904	.24122	.01260	.31319	.00328	.38199		
.58	.02365	.24542	.01737	.31910	.00785	.38950		
.60	.02853	.24948	.02242	.32492	.01266	.39697	.00097	.46541
.62	.03370	.25338	.02777	.33063	.01773	.40439	.00553	.47436
.64	.03918	.25710	.03344	.33621	.02307	.41177	.01031	.48331
.66	.04499	.26059	.03945	.34165	.02871	.41907	.01532	.49227
.68	.05115	.26383	.04584	.34691	.03467	.42630	.02057	.50121
.70	.05768	.26679	.05264	.35196	.04099	.43342	.02607	.51014
.72	.06463	.26940	.05991	.35677	.04769	.44042	.03186	.51904
.74	.07203	.27163	.06768	.36128	.05483	.44727	.03795	.52790
.76	.07991	.27340	.07604	.36544	.06245	.45393	.04437	.53672
.78	.08831	.27465	.08505	.36918	.07062	.46038	.05117	.54547
.80	.09729	.27527	.09481	.37240	.07942	.46655	.05837	.55415
.82	.10689	.27518	.10546	.37498	.08895	.47239	.06603	.56272
.84	.11715	.27424	.11716	.37677	.09935	.47780	.07423	.57116
.86	.12813	.27230	.13012	.37753	.11081	.48266	.08303	.57943
.88	.13984	.26921	.14463	.37696	.12359	.48681	.09255	.58749
.90	.15228	.26478	.16112	.37460	.13808	.48998	.10294	.59528
.92	.16538	.25884	.18016	.36973	.15488	.49176	.11439	.60269
.94	.17899	.25126	.20255	.36125	.17506	.49140	.12721	.60959
.96	.19281	.24203	.22934	.34737	.20081	.48733	.14183	.61578
.98	.20642	.23127	.26111	.32571	.23811	.47506	.15905	.62087
1.00	.21933	.21933	.29544	.29544	.37634	.37634	.46346	.46346

Table D–3: $0.586 \leq \hat{\rho}_4 \leq 0.634$ and $\lambda_3, \ \lambda_4 > 0$

$\hat{\rho}_3$	$\hat{\rho}_4 = 0.586$		$\hat{\rho}_4 = 0.598$		$\hat{\rho}_4 = 0.610$		$\hat{\rho}_4 = 0.622$	
	λ_3	λ_4	λ_3	λ_4	λ_3	λ_4	λ_3	λ_4
.54					2.03531	.63682		
.56					1.77814	.51340		
.58					1.65593	.47138		
.60					1.56598	.45088		
.62					1.49268	.44293		
.64					1.42962	.44349		
.66					1.37344	.45007		
.68					1.32222	.46094	1.58799	.73803
.70					1.27479	.47491	1.51059	.72949
.72					1.23040	.49112	1.44543	.72855
.74					1.18857	.50897	1.38847	.73267
.76					1.14896	.52805	1.33755	.74044
.78					1.11133	.54805	1.29132	.75098
.80			.96798	.36556	1.07548	.56876	1.24890	.76368
.82			.93249	.40578	1.04128	.59000	1.20967	.77811
.84			.89864	.44062	1.00859	.61166	1.17318	.79396
.86			.86610	.47263	.97733	.63364	1.13909	.81097
.88			.83472	.50286	.94739	.65587	1.10713	.82897
.90	.73415	.35942	.80440	.53185	.91871	.67829	1.07710	.84779
.92	.69452	.41179	.77507	.55993	.89122	.70086	1.04883	.86732
.94	.65865	.45379	.74666	.58730	.86485	.72354	1.02218	.88746
.96	.62454	.49119	.71909	.61408	.83954	.74629	.99703	.90814
.98	.59146	.52592	.69232	.64039	.81525	.76910	.97329	.92929
1.00	.55894	.55894	.66628	.66628	.79194	.79194	.95087	.95087

$\hat{\rho}_3$	$\hat{\rho}_4 = 0.634$	
	λ_3	λ_4
.82	1.57391	1.10986
.84	1.49927	1.09761
.86	1.44174	1.09742
.88	1.39373	1.10368
.90	1.35223	1.11420
.92	1.31565	1.12786
.94	1.28306	1.14399
.96	1.25382	1.16220
.98	1.22752	1.18221
1.00	1.20385	1.20385

Table D–4: $0.586 \leq \hat{\rho}_4 \leq 0.622$ and $\lambda_3, \lambda_4 > 0$

$\hat{\rho}_3$	$\hat{\rho}_4 = 0.586$		$\hat{\rho}_4 = 0.598$		$\hat{\rho}_4 = 0.610$		$\hat{\rho}_4 = 0.622$	
	λ_3	λ_4	λ_3	λ_4	λ_3	λ_4	λ_3	λ_4
.66	.00096	.56123						
.68	.00544	.57161						
.70	.01009	.58203						
.72	.01492	.59248						
.74	.01995	.60297	.00231	.67288				
.76	.02518	.61348	.00651	.68471				
.78	.03063	.62402	.01084	.69660				
.80	.03632	.63459	.01529	.70857				
.82	.04226	.64517	.01987	.72060				
.84	.04848	.65577	.02458	.73271	.00309	.80327		
.86	.05501	.66638	.02943	.74488	.00674	.81653		
.88	.06188	.67699	.03444	.75714	.01044	.82989		
.90	.06914	.68759	.03959	.76947	.01419	.84334		
.92	.07682	.69817	.04492	.78187	.01800	.85688		
.94	.08500	.70871	.05042	.79436	.02185	.87053		
.96	.09376	.71920	.05611	.80693	.02575	.88427	.00061	.95405
.98	.10321	.72960	.06201	.81958	.02970	.89812	.00336	.96862
1.00	.11349	.73988	.06813	.83232	.03369	.91208	.00610	.98327

Table D-5: $0.005 \leq \hat{\rho}_4 \leq 0.008$ and λ_3, $\lambda_4 < 0$

$\hat{\rho}_3$	$\hat{\rho}_4 = 0.005$		$\hat{\rho}_4 = 0.006$		$\hat{\rho}_4 = 0.007$		$\hat{\rho}_4 = 0.008$	
	λ_3	λ_4	λ_3	λ_4	λ_3	λ_4	λ_3	λ_4
.02	-4.1567	-5.8568	-3.9567	-5.6573	-3.7873	-5.4884	-3.6403	-5.3419
.04	-4.4853	-5.8837	-4.2855	-5.6842	-4.1164	-5.5153	-3.9698	-5.3689
.06	-4.6782	-5.9004	-4.4786	-5.7008	-4.3096	-5.5320	-4.1631	-5.3856
.08	-4.8145	-5.9116	-4.6149	-5.7121	-4.4460	-5.5433	-4.2996	-5.3970
.10	-4.9193	-5.9194	-4.7197	-5.7200	-4.5509	-5.5512	-4.4045	-5.4048
.12	-5.0039	-5.9249	-4.8044	-5.7254	-4.6355	-5.5566	-4.4892	-5.4103
.14	-5.0745	-5.9285	-4.8750	-5.7290	-4.7062	-5.5603	-4.5598	-5.4139
.16	-5.1348	-5.9307	-4.9353	-5.7313	-4.7665	-5.5625	-4.6202	-5.4162
.18	-5.1871	-5.9319	-4.9876	-5.7325	-4.8189	-5.5637	-4.6725	-5.4174
.20	-5.2332	-5.9322	-5.0337	-5.7328	-4.8649	-5.5640	-4.7186	-5.4177
.22	-5.2741	-5.9318	-5.0747	-5.7324	-4.9059	-5.5636	-4.7596	-5.4173
.24	-5.3109	-5.9308	-5.1115	-5.7313	-4.9427	-5.5626	-4.7964	-5.4163
.26	-5.3441	-5.9292	-5.1447	-5.7298	-4.9760	-5.5611	-4.8297	-5.4148
.28	-5.3744	-5.9273	-5.1750	-5.7279	-5.0062	-5.5591	-4.8599	-5.4129
.30	-5.4020	-5.9249	-5.2026	-5.7255	-5.0339	-5.5568	-4.8876	-5.4106
.32	-5.4274	-5.9223	-5.2280	-5.7229	-5.0593	-5.5542	-4.9130	-5.4079
.34	-5.4509	-5.9194	-5.2515	-5.7200	-5.0827	-5.5513	-4.9365	-5.4050
.36	-5.4726	-5.9163	-5.2732	-5.7169	-5.1044	-5.5482	-4.9582	-5.4019
.38	-5.4927	-5.9130	-5.2933	-5.7136	-5.1246	-5.5449	-4.9783	-5.3986
.40	-5.5115	-5.9095	-5.3121	-5.7101	-5.1434	-5.5414	-4.9971	-5.3951
.42	-5.5291	-5.9058	-5.3297	-5.7064	-5.1609	-5.5377	-5.0147	-5.3915
.44	-5.5455	-5.9020	-5.3461	-5.7027	-5.1774	-5.5339	-5.0311	-5.3877
.46	-5.5609	-5.8982	-5.3615	-5.6988	-5.1928	-5.5301	-5.0465	-5.3838
.48	-5.5754	-5.8942	-5.3760	-5.6948	-5.2073	-5.5261	-5.0610	-5.3798
.50	-5.5891	-5.8901	-5.3897	-5.6907	-5.2209	-5.5220	-5.0747	-5.3758
.52	-5.6019	-5.8860	-5.4026	-5.6866	-5.2338	-5.5179	-5.0876	-5.3716
.54	-5.6141	-5.8818	-5.4147	-5.6824	-5.2460	-5.5137	-5.0998	-5.3674
.56	-5.6257	-5.8775	-5.4263	-5.6781	-5.2576	-5.5094	-5.1113	-5.3632
.58	-5.6366	-5.8732	-5.4372	-5.6738	-5.2685	-5.5051	-5.1223	-5.3589
.60	-5.6470	-5.8689	-5.4476	-5.6695	-5.2789	-5.5008	-5.1327	-5.3545
.62	-5.6569	-5.8645	-5.4575	-5.6651	-5.2888	-5.4964	-5.1425	-5.3502
.64	-5.6662	-5.8601	-5.4669	-5.6607	-5.2982	-5.4920	-5.1519	-5.3458
.66	-5.6752	-5.8557	-5.4758	-5.6563	-5.3071	-5.4876	-5.1609	-5.3414
.68	-5.6837	-5.8512	-5.4843	-5.6518	-5.3156	-5.4831	-5.1694	-5.3369
.70	-5.6919	-5.8468	-5.4925	-5.6474	-5.3238	-5.4787	-5.1775	-5.3325
.72	-5.6996	-5.8423	-5.5003	-5.6429	-5.3316	-5.4742	-5.1853	-5.3280
.74	-5.7071	-5.8378	-5.5077	-5.6385	-5.3390	-5.4698	-5.1928	-5.3235
.76	-5.7142	-5.8334	-5.5148	-5.6340	-5.3461	-5.4653	-5.1999	-5.3191
.78	-5.7210	-5.8289	-5.5216	-5.6295	-5.3529	-5.4608	-5.2067	-5.3146
.80	-5.7275	-5.8244	-5.5282	-5.6251	-5.3595	-5.4564	-5.2132	-5.3102
.82	-5.7338	-5.8200	-5.5344	-5.6206	-5.3657	-5.4519	-5.2195	-5.3057
.84	-5.7398	-5.8155	-5.5404	-5.6162	-5.3717	-5.4475	-5.2255	-5.3012
.86	-5.7456	-5.8111	-5.5462	-5.6117	-5.3775	-5.4430	-5.2313	-5.2968
.88	-5.7511	-5.8067	-5.5518	-5.6073	-5.3831	-5.4386	-5.2368	-5.2924
.90	-5.7565	-5.8022	-5.5571	-5.6029	-5.3884	-5.4342	-5.2422	-5.2879
.92	-5.7616	-5.7978	-5.5622	-5.5984	-5.3935	-5.4297	-5.2473	-5.2835
.94	-5.7665	-5.7934	-5.5672	-5.5940	-5.3985	-5.4253	-5.2522	-5.2791
.96	-5.7713	-5.7890	-5.5719	-5.5897	-5.4032	-5.4210	-5.2570	-5.2747
.98	-5.7759	-5.7847	-5.5765	-5.5853	-5.4078	-5.4166	-5.2616	-5.2704
1.00	-5.7803	-5.7803	-5.5809	-5.5809	-5.4122	-5.4122	-5.2660	-5.2660

Table D–5: $0.009 \leq \hat{\rho}_4 \leq 0.012$ and $\lambda_3, \lambda_4 < 0$

	$\hat{\rho}_4 = 0.009$		$\hat{\rho}_4 = 0.010$		$\hat{\rho}_4 = 0.011$		$\hat{\rho}_4 = 0.012$	
$\hat{\rho}_3$	λ_3	λ_4	λ_3	λ_4	λ_3	λ_4	λ_3	λ_4
.02	-3.5104	-5.2126	-3.3940	-5.0969	-3.2885	-4.9921	-3.1920	-4.8963
.04	-3.8403	-5.2397	-3.7243	-5.1240	-3.6193	-5.0192	-3.5233	-4.9235
.06	-4.0337	-5.2564	-3.9179	-5.1408	-3.8130	-5.0361	-3.7171	-4.9404
.08	-4.1703	-5.2678	-4.0545	-5.1521	-3.9497	-5.0475	-3.8539	-4.9518
.10	-4.2752	-5.2757	-4.1595	-5.1600	-4.0547	-5.0554	-3.9590	-4.9598
.12	-4.3600	-5.2811	-4.2443	-5.1655	-4.1396	-5.0609	-4.0439	-4.9653
.14	-4.4306	-5.2848	-4.3150	-5.1692	-4.2103	-5.0646	-4.1146	-4.9690
.16	-4.4910	-5.2871	-4.3753	-5.1715	-4.2707	-5.0669	-4.1750	-4.9713
.18	-4.5434	-5.2883	-4.4278	-5.1727	-4.3231	-5.0681	-4.2275	-4.9725
.20	-4.5895	-5.2886	-4.4739	-5.1731	-4.3692	-5.0685	-4.2736	-4.9729
.22	-4.6305	-5.2882	-4.5149	-5.1727	-4.4103	-5.0681	-4.3147	-4.9725
.24	-4.6673	-5.2872	-4.5517	-5.1717	-4.4471	-5.0671	-4.3515	-4.9715
.26	-4.7006	-5.2857	-4.5850	-5.1702	-4.4804	-5.0656	-4.3848	-4.9700
.28	-4.7308	-5.2838	-4.6152	-5.1682	-4.5106	-5.0636	-4.4151	-4.9681
.30	-4.7585	-5.2815	-4.6429	-5.1659	-4.5383	-5.0613	-4.4427	-4.9658
.32	-4.7839	-5.2789	-4.6683	-5.1633	-4.5637	-5.0587	-4.4682	-4.9632
.34	-4.8074	-5.2760	-4.6918	-5.1604	-4.5872	-5.0559	-4.4917	-4.9603
.36	-4.8291	-5.2729	-4.7135	-5.1573	-4.6089	-5.0527	-4.5134	-4.9572
.38	-4.8492	-5.2695	-4.7337	-5.1540	-4.6291	-5.0494	-4.5336	-4.9539
.40	-4.8680	-5.2660	-4.7525	-5.1505	-4.6479	-5.0460	-4.5524	-4.9504
.42	-4.8856	-5.2624	-4.7701	-5.1469	-4.6655	-5.0423	-4.5699	-4.9468
.44	-4.9020	-5.2586	-4.7865	-5.1431	-4.6819	-5.0386	-4.5864	-4.9430
.46	-4.9175	-5.2548	-4.8019	-5.1392	-4.6974	-5.0347	-4.6018	-4.9392
.48	-4.9320	-5.2508	-4.8164	-5.1353	-4.7119	-5.0307	-4.6163	-4.9352
.50	-4.9456	-5.2467	-4.8301	-5.1312	-4.7255	-5.0266	-4.6300	-4.9311
.52	-4.9585	-5.2426	-4.8430	-5.1271	-4.7384	-5.0225	-4.6429	-4.9270
.54	-4.9707	-5.2384	-4.8552	-5.1229	-4.7506	-5.0183	-4.6551	-4.9228
.56	-4.9823	-5.2341	-4.8668	-5.1186	-4.7622	-5.0141	-4.6667	-4.9185
.58	-4.9932	-5.2298	-4.8777	-5.1143	-4.7731	-5.0098	-4.6776	-4.9142
.60	-5.0036	-5.2255	-4.8881	-5.1100	-4.7835	-5.0054	-4.6880	-4.9099
.62	-5.0135	-5.2211	-4.8980	-5.1056	-4.7934	-5.0011	-4.6979	-4.9055
.64	-5.0229	-5.2167	-4.9074	-5.1012	-4.8028	-4.9967	-4.7073	-4.9011
.66	-5.0318	-5.2123	-4.9163	-5.0968	-4.8117	-4.9922	-4.7162	-4.8967
.68	-5.0403	-5.2079	-4.9248	-5.0924	-4.8203	-4.9878	-4.7248	-4.8923
.70	-5.0485	-5.2034	-4.9330	-5.0879	-4.8284	-4.9834	-4.7329	-4.8879
.72	-5.0563	-5.1990	-4.9408	-5.0835	-4.8362	-4.9789	-4.7407	-4.8834
.74	-5.0637	-5.1945	-4.9482	-5.0790	-4.8436	-4.9744	-4.7481	-4.8789
.76	-5.0708	-5.1900	-4.9553	-5.0745	-4.8508	-4.9700	-4.7553	-4.8745
.78	-5.0776	-5.1856	-4.9621	-5.0701	-4.8576	-4.9655	-4.7621	-4.8700
.80	-5.0842	-5.1811	-4.9687	-5.0656	-4.8641	-4.9610	-4.7686	-4.8655
.82	-5.0904	-5.1766	-4.9749	-5.0611	-4.8704	-4.9566	-4.7749	-4.8611
.84	-5.0965	-5.1722	-4.9809	-5.0567	-4.8764	-4.9521	-4.7809	-4.8566
.86	-5.1022	-5.1677	-4.9867	-5.0522	-4.8822	-4.9477	-4.7867	-4.8522
.88	-5.1078	-5.1633	-4.9923	-5.0478	-4.8877	-4.9433	-4.7922	-4.8477
.90	-5.1131	-5.1589	-4.9976	-5.0434	-4.8931	-4.9388	-4.7976	-4.8433
.92	-5.1183	-5.1545	-5.0027	-5.0390	-4.8982	-4.9344	-4.8027	-4.8389
.94	-5.1232	-5.1501	-5.0077	-5.0346	-4.9031	-4.9300	-4.8076	-4.8345
.96	-5.1280	-5.1457	-5.0124	-5.0302	-4.9079	-4.9256	-4.8124	-4.8301
.98	-5.1325	-5.1413	-5.0170	-5.0258	-4.9125	-4.9213	-4.8170	-4.8258
1.00	-5.1370	-5.1370	-5.0215	-5.0215	-4.9169	-4.9169	-4.8214	-4.8214

Table D–5: $0.0135 \leq \hat{\rho}_4 \leq 0.0180$ and $\lambda_3, \ \lambda_4 < 0$

	$\hat{\rho}_4 = 0.0135$		$\hat{\rho}_4 = 0.0150$		$\hat{\rho}_4 = 0.0165$		$\hat{\rho}_4 = 0.0180$	
$\hat{\rho}_3$	λ_3	λ_4	λ_3	λ_4	λ_3	λ_4	λ_3	λ_4
.02	-3.0609	-4.7665	-2.9434	-4.6503	-2.8367	-4.5450	-2.7388	-4.4487
.04	-3.3930	-4.7938	-3.2763	-4.6777	-3.1705	-4.5725	-3.0737	-4.4763
.06	-3.5872	-4.8107	-3.4708	-4.6947	-3.3652	-4.5895	-3.2687	-4.4934
.08	-3.7241	-4.8222	-3.6078	-4.7062	-3.5025	-4.6011	-3.4061	-4.5050
.10	-3.8293	-4.8302	-3.7131	-4.7142	-3.6078	-4.6091	-3.5116	-4.5131
.12	-3.9142	-4.8357	-3.7980	-4.7197	-3.6929	-4.6147	-3.5967	-4.5187
.14	-3.9850	-4.8394	-3.8689	-4.7235	-3.7638	-4.6185	-3.6676	-4.5225
.16	-4.0454	-4.8418	-3.9294	-4.7258	-3.8243	-4.6208	-3.7282	-4.5249
.18	-4.0979	-4.8430	-3.9819	-4.7271	-3.8768	-4.6221	-3.7807	-4.5262
.20	-4.1440	-4.8434	-4.0281	-4.7275	-3.9230	-4.6225	-3.8270	-4.5265
.22	-4.1851	-4.8430	-4.0691	-4.7271	-3.9641	-4.6221	-3.8681	-4.5262
.24	-4.2219	-4.8420	-4.1060	-4.7261	-4.0010	-4.6212	-3.9050	-4.5253
.26	-4.2553	-4.8405	-4.1393	-4.7247	-4.0343	-4.6197	-3.9383	-4.5238
.28	-4.2855	-4.8386	-4.1696	-4.7227	-4.0646	-4.6178	-3.9687	-4.5219
.30	-4.3132	-4.8363	-4.1973	-4.7205	-4.0924	-4.6155	-3.9964	-4.5196
.32	-4.3387	-4.8337	-4.2228	-4.7179	-4.1178	-4.6129	-4.0219	-4.5171
.34	-4.3622	-4.8309	-4.2463	-4.7150	-4.1413	-4.6101	-4.0454	-4.5142
.36	-4.3839	-4.8278	-4.2680	-4.7119	-4.1631	-4.6070	-4.0672	-4.5111
.38	-4.4041	-4.8245	-4.2882	-4.7086	-4.1833	-4.6037	-4.0874	-4.5078
.40	-4.4229	-4.8210	-4.3070	-4.7051	-4.2021	-4.6002	-4.1062	-4.5044
.42	-4.4405	-4.8174	-4.3246	-4.7015	-4.2197	-4.5966	-4.1238	-4.5007
.44	-4.4569	-4.8136	-4.3411	-4.6978	-4.2361	-4.5929	-4.1403	-4.4970
.46	-4.4724	-4.8097	-4.3565	-4.6939	-4.2516	-4.5890	-4.1557	-4.4931
.48	-4.4869	-4.8058	-4.3710	-4.6899	-4.2661	-4.5850	-4.1702	-4.4892
.50	-4.5006	-4.8017	-4.3847	-4.6859	-4.2798	-4.5810	-4.1839	-4.4851
.52	-4.5135	-4.7976	-4.3976	-4.6817	-4.2927	-4.5768	-4.1969	-4.4810
.54	-4.5257	-4.7934	-4.4098	-4.6775	-4.3049	-4.5727	-4.2091	-4.4768
.56	-4.5372	-4.7891	-4.4214	-4.6733	-4.3165	-4.5684	-4.2206	-4.4726
.58	-4.5482	-4.7848	-4.4323	-4.6690	-4.3274	-4.5641	-4.2316	-4.4683
.60	-4.5586	-4.7805	-4.4427	-4.6647	-4.3378	-4.5598	-4.2420	-4.4639
.62	-4.5685	-4.7761	-4.4526	-4.6603	-4.3477	-4.5554	-4.2519	-4.4596
.64	-4.5779	-4.7717	-4.4620	-4.6559	-4.3571	-4.5510	-4.2613	-4.4552
.66	-4.5868	-4.7673	-4.4710	-4.6515	-4.3661	-4.5466	-4.2702	-4.4508
.68	-4.5953	-4.7629	-4.4795	-4.6471	-4.3746	-4.5422	-4.2788	-4.4463
.70	-4.6035	-4.7584	-4.4877	-4.6426	-4.3828	-4.5377	-4.2869	-4.4419
.72	-4.6113	-4.7540	-4.4955	-4.6382	-4.3906	-4.5333	-4.2947	-4.4374
.74	-4.6187	-4.7495	-4.5029	-4.6337	-4.3980	-4.5288	-4.3022	-4.4330
.76	-4.6258	-4.7451	-4.5100	-4.6292	-4.4051	-4.5244	-4.3093	-4.4285
.78	-4.6327	-4.7406	-4.5168	-4.6248	-4.4120	-4.5199	-4.3161	-4.4241
.80	-4.6392	-4.7361	-4.5234	-4.6203	-4.4185	-4.5154	-4.3226	-4.4196
.82	-4.6455	-4.7317	-4.5296	-4.6159	-4.4248	-4.5110	-4.3289	-4.4151
.84	-4.6515	-4.7272	-4.5357	-4.6114	-4.4308	-4.5065	-4.3349	-4.4107
.86	-4.6573	-4.7228	-4.5414	-4.6070	-4.4366	-4.5021	-4.3407	-4.4062
.88	-4.6628	-4.7183	-4.5470	-4.6025	-4.4421	-4.4977	-4.3463	-4.4018
.90	-4.6681	-4.7139	-4.5523	-4.5981	-4.4475	-4.4932	-4.3516	-4.3974
.92	-4.6733	-4.7095	-4.5575	-4.5937	-4.4526	-4.4888	-4.3567	-4.3930
.94	-4.6782	-4.7051	-4.5624	-4.5893	-4.4575	-4.4844	-4.3617	-4.3886
.96	-4.6830	-4.7007	-4.5672	-4.5849	-4.4623	-4.4800	-4.3665	-4.3842
.98	-4.6876	-4.6963	-4.5718	-4.5805	-4.4669	-4.4757	-4.3710	-4.3798
1.00	-4.6920	-4.6920	-4.5762	-4.5762	-4.4713	-4.4713	-4.3755	-4.3755

Table D–5: $0.0195 \leq \hat{\rho}_4 \leq 0.0240$ and $\lambda_3, \lambda_4 < 0$

| | $\hat{\rho}_4 = 0.0195$ | | $\hat{\rho}_4 = 0.0210$ | | $\hat{\rho}_4 = 0.0225$ | | $\hat{\rho}_4 = 0.0240$ | |
$\hat{\rho}_3$	λ_3	λ_4	λ_3	λ_4	λ_3	λ_4	λ_3	λ_4
.02	-2.6485	-4.3600	-2.5645	-4.2777	-2.4860	-4.2010	-2.4122	-4.1291
.04	-2.9844	-4.3877	-2.9015	-4.3056	-2.8242	-4.2290	-2.7516	-4.1572
.06	-3.1798	-4.4049	-3.0973	-4.3228	-3.0203	-4.2463	-2.9481	-4.1746
.08	-3.3173	-4.4165	-3.2350	-4.3345	-3.1582	-4.2580	-3.0862	-4.1864
.10	-3.4229	-4.4246	-3.3407	-4.3426	-3.2640	-4.2662	-3.1922	-4.1946
.12	-3.5081	-4.4302	-3.4259	-4.3483	-3.3493	-4.2719	-3.2776	-4.2003
.14	-3.5791	-4.4341	-3.4970	-4.3521	-3.4204	-4.2757	-3.3487	-4.2042
.16	-3.6397	-4.4365	-3.5576	-4.3545	-3.4811	-4.2782	-3.4095	-4.2067
.18	-3.6923	-4.4378	-3.6102	-4.3559	-3.5338	-4.2795	-3.4622	-4.2080
.20	-3.7385	-4.4382	-3.6565	-4.3563	-3.5801	-4.2799	-3.5085	-4.2085
.22	-3.7797	-4.4379	-3.6977	-4.3560	-3.6213	-4.2796	-3.5497	-4.2082
.24	-3.8166	-4.4369	-3.7346	-4.3550	-3.6583	-4.2787	-3.5867	-4.2073
.26	-3.8500	-4.4355	-3.7680	-4.3536	-3.6917	-4.2773	-3.6201	-4.2058
.28	-3.8803	-4.4336	-3.7984	-4.3517	-3.7220	-4.2754	-3.6505	-4.2040
.30	-3.9080	-4.4313	-3.8261	-4.3495	-3.7498	-4.2732	-3.6783	-4.2017
.32	-3.9335	-4.4287	-3.8516	-4.3469	-3.7753	-4.2706	-3.7038	-4.1992
.34	-3.9570	-4.4259	-3.8752	-4.3441	-3.7988	-4.2678	-3.7274	-4.1964
.36	-3.9788	-4.4228	-3.8969	-4.3410	-3.8206	-4.2647	-3.7492	-4.1933
.38	-3.9990	-4.4195	-3.9172	-4.3377	-3.8409	-4.2614	-3.7694	-4.1900
.40	-4.0179	-4.4161	-3.9360	-4.3342	-3.8597	-4.2580	-3.7883	-4.1866
.42	-4.0355	-4.4125	-3.9536	-4.3306	-3.8773	-4.2544	-3.8059	-4.1830
.44	-4.0519	-4.4087	-3.9701	-4.3269	-3.8938	-4.2506	-3.8224	-4.1792
.46	-4.0674	-4.4049	-3.9856	-4.3230	-3.9093	-4.2468	-3.8379	-4.1754
.48	-4.0819	-4.4009	-4.0001	-4.3191	-3.9238	-4.2428	-3.8524	-4.1714
.50	-4.0956	-4.3968	-4.0138	-4.3150	-3.9375	-4.2388	-3.8661	-4.1674
.52	-4.1086	-4.3927	-4.0267	-4.3109	-3.9505	-4.2347	-3.8791	-4.1633
.54	-4.1208	-4.3885	-4.0390	-4.3067	-3.9627	-4.2305	-3.8913	-4.1591
.56	-4.1323	-4.3843	-4.0505	-4.3025	-3.9743	-4.2263	-3.9029	-4.1549
.58	-4.1433	-4.3800	-4.0615	-4.2982	-3.9852	-4.2220	-3.9138	-4.1506
.60	-4.1537	-4.3757	-4.0719	-4.2939	-3.9956	-4.2176	-3.9242	-4.1463
.62	-4.1636	-4.3713	-4.0818	-4.2895	-4.0055	-4.2133	-3.9341	-4.1419
.64	-4.1730	-4.3669	-4.0912	-4.2851	-4.0149	-4.2089	-3.9435	-4.1375
.66	-4.1820	-4.3625	-4.1002	-4.2807	-4.0239	-4.2045	-3.9525	-4.1331
.68	-4.1905	-4.3581	-4.1087	-4.2763	-4.0325	-4.2001	-3.9611	-4.1287
.70	-4.1987	-4.3536	-4.1169	-4.2718	-4.0406	-4.1956	-3.9692	-4.1242
.72	-4.2064	-4.3492	-4.1246	-4.2674	-4.0484	-4.1912	-3.9770	-4.1198
.74	-4.2139	-4.3447	-4.1321	-4.2629	-4.0559	-4.1867	-3.9845	-4.1153
.76	-4.2210	-4.3403	-4.1392	-4.2585	-4.0630	-4.1823	-3.9916	-4.1109
.78	-4.2278	-4.3358	-4.1460	-4.2540	-4.0698	-4.1778	-3.9984	-4.1064
.80	-4.2344	-4.3313	-4.1526	-4.2496	-4.0764	-4.1733	-4.0050	-4.1020
.82	-4.2407	-4.3269	-4.1589	-4.2451	-4.0826	-4.1689	-4.0112	-4.0975
.84	-4.2467	-4.3224	-4.1649	-4.2406	-4.0886	-4.1644	-4.0173	-4.0930
.86	-4.2525	-4.3180	-4.1707	-4.2362	-4.0944	-4.1600	-4.0230	-4.0886
.88	-4.2580	-4.3136	-4.1762	-4.2318	-4.1000	-4.1555	-4.0286	-4.0842
.90	-4.2634	-4.3091	-4.1816	-4.2273	-4.1053	-4.1511	-4.0339	-4.0797
.92	-4.2685	-4.3047	-4.1867	-4.2229	-4.1105	-4.1467	-4.0391	-4.0753
.94	-4.2734	-4.3003	-4.1916	-4.2185	-4.1154	-4.1423	-4.0440	-4.0709
.96	-4.2782	-4.2959	-4.1964	-4.2141	-4.1202	-4.1379	-4.0488	-4.0665
.98	-4.2828	-4.2916	-4.2010	-4.2098	-4.1248	-4.1336	-4.0534	-4.0622
1.00	-4.2872	-4.2872	-4.2054	-4.2054	-4.1292	-4.1292	-4.0578	-4.0578

Table D–5: $0.026 \leq \hat{\rho}_4 \leq 0.032$ and $\lambda_3, \ \lambda_4 < 0$

	$\hat{\rho}_4 = 0.026$		$\hat{\rho}_4 = 0.028$		$\hat{\rho}_4 = 0.030$		$\hat{\rho}_4 = 0.032$	
$\hat{\rho}_3$	λ_3	λ_4	λ_3	λ_4	λ_3	λ_4	λ_3	λ_4
.02	-2.3201	-4.0398	-2.2344	-3.9569	-2.1540	-3.8795	-2.0782	-3.8069
.04	-2.6613	-4.0681	-2.5774	-3.9853	-2.4990	-3.9082	-2.4254	-3.8359
.06	-2.8584	-4.0855	-2.7751	-4.0029	-2.6973	-3.9259	-2.6243	-3.8537
.08	-2.9968	-4.0974	-2.9137	-4.0148	-2.8363	-3.9378	-2.7636	-3.8657
.10	-3.1029	-4.1056	-3.0200	-4.0232	-2.9427	-3.9462	-2.8702	-3.8741
.12	-3.1884	-4.1114	-3.1056	-4.0289	-3.0285	-3.9521	-2.9561	-3.8800
.14	-3.2597	-4.1153	-3.1770	-4.0329	-3.0999	-3.9560	-3.0276	-3.8840
.16	-3.3204	-4.1178	-3.2379	-4.0354	-3.1608	-3.9586	-3.0886	-3.8866
.18	-3.3732	-4.1192	-3.2907	-4.0368	-3.2137	-3.9600	-3.1415	-3.8881
.20	-3.4196	-4.1197	-3.3371	-4.0373	-3.2602	-3.9605	-3.1881	-3.8886
.22	-3.4608	-4.1194	-3.3784	-4.0371	-3.3015	-3.9603	-3.2294	-3.8884
.24	-3.4978	-4.1185	-3.4154	-4.0362	-3.3386	-3.9594	-3.2665	-3.8876
.26	-3.5313	-4.1171	-3.4489	-4.0348	-3.3720	-3.9581	-3.3001	-3.8862
.28	-3.5617	-4.1152	-3.4793	-4.0330	-3.4025	-3.9562	-3.3305	-3.8844
.30	-3.5895	-4.1130	-3.5072	-4.0308	-3.4303	-3.9540	-3.3584	-3.8822
.32	-3.6150	-4.1105	-3.5327	-4.0282	-3.4559	-3.9515	-3.3840	-3.8797
.34	-3.6386	-4.1077	-3.5563	-4.0254	-3.4795	-3.9487	-3.4076	-3.8769
.36	-3.6604	-4.1046	-3.5781	-4.0224	-3.5013	-3.9457	-3.4295	-3.8738
.38	-3.6807	-4.1013	-3.5984	-4.0191	-3.5216	-3.9424	-3.4497	-3.8706
.40	-3.6995	-4.0979	-3.6173	-4.0157	-3.5405	-3.9390	-3.4686	-3.8672
.42	-3.7172	-4.0943	-3.6349	-4.0121	-3.5582	-3.9354	-3.4863	-3.8636
.44	-3.7337	-4.0906	-3.6514	-4.0084	-3.5747	-3.9317	-3.5028	-3.8599
.46	-3.7491	-4.0867	-3.6669	-4.0045	-3.5902	-3.9279	-3.5183	-3.8561
.48	-3.7637	-4.0828	-3.6815	-4.0006	-3.6048	-3.9239	-3.5329	-3.8521
.50	-3.7774	-4.0787	-3.6952	-3.9965	-3.6185	-3.9199	-3.5467	-3.8481
.52	-3.7904	-4.0746	-3.7081	-3.9924	-3.6314	-3.9158	-3.5596	-3.8440
.54	-3.8026	-4.0705	-3.7204	-3.9883	-3.6437	-3.9116	-3.5719	-3.8398
.56	-3.8142	-4.0662	-3.7320	-3.9840	-3.6553	-3.9074	-3.5835	-3.8356
.58	-3.8252	-4.0619	-3.7429	-3.9797	-3.6663	-3.9031	-3.5945	-3.8313
.60	-3.8356	-4.0576	-3.7534	-3.9754	-3.6767	-3.8988	-3.6049	-3.8270
.62	-3.8455	-4.0533	-3.7633	-3.9711	-3.6866	-3.8944	-3.6148	-3.8227
.64	-3.8549	-4.0489	-3.7727	-3.9667	-3.6960	-3.8901	-3.6242	-3.8183
.66	-3.8639	-4.0445	-3.7817	-3.9623	-3.7050	-3.8857	-3.6332	-3.8139
.68	-3.8724	-4.0400	-3.7902	-3.9579	-3.7136	-3.8812	-3.6418	-3.8095
.70	-3.8806	-4.0356	-3.7984	-3.9534	-3.7217	-3.8768	-3.6499	-3.8050
.72	-3.8884	-4.0312	-3.8062	-3.9490	-3.7295	-3.8723	-3.6577	-3.8006
.74	-3.8958	-4.0267	-3.8136	-3.9445	-3.7370	-3.8679	-3.6652	-3.7961
.76	-3.9030	-4.0222	-3.8208	-3.9401	-3.7441	-3.8634	-3.6723	-3.7917
.78	-3.9098	-4.0178	-3.8276	-3.9356	-3.7509	-3.8590	-3.6792	-3.7872
.80	-3.9163	-4.0133	-3.8341	-3.9311	-3.7575	-3.8545	-3.6857	-3.7827
.82	-3.9226	-4.0089	-3.8404	-3.9267	-3.7638	-3.8501	-3.6920	-3.7783
.84	-3.9286	-4.0044	-3.8464	-3.9222	-3.7698	-3.8456	-3.6980	-3.7738
.86	-3.9344	-4.0000	-3.8522	-3.9178	-3.7756	-3.8412	-3.7038	-3.7694
.88	-3.9400	-3.9955	-3.8578	-3.9134	-3.7812	-3.8367	-3.7094	-3.7650
.90	-3.9453	-3.9911	-3.8631	-3.9089	-3.7865	-3.8323	-3.7147	-3.7605
.92	-3.9505	-3.9867	-3.8683	-3.9045	-3.7916	-3.8279	-3.7199	-3.7561
.94	-3.9554	-3.9823	-3.8732	-3.9001	-3.7966	-3.8235	-3.7248	-3.7517
.96	-3.9602	-3.9779	-3.8780	-3.8957	-3.8014	-3.8191	-3.7296	-3.7473
.98	-3.9648	-3.9735	-3.8826	-3.8914	-3.8059	-3.8147	-3.7342	-3.7430
1.00	-3.9692	-3.9692	-3.8870	-3.8870	-3.8104	-3.8104	-3.7386	-3.7386

Table D–5: $0.034 \leq \hat{\rho}_4 \leq 0.040$ and $\lambda_3, \lambda_4 < 0$

$\hat{\rho}_3$	$\hat{\rho}_4 = 0.034$		$\hat{\rho}_4 = 0.036$		$\hat{\rho}_4 = 0.038$		$\hat{\rho}_4 = 0.040$	
	λ_3	λ_4	λ_3	λ_4	λ_3	λ_4	λ_3	λ_4
.02	-2.0065	-3.7386	-1.9383	-3.6740	-1.8733	-3.6128	-1.8111	-3.5545
.04	-2.3559	-3.7678	-2.2901	-3.7034	-2.2276	-3.6424	-2.1680	-3.5844
.06	-2.5555	-3.7857	-2.4904	-3.7215	-2.4286	-3.6606	-2.3698	-3.6027
.08	-2.6951	-3.7978	-2.6304	-3.7337	-2.5690	-3.6729	-2.5106	-3.6151
.10	-2.8020	-3.8063	-2.7374	-3.7422	-2.6762	-3.6815	-2.6180	-3.6237
.12	-2.8880	-3.8122	-2.8236	-3.7482	-2.7625	-3.6875	-2.7045	-3.6298
.14	-2.9596	-3.8163	-2.8953	-3.7523	-2.8344	-3.6916	-2.7764	-3.6340
.16	3.0207	-3.8189	-2.9565	-3.7549	-2.8956	-3.6943	-2.8377	-3.6367
.18	-3.0736	-3.8204	-3.0095	-3.7564	-2.9487	-3.6958	-2.8909	-3.6382
.20	-3.1202	-3.8209	-3.0561	-3.7570	-2.9954	-3.6964	-2.9376	-3.6389
.22	-3.1616	-3.8207	-3.0976	-3.7568	-3.0369	-3.6963	-2.9791	-3.6387
.24	-3.1988	-3.8199	-3.1347	-3.7560	-3.0741	-3.6955	-3.0164	-3.6379
.26	-3.2323	-3.8186	-3.1683	-3.7547	-3.1077	-3.6942	-3.0500	-3.6366
.28	-3.2628	-3.8168	-3.1988	-3.7529	-3.1382	-3.6924	-3.0806	-3.6349
.30	-3.2907	-3.8146	-3.2267	-3.7507	-3.1661	-3.6902	-3.1086	-3.6327
.32	-3.3163	-3.8121	-3.2524	-3.7482	-3.1918	-3.6877	-3.1342	-3.6303
.34	-3.3399	-3.8093	-3.2760	-3.7455	-3.2155	-3.6850	-3.1579	-3.6275
.36	-3.3618	-3.8063	-3.2979	-3.7424	-3.2374	-3.6820	-3.1798	-3.6245
.38	-3.3821	-3.8030	-3.3182	-3.7392	-3.2577	-3.6788	-3.2002	-3.6213
.40	-3.4010	-3.7996	-3.3371	-3.7358	-3.2766	-3.6754	-3.2191	-3.6179
.42	-3.4187	-3.7960	-3.3548	-3.7322	-3.2943	-3.6718	-3.2368	-3.6144
.44	-3.4352	-3.7923	-3.3714	-3.7285	-3.3109	-3.6681	-3.2534	-3.6107
.46	-3.4507	-3.7885	-3.3869	-3.7247	-3.3264	-3.6643	-3.2689	-3.6068
.48	-3.4653	-3.7846	-3.4015	-3.7208	-3.3410	-3.6604	-3.2835	-3.6029
.50	-3.4791	-3.7806	-3.4152	-3.7168	-3.3548	-3.6563	-3.2973	-3.5989
.52	-3.4920	-3.7765	-3.4282	-3.7127	-3.3678	-3.6523	-3.3103	-3.5948
.54	-3.5043	-3.7723	-3.4405	-3.7085	-3.3800	-3.6481	-3.3226	-3.5907
.56	-3.5159	-3.7681	-3.4521	-3.7043	-3.3916	-3.6439	-3.3342	-3.5865
.58	-3.5269	-3.7638	-3.4631	-3.7000	-3.4026	-3.6396	-3.3452	-3.5822
.60	-3.5373	-3.7595	-3.4735	-3.6957	-3.4131	-3.6353	-3.3556	-3.5779
.62	-3.5472	-3.7551	-3.4834	-3.6914	-3.4230	-3.6310	-3.3656	-3.5736
.64	-3.5567	-3.7508	-3.4929	-3.6870	-3.4324	-3.6266	-3.3750	-3.5692
.66	-3.5557	-3.7464	-3.5019	-3.6826	-3.4414	-3.6222	-3.3840	-3.5648
.68	-3.5742	-3.7419	-3.5104	-3.6782	-3.4500	-3.6178	-3.3926	-3.5604
.70	-3.5824	-3.7375	-3.5186	-3.6737	-3.4582	-3.6133	-3.4008	-3.5559
.72	-3.5902	-3.7331	-3.5264	-3.6693	-3.4660	-3.6089	-3.4086	-3.5515
.74	-3.5977	-3.7286	-3.5339	-3.6648	-3.4735	-3.6044	-3.4161	-3.5471
.76	-3.6048	-3.7241	-3.5410	-3.6604	-3.4806	-3.6000	-3.4232	-3.5426
.78	-3.6116	-3.7197	-3.5479	-3.6559	-3.4875	-3.5955	-3.4301	-3.5381
.80	-3.6182	-3.7152	-3.5544	-3.6515	-3.4940	-3.5911	-3.4366	-3.5337
.82	-3.6245	-3.7108	-3.5607	-3.6470	-3.5003	-3.5866	-3.4429	-3.5292
.84	-3.6305	-3.7063	-3.5667	-3.6426	-3.5063	-3.5822	-3.4489	-3.5248
.86	-3.6363	-3.7019	-3.5725	-3.6381	-3.5121	-3.5777	-3.4547	-3.5203
.88	-3.6419	-3.6974	-3.5781	-3.6337	-3.5177	-3.5733	-3.4603	-3.5159
.90	-3.6472	-3.6930	-3.5834	-3.6293	-3.5230	-3.5689	-3.4656	-3.5115
.92	-3.6524	-3.6886	-3.5886	-3.6249	-3.5282	-3.5645	-3.4708	-3.5071
.94	-3.6573	-3.6842	-3.5935	-3.6205	-3.5331	-3.5601	-3.4757	-3.5027
.96	-3.6621	-3.6798	-3.5983	-3.6161	-3.5379	-3.5557	-3.4805	-3.4983
.98	-3.6667	-3.6754	-3.6029	-3.6117	-3.5425	-3.5513	-3.4851	-3.4939
1.00	-3.6711	-3.6711	-3.6073	-3.6073	-3.5469	-3.5469	-3.4895	-3.4895

Table D–5: $0.0425 \leq \hat{\rho}_4 \leq 0.0500$ and $\lambda_3,\ \lambda_4 < 0$

	$\hat{\rho}_4 = 0.0425$		$\hat{\rho}_4 = 0.0450$		$\hat{\rho}_4 = 0.0475$		$\hat{\rho}_4 = 0.0500$	
$\hat{\rho}_3$	λ_3	λ_4	λ_3	λ_4	λ_3	λ_4	λ_3	λ_4
.02	-1.7368	-3.4854	-1.6660	-3.4201	-1.5981	-3.3580	-1.5329	-3.2989
.04	-2.0972	-3.5157	-2.0301	-3.4507	-1.9662	-3.3890	-1.9051	-3.3303
.06	-2.3001	-3.5341	-2.2340	-3.4693	-2.1712	-3.4078	-2.1113	-3.3492
.08	-2.4413	-3.5466	-2.3757	-3.4819	-2.3134	-3.4205	-2.2541	-3.3621
.10	-2.5490	-3.5553	-2.4837	-3.4907	-2.4218	-3.4294	-2.3628	-3.3711
.12	-2.6357	-3.5615	-2.5706	-3.4969	-2.5089	-3.4356	-2.4501	-3.3774
.14	-2.7078	-3.5657	-2.6428	-3.5011	-2.5812	-3.4400	-2.5226	-3.3818
.16	-2.7692	-3.5684	-2.7043	-3.5039	-2.6429	-3.4428	-2.5843	-3.3847
.18	-2.8224	-3.5700	-2.7577	-3.5056	-2.6963	-3.4445	-2.6378	-3.3864
.20	-2.8692	-3.5707	-2.8045	-3.5063	-2.7432	-3.4452	-2.6849	-3.3871
.22	-2.9108	-3.5706	-2.8462	-3.5062	-2.7849	-3.4451	-2.7266	-3.3871
.24	-2.9481	-3.5698	-2.8835	-3.5054	-2.8223	-3.4444	-2.7640	-3.3864
.26	-2.9817	-3.5685	-2.9172	-3.5042	-2.8560	-3.4432	-2.7978	-3.3852
.28	-3.0124	-3.5668	-2.9479	-3.5024	-2.8867	-3.4415	-2.8286	-3.3835
.30	-3.0403	-3.5647	-2.9759	-3.5003	-2.9148	-3.4394	-2.8566	-3.3814
.32	-3.0660	-3.5622	-3.0016	-3.4979	-2.9405	-3.4369	-2.8824	-3.3790
.34	-3.0897	-3.5595	-3.0253	-3.4952	-2.9643	-3.4342	-2.9062	-3.3763
.36	-3.1117	-3.5565	-3.0473	-3.4922	-2.9862	-3.4313	-2.9282	-3.3733
.38	-3.1320	-3.5533	-3.0677	-3.4890	-3.0066	-3.4281	-2.9486	-3.3702
.40	-3.1510	-3.5499	-3.0866	-3.4856	-3.0256	-3.4247	-2.9676	-3.3668
.42	-3.1687	-3.5463	-3.1044	-3.4821	-3.0434	-3.4212	-2.9854	-3.3633
.44	-3.1853	-3.5427	-3.1210	-3.4784	-3.0600	-3.4175	-3.0020	-3.3596
.46	-3.2009	-3.5389	-3.1365	-3.4746	-3.0756	-3.4137	-3.0176	-3.3558
.48	-3.2155	-3.5349	-3.1512	-3.4707	-3.0902	-3.4098	-3.0323	-3.3519
.50	-3.2293	-3.5309	-3.1650	-3.4667	-3.1040	-3.4058	-3.0461	-3.3480
.52	-3.2423	-3.5269	-3.1780	-3.4626	-3.1170	-3.4018	-3.0591	-3.3439
.54	-3.2545	-3.5227	-3.1903	-3.4585	-3.1293	-3.3976	-3.0714	-3.3398
.56	-3.2662	-3.5185	-3.2019	-3.4543	-3.1410	-3.3934	-3.0831	-3.3356
.58	-3.2772	-3.5142	-3.2129	-3.4500	-3.1520	-3.3892	-3.0941	-3.3313
.60	-3.2876	-3.5099	-3.2234	-3.4457	-3.1625	-3.3849	-3.1046	-3.3270
.62	-3.2976	-3.5056	-3.2333	-3.4414	-3.1724	-3.3805	-3.1145	-3.3227
.64	-3.3070	-3.5012	-3.2428	-3.4370	-3.1819	-3.3762	-3.1240	-3.3183
.66	-3.3160	-3.4968	-3.2518	-3.4326	-3.1909	-3.3718	-3.1330	-3.3139
.68	-3.3246	-3.4924	-3.2604	-3.4282	-3.1995	-3.3674	-3.1416	-3.3095
.70	-3.3328	-3.4880	-3.2685	-3.4238	-3.2077	-3.3629	-3.1498	-3.3051
.72	-3.3406	-3.4836	-3.2764	-3.4193	-3.2155	-3.3585	-3.1576	-3.3007
.74	-3.3481	-3.4791	-3.2838	-3.4149	-3.2230	-3.3541	-3.1651	-3.2962
.76	-3.3552	-3.4746	-3.2910	-3.4104	-3.2301	-3.3496	-3.1723	-3.2918
.78	-3.3621	-3.4702	-3.2979	-3.4060	-3.2370	-3.3452	-3.1791	-3.2873
.80	-3.3686	-3.4657	-3.3044	-3.4015	-3.2436	-3.3407	-3.1857	-3.2829
.82	-3.3749	-3.4613	-3.3107	-3.3971	-3.2499	-3.3362	-3.1920	-3.2784
.84	-3.3810	-3.4568	-3.3167	-3.3926	-3.2559	-3.3318	-3.1981	-3.2740
.86	-3.3868	-3.4524	-3.3225	-3.3882	-3.2617	-3.3274	-3.2039	-3.2695
.88	-3.3923	-3.4480	-3.3281	-3.3837	-3.2673	-3.3229	-3.2094	-3.2651
.90	-3.3977	-3.4435	-3.3335	-3.3793	-3.2726	-3.3185	-3.2148	-3.2607
.92	-3.4028	-3.4391	-3.3386	-3.3749	-3.2778	-3.3141	-3.2199	-3.2562
.94	-3.4078	-3.4347	-3.3436	-3.3705	-3.2827	-3.3097	-3.2249	-3.2518
.96	-3.4126	-3.4303	-3.3484	-3.3661	-3.2875	-3.3053	-3.2297	-3.2474
.98	-3.4172	-3.4259	-3.3529	-3.3617	-3.2921	-3.3009	-3.2343	-3.2431
1.00	-3.4216	-3.4216	-3.3574	-3.3574	-3.2965	-3.2965	-3.2387	-3.2387

Table D–5: $0.053 \le \hat{\rho}_4 \le 0.062$ and $\lambda_3, \lambda_4 < 0$

$\hat{\rho}_3$	$\hat{\rho}_4 = 0.053$		$\hat{\rho}_4 = 0.056$		$\hat{\rho}_4 = 0.059$		$\hat{\rho}_4 = 0.062$	
	λ_3	λ_4	λ_3	λ_4	λ_3	λ_4	λ_3	λ_4
.02	-1.4577	-3.2315	-1.3854	-3.1675	-1.3155	-3.1065	-1.2478	-3.0483
.04	-1.8353	-3.2633	-1.7687	-3.1998	-1.7051	-3.1394	-1.6441	-3.0818
.06	-2.0429	-3.2825	-1.9779	-3.2193	-1.9159	-3.1591	-1.8567	-3.1017
.08	-2.1864	-3.2955	-2.1221	-3.2324	-2.0609	-3.1724	-2.0024	-3.1152
.10	-2.2955	-3.3046	-2.2317	-3.2416	-2.1709	-3.1817	-2.1128	-3.1246
.12	-2.3831	-3.3110	-2.3195	-3.2481	-2.2590	-3.1883	-2.2012	-3.1313
.14	-2.4558	-3.3155	-2.3924	-3.2526	-2.3321	-3.1929	-2.2746	-3.1360
.16	-2.5177	-3.3184	-2.4544	-3.2556	-2.3943	-3.1960	-2.3369	-3.1391
.18	-2.5713	-3.3202	-2.5082	-3.2574	-2.4481	-3.1978	-2.3909	-3.1410
.20	-2.6184	-3.3210	-2.5554	-3.2583	-2.4954	-3.1987	-2.4383	-3.1419
.22	-2.6602	-3.3210	-2.5973	-3.2583	-2.5374	-3.1988	-2.4803	-3.1420
.24	-2.6977	-3.3203	-2.6348	-3.2577	-2.5750	-3.1982	-2.5180	-3.1414
.26	-2.7316	-3.3191	-2.6687	-3.2565	-2.6090	-3.1970	-2.5520	-3.1403
.28	-2.7623	-3.3174	-2.6995	-3.2549	-2.6399	-3.1954	-2.5830	-3.1387
.30	-2.7904	-3.3154	-2.7277	-3.2528	-2.6681	-3.1934	-2.6112	-3.1367
.32	-2.8163	-3.3130	-2.7535	-3.2504	-2.6940	-3.1910	-2.6371	-3.1344
.34	-2.8401	-3.3103	-2.7774	-3.2478	-2.7178	-3.1884	-2.6610	-3.1318
.36	-2.8621	-3.3074	-2.7994	-3.2449	-2.7399	-3.1855	-2.6831	-3.1289
.38	-2.8825	-3.3042	-2.8199	-3.2417	-2.7604	-3.1823	-2.7037	-3.1257
.40	-2.9016	-3.3009	-2.8390	-3.2384	-2.7795	-3.1790	-2.7228	-3.1224
.42	-2.9194	-3.2973	-2.8568	-3.2349	-2.7973	-3.1755	-2.7406	-3.1190
.44	-2.9360	-3.2937	-2.8734	-3.2312	-2.8140	-3.1719	-2.7573	-3.1153
.46	-2.9516	-3.2899	-2.8891	-3.2275	-2.8296	-3.1681	-2.7730	-3.1116
.48	-2.9663	-3.2860	-2.9038	-3.2236	-2.8443	-3.1643	-2.7877	-3.1077
.50	-2.9801	-3.2821	-2.9176	-3.2196	-2.8582	-3.1603	-2.8016	-3.1038
.52	-2.9931	-3.2780	-2.9306	-3.2156	-2.8712	-3.1562	-2.8146	-3.0997
.54	-3.0055	-3.2739	-2.9430	-3.2114	-2.8836	-3.1521	-2.8270	-3.0956
.56	-3.0171	-3.2697	-2.9546	-3.2073	-2.8953	-3.1479	-2.8387	-3.0914
.58	-3.0282	-3.2654	-2.9657	-3.2030	-2.9063	-3.1437	-2.8497	-3.0872
.60	-3.0386	-3.2611	-2.9762	-3.1987	-2.9168	-3.1394	-2.8603	-3.0829
.62	-3.0486	-3.2568	-2.9862	-3.1944	-2.9268	-3.1351	-2.8702	-3.0786
.64	-3.0581	-3.2525	-2.9956	-3.1901	-2.9363	-3.1308	-2.8797	-3.0743
.66	-3.0671	-3.2481	-3.0047	-3.1857	-2.9453	-3.1264	-2.8888	-3.0699
.68	-3.0757	-3.2437	-3.0133	-3.1813	-2.9539	-3.1220	-2.8974	-3.0655
.70	-3.0839	-3.2392	-3.0215	-3.1768	-2.9622	-3.1176	-2.9056	-3.0611
.72	-3.0917	-3.2348	-3.0293	-3.1724	-2.9700	-3.1131	-2.9135	-3.0566
.74	-3.0992	-3.2304	-3.0368	-3.1680	-2.9775	-3.1087	-2.9210	-3.0522
.76	-3.1064	-3.2259	-3.0440	-3.1635	-2.9847	-3.1042	-2.9282	-3.0478
.78	-3.1133	-3.2215	-3.0508	-3.1591	-2.9915	-3.0998	-2.9350	-3.0433
.80	-3.1198	-3.2170	-3.0574	-3.1546	-2.9981	-3.0953	-2.9416	-3.0389
.82	-3.1261	-3.2126	-3.0637	-3.1502	-3.0044	-3.0909	-2.9479	-3.0344
.84	-3.1322	-3.2081	-3.0698	-3.1457	-3.0105	-3.0864	-2.9540	-3.0300
.86	-3.1380	-3.2037	-3.0756	-3.1413	-3.0163	-3.0820	-2.9598	-3.0255
.88	-3.1436	-3.1992	-3.0812	-3.1368	-3.0219	-3.0776	-2.9654	-3.0211
.90	-3.1489	-3.1948	-3.0865	-3.1324	-3.0272	-3.0731	-2.9707	-3.0166
.92	-3.1541	-3.1904	-3.0917	-3.1280	-3.0324	-3.0687	-2.9759	-3.0122
.94	-3.1590	-3.1860	-3.0966	-3.1236	-3.0374	-3.0643	-2.9809	-3.0078
.96	-3.1638	-3.1816	-3.1014	-3.1192	-3.0421	-3.0599	-2.9856	-3.0034
.98	-3.1684	-3.1772	-3.1060	-3.1148	-3.0467	-3.0555	-2.9902	-2.9991
1.00	-3.1728	-3.1728	-3.1104	-3.1104	-3.0512	-3.0512	-2.9947	-2.9947

Table D–5: $0.0655 \le \hat{\rho}_4 \le 0.0760$ and $\lambda_3,\ \lambda_4 < 0$

	$\hat{\rho}_4 = 0.0655$		$\hat{\rho}_4 = 0.0690$		$\hat{\rho}_4 = 0.0725$		$\hat{\rho}_4 = 0.0760$	
$\hat{\rho}_3$	λ_3	λ_4	λ_3	λ_4	λ_3	λ_4	λ_3	λ_4
.02	-1.1711	-2.9834	-1.0964	-2.9215	-1.0233	-2.8623	-.9515	-2.8055
.04	-1.5759	-3.0176	-1.5105	-2.9565	-1.4476	-2.8981	-1.3870	-2.8422
.06	-1.7905	-3.0379	-1.7274	-2.9771	-1.6669	-2.9191	-1.6087	-2.8635
.08	-1.9372	-3.0516	-1.8751	-2.9910	-1.8156	-2.9332	-1.7586	-2.8778
.10	-2.0482	-3.0611	-1.9866	-3.0007	-1.9278	-2.9430	-1.8713	-2.8879
.12	-2.1370	-3.0679	-2.0758	-3.0076	-2.0173	-2.9501	-1.9613	-2.8950
.14	-2.2106	-3.0727	-2.1496	-3.0125	-2.0914	-2.9550	-2.0357	-2.9000
.16	-2.2731	-3.0759	-2.2124	-3.0157	-2.1544	-2.9583	-2.0988	-2.9035
.18	-2.3272	-3.0778	-2.2666	-3.0177	-2.2088	-2.9604	-2.1534	-2.9056
.20	-2.3747	-3.0788	-2.3143	-3.0188	-2.2565	-2.9615	-2.2013	-2.9067
.22	-2.4169	-3.0789	-2.3565	-3.0190	-2.2989	-2.9617	-2.2438	-2.9070
.24	-2.4547	-3.0784	-2.3944	-3.0185	-2.3369	-2.9613	-2.2818	-2.9066
.26	-2.4888	-3.0773	-2.4286	-3.0174	-2.3711	-2.9603	-2.3161	-2.9056
.28	-2.5197	-3.0758	-2.4596	-3.0159	-2.4022	-2.9588	-2.3473	-2.9042
.30	-2.5480	-3.0738	-2.4879	-3.0139	-2.4306	-2.9569	-2.3758	-2.9023
.32	-2.5740	-3.0715	-2.5140	-3.0117	-2.4567	-2.9546	-2.4019	-2.9000
.34	-2.5979	-3.0689	-2.5379	-3.0091	-2.4807	-2.9520	-2.4259	-2.8975
.36	-2.6201	-3.0660	-2.5601	-3.0062	-2.5029	-2.9492	-2.4482	-2.8947
.38	-2.6406	-3.0629	-2.5807	-3.0031	-2.5235	-2.9461	-2.4689	-2.8916
.40	-2.6598	-3.0596	-2.5999	-2.9998	-2.5427	-2.9429	-2.4881	-2.8884
.42	-2.6777	-3.0561	-2.6178	-2.9964	-2.5606	-2.9394	-2.5060	-2.8850
.44	-2.6944	-3.0525	-2.6345	-2.9928	-2.5774	-2.9358	-2.5228	-2.8814
.46	-2.7101	-3.0488	-2.6502	-2.9891	-2.5931	-2.9321	-2.5386	-2.8777
.48	-2.7248	-3.0449	-2.6650	-2.9852	-2.6079	-2.9283	-2.5534	-2.8739
.50	-2.7387	-3.0410	-2.6789	-2.9813	-2.6218	-2.9244	-2.5673	-2.8699
.52	-2.7518	-3.0370	-2.6920	-2.9773	-2.6349	-2.9203	-2.5804	-2.8659
.54	-2.7641	-3.0328	-2.7044	-2.9732	-2.6473	-2.9162	-2.5928	-2.8618
.56	-2.7758	-3.0287	-2.7161	-2.9690	-2.6591	-2.9121	-2.6046	-2.8577
.58	-2.7869	-3.0245	-2.7272	-2.9648	-2.6702	-2.9079	-2.6157	-2.8535
.60	-2.7974	-3.0202	-2.7377	-2.9605	-2.6807	-2.9036	-2.6263	-2.8492
.62	-2.8074	-3.0159	-2.7477	-2.9562	-2.6907	-2.8993	-2.6363	-2.8449
.64	-2.8169	-3.0115	-2.7572	-2.9519	-2.7003	-2.8950	-2.6458	-2.8406
.66	-2.8260	-3.0072	-2.7663	-2.9475	-2.7093	-2.8906	-2.6549	-2.8362
.68	-2.8346	-3.0028	-2.7749	-2.9431	-2.7180	-2.8862	-2.6635	-2.8319
.70	-2.8428	-2.9983	-2.7831	-2.9387	-2.7262	-2.8818	-2.6718	-2.8275
.72	-2.8507	-2.9939	-2.7910	-2.9343	-2.7341	-2.8774	-2.6797	-2.8230
.74	-2.8582	-2.9895	-2.7985	-2.9298	-2.7416	-2.8730	-2.6872	-2.8186
.76	-2.8654	-2.9850	-2.8057	-2.9254	-2.7488	-2.8685	-2.6944	-2.8142
.78	-2.8723	-2.9806	-2.8126	-2.9209	-2.7557	-2.8641	-2.7013	-2.8097
.80	-2.8789	-2.9761	-2.8192	-2.9165	-2.7623	-2.8596	-2.7079	-2.8053
.82	-2.8852	-2.9717	-2.8255	-2.9120	-2.7686	-2.8552	-2.7142	-2.8008
.84	-2.8912	-2.9672	-2.8316	-2.9076	-2.7747	-2.8507	-2.7203	-2.7964
.86	-2.8970	-2.9628	-2.8374	-2.9032	-2.7805	-2.8463	-2.7261	-2.7919
.88	-2.9026	-2.9584	-2.8430	-2.8987	-2.7861	-2.8418	-2.7317	-2.7875
.90	-2.9080	-2.9539	-2.8483	-2.8943	-2.7915	-2.8374	-2.7371	-2.7831
.92	-2.9132	-2.9495	-2.8535	-2.8899	-2.7966	-2.8330	-2.7423	-2.7786
.94	-2.9181	-2.9451	-2.8585	-2.8855	-2.8016	-2.8286	-2.7472	-2.7742
.96	-2.9229	-2.9407	-2.8633	-2.8811	-2.8064	-2.8242	-2.7520	-2.7698
.98	-2.9275	-2.9363	-2.8679	-2.8767	-2.8110	-2.8198	-2.7566	-2.7655
1.00	-2.9320	-2.9320	-2.8723	-2.8723	-2.8154	-2.8154	-2.7611	-2.7611

Table D-5: $0.080 \leq \hat{\rho}_4 \leq 0.092$ and λ_3, $\lambda_4 < 0$

	$\hat{\rho}_4 = 0.080$		$\hat{\rho}_4 = 0.084$		$\hat{\rho}_4 = 0.088$		$\hat{\rho}_4 = 0.092$	
$\hat{\rho}_3$	λ_3	λ_4	λ_3	λ_4	λ_3	λ_4	λ_3	λ_4
.02	-.8704	-2.7431	-.7897	-2.6833	-.7090	-2.6257	-.6273	-2.5700
.04	-1.3201	-2.7810	-1.2556	-2.7224	-1.1931	-2.6661	-1.1324	-2.6120
.06	-1.5449	-2.8027	-1.4836	-2.7446	-1.4245	-2.6888	-1.3675	-2.6352
.08	-1.6961	-2.8173	-1.6361	-2.7595	-1.5785	-2.7040	-1.5230	-2.6507
.10	-1.8096	-2.8276	-1.7504	-2.7699	-1.6936	-2.7146	-1.6389	-2.6615
.12	-1.9000	-2.8348	-1.8413	-2.7773	-1.7850	-2.7222	-1.7309	-2.6693
.14	-1.9747	-2.8400	-1.9164	-2.7826	-1.8605	-2.7276	-1.8067	-2.6748
.16	-2.0381	-2.8435	-1.9801	-2.7862	-1.9244	-2.7313	-1.8709	-2.6786
.18	-2.0929	-2.8457	-2.0350	-2.7885	-1.9796	-2.7337	-1.9263	-2.6811
.20	-2.1409	-2.8469	-2.0832	-2.7897	-2.0279	-2.7350	-1.9748	-2.6825
.22	-2.1836	-2.8472	-2.1260	-2.7902	-2.0708	-2.7355	-2.0178	-2.6830
.24	-2.2217	-2.8469	-2.1642	-2.7899	-2.1092	-2.7352	-2.0563	-2.6828
.26	-2.2561	-2.8460	-2.1987	-2.7890	-2.1437	-2.7344	-2.0910	-2.6820
.28	-2.2873	-2.8445	-2.2300	-2.7876	-2.1751	-2.7330	-2.1224	-2.6807
.30	-2.3159	-2.8427	-2.2586	-2.7857	-2.2038	-2.7312	-2.1512	-2.6789
.32	-2.3420	-2.8405	-2.2848	-2.7836	-2.2301	-2.7291	-2.1775	-2.6768
.34	-2.3661	-2.8379	-2.3090	-2.7811	-2.2543	-2.7266	-2.2018	-2.6744
.36	-2.3884	-2.8351	-2.3314	-2.7783	-2.2767	-2.7239	-2.2242	-2.6716
.38	-2.4091	-2.8321	-2.3521	-2.7753	-2.2975	-2.7209	-2.2451	-2.6687
.40	-2.4284	-2.8289	-2.3714	-2.7721	-2.3168	-2.7177	-2.2644	-2.6655
.42	-2.4464	-2.8255	-2.3894	-2.7687	-2.3349	-2.7143	-2.2825	-2.6622
.44	-2.4632	-2.8219	-2.4063	-2.7652	-2.3517	-2.7108	-2.2994	-2.6587
.46	-2.4790	-2.8182	-2.4221	-2.7615	-2.3676	-2.7071	-2.3153	-2.6550
.48	-2.4938	-2.8144	-2.4369	-2.7577	-2.3824	-2.7034	-2.3302	-2.6512
.50	-2.5077	-2.8105	-2.4509	-2.7538	-2.3964	-2.6995	-2.3442	-2.6474
.52	-2.5209	-2.8065	-2.4640	-2.7498	-2.4096	-2.6955	-2.3574	-2.6434
.54	-2.5333	-2.8024	-2.4765	-2.7457	-2.4221	-2.6914	-2.3699	-2.6394
.56	-2.5451	-2.7983	-2.4883	-2.7416	-2.4339	-2.6873	-2.3817	-2.6352
.58	-2.5562	-2.7941	-2.4994	-2.7374	-2.4451	-2.6831	-2.3929	-2.6311
.60	-2.5668	-2.7899	-2.5100	-2.7332	-2.4556	-2.6789	-2.4035	-2.6268
.62	-2.5768	-2.7856	-2.5201	-2.7289	-2.4657	-2.6746	-2.4136	-2.6226
.64	-2.5864	-2.7812	-2.5296	-2.7246	-2.4753	-2.6703	-2.4231	-2.6182
.66	-2.5955	-2.7769	-2.5387	-2.7202	-2.4844	-2.6659	-2.4323	-2.6139
.68	-2.6041	-2.7725	-2.5474	-2.7158	-2.4931	-2.6616	-2.4409	-2.6095
.70	-2.6124	-2.7681	-2.5556	-2.7114	-2.5013	-2.6572	-2.4492	-2.6051
.72	-2.6203	-2.7637	-2.5635	-2.7070	-2.5092	-2.6528	-2.4571	-2.6007
.74	-2.6278	-2.7592	-2.5711	-2.7026	-2.5168	-2.6483	-2.4647	-2.5963
.76	-2.6350	-2.7548	-2.5783	-2.6981	-2.5240	-2.6439	-2.4719	-2.5919
.78	-2.6419	-2.7504	-2.5852	-2.6937	-2.5309	-2.6395	-2.4788	-2.5874
.80	-2.6485	-2.7459	-2.5918	-2.6893	-2.5375	-2.6350	-2.4855	-2.5830
.82	-2.6548	-2.7415	-2.5981	-2.6848	-2.5439	-2.6306	-2.4918	-2.5785
.84	-2.6609	-2.7370	-2.6042	-2.6804	-2.5500	-2.6261	-2.4979	-2.5741
.86	-2.6667	-2.7326	-2.6101	-2.6759	-2.5558	-2.6217	-2.5037	-2.5697
.88	-2.6723	-2.7281	-2.6157	-2.6715	-2.5614	-2.6172	-2.5094	-2.5652
.90	-2.6777	-2.7237	-2.6210	-2.6671	-2.5668	-2.6128	-2.5147	-2.5608
.92	-2.6829	-2.7193	-2.6262	-2.6626	-2.5720	-2.6084	-2.5199	-2.5564
.94	-2.6879	-2.7149	-2.6312	-2.6582	-2.5770	-2.6040	-2.5249	-2.5520
.96	-2.6927	-2.7105	-2.6360	-2.6538	-2.5818	-2.5996	-2.5297	-2.5476
.98	-2.6973	-2.7061	-2.6406	-2.6494	-2.5864	-2.5952	-2.5343	-2.5432
1.00	-2.7017	-2.7017	-2.6451	-2.6451	-2.5908	-2.5908	-2.5388	-2.5388

Table D–5: $0.0965 \le \hat{\rho}_4 \le 0.1100$ and $\lambda_3,\ \lambda_4 < 0$

	$\hat{\rho}_4 = 0.0965$		$\hat{\rho}_4 = 0.1010$		$\hat{\rho}_4 = 0.1055$		$\hat{\rho}_4 = 0.1100$	
$\hat{\rho}_3$	λ_3	λ_4	λ_3	λ_4	λ_3	λ_4	λ_3	λ_4
.02	-.5333	-2.5095	-.4354	-2.4509	-.3310	-2.3940	-.2158	-2.3383
.04	-1.0660	-2.5534	-1.0013	-2.4970	-.9381	-2.4426	-.8762	-2.3900
.06	-1.3055	-2.5773	-1.2456	-2.5216	-1.1875	-2.4679	-1.1311	-2.4160
.08	-1.4628	-2.5931	-1.4048	-2.5377	-1.3488	-2.4845	-1.2945	-2.4330
.10	-1.5797	-2.6042	-1.5227	-2.5491	-1.4678	-2.4961	-1.4146	-2.4450
.12	-1.6723	-2.6121	-1.6160	-2.5572	-1.5617	-2.5044	-1.5092	-2.4535
.14	-1.7485	-2.6178	-1.6927	-2.5630	-1.6388	-2.5104	-1.5869	-2.4596
.16	-1.8130	-2.6217	-1.7575	-2.5671	-1.7040	-2.5145	-1.6524	-2.4639
.18	-1.8687	-2.6243	-1.8134	-2.5697	-1.7602	-2.5173	-1.7088	-2.4668
.20	-1.9174	-2.6257	-1.8623	-2.5713	-1.8093	-2.5190	-1.7582	-2.4685
.22	-1.9606	-2.6263	-1.9056	-2.5720	-1.8528	-2.5197	-1.8018	-2.4694
.24	-1.9992	-2.6262	-1.9444	-2.5719	-1.8917	-2.5197	-1.8408	-2.4694
.26	-2.0340	-2.6255	-1.9793	-2.5712	-1.9267	-2.5191	-1.8760	-2.4689
.28	-2.0656	-2.6242	-2.0110	-2.5700	-1.9585	-2.5179	-1.9079	-2.4678
.30	-2.0944	-2.6225	-2.0399	-2.5683	-1.9874	-2.5163	-1.9369	-2.4662
.32	-2.1208	-2.6204	-2.0663	-2.5663	-2.0140	-2.5143	-1.9636	-2.4642
.34	-2.1451	-2.6180	-2.0907	-2.5639	-2.0385	-2.5120	-1.9881	-2.4619
.36	-2.1676	-2.6153	-2.1133	-2.5613	-2.0611	-2.5093	-2.0108	-2.4593
.38	-2.1885	-2.6124	-2.1342	-2.5584	-2.0820	-2.5064	-2.0318	-2.4565
.40	-2.2079	-2.6092	-2.1537	-2.5552	-2.1016	-2.5034	-2.0513	-2.4534
.42	-2.2260	-2.6059	-2.1718	-2.5519	-2.1198	-2.5001	-2.0696	-2.4501
.44	-2.2430	-2.6024	-2.1888	-2.5485	-2.1368	-2.4966	-2.0866	-2.4467
.46	-2.2588	-2.5988	-2.2047	-2.5448	-2.1527	-2.4930	-2.1026	-2.4431
.48	-2.2738	-2.5950	-2.2197	-2.5411	-2.1677	-2.4893	-2.1176	-2.4394
.50	-2.2878	-2.5912	-2.2338	-2.5373	-2.1818	-2.4855	-2.1318	-2.4356
.52	-2.3011	-2.5872	-2.2470	-2.5333	-2.1951	-2.4815	-2.1451	-2.4317
.54	-2.3136	-2.5832	-2.2595	-2.5293	-2.2076	-2.4775	-2.1577	-2.4277
.56	-2.3254	-2.5791	-2.2714	-2.5252	-2.2195	-2.4734	-2.1695	-2.4236
.58	-2.3366	-2.5749	-2.2826	-2.5210	-2.2308	-2.4693	-2.1808	-2.4195
.60	-2.3472	-2.5707	-2.2933	-2.5168	-2.2414	-2.4651	-2.1915	-2.4153
.62	-2.3573	-2.5664	-2.3034	-2.5126	-2.2515	-2.4608	-2.2016	-2.4110
.64	-2.3669	-2.5621	-2.3130	-2.5083	-2.2612	-2.4565	-2.2113	-2.4067
.66	-2.3760	-2.5578	-2.3221	-2.5039	-2.2703	-2.4522	-2.2204	-2.4024
.68	-2.3847	-2.5534	-2.3308	-2.4996	-2.2790	-2.4479	-2.2292	-2.3981
.70	-2.3930	-2.5490	-2.3391	-2.4952	-2.2874	-2.4435	-2.2375	-2.3937
.72	-2.4009	-2.5446	-2.3471	-2.4908	-2.2953	-2.4391	-2.2454	-2.3893
.74	-2.4085	-2.5402	-2.3546	-2.4864	-2.3029	-2.4347	-2.2530	-2.3849
.76	-2.4157	-2.5357	-2.3619	-2.4819	-2.3101	-2.4302	-2.2603	-2.3805
.78	-2.4227	-2.5313	-2.3688	-2.4775	-2.3171	-2.4258	-2.2672	-2.3760
.80	-2.4293	-2.5269	-2.3754	-2.4731	-2.3237	-2.4214	-2.2739	-2.3716
.82	-2.4357	-2.5224	-2.3818	-2.4686	-2.3301	-2.4169	-2.2803	-2.3671
.84	-2.4417	-2.5180	-2.3879	-2.4642	-2.3362	-2.4125	-2.2864	-2.3627
.86	-2.4476	-2.5135	-2.3938	-2.4597	-2.3420	-2.4080	-2.2922	-2.3583
.88	-2.4532	-2.5091	-2.3994	-2.4553	-2.3477	-2.4036	-2.2979	-2.3538
.90	-2.4586	-2.5047	-2.4048	-2.4509	-2.3531	-2.3992	-2.3033	-2.3494
.92	-2.4638	-2.5002	-2.4100	-2.4464	-2.3583	-2.3947	-2.3085	-2.3450
.94	-2.4688	-2.4958	-2.4150	-2.4420	-2.3633	-2.3903	-2.3135	-2.3406
.96	-2.4736	-2.4914	-2.4198	-2.4376	-2.3681	-2.3859	-2.3183	-2.3361
.98	-2.4782	-2.4870	-2.4244	-2.4332	-2.3727	-2.3815	-2.3229	-2.3317
1.00	-2.4827	-2.4827	-2.4288	-2.4288	-2.3771	-2.3771	-2.3274	-2.3274

Table D–5: $0.115 \leq \hat{\rho}_4 \leq 0.130$ and λ_3, $\lambda_4 < 0$

	$\hat{\rho}_4 = 0.115$		$\hat{\rho}_4 = 0.120$		$\hat{\rho}_4 = 0.125$		$\hat{\rho}_4 = 0.130$	
$\hat{\rho}_3$	λ_3	λ_4	λ_3	λ_4	λ_3	λ_4	λ_3	λ_4
.02	-.0648	-2.2773						
.04	-.8086	-2.3335	-.7420	-2.2787	-.6761	-2.2256	-.6108	-2.1740
.06	-1.0703	-2.3604	-1.0110	-2.3067	-.9533	-2.2547	-.8969	-2.2042
.08	-1.2361	-2.3779	-1.1795	-2.3247	-1.1246	-2.2733	-1.0712	-2.2234
.10	-1.3576	-2.3902	-1.3023	-2.3373	-1.2489	-2.2862	-1.1969	-2.2367
.12	-1.4530	-2.3989	-1.3986	-2.3463	-1.3460	-2.2955	-1.2949	-2.2463
.14	-1.5312	-2.4053	-1.4774	-2.3528	-1.4253	-2.3022	-1.3749	-2.2532
.16	-1.5971	-2.4097	-1.5437	-2.3575	-1.4921	-2.3070	-1.4421	-2.2582
.18	-1.6538	-2.4127	-1.6008	-2.3606	-1.5495	-2.3103	-1.4998	-2.2616
.20	-1.7034	-2.4146	-1.6506	-2.3626	-1.5996	-2.3123	-1.5502	-2.2638
.22	-1.7473	-2.4155	-1.6946	-2.3636	-1.6438	-2.3134	-1.5947	-2.2650
.24	-1.7864	-2.4156	-1.7340	-2.3638	-1.6834	-2.3138	-1.6344	-2.2654
.26	-1.8217	-2.4151	-1.7694	-2.3633	-1.7189	-2.3134	-1.6701	-2.2651
.28	-1.8537	-2.4141	-1.8015	-2.3624	-1.7511	-2.3125	-1.7024	-2.2642
.30	-1.8829	-2.4125	-1.8308	-2.3609	-1.7805	-2.3110	-1.7319	-2.2629
.32	-1.9096	-2.4106	-1.8576	-2.3590	-1.8074	-2.3092	-1.7589	-2.2611
.34	-1.9342	-2.4084	-1.8823	-2.3568	-1.8322	-2.3070	-1.7838	-2.2590
.36	-1.9570	-2.4058	-1.9051	-2.3543	-1.8551	-2.3046	-1.8067	-2.2565
.38	-1.9780	-2.4030	-1.9262	-2.3515	-1.8763	-2.3018	-1.8280	-2.2538
.40	-1.9976	-2.3999	-1.9459	-2.3485	-1.8960	-2.2988	-1.8478	-2.2509
.42	-2.0159	-2.3967	-1.9642	-2.3453	-1.9144	-2.2957	-1.8662	-2.2477
.44	-2.0330	-2.3933	-1.9814	-2.3419	-1.9316	-2.2923	-1.8834	-2.2444
.46	-2.0490	-2.3897	-1.9974	-2.3384	-1.9477	-2.2888	-1.8996	-2.2409
.48	-2.0641	-2.3861	-2.0125	-2.3347	-1.9628	-2.2851	-1.9147	-2.2373
.50	-2.0783	-2.3823	-2.0267	-2.3309	-1.9770	-2.2814	-1.9290	-2.2335
.52	-2.0916	-2.3784	-2.0401	-2.3270	-1.9904	-2.2775	-1.9424	-2.2297
.54	-2.1042	-2.3744	-2.0527	-2.3230	-2.0031	-2.2735	-1.9551	-2.2257
.56	-2.1161	-2.3703	-2.0647	-2.3190	-2.0150	-2.2695	-1.9671	-2.2217
.58	-2.1274	-2.3662	-2.0760	-2.3149	-2.0264	-2.2654	-1.9784	-2.2176
.60	-2.1381	-2.3620	-2.0867	-2.3107	-2.0371	-2.2612	-1.9892	-2.2135
.62	-2.1482	-2.3578	-2.0969	-2.3065	-2.0473	-2.2570	-1.9994	-2.2092
.64	-2.1579	-2.3535	-2.1065	-2.3022	-2.0570	-2.2528	-2.0091	-2.2050
.66	-2.1671	-2.3492	-2.1157	-2.2979	-2.0662	-2.2485	-2.0183	-2.2007
.68	-2.1758	-2.3448	-2.1245	-2.2936	-2.0750	-2.2441	-2.0271	-2.1964
.70	-2.1842	-2.3404	-2.1328	-2.2892	-2.0833	-2.2398	-2.0355	-2.1920
.72	-2.1921	-2.3361	-2.1408	-2.2848	-2.0913	-2.2354	-2.0435	-2.1876
.74	-2.1997	-2.3317	-2.1484	-2.2804	-2.0989	-2.2310	-2.0511	-2.1833
.76	-2.2070	-2.3272	-2.1557	-2.2760	-2.1062	-2.2266	-2.0584	-2.1788
.78	-2.2140	-2.3228	-2.1627	-2.2716	-2.1132	-2.2222	-2.0654	-2.1744
.80	-2.2206	-2.3184	-2.1693	-2.2671	-2.1199	-2.2177	-2.0721	-2.1700
.82	-2.2270	-2.3139	-2.1757	-2.2627	-2.1263	-2.2133	-2.0785	-2.1656
.84	-2.2331	-2.3095	-2.1818	-2.2583	-2.1324	-2.2088	-2.0846	-2.1611
.86	-2.2390	-2.3050	-2.1877	-2.2538	-2.1383	-2.2044	-2.0905	-2.1567
.88	-2.2446	-2.3006	-2.1934	-2.2494	-2.1439	-2.2000	-2.0962	-2.1522
.90	-2.2500	-2.2962	-2.1988	-2.2449	-2.1493	-2.1955	-2.1016	-2.1478
.92	-2.2552	-2.2918	-2.2040	-2.2405	-2.1546	-2.1911	-2.1068	-2.1434
.94	-2.2602	-2.2873	-2.2090	-2.2361	-2.1596	-2.1867	-2.1118	-2.1390
.96	-2.2650	-2.2829	-2.2138	-2.2317	-2.1644	-2.1823	-2.1166	-2.1346
.98	-2.2697	-2.2785	-2.2184	-2.2273	-2.1690	-2.1779	-2.1213	-2.1301
1.00	-2.2741	-2.2741	-2.2229	-2.2229	-2.1735	-2.1735	-2.1258	-2.1258

Table D–5: $0.136 \leq \hat{\rho}_4 \leq 0.154$ and $\lambda_3,\ \lambda_4 < 0$

	$\hat{\rho}_4 = 0.136$		$\hat{\rho}_4 = 0.142$		$\hat{\rho}_4 = 0.148$		$\hat{\rho}_4 = 0.154$	
$\hat{\rho}_3$	λ_3	λ_4	λ_3	λ_4	λ_3	λ_4	λ_3	λ_4
.04	-.5325	-2.1138	-.4539	-2.0551	-.3744	-1.9979	-.2930	-1.9419
.06	-.8307	-2.1455	-.7659	-2.0886	-.7024	-2.0334	-.6399	-1.9797
.08	-1.0088	-2.1655	-.9482	-2.1094	-.8892	-2.0551	-.8315	-2.0023
.10	-1.1365	-2.1793	-1.0779	-2.1238	-1.0210	-2.0700	-.9656	-2.0178
.12	-1.2356	-2.1892	-1.1782	-2.1341	-1.1226	-2.0807	-1.0685	-2.0289
.14	-1.3164	-2.1964	-1.2598	-2.1415	-1.2049	-2.0884	-1.1517	-2.0370
.16	-1.3841	-2.2016	-1.3281	-2.1469	-1.2738	-2.0941	-1.2212	-2.0428
.18	-1.4423	-2.2051	-1.3866	-2.1507	-1.3328	-2.0980	-1.2806	-2.0470
.20	-1.4929	-2.2075	-1.4376	-2.1532	-1.3842	-2.1007	-1.3323	-2.0498
.22	-1.5377	-2.2088	-1.4827	-2.1546	-1.4294	-2.1023	-1.3779	-2.0515
.24	-1.5776	-2.2093	-1.5228	-2.1552	-1.4698	-2.1030	-1.4185	-2.0524
.26	-1.6135	-2.2091	-1.5589	-2.1551	-1.5061	-2.1030	-1.4549	-2.0525
.28	-1.6460	-2.2083	-1.5915	-2.1544	-1.5389	-2.1024	-1.4879	-2.0520
.30	-1.6756	-2.2070	-1.6213	-2.1532	-1.5688	-2.1013	-1.5180	-2.0509
.32	-1.7027	-2.2053	-1.6485	-2.1516	-1.5961	-2.0997	-1.5454	-2.0494
.34	-1.7276	-2.2033	-1.6736	-2.1496	-1.6213	-2.0977	-1.5707	-2.0475
.36	-1.7507	-2.2009	-1.6967	-2.1472	-1.6445	-2.0954	-1.5940	-2.0453
.38	-1.7720	-2.1982	-1.7181	-2.1446	-1.6660	-2.0929	-1.6156	-2.0428
.40	-1.7919	-2.1953	-1.7380	-2.1418	-1.6860	-2.0900	-1.6357	-2.0400
.42	-1.8104	-2.1922	-1.7566	-2.1387	-1.7046	-2.0870	-1.6544	-2.0370
.44	-1.8277	-2.1889	-1.7739	-2.1354	-1.7220	-2.0838	-1.6718	-2.0338
.46	-1.8438	-2.1854	-1.7902	-2.1320	-1.7383	-2.0804	-1.6882	-2.0305
.48	-1.8590	-2.1818	-1.8054	-2.1284	-1.7536	-2.0768	-1.7035	-2.0269
.50	-1.8733	-2.1781	-1.8198	-2.1247	-1.7680	-2.0732	-1.7179	-2.0233
.52	-1.8868	-2.1743	-1.8333	-2.1209	-1.7815	-2.0694	-1.7315	-2.0195
.54	-1.8995	-2.1703	-1.8460	-2.1170	-1.7943	-2.0655	-1.7443	-2.0157
.56	-1.9115	-2.1663	-1.8581	-2.1130	-1.8064	-2.0615	-1.7564	-2.0117
.58	-1.9229	-2.1623	-1.8695	-2.1090	-1.8178	-2.0575	-1.7679	-2.0077
.60	-1.9337	-2.1581	-1.8803	-2.1048	-1.8287	-2.0534	-1.7788	-2.0036
.62	-1.9439	-2.1539	-1.8905	-2.1006	-1.8390	-2.0492	-1.7891	-1.9994
.64	-1.9537	-2.1497	-1.9003	-2.0964	-1.8487	-2.0450	-1.7989	-1.9952
.66	-1.9629	-2.1454	-1.9095	-2.0921	-1.8580	-2.0407	-1.8082	-1.9910
.68	-1.9717	-2.1411	-1.9184	-2.0878	-1.8669	-2.0364	-1.8170	-1.9867
.70	-1.9801	-2.1367	-1.9268	-2.0835	-1.8753	-2.0321	-1.8255	-1.9824
.72	-1.9881	-2.1324	-1.9348	-2.0791	-1.8833	-2.0277	-1.8335	-1.9780
.74	-1.9958	-2.1280	-1.9425	-2.0748	-1.8910	-2.0234	-1.8412	-1.9737
.76	-2.0031	-2.1236	-1.9498	-2.0704	-1.8983	-2.0190	-1.8486	-1.9693
.78	-2.0101	-2.1192	-1.9568	-2.0659	-1.9054	-2.0146	-1.8556	-1.9649
.80	-2.0168	-2.1147	-1.9635	-2.0615	-1.9121	-2.0101	-1.8623	-1.9605
.82	-2.0232	-2.1103	-1.9699	-2.0571	-1.9185	-2.0057	-1.8688	-1.9560
.84	-2.0293	-2.1059	-1.9761	-2.0527	-1.9247	-2.0013	-1.8749	-1.9516
.86	-2.0352	-2.1014	-1.9820	-2.0482	-1.9306	-1.9968	-1.8808	-1.9472
.88	-2.0409	-2.0970	-1.9876	-2.0438	-1.9362	-1.9924	-1.8865	-1.9427
.90	-2.0463	-2.0926	-1.9931	-2.0393	-1.9417	-1.9880	-1.8920	-1.9383
.92	-2.0515	-2.0881	-1.9983	-2.0349	-1.9469	-1.9835	-1.8972	-1.9339
.94	-2.0566	-2.0837	-2.0033	-2.0305	-1.9519	-1.9791	-1.9022	-1.9294
.96	-2.0614	-2.0793	-2.0082	-2.0261	-1.9568	-1.9747	-1.9071	-1.9250
.98	-2.0660	-2.0749	-2.0128	-2.0217	-1.9614	-1.9703	-1.9117	-1.9206
1.00	-2.0705	-2.0705	-2.0173	-2.0173	-1.9659	-1.9659	-1.9162	-1.9162

Table D–5: $0.161 \leq \hat{\rho}_4 \leq 0.182$ and $\lambda_3, \lambda_4 < 0$

	$\hat{\rho}_4 = 0.161$		$\hat{\rho}_4 = 0.168$		$\hat{\rho}_4 = 0.175$		$\hat{\rho}_4 = 0.182$	
$\hat{\rho}_3$	λ_3	λ_4	λ_3	λ_4	λ_3	λ_4	λ_3	λ_4
.04	-.1943	-1.8777	-.0889	-1.8143				
.06	-.5681	-1.9186	-.4971	-1.8591	-.4268	-1.8009	-.3567	-1.7439
.08	-.7659	-1.9425	-.7018	-1.8843	-.6389	-1.8277	-.5773	-1.7723
.10	-.9027	-1.9587	-.8415	-1.9013	-.7818	-1.8455	-.7236	-1.7912
.12	-1.0072	-1.9703	-.9476	-1.9135	-.8897	-1.8583	-.8333	-1.8045
.14	-1.0914	-1.9788	-1.0330	-1.9224	-.9762	-1.8676	-.9209	-1.8143
.16	-1.1616	-1.9849	-1.1040	-1.9289	-1.0480	-1.8744	-.9935	-1.8215
.18	-1.2216	-1.9893	-1.1645	-1.9335	-1.1091	-1.8794	-1.0552	-1.8267
.20	-1.2738	-1.9924	-1.2171	-1.9368	-1.1621	-1.8828	-1.1087	-1.8304
.22	-1.3197	-1.9943	-1.2633	-1.9389	-1.2087	-1.8851	-1.1557	-1.8329
.24	-1.3605	-1.9953	-1.3045	-1.9400	-1.2502	-1.8864	-1.1974	-1.8344
.26	-1.3972	-1.9955	-1.3414	-1.9404	-1.2873	-1.8870	-1.2348	-1.8351
.28	-1.4304	-1.9951	-1.3748	-1.9401	-1.3209	-1.8868	-1.2686	-1.8350
.30	-1.4606	-1.9941	-1.4052	-1.9392	-1.3515	-1.8860	-1.2993	-1.8344
.32	-1.4882	-1.9927	-1.4329	-1.9379	-1.3794	-1.8848	-1.3274	-1.8333
.34	-1.5136	-1.9909	-1.4585	-1.9362	-1.4050	-1.8832	-1.3532	-1.8317
.36	-1.5370	-1.9888	-1.4820	-1.9341	-1.4287	-1.8812	-1.3770	-1.8298
.38	-1.5587	-1.9863	-1.5038	-1.9317	-1.4506	-1.8788	-1.3990	-1.8275
.40	-1.5789	-1.9836	-1.5240	-1.9290	-1.4709	-1.8762	-1.4194	-1.8250
.42	-1.5976	-1.9806	-1.5429	-1.9261	-1.4898	-1.8734	-1.4384	-1.8222
.44	-1.6152	-1.9775	-1.5605	-1.9230	-1.5075	-1.8703	-1.4561	-1.8192
.46	-1.6316	-1.9741	-1.5769	-1.9198	-1.5240	-1.8671	-1.4727	-1.8160
.48	-1.6470	-1.9707	-1.5924	-1.9163	-1.5396	-1.8637	-1.4883	-1.8126
.50	-1.6615	-1.9671	-1.6069	-1.9127	-1.5541	-1.8601	-1.5029	-1.8091
.52	-1.6751	-1.9633	-1.6206	-1.9090	-1.5679	-1.8565	-1.5167	-1.8055
.54	-1.6880	-1.9595	-1.6335	-1.9052	-1.5808	-1.8527	-1.5297	-1.8018
.56	-1.7001	-1.9556	-1.6457	-1.9013	-1.5930	-1.8488	-1.5420	-1.7979
.58	-1.7116	-1.9516	-1.6572	-1.8973	-1.6046	-1.8449	-1.5536	-1.7940
.60	-1.7225	-1.9475	-1.6682	-1.8933	-1.6156	-1.8408	-1.5646	-1.7900
.62	-1.7328	-1.9433	-1.6785	-1.8892	-1.6260	-1.8367	-1.5750	-1.7859
.64	-1.7427	-1.9392	-1.6884	-1.8850	-1.6358	-1.8326	-1.5849	-1.7817
.66	-1.7520	-1.9349	-1.6977	-1.8808	-1.6452	-1.8284	-1.5943	-1.7776
.68	-1.7609	-1.9306	-1.7066	-1.8765	-1.6541	-1.8241	-1.6033	-1.7733
.70	-1.7693	-1.9263	-1.7151	-1.8722	-1.6626	-1.8198	-1.6118	-1.7690
.72	-1.7774	-1.9220	-1.7232	-1.8679	-1.6708	-1.8155	-1.6199	-1.7647
.74	-1.7851	-1.9176	-1.7309	-1.8635	-1.6785	-1.8112	-1.6277	-1.7604
.76	-1.7925	-1.9132	-1.7383	-1.8591	-1.6859	-1.8068	-1.6351	-1.7560
.78	-1.7995	-1.9089	-1.7454	-1.8548	-1.6930	-1.8024	-1.6421	-1.7517
.80	-1.8063	-1.9044	-1.7521	-1.8504	-1.6997	-1.7980	-1.6489	-1.7473
.82	-1.8127	-1.9000	-1.7586	-1.8459	-1.7062	-1.7936	-1.6554	-1.7429
.84	-1.8189	-1.8956	-1.7648	-1.8415	-1.7124	-1.7892	-1.6616	-1.7384
.86	-1.8248	-1.8912	-1.7707	-1.8371	-1.7183	-1.7848	-1.6676	-1.7340
.88	-1.8305	-1.8867	-1.7764	-1.8327	-1.7240	-1.7803	-1.6733	-1.7296
.90	-1.8359	-1.8823	-1.7818	-1.8282	-1.7295	-1.7759	-1.6787	-1.7252
.92	-1.8412	-1.8779	-1.7871	-1.8238	-1.7347	-1.7715	-1.6840	-1.7207
.94	-1.8462	-1.8734	-1.7921	-1.8194	-1.7398	-1.7670	-1.6891	-1.7163
.96	-1.8511	-1.8690	-1.7970	-1.8149	-1.7446	-1.7626	-1.6939	-1.7119
.98	-1.8557	-1.8646	-1.8016	-1.8105	-1.7493	-1.7582	-1.6986	-1.7075
1.00	-1.8602	-1.8602	-1.8061	-1.8061	-1.7538	-1.7538	-1.7031	-1.7031

Table D–5: $0.190 \leq \hat{\rho}_4 \leq 0.214$ and $\lambda_3, \ \lambda_4 < 0$

	$\hat{\rho}_4 = 0.190$		$\hat{\rho}_4 = 0.198$		$\hat{\rho}_4 = 0.206$		$\hat{\rho}_4 = 0.214$	
$\hat{\rho}_3$	λ_3	λ_4	λ_3	λ_4	λ_3	λ_4	λ_3	λ_4
.06	-.2767	-1.6799	-.1962	-1.6169	-.1148	-1.5545	-.0319	-1.4925
.08	-.5082	-1.7105	-.4403	-1.6501	-.3734	-1.5907	-.3075	-1.5322
.10	-.6586	-1.7305	-.5950	-1.6713	-.5329	-1.6134	-.4721	-1.5565
.12	-.7705	-1.7447	-.7092	-1.6863	-.6495	-1.6293	-.5911	-1.5735
.14	-.8594	-1.7550	-.7996	-1.6973	-.7413	-1.6409	-.6844	-1.5858
.16	-.9330	-1.7627	-.8741	-1.7054	-.8168	-1.6495	-.7609	-1.5949
.18	-.9954	-1.7682	-.9372	-1.7114	-.8807	-1.6559	-.8255	-1.6017
.20	-1.0494	-1.7722	-.9918	-1.7157	-.9358	-1.6605	-.8812	-1.6067
.22	-1.0968	-1.7750	-1.0396	-1.7186	-.9840	-1.6638	-.9299	-1.6103
.24	-1.1389	-1.7767	-1.0821	-1.7206	-1.0268	-1.6659	-.9730	-1.6126
.26	-1.1766	-1.7775	-1.1201	-1.7216	-1.0651	-1.6672	-1.0116	-1.6141
.28	-1.2106	-1.7776	-1.1543	-1.7219	-1.0997	-1.6676	-1.0464	-1.6147
.30	-1.2415	-1.7771	-1.1855	-1.7215	-1.1310	-1.6674	-1.0780	-1.6147
.32	-1.2698	-1.7761	-1.2139	-1.7206	-1.1596	-1.6667	-1.1068	-1.6141
.34	-1.2957	-1.7747	-1.2400	-1.7193	-1.1859	-1.6655	-1.1332	-1.6130
.36	-1.3196	-1.7728	-1.2641	-1.7176	-1.2101	-1.6638	-1.1575	-1.6115
.38	-1.3418	-1.7707	-1.2863	-1.7155	-1.2324	-1.6618	-1.1800	-1.6096
.40	-1.3623	-1.7682	-1.3069	-1.7131	-1.2532	-1.6595	-1.2009	-1.6074
.42	-1.3814	-1.7655	-1.3261	-1.7104	-1.2725	-1.6570	-1.2202	-1.6049
.44	-1.3992	-1.7625	-1.3441	-1.7076	-1.2905	-1.6542	-1.2383	-1.6021
.46	-1.4159	-1.7594	-1.3608	-1.7045	-1.3073	-1.6511	-1.2553	-1.5992
.48	-1.4315	-1.7561	-1.3765	-1.7012	-1.3231	-1.6479	-1.2711	-1.5961
.50	-1.4462	-1.7526	-1.3913	-1.6978	-1.3379	-1.6446	-1.2860	-1.5928
.52	-1.4601	-1.7490	-1.4052	-1.6943	-1.3519	-1.6411	-1.3000	-1.5893
.54	-1.4731	-1.7453	-1.4183	-1.6906	-1.3650	-1.6374	-1.3132	-1.5857
.56	-1.4854	-1.7415	-1.4306	-1.6868	-1.3774	-1.6337	-1.3256	-1.5820
.58	-1.4971	-1.7376	-1.4423	-1.6830	-1.3891	-1.6299	-1.3374	-1.5782
.60	-1.5081	-1.7336	-1.4534	-1.6790	-1.4002	-1.6259	-1.3486	-1.5743
.62	-1.5186	-1.7296	-1.4639	-1.6750	-1.4108	-1.6219	-1.3591	-1.5703
.64	-1.5285	-1.7254	-1.4738	-1.6709	-1.4208	-1.6179	-1.3691	-1.5663
.66	-1.5379	-1.7213	-1.4833	-1.6667	-1.4303	-1.6137	-1.3787	-1.5622
.68	-1.5469	-1.7171	-1.4923	-1.6625	-1.4393	-1.6096	-1.3877	-1.5580
.70	-1.5555	-1.7128	-1.5009	-1.6583	-1.4479	-1.6053	-1.3963	-1.5538
.72	-1.5636	-1.7085	-1.5090	-1.6540	-1.4561	-1.6011	-1.4045	-1.5496
.74	-1.5714	-1.7042	-1.5168	-1.6497	-1.4639	-1.5968	-1.4124	-1.5453
.76	-1.5788	-1.6998	-1.5243	-1.6454	-1.4713	-1.5925	-1.4198	-1.5410
.78	-1.5859	-1.6955	-1.5314	-1.6410	-1.4785	-1.5881	-1.4270	-1.5367
.80	-1.5927	-1.6911	-1.5382	-1.6366	-1.4853	-1.5837	-1.4338	-1.5323
.82	-1.5992	-1.6867	-1.5447	-1.6322	-1.4918	-1.5794	-1.4403	-1.5279
.84	-1.6054	-1.6823	-1.5509	-1.6278	-1.4980	-1.5750	-1.4466	-1.5235
.86	-1.6113	-1.6778	-1.5569	-1.6234	-1.5040	-1.5705	-1.4526	-1.5191
.88	-1.6171	-1.6734	-1.5626	-1.6190	-1.5097	-1.5661	-1.4583	-1.5147
.90	-1.6225	-1.6690	-1.5681	-1.6146	-1.5152	-1.5617	-1.4638	-1.5103
.92	-1.6278	-1.6646	-1.5734	-1.6101	-1.5205	-1.5573	-1.4691	-1.5059
.94	-1.6329	-1.6601	-1.5784	-1.6057	-1.5256	-1.5529	-1.4742	-1.5015
.96	-1.6377	-1.6557	-1.5833	-1.6013	-1.5304	-1.5484	-1.4790	-1.4970
.98	-1.6424	-1.6513	-1.5880	-1.5969	-1.5351	-1.5440	-1.4837	-1.4926
1.00	-1.6469	-1.6469	-1.5925	-1.5925	-1.5396	-1.5396	-1.4882	-1.4882

Table D–5: $0.223 \leq \hat{\rho}_4 \leq 0.250$ and λ_3, $\lambda_4 < 0$

$\hat{\rho}_3$	$\hat{\rho}_4 = 0.223$		$\hat{\rho}_4 = 0.232$		$\hat{\rho}_4 = 0.241$		$\hat{\rho}_4 = 0.250$	
	λ_3	λ_4	λ_3	λ_4	λ_3	λ_4	λ_3	λ_4
.08	-.2346	-1.4674	-.1628	-1.4032	-.0925	-1.3395	-.0241	-1.2761
.10	-.4051	-1.4937	-.3397	-1.4318	-.2759	-1.3707	-.2139	-1.3102
.12	-.5270	-1.5118	-.4645	-1.4513	-.4036	-1.3917	-.3442	-1.3328
.14	-.6220	-1.5250	-.5612	-1.4654	-.5019	-1.4068	-.4441	-1.3491
.16	-.6997	-1.5348	-.6400	-1.4759	-.5818	-1.4181	-.5250	-1.3612
.18	-.7651	-1.5421	-.7062	-1.4838	-.6488	-1.4266	-.5928	-1.3703
.20	-.8214	-1.5475	-.7631	-1.4897	-.7063	-1.4329	-.6509	-1.3772
.22	-.8706	-1.5514	-.8128	-1.4939	-.7565	-1.4376	-.7015	-1.3823
.24	-.9141	-1.5541	-.8567	-1.4970	-.8008	-1.4410	-.7462	-1.3860
.26	-.9530	-1.5558	-.8960	-1.4989	-.8404	-1.4432	-.7861	-1.3886
.28	-.9881	-1.5567	-.9313	-1.5000	-.8760	-1.4446	-.8220	-1.3902
.30	-1.0199	-1.5568	-.9634	-1.5004	-.9083	-1.4452	-.8544	-1.3910
.32	-1.0489	-1.5564	-.9926	-1.5001	-.9377	-1.4451	-.8841	-1.3912
.34	-1.0755	-1.5555	-1.0194	-1.4993	-.9646	-1.4445	-.9112	-1.3908
.36	-1.1000	-1.5541	-1.0440	-1.4981	-.9894	-1.4434	-.9361	-1.3898
.38	-1.1226	-1.5523	-1.0668	-1.4965	-1.0123	-1.4419	-.9591	-1.3885
.40	-1.1436	-1.5502	-1.0879	-1.4945	-1.0335	-1.4400	-.9805	-1.3867
.42	-1.1631	-1.5478	-1.1075	-1.4922	-1.0533	-1.4378	-1.0003	-1.3847
.44	-1.1813	-1.5452	-1.1258	-1.4896	-1.0716	-1.4354	-1.0188	-1.3823
.46	-1.1983	-1.5423	-1.1429	-1.4868	-1.0888	-1.4327	-1.0360	-1.3797
.48	-1.2142	-1.5392	-1.1589	-1.4838	-1.1049	-1.4298	-1.0522	-1.3769
.50	-1.2292	-1.5360	-1.1739	-1.4807	-1.1200	-1.4267	-1.0674	-1.3739
.52	-1.2432	-1.5326	-1.1880	-1.4773	-1.1342	-1.4234	-1.0816	-1.3707
.54	-1.2565	-1.5290	-1.2013	-1.4739	-1.1475	-1.4200	-1.0950	-1.3673
.56	-1.2690	-1.5254	-1.2139	-1.4702	-1.1602	-1.4164	-1.1077	-1.3638
.58	-1.2808	-1.5216	-1.2257	-1.4665	-1.1721	-1.4128	-1.1197	-1.3602
.60	-1.2920	-1.5178	-1.2370	-1.4627	-1.1833	-1.4090	-1.1310	-1.3565
.62	-1.3026	-1.5138	-1.2476	-1.4588	-1.1940	-1.4052	-1.1417	-1.3527
.64	-1.3127	-1.5098	-1.2577	-1.4548	-1.2041	-1.4012	-1.1518	-1.3488
.66	-1.3222	-1.5058	-1.2673	-1.4508	-1.2137	-1.3972	-1.1615	-1.3448
.68	-1.3313	-1.5016	-1.2764	-1.4467	-1.2229	-1.3931	-1.1706	-1.3408
.70	-1.3399	-1.4974	-1.2851	-1.4425	-1.2316	-1.3890	-1.1793	-1.3367
.72	-1.3482	-1.4932	-1.2933	-1.4383	-1.2398	-1.3848	-1.1876	-1.3325
.74	-1.3560	-1.4890	-1.3012	-1.4341	-1.2477	-1.3806	-1.1955	-1.3283
.76	-1.3635	-1.4847	-1.3087	-1.4298	-1.2553	-1.3764	-1.2031	-1.3241
.78	-1.3707	-1.4803	-1.3159	-1.4255	-1.2624	-1.3721	-1.2103	-1.3198
.80	-1.3775	-1.4760	-1.3227	-1.4212	-1.2693	-1.3678	-1.2172	-1.3155
.82	-1.3840	-1.4716	-1.3293	-1.4168	-1.2759	-1.3634	-1.2237	-1.3112
.84	-1.3903	-1.4673	-1.3355	-1.4125	-1.2821	-1.3591	-1.2300	-1.3069
.86	-1.3963	-1.4629	-1.3415	-1.4081	-1.2882	-1.3547	-1.2360	-1.3025
.88	-1.4020	-1.4585	-1.3473	-1.4037	-1.2939	-1.3503	-1.2418	-1.2981
.90	-1.4075	-1.4540	-1.3528	-1.3993	-1.2994	-1.3459	-1.2473	-1.2938
.92	-1.4128	-1.4496	-1.3581	-1.3949	-1.3047	-1.3415	-1.2526	-1.2894
.94	-1.4179	-1.4452	-1.3632	-1.3905	-1.3098	-1.3371	-1.2577	-1.2850
.96	-1.4228	-1.4408	-1.3680	-1.3861	-1.3147	-1.3327	-1.2626	-1.2806
.98	-1.4275	-1.4364	-1.3727	-1.3816	-1.3194	-1.3283	-1.2672	-1.2761
1.00	-1.4320	-1.4320	-1.3772	-1.3772	-1.3239	-1.3239	-1.2717	-1.2717

Table D–5: $0.26 \leq \hat{\rho}_4 \leq 0.29$ and $\lambda_3,\ \lambda_4 < 0$

	$\hat{\rho}_4 = 0.26$		$\hat{\rho}_4 = 0.27$		$\hat{\rho}_4 = 0.28$		$\hat{\rho}_4 = 0.29$	
$\hat{\rho}_3$	λ_3	λ_4	λ_3	λ_4	λ_3	λ_4	λ_3	λ_4
.10	-.1474	-1.2434	-.0839	-1.1771	-.0245	-1.1110		
.12	-.2803	-1.2681	-.2188	-1.2040	-.1600	-1.1402	-.1047	-1.0767
.14	-.3817	-1.2858	-.3214	-1.2231	-.2632	-1.1609	-.2076	-1.0991
.16	-.4636	-1.2989	-.4041	-1.2373	-.3464	-1.1763	-.2908	-1.1157
.18	-.5322	-1.3088	-.4732	-1.2480	-.4160	-1.1879	-.3605	-1.1283
.20	-.5908	-1.3163	-.5323	-1.2562	-.4755	-1.1969	-.4202	-1.1381
.22	-.6419	-1.3219	-.5838	-1.2624	-.5273	-1.2037	-.4722	-1.1455
.24	-.6870	-1.3261	-.6293	-1.2670	-.5730	-1.2088	-.5181	-1.1512
.26	-.7272	-1.3290	-.6698	-1.2704	-.6137	-1.2126	-.5590	-1.1555
.28	-.7633	-1.3309	-.7062	-1.2727	-.6503	-1.2152	-.5957	-1.1585
.30	-.7961	-1.3320	-.7391	-1.2741	-.6834	-1.2170	-.6290	-1.1606
.32	-.8259	-1.3324	-.7691	-1.2747	-.7136	-1.2179	-.6593	-1.1619
.34	-.8532	-1.3322	-.7966	-1.2747	-.7412	-1.2182	-.6871	-1.1624
.36	-.8783	-1.3315	-.8218	-1.2742	-.7666	-1.2179	-.7125	-1.1624
.38	-.9015	-1.3303	-.8451	-1.2732	-.7900	-1.2171	-.7361	-1.1618
.40	-.9229	-1.3287	-.8667	-1.2718	-.8117	-1.2159	-.7578	-1.1607
.42	-.9429	-1.3268	-.8867	-1.2700	-.8318	-1.2143	-.7780	-1.1593
.44	-.9614	-1.3246	-.9054	-1.2679	-.8506	-1.2123	-.7969	-1.1575
.46	-.9788	-1.3221	-.9228	-1.2656	-.8681	-1.2101	-.8145	-1.1554
.48	-.9950	-1.3194	-.9392	-1.2630	-.8845	-1.2076	-.8309	-1.1531
.50	-1.0103	-1.3164	-.9545	-1.2601	-.8998	-1.2049	-.8463	-1.1505
.52	-1.0246	-1.3133	-.9688	-1.2571	-.9143	-1.2019	-.8608	-1.1477
.54	-1.0381	-1.3100	-.9824	-1.2539	-.9279	-1.1989	-.8744	-1.1447
.56	-1.0508	-1.3066	-.9951	-1.2506	-.9407	-1.1956	-.8873	-1.1415
.58	-1.0628	-1.3031	-1.0072	-1.2471	-.9527	-1.1922	-.8994	-1.1382
.60	-1.0741	-1.2994	-1.0186	-1.2435	-.9642	-1.1887	-.9108	-1.1347
.62	-1.0849	-1.2957	-1.0293	-1.2398	-.9750	-1.1850	-.9217	-1.1312
.64	-1.0951	-1.2918	-1.0396	-1.2360	-.9852	-1.1813	-.9319	-1.1275
.66	-1.1047	-1.2879	-1.0493	-1.2321	-.9949	-1.1775	-.9417	-1.1237
.68	-1.1139	-1.2839	-1.0585	-1.2282	-1.0042	-1.1736	-.9509	-1.1199
.70	-1.1226	-1.2798	-1.0672	-1.2242	-1.0130	-1.1696	-.9597	-1.1160
.72	-1.1310	-1.2757	-1.0756	-1.2201	-1.0213	-1.1656	-.9681	-1.1120
.74	-1.1389	-1.2715	-1.0835	-1.2160	-1.0293	-1.1615	-.9761	-1.1079
.76	-1.1464	-1.2673	-1.0911	-1.2118	-1.0368	-1.1573	-.9836	-1.1038
.78	-1.1537	-1.2631	-1.0983	-1.2076	-1.0441	-1.1531	-.9909	-1.0997
.80	-1.1605	-1.2588	-1.1052	-1.2033	-1.0510	-1.1489	-.9978	-1.0955
.82	-1.1671	-1.2545	-1.1118	-1.1990	-1.0576	-1.1447	-1.0044	-1.0913
.84	-1.1734	-1.2502	-1.1181	-1.1947	-1.0639	-1.1404	-1.0107	-1.0870
.86	-1.1794	-1.2459	-1.1241	-1.1904	-1.0699	-1.1361	-1.0167	-1.0827
.88	-1.1852	-1.2415	-1.1299	-1.1861	-1.0757	-1.1318	-1.0225	-1.0784
.90	-1.1907	-1.2371	-1.1354	-1.1817	-1.0812	-1.1274	-1.0280	-1.0741
.92	-1.1960	-1.2327	-1.1407	-1.1773	-1.0865	-1.1231	-1.0333	-1.0698
.94	-1.2011	-1.2283	-1.1458	-1.1730	-1.0915	-1.1187	-1.0383	-1.0654
.96	-1.2060	-1.2239	-1.1506	-1.1686	-1.0964	-1.1143	-1.0432	-1.0611
.98	-1.2106	-1.2195	-1.1553	-1.1642	-1.1011	-1.1099	-1.0478	-1.0567
1.00	-1.2151	-1.2151	-1.1598	-1.1598	-1.1056	-1.1056	-1.0523	-1.0523

Table D–5: $0.301 \leq \hat{\rho}_4 \leq 0.334$ and $\lambda_3, \lambda_4 < 0$

	$\hat{\rho}_4 = 0.301$		$\hat{\rho}_4 = 0.312$		$\hat{\rho}_4 = 0.323$		$\hat{\rho}_4 = 0.334$	
$\hat{\rho}_3$	λ_3	λ_4	λ_3	λ_4	λ_3	λ_4	λ_3	λ_4
.12	-.0486	-1.0071						
.14	-.1499	-1.0314	-.0965	-.9638	-.0486	-.8965	-.0070	-.8295
.16	-.2324	-1.0494	-.1773	-.9833	-.1263	-.9174	-.0800	-.8515
.18	-.3018	-1.0632	-.2458	-.9983	-.1930	-.9336	-.1439	-.8688
.20	-.3614	-1.0739	-.3049	-1.0100	-.2510	-.9462	-.2001	-.8825
.22	-.4134	-1.0821	-.3566	-1.0190	-.3021	-.9562	-.2501	-.8933
.24	-.4593	-1.0884	-.4024	-1.0261	-.3474	-.9639	-.2947	-.9019
.26	-.5003	-1.0932	-.4433	-1.0315	-.3881	-.9700	-.3348	-.9087
.28	-.5372	-1.0968	-.4802	-1.0356	-.4248	-.9746	-.3711	-.9139
.30	-.5706	-1.0993	-.5136	-1.0385	-.4581	-.9781	-.4041	-.9179
.32	-.6010	-1.1009	-.5440	-1.0405	-.4884	-.9806	-.4343	-.9208
.34	-.6288	-1.1018	-.5718	-1.0418	-.5162	-.9822	-.4620	-.9229
.36	-.6544	-1.1020	-.5975	-1.0423	-.5418	-.9831	-.4875	-.9241
.38	-.6780	-1.1017	-.6211	-1.0423	-.5655	-.9833	-.5110	-.9247
.40	-.6998	-1.1009	-.6430	-1.0417	-.5874	-.9831	-.5329	-.9248
.42	-.7201	-1.0997	-.6633	-1.0407	-.6077	-.9823	-.5532	-.9243
.44	-.7390	-1.0981	-.6823	-1.0394	-.6266	-.9812	-.5721	-.9234
.46	-.7566	-1.0962	-.6999	-1.0376	-.6443	-.9797	-.5897	-.9221
.48	-.7731	-1.0940	-.7165	-1.0356	-.6609	-.9778	-.6062	-.9205
.50	-.7886	-1.0915	-.7320	-1.0333	-.6764	-.9757	-.6217	-.9185
.52	-.8031	-1.0888	-.7465	-1.0308	-.6909	-.9733	-.6363	-.9163
.54	-.8168	-1.0860	-.7602	-1.0280	-.7046	-.9707	-.6500	-.9139
.56	-.8297	-1.0829	-.7731	-1.0251	-.7175	-.9679	-.6629	-.9112
.58	-.8418	-1.0797	-.7853	-1.0220	-.7297	-.9649	-.6750	-.9084
.60	-.8533	-1.0763	-.7968	-1.0187	-.7412	-.9618	-.6865	-.9054
.62	-.8641	-1.0728	-.8076	-1.0153	-.7521	-.9585	-.6974	-.9022
.64	-.8744	-1.0693	-.8179	-1.0118	-.7624	-.9551	-.7077	-.8989
.66	-.8842	-1.0656	-.8277	-1.0082	-.7722	-.9516	-.7174	-.8955
.68	-.8935	-1.0618	-.8370	-1.0045	-.7814	-.9479	-.7267	-.8920
.70	-.9023	-1.0579	-.8458	-1.0007	-.7902	-.9442	-.7354	-.8883
.72	-.9106	-1.0540	-.8542	-.9968	-.7986	-.9404	-.7438	-.8846
.74	-.9186	-1.0500	-.8621	-.9929	-.8065	-.9366	-.7517	-.8808
.76	-.9262	-1.0459	-.8697	-.9889	-.8141	-.9326	-.7593	-.8769
.78	-.9334	-1.0418	-.8769	-.9849	-.8213	-.9286	-.7665	-.8730
.80	-.9403	-1.0377	-.8838	-.9808	-.8282	-.9246	-.7734	-.8690
.82	-.9469	-1.0335	-.8904	-.9766	-.8348	-.9205	-.7799	-.8650
.84	-.9532	-1.0293	-.8967	-.9724	-.8411	-.9164	-.7862	-.8609
.86	-.9592	-1.0250	-.9027	-.9682	-.8471	-.9122	-.7921	-.8568
.88	-.9650	-1.0208	-.9085	-.9640	-.8528	-.9080	-.7978	-.8526
.90	-.9705	-1.0165	-.9140	-.9597	-.8583	-.9038	-.8033	-.8484
.92	-.9758	-1.0121	-.9192	-.9554	-.8635	-.8995	-.8085	-.8442
.94	-.9808	-1.0078	-.9243	-.9511	-.8685	-.8952	-.8135	-.8400
.96	-.9857	-1.0035	-.9291	-.9468	-.8733	-.8909	-.8183	-.8358
.98	-.9903	-.9991	-.9337	-.9425	-.8779	-.8866	-.8228	-.8315
1.00	-.9948	-.9948	-.9381	-.9381	-.8823	-.8823	-.8272	-.8272

Table D–5: $0.346 \leq \hat{\rho}_4 \leq 0.382$ and λ_3, $\lambda_4 < 0$

$\hat{\rho}_3$	$\hat{\rho}_4 = 0.346$ λ_3	λ_4	$\hat{\rho}_4 = 0.358$ λ_3	λ_4	$\hat{\rho}_4 = 0.370$ λ_3	λ_4	$\hat{\rho}_4 = 0.382$ λ_3	λ_4
.16	-.0360	-.7799						
.18	-.0954	-.7982	-.0533	-.7277	-.0184	-.6575		
.20	-.1488	-.8130	-.1025	-.7433	-.0624	-.6736	-.0291	-.6042
.22	-.1967	-.8247	-.1476	-.7560	-.1036	-.6870	-.0655	-.6179
.24	-.2400	-.8342	-.1888	-.7662	-.1419	-.6980	-.1002	-.6294
.26	-.2791	-.8417	-.2265	-.7745	-.1775	-.7070	-.1328	-.6391
.28	-.3147	-.8476	-.2610	-.7812	-.2104	-.7144	-.1635	-.6471
.30	-.3472	-.8522	-.2926	-.7864	-.2408	-.7203	-.1922	-.6536
.32	-.3770	-.8557	-.3218	-.7905	-.2689	-.7250	-.2190	-.6589
.34	-.4044	-.8583	-.3487	-.7936	-.2951	-.7286	-.2440	-.6631
.36	-.4296	-.8600	-.3735	-.7958	-.3193	-.7313	-.2673	-.6663
.38	-.4530	-.8610	-.3966	-.7972	-.3419	-.7332	-.2891	-.6688
.40	-.4747	-.8614	-.4180	-.7980	-.3629	-.7344	-.3095	-.6704
.42	-.4949	-.8612	-.4380	-.7982	-.3825	-.7350	-.3286	-.6715
.44	-.5137	-.8606	-.4566	-.7979	-.4008	-.7351	-.3465	-.6719
.46	-.5313	-.8596	-.4740	-.7972	-.4180	-.7347	-.3634	-.6719
.48	-.5477	-.8582	-.4904	-.7961	-.4341	-.7339	-.3791	-.6715
.50	-.5632	-.8565	-.5057	-.7947	-.4493	-.7328	-.3940	-.6706
.52	-.5777	-.8545	-.5201	-.7929	-.4635	-.7313	-.4080	-.6694
.54	-.5913	-.8523	-.5336	-.7909	-.4769	-.7295	-.4212	-.6679
.56	-.6042	-.8498	-.5464	-.7886	-.4895	-.7274	-.4336	-.6661
.58	-.6163	-.8471	-.5584	-.7861	-.5015	-.7252	-.4453	-.6640
.60	-.6277	-.8443	-.5698	-.7834	-.5127	-.7227	-.4565	-.6618
.62	-.6386	-.8412	-.5806	-.7806	-.5234	-.7200	-.4670	-.6593
.64	-.6488	-.8381	-.5908	-.7775	-.5335	-.7171	-.4769	-.6566
.66	-.6585	-.8348	-.6004	-.7744	-.5431	-.7141	-.4864	-.6538
.68	-.6678	-.8314	-.6096	-.7711	-.5521	-.7110	-.4953	-.6509
.70	-.6765	-.8278	-.6183	-.7677	-.5608	-.7077	-.5038	-.6478
.72	-.6848	-.8242	-.6266	-.7642	-.5689	-.7044	-.5119	-.6445
.74	-.6927	-.8205	-.6344	-.7606	-.5767	-.7009	-.5196	-.6412
.76	-.7003	-.8167	-.6419	-.7569	-.5841	-.6973	-.5269	-.6378
.78	-.7074	-.8129	-.6490	-.7532	-.5912	-.6937	-.5338	-.6343
.80	-.7142	-.8090	-.6558	-.7493	-.5979	-.6900	-.5404	-.6307
.82	-.7208	-.8050	-.6623	-.7455	-.6043	-.6862	-.5467	-.6270
.84	-.7270	-.8010	-.6684	-.7415	-.6104	-.6823	-.5527	-.6233
.86	-.7329	-.7969	-.6743	-.7375	-.6162	-.6785	-.5585	-.6195
.88	-.7386	-.7928	-.6799	-.7335	-.6217	-.6745	-.5639	-.6157
.90	-.7440	-.7887	-.6853	-.7295	-.6270	-.6705	-.5691	-.6118
.92	-.7492	-.7846	-.6904	-.7254	-.6321	-.6665	-.5741	-.6079
.94	-.7541	-.7804	-.6953	-.7213	-.6369	-.6625	-.5789	-.6039
.96	-.7588	-.7762	-.7000	-.7171	-.6416	-.6584	-.5834	-.5999
.98	-.7634	-.7720	-.7045	-.7129	-.6460	-.6543	-.5877	-.5959
1.00	-.7677	-.7677	-.7088	-.7088	-.6502	-.6502	-.5919	-.5919

Table D-5: $0.394 \leq \hat{\rho}_4 \leq 0.430$ and $\lambda_3,\ \lambda_4 < 0$

$\hat{\rho}_3$	$\hat{\rho}_4 = 0.394$		$\hat{\rho}_4 = 0.406$		$\hat{\rho}_4 = 0.418$		$\hat{\rho}_4 = 0.430$	
	λ_3	λ_4	λ_3	λ_4	λ_3	λ_4	λ_3	λ_4
.20	-.0032	-.5353						
.22	-.0341	-.5489	-.0099	-.4805				
.24	-.0643	-.5607	-.0350	-.4921	-.0127	-.4240		
.26	-.0934	-.5708	-.0599	-.5023	-.0330	-.4339	-.0130	-.3659
.28	-.1212	-.5793	-.0842	-.5111	-.0533	-.4426	-.0290	-.3742
.30	-.1475	-.5864	-.1076	-.5185	-.0732	-.4502	-.0452	-.3816
.32	-.1724	-.5922	-.1301	-.5248	-.0928	-.4568	-.0613	-.3882
.34	-.1959	-.5970	-.1516	-.5301	-.1117	-.4623	-.0772	-.3938
.36	-.2180	-.6007	-.1720	-.5343	-.1300	-.4670	-.0929	-.3987
.38	-.2388	-.6037	-.1914	-.5377	-.1475	-.4708	-.1081	-.4029
.40	-.2583	-.6058	-.2097	-.5404	-.1643	-.4739	-.1229	-.4063
.42	-.2767	-.6073	-.2270	-.5423	-.1803	-.4763	-.1371	-.4091
.44	-.2940	-.6082	-.2435	-.5436	-.1955	-.4781	-.1508	-.4112
.46	-.3102	-.6086	-.2590	-.5444	-.2100	-.4793	-.1639	-.4128
.48	-.3256	-.6085	-.2736	-.5447	-.2238	-.4800	-.1765	-.4140
.50	-.3400	-.6080	-.2875	-.5446	-.2368	-.4802	-.1885	-.4146
.52	-.3536	-.6071	-.3006	-.5441	-.2493	-.4801	-.2000	-.4149
.54	-.3665	-.6059	-.3130	-.5432	-.2610	-.4796	-.2109	-.4147
.56	-.3786	-.6044	-.3248	-.5420	-.2722	-.4787	-.2213	-.4142
.58	-.3901	-.6026	-.3359	-.5405	-.2829	-.4776	-.2313	-.4134
.60	-.4010	-.6005	-.3464	-.5388	-.2929	-.4761	-.2407	-.4124
.62	-.4113	-.5983	-.3564	-.5368	-.3025	-.4744	-.2498	-.4110
.64	-.4211	-.5959	-.3659	-.5346	-.3116	-.4726	-.2583	-.4095
.66	-.4303	-.5933	-.3749	-.5322	-.3203	-.4705	-.2665	-.4077
.68	-.4391	-.5905	-.3835	-.5297	-.3285	-.4682	-.2743	-.4057
.70	-.4474	-.5876	-.3916	-.5270	-.3363	-.4657	-.2817	-.4035
.72	-.4554	-.5845	-.3993	-.5242	-.3437	-.4631	-.2887	-.4012
.74	-.4629	-.5814	-.4066	-.5212	-.3508	-.4604	-.2954	-.3988
.76	-.4700	-.5781	-.4136	-.5181	-.3575	-.4576	-.3018	-.3962
.78	-.4769	-.5748	-.4202	-.5149	-.3639	-.4546	-.3079	-.3935
.80	-.4833	-.5713	-.4266	-.5117	-.3700	-.4515	-.3137	-.3906
.82	-.4895	-.5678	-.4326	-.5083	-.3758	-.4484	-.3192	-.3877
.84	-.4954	-.5642	-.4383	-.5049	-.3813	-.4451	-.3245	-.3847
.86	-.5010	-.5605	-.4438	-.5014	-.3866	-.4418	-.3295	-.3816
.88	-.5064	-.5568	-.4490	-.4978	-.3916	-.4384	-.3343	-.3785
.90	-.5115	-.5531	-.4539	-.4942	-.3964	-.4350	-.3388	-.3753
.92	-.5163	-.5493	-.4587	-.4905	-.4010	-.4315	-.3431	-.3720
.94	-.5210	-.5454	-.4632	-.4868	-.4053	-.4280	-.3472	-.3687
.96	-.5254	-.5415	-.4675	-.4831	-.4094	-.4244	-.3511	-.3653
.98	-.5297	-.5376	-.4716	-.4793	-.4134	-.4208	-.3549	-.3619
1.00	-.5337	-.5337	-.4755	-.4755	-.4171	-.4171	-.3584	-.3584

Table D–5: $0.442 \leq \hat{\rho}_4 \leq 0.478$ and $\lambda_3,\ \lambda_4 < 0$

	$\hat{\rho}_4 = 0.442$		$\hat{\rho}_4 = 0.454$		$\hat{\rho}_4 = 0.466$		$\hat{\rho}_4 = 0.478$	
$\hat{\rho}_3$	λ_3	λ_4	λ_3	λ_4	λ_3	λ_4	λ_3	λ_4
.28	-.0114	-.3064	-.0003	-.2397				
.30	-.0237	-.3132	-.0089	-.2455	-.0002	-.1789	.0033	-.1140
.32	-.0363	-.3194	-.0179	-.2509	-.0060	-.1832	.0001	-.1168
.34	-.0489	-.3249	-.0272	-.2558	-.0121	-.1872	-.0032	-.1196
.36	-.0615	-.3297	-.0366	-.2603	-.0184	-.1909	-.0068	-.1222
.38	-.0740	-.3339	-.0461	-.2643	-.0250	-.1943	-.0106	-.1247
.40	-.0863	-.3375	-.0557	-.2678	-.0316	-.1974	-.0145	-.1270
.42	-.0984	-.3406	-.0652	-.2708	-.0384	-.2002	-.0186	-.1291
.44	-.1102	-.3430	-.0746	-.2734	-.0452	-.2026	-.0227	-.1310
.46	-.1215	-.3450	-.0839	-.2756	-.0520	-.2047	-.0269	-.1327
.48	-.1326	-.3465	-.0930	-.2773	-.0588	-.2065	-.0312	-.1342
.50	-.1432	-.3475	-.1018	-.2786	-.0655	-.2079	-.0355	-.1355
.52	-.1534	-.3481	-.1104	-.2796	-.0721	-.2090	-.0398	-.1366
.54	-.1632	-.3484	-.1188	-.2802	-.0786	-.2099	-.0441	-.1374
.56	-.1726	-.3483	-.1268	-.2804	-.0850	-.2104	-.0484	-.1381
.58	-.1816	-.3478	-.1346	-.2804	-.0912	-.2107	-.0526	-.1386
.60	-.1903	-.3471	-.1421	-.2801	-.0972	-.2107	-.0568	-.1388
.62	-.1985	-.3462	-.1493	-.2795	-.1031	-.2105	-.0609	-.1389
.64	-.2064	-.3449	-.1563	-.2786	-.1087	-.2101	-.0649	-.1388
.66	-.2139	-.3435	-.1629	-.2776	-.1142	-.2095	-.0688	-.1386
.68	-.2211	-.3419	-.1693	-.2764	-.1194	-.2086	-.0726	-.1382
.70	-.2279	-.3401	-.1754	-.2749	-.1245	-.2076	-.0763	-.1376
.72	-.2345	-.3381	-.1812	-.2733	-.1294	-.2065	-.0799	-.1369
.74	-.2407	-.3359	-.1868	-.2716	-.1341	-.2051	-.0833	-.1360
.76	-.2466	-.3337	-.1921	-.2696	-.1385	-.2037	-.0866	-.1351
.78	-.2523	-.3313	-.1972	-.2676	-.1428	-.2021	-.0898	-.1340
.80	-.2577	-.3287	-.2020	-.2655	-.1469	-.2003	-.0929	-.1328
.82	-.2628	-.3261	-.2066	-.2632	-.1508	-.1985	-.0959	-.1315
.84	-.2677	-.3234	-.2110	-.2608	-.1546	-.1966	-.0987	-.1302
.86	-.2724	-.3206	-.2152	-.2584	-.1582	-.1946	-.1014	-.1287
.88	-.2768	-.3177	-.2192	-.2559	-.1616	-.1925	-.1040	-.1272
.90	-.2810	-.3148	-.2230	-.2533	-.1648	-.1904	-.1064	-.1256
.92	-.2850	-.3118	-.2266	-.2506	-.1679	-.1881	-.1087	-.1240
.94	-.2889	-.3087	-.2301	-.2479	-.1708	-.1859	-.1110	-.1223
.96	-.2925	-.3056	-.2334	-.2451	-.1736	-.1835	-.1131	-.1205
.98	-.2960	-.3025	-.2365	-.2423	-.1762	-.1811	-.1151	-.1188
1.00	-.2993	-.2993	-.2394	-.2394	-.1787	-.1787	-.1169	-.1169

Appendix E

The Normal Distribution

A random variable X is said to have the normal distribution $N(\mu, \sigma^2)$ if (for some $-\infty < \mu < +\infty$, $\sigma^2 > 0$)

$$f_X(x) = \frac{1}{\sqrt{2\pi}\,\sigma}\, e^{-\frac{1}{2}(\frac{x-\mu}{\sigma})^2} \qquad -\infty < x < +\infty.$$

The table on the following two pages gives for various values of y (when $\mu = 0$ and $\sigma^2 = 1$) the probability $P[X \le y] = \int_{-\infty}^{y} f_X(x)\,dx$. A table entry is either a number, a, or a number, a, and an exponent, b. If an exponent is not present, then a is the value of the integral to 4 significant digits. If an exponent, b, is present, then it signifies the number of 0s (in case $y < 0$) or the number of 9s (in case $y > 0$) that must be inserted immediately following the decimal point of a. For example, if $y = -3.26$, $a = .5771$ and $b = 3$; hence the value of the integral is .0005771. If $y = 4.52$, then $a = .6908$ and $b = 5$; hence the value of the integral is .999996908.

This table is from Appendix C of *Modern Statistical, Systems, and GPSS Simulation*, Second Edition, by Z.A. Karian and E.J. Dudewicz, CRC Press (1999), pp. 470–471. Reprinted with the permission of CRC Press.

The Normal Distribution (continues)

y	9	8	7	6	5	4	3	2	1	0
-4.90	$.3019^6$	$.3179^6$	$.3347^6$	$.3524^6$	$.3710^6$	$.3906^6$	$.4111^6$	$.4327^6$	$.4554^6$	$.4792^6$
-4.80	$.5042^6$	$.5304^6$	$.5580^6$	$.5869^6$	$.6173^6$	$.6492^6$	$.6826^6$	$.7178^6$	$.7546^6$	$.7933^6$
-4.70	$.8339^6$	$.8765^6$	$.9211^6$	$.9679^6$	$.1017^5$	$.1069^5$	$.1123^5$	$.1179^5$	$.1239^5$	$.1301^5$
-4.60	$.1366^5$	$.1434^5$	$.1506^5$	$.1581^5$	$.1660^5$	$.1742^5$	$.1828^5$	$.1919^5$	$.2013^5$	$.2112^5$
-4.50	$.2216^5$	$.2325^5$	$.2439^5$	$.2558^5$	$.2682^5$	$.2813^5$	$.2949^5$	$.3092^5$	$.3241^5$	$.3398^5$
-4.40	$.3561^5$	$.3732^5$	$.3911^5$	$.4098^5$	$.4294^5$	$.4498^5$	$.4712^5$	$.4935^5$	$.5169^5$	$.5413^5$
-4.30	$.5668^5$	$.5934^5$	$.6212^5$	$.6503^5$	$.6807^5$	$.7124^5$	$.7455^5$	$.7801^5$	$.8163^5$	$.8540^5$
-4.20	$.8934^5$	$.9345^5$	$.9774^5$	$.1022^4$	$.1069^4$	$.1118^4$	$.1168^4$	$.1222^4$	$.1277^4$	$.1335^4$
-4.10	$.1395^4$	$.1458^4$	$.1523^4$	$.1591^4$	$.1662^4$	$.1737^4$	$.1814^4$	$.1894^4$	$.1978^4$	$.2066^4$
-4.00	$.2157^4$	$.2252^4$	$.2351^4$	$.2454^4$	$.2561^4$	$.2673^4$	$.2789^4$	$.2910^4$	$.3036^4$	$.3167^4$
-3.90	$.3304^4$	$.3446^4$	$.3594^4$	$.3747^4$	$.3908^4$	$.4074^4$	$.4247^4$	$.4427^4$	$.4615^4$	$.4810^4$
-3.80	$.5012^4$	$.5223^4$	$.5442^4$	$.5669^4$	$.5906^4$	$.6152^4$	$.6407^4$	$.6673^4$	$.6948^4$	$.7235^4$
-3.70	$.7532^4$	$.7841^4$	$.8162^4$	$.8496^4$	$.8842^4$	$.9201^4$	$.9574^4$	$.9961^4$	$.1036^3$	$.1078^3$
-3.60	$.1121^3$	$.1166^3$	$.1213^3$	$.1261^3$	$.1311^3$	$.1363^3$	$.1417^3$	$.1473^3$	$.1531^3$	$.1591^3$
-3.50	$.1653^3$	$.1718^3$	$.1785^3$	$.1854^3$	$.1926^3$	$.2001^3$	$.2078^3$	$.2158^3$	$.2241^3$	$.2326^3$
-3.40	$.2415^3$	$.2507^3$	$.2602^3$	$.2701^3$	$.2803^3$	$.2909^3$	$.3018^3$	$.3131^3$	$.3248^3$	$.3369^3$
-3.30	$.3495^3$	$.3624^3$	$.3758^3$	$.3897^3$	$.4041^3$	$.4189^3$	$.4342^3$	$.4501^3$	$.4665^3$	$.4834^3$
-3.20	$.5009^3$	$.5190^3$	$.5377^3$	$.5571^3$	$.5770^3$	$.5976^3$	$.6190^3$	$.6410^3$	$.6637^3$	$.6871^3$
-3.10	$.7114^3$	$.7364^3$	$.7622^3$	$.7888^3$	$.8164^3$	$.8447^3$	$.8740^3$	$.9043^3$	$.9354^3$	$.9676^3$
-3.00	$.1001^2$	$.1035^2$	$.1070^2$	$.1107^2$	$.1144^2$	$.1183^2$	$.1223^2$	$.1264^2$	$.1306^2$	$.1350^2$
-2.90	$.1395^2$	$.1441^2$	$.1489^2$	$.1538^2$	$.1589^2$	$.1641^2$	$.1695^2$	$.1750^2$	$.1807^2$	$.1866^2$
-2.80	$.1926^2$	$.1988^2$	$.2052^2$	$.2118^2$	$.2186^2$	$.2256^2$	$.2327^2$	$.2401^2$	$.2477^2$	$.2555^2$
-2.70	$.2635^2$	$.2718^2$	$.2803^2$	$.2890^2$	$.2980^2$	$.3072^2$	$.3167^2$	$.3264^2$	$.3364^2$	$.3467^2$
-2.60	$.3573^2$	$.3681^2$	$.3793^2$	$.3907^2$	$.4025^2$	$.4145^2$	$.4269^2$	$.4396^2$	$.4527^2$	$.4661^2$
-2.50	$.4799^2$	$.4940^2$	$.5085^2$	$.5234^2$	$.5386^2$	$.5543^2$	$.5703^2$	$.5868^2$	$.6037^2$	$.6210^2$
-2.40	$.6387^2$	$.6569^2$	$.6756^2$	$.6947^2$	$.7143^2$	$.7344^2$	$.7549^2$	$.7760^2$	$.7976^2$	$.8198^2$
-2.30	$.8424^2$	$.8656^2$	$.8894^2$	$.9137^2$	$.9387^2$	$.9642^2$	$.9903^2$	$.1017^1$	$.1044^1$	$.1072^1$
-2.20	$.1101^1$	$.1130^1$	$.1160^1$	$.1191^1$	$.1222^1$	$.1255^1$	$.1287^1$	$.1321^1$	$.1355^1$	$.1390^1$
-2.10	$.1426^1$	$.1463^1$	$.1500^1$	$.1539^1$	$.1578^1$	$.1618^1$	$.1659^1$	$.1700^1$	$.1743^1$	$.1786^1$
-2.00	$.1831^1$	$.1876^1$	$.1923^1$	$.1970^1$	$.2018^1$	$.2068^1$	$.2118^1$	$.2169^1$	$.2222^1$	$.2275^1$
-1.90	$.2330^1$	$.2385^1$	$.2442^1$	$.2500^1$	$.2559^1$	$.2619^1$	$.2680^1$	$.2743^1$	$.2807^1$	$.2872^1$
-1.80	$.2938^1$	$.3005^1$	$.3074^1$	$.3144^1$	$.3216^1$	$.3288^1$	$.3362^1$	$.3438^1$	$.3515^1$	$.3593^1$
-1.70	$.3673^1$	$.3754^1$	$.3836^1$	$.3920^1$	$.4006^1$	$.4093^1$	$.4182^1$	$.4272^1$	$.4363^1$	$.4457^1$
-1.60	$.4551^1$	$.4648^1$	$.4746^1$	$.4846^1$	$.4947^1$	$.5050^1$	$.5155^1$	$.5262^1$	$.5370^1$	$.5480^1$
-1.50	$.5592^1$	$.5705^1$	$.5821^1$	$.5938^1$	$.6057^1$	$.6178^1$	$.6301^1$	$.6426^1$	$.6552^1$	$.6681^1$
-1.40	$.6811^1$	$.6944^1$	$.7078^1$	$.7215^1$	$.7353^1$	$.7493^1$	$.7636^1$	$.7780^1$	$.7927^1$	$.8076^1$
-1.30	$.8226^1$	$.8379^1$	$.8534^1$	$.8691^1$	$.8851^1$	$.9012^1$	$.9176^1$	$.9342^1$	$.9510^1$	$.9680^1$
-1.20	$.9853^1$	.1003	.1020	.1038	.1056	.1075	.1093	.1112	.1131	.1151
-1.10	.1170	.1190	.1210	.1230	.1251	.1271	.1292	.1314	.1335	.1357
-1.00	.1379	.1401	.1423	.1446	.1469	.1492	.1515	.1539	.1562	.1587
-.90	.1611	.1635	.1660	.1685	.1711	.1736	.1762	.1788	.1814	.1841
-.80	.1867	.1894	.1922	.1949	.1977	.2005	.2033	.2061	.2090	.2119
-.70	.2148	.2177	.2206	.2236	.2266	.2296	.2327	.2358	.2389	.2420
-.60	.2451	.2483	.2514	.2546	.2578	.2611	.2643	.2676	.2709	.2743
-.50	.2776	.2810	.2843	.2877	.2912	.2946	.2981	.3015	.3050	.3085
-.40	.3121	.3156	.3192	.3228	.3264	.3300	.3336	.3372	.3409	.3446
-.30	.3483	.3520	.3557	.3594	.3632	.3669	.3707	.3745	.3783	.3821
-.20	.3859	.3897	.3936	.3974	.4013	.4052	.4090	.4129	.4168	.4207
-.10	.4247	.4286	.4325	.4364	.4404	.4443	.4483	.4522	.4562	.4602
-0.00	.4641	.4681	.4721	.4761	.4801	.4840	.4880	.4920	.4960	.5000

The Normal Distribution (concluded)

y	0	1	2	3	4	5	6	7	8	9
.00	.5000	.5040	.5080	.5120	.5160	.5199	.5239	.5279	.5319	.5359
.10	.5398	.5438	.5478	.5517	.5557	.5596	.5636	.5675	.5714	.5753
.20	.5793	.5832	.5871	.5910	.5948	.5987	.6026	.6064	.6103	.6141
.30	.6179	.6217	.6255	.6293	.6331	.6368	.6406	.6443	.6480	.6517
.40	.6554	.6591	.6628	.6664	.6700	.6736	.6772	.6808	.6844	.6879
.50	.6915	.6950	.6985	.7019	.7054	.7088	.7123	.7157	.7190	.7224
.60	.7257	.7291	.7324	.7357	.7389	.7422	.7454	.7486	.7517	.7549
.70	.7580	.7611	.7642	.7673	.7704	.7734	.7764	.7794	.7823	.7852
.80	.7881	.7910	.7939	.7967	.7995	.8023	.8051	.8078	.8106	.8133
.9	.8159	.8186	.8212	.8238	.8264	.8289	.8315	.8340	.8365	.8389
1.00	.8413	.8438	.8461	.8485	.8508	.8531	.8554	.8577	.8599	.8621
1.10	.8643	.8665	.8686	.8708	.8729	.8749	.8770	.8790	.8810	.8830
1.20	.8849	.8869	.8888	.8907	.8925	.8944	.8962	.8980	.8997	$.0148^1$
1.30	$.0320^1$	$.0490^1$	$.0658^1$	$.0824^1$	$.0988^1$	$.1149^1$	$.1309^1$	$.1466^1$	$.1621^1$	$.1774^1$
1.40	$.1924^1$	$.2073^1$	$.2220^1$	$.2364^1$	$.2507^1$	$.2647^1$	$.2785^1$	$.2922^1$	$.3056^1$	$.3189^1$
1.50	$.3319^1$	$.3448^1$	$.3574^1$	$.3699^1$	$.3822^1$	$.3943^1$	$.4062^1$	$.4179^1$	$.4295^1$	$.4408^1$
1.60	$.4520^1$	$.4630^1$	$.4738^1$	$.4845^1$	$.4950^1$	$.5053^1$	$.5154^1$	$.5254^1$	$.5352^1$	$.5449^1$
1.70	$.5543^1$	$.5637^1$	$.5728^1$	$.5818^1$	$.5907^1$	$.5994^1$	$.6080^1$	$.6164^1$	$.6246^1$	$.6327^1$
1.80	$.6407^1$	$.6485^1$	$.6562^1$	$.6638^1$	$.6712^1$	$.6784^1$	$.6856^1$	$.6926^1$	$.6995^1$	$.7062^1$
1.90	$.7128^1$	$.7193^1$	$.7257^1$	$.7320^1$	$.7381^1$	$.7441^1$	$.7500^1$	$.7558^1$	$.7615^1$	$.7670^1$
2.00	$.7725^1$	$.7778^1$	$.7831^1$	$.7882^1$	$.7932^1$	$.7982^1$	$.8030^1$	$.8077^1$	$.8124^1$	$.8169^1$
2.10	$.8214^1$	$.8257^1$	$.8300^1$	$.8341^1$	$.8382^1$	$.8422^1$	$.8461^1$	$.8500^1$	$.8537^1$	$.8574^1$
2.20	$.8610^1$	$.8645^1$	$.8679^1$	$.8713^1$	$.8745^1$	$.8778^1$	$.8809^1$	$.8840^1$	$.8870^1$	$.8899^1$
2.30	$.8928^1$	$.8956^1$	$.8983^1$	$.0097^2$	$.0358^2$	$.0613^2$	$.0863^2$	$.1106^2$	$.1344^2$	$.1576^2$
2.40	$.1802^2$	$.2024^2$	$.2240^2$	$.2451^2$	$.2656^2$	$.2857^2$	$.3053^2$	$.3244^2$	$.3431^2$	$.3613^2$
2.50	$.3790^2$	$.3963^2$	$.4132^2$	$.4297^2$	$.4457^2$	$.4614^2$	$.4766^2$	$.4915^2$	$.5060^2$	$.5201^2$
2.60	$.5339^2$	$.5473^2$	$.5604^2$	$.5731^2$	$.5855^2$	$.5975^2$	$.6093^2$	$.6207^2$	$.6319^2$	$.6427^2$
2.70	$.6533^2$	$.6636^2$	$.6736^2$	$.6833^2$	$.6928^2$	$.7020^2$	$.7110^2$	$.7197^2$	$.7282^2$	$.7365^2$
2.80	$.7445^2$	$.7523^2$	$.7599^2$	$.7673^2$	$.7744^2$	$.7814^2$	$.7882^2$	$.7948^2$	$.8012^2$	$.8074^2$
2.90	$.8134^2$	$.8193^2$	$.8250^2$	$.8305^2$	$.8359^2$	$.8411^2$	$.8462^2$	$.8511^2$	$.8559^2$	$.8605^2$
3.00	$.8650^2$	$.8694^2$	$.8736^2$	$.8777^2$	$.8817^2$	$.8856^2$	$.8893^2$	$.8930^2$	$.8965^2$	$.8999^2$
3.10	$.0324^3$	$.0646^3$	$.0957^3$	$.1260^3$	$.1553^3$	$.1836^3$	$.2112^3$	$.2378^3$	$.2636^3$	$.2886^3$
3.20	$.3129^3$	$.3363^3$	$.3590^3$	$.3810^3$	$.4024^3$	$.4230^3$	$.4429^3$	$.4623^3$	$.4810^3$	$.4991^3$
3.30	$.5166^3$	$.5335^3$	$.5499^3$	$.5658^3$	$.5811^3$	$.5959^3$	$.6103^3$	$.6242^3$	$.6376^3$	$.6505^3$
3.40	$.6631^3$	$.6752^3$	$.6869^3$	$.6982^3$	$.7091^3$	$.7197^3$	$.7299^3$	$.7398^3$	$.7493^3$	$.7585^3$
3.50	$.7674^3$	$.7759^3$	$.7842^3$	$.7922^3$	$.7999^3$	$.8074^3$	$.8146^3$	$.8215^3$	$.8282^3$	$.8347^3$
3.60	$.8409^3$	$.8469^3$	$.8527^3$	$.8583^3$	$.8637^3$	$.8689^3$	$.8739^3$	$.8787^3$	$.8834^3$	$.8879^3$
3.70	$.8922^3$	$.8964^3$	$.0039^4$	$.0426^4$	$.0799^4$	$.1158^4$	$.1504^4$	$.1838^4$	$.2159^4$	$.2468^4$
3.80	$.2765^4$	$.3052^4$	$.3327^4$	$.3593^4$	$.3848^4$	$.4094^4$	$.4331^4$	$.4558^4$	$.4777^4$	$.4988^4$
3.90	$.5190^4$	$.5385^4$	$.5573^4$	$.5753^4$	$.5926^4$	$.6092^4$	$.6253^4$	$.6406^4$	$.6554^4$	$.6696^4$
4.00	$.6833^4$	$.6964^4$	$.7090^4$	$.7211^4$	$.7327^4$	$.7439^4$	$.7546^4$	$.7649^4$	$.7748^4$	$.7843^4$
4.10	$.7934^4$	$.8022^4$	$.8106^4$	$.8186^4$	$.8263^4$	$.8338^4$	$.8409^4$	$.8477^4$	$.8542^4$	$.8605^4$
4.20	$.8665^4$	$.8723^4$	$.8778^4$	$.8832^4$	$.8882^4$	$.8931^4$	$.8978^4$	$.0226^5$	$.0655^5$	$.1066^5$
4.30	$.1460^5$	$.1837^5$	$.2199^5$	$.2545^5$	$.2876^5$	$.3193^5$	$.3497^5$	$.3788^5$	$.4066^5$	$.4333^5$
4.40	$.4588^5$	$.4832^5$	$.5065^5$	$.5285^5$	$.5502^5$	$.5707^5$	$.5902^5$	$.6089^5$	$.6268^5$	$.6439^5$
4.50	$.6602^5$	$.6759^5$	$.6908^5$	$.7051^5$	$.7187^5$	$.7318^5$	$.7442^5$	$.7561^5$	$.7675^5$	$.7784^5$
4.60	$.7888^5$	$.7987^5$	$.8081^5$	$.8172^5$	$.8258^5$	$.8340^5$	$.8419^5$	$.8494^5$	$.8566^5$	$.8634^5$
4.70	$.8699^5$	$.8761^5$	$.8821^5$	$.8877^5$	$.8931^5$	$.8983^5$	$.0321^6$	$.0789^6$	$.1235^6$	$.1661^6$
4.80	$.2067^6$	$.2456^6$	$.2822^6$	$.3174^6$	$.3508^6$	$.3827^6$	$.4131^6$	$.4420^6$	$.4696^6$	$.4958^6$
4.90	$.5208^6$	$.5446^6$	$.5673^6$	$.5889^6$	$.6094^6$	$.6290^6$	$.6476^6$	$.6653^6$	$.6821^6$	$.6981^6$

References and Author Index†

Abramowitz, M. and Stegun, I. A. (Editors) (1964). *Handbook of Mathematical Functions With Formulas, Graphs, and Mathematical Tables,* National Bureau of Standards Applied Mathematics Series • 55, U.S. Government Printing Office, Washington, D.C. 20402. [177]

Anderson, T. W., Fang, K. T., and Olkin, I. (Editors) (1994). *Multivariate Analysis and Its Applications,* Institute of Mathematical Statistics Lecture Notes–Monograph Series, Vol. 24, Institute of Mathematical Statistics, Hayward, California. [271]

Artin, E. (1964). *The Gamma Function,* Holt, Rinehart and Winston, New York. [70]

Beckwith, N. B. and Dudewicz, E. J. (1996). "A bivariate Generalized Lambda Distribution (GLD–2) using Plackett's method of construction: distribution, examples and applications," *American Journal of Mathematical and Management Sciences,* 16, 333–393. [vi, 217, 219, 232, 233, 242, 252, 253, 270, 301]

Bowman, A. W. and Azzalini, A. (1997). *Applied Smoothing Techniques for Data Analysis, The Kernel Approach with S-Plus Illustrations,* Clarendon Press, Oxford. [1]

Burr, I. W. (1973). "Parameters for a general system of distributions to match a grid of α_3 and α_4," *Comm. Statist.,* 2, 1–21. [54]

Chernick, M. R. (1999). *Bootstrap Methods, A Practitioner's Guide,* John Wiley & Sons, Inc., New York. [275, 282]

---

†The list in the braces [ ] consists of page numbers where the reference either is cited or can be referred to for further detail.

Christian, J. C., Carmelli, D., Castelli, W. P., Fabsitz, R., Grim, C. E., Meaney, F. J., Norton, J. A. Jr., Reed, T., Williams, C. J., and Wood, P. D. (1990). "High density lipoprotein cholesterol: A 16–year longitudinal study in aging male twins," *Arteriosclerosis*, 10, 1020–1025. [99]

Christian, J. C., Kang, K. W., and Norton, J. A. Jr. (1974). "Choice of an estimate of genetic variance from twin data," *American Journal of Human Genetics*, 26, 154–161. [99]

Cooley, C. A. (1991). *The Generalized Lambda Distribution: Applications and Parameter Estimation*, Honors Project, Denison University. Available from: Doan Library, Denison University, Granville, Ohio 43023. [63]

Cramér, H. (1946). *Mathematical Methods of Statistics*, Princeton University Press, Princeton, New Jersey. [108]

Csörgö, M. and Szyszkowicz, B. (1994). "Weighted multivariate empirical processes and continguous change-point analysis," *Change-point Problem*, Institute of Mathematical Statistics Lecture Notes–Monograph Series, Vol. 23, Carlstein, E., Müller, H.-G., and Siegmund, D., Editors, pp. 93–98. [270]

Dahl-Jensen, D., Mosegaard, K., Gundestrup, N., Clow, G. D., Johnsen, S. J., Hansen, A. W., and Balling, N. (1998). "Past temperatures from the Greenland ice sheet," *Science*, 282, 268–271. [104]

Dall'Aglio, G., Kotz, S., and Salinetti, G. (Editors) (1991). *Advances in Probability Distributions with Given Marginals, Beyond the Copulas*, Kluwer Academic Publishers, Dordrecht, The Netherlands. [271]

Dudewicz, E. J. (1992). "The Generalized Bootstrap," *Bootstrapping and Related Techniques* (K.-H. Jöckel, G. Rothe, and W. Sendler, eds.), V. 376 of Lecture Notes in Economics and Mathematical Systems, Springer-Verlag, Berlin, pp. 31–37. [274, 278]

Dudewicz, E. J. (Editor) (1997). *Modern Digital Simulation Methodology, III: Advances in Theory, Application, and Design–Electric Power Systems, Spare Parts Inventory, Purchase Interval and Incidence Modeling, Automobile Insurance Bonus-Malus Systems, Genetic Algorithms–DNA Sequence Assembly, Education, & Water Resources Case Studies*, American Sciences Press, Inc., Columbus, Ohio. [273]

Dudewicz, E. J. (1999). "Basic statistical methods," Chapter 44, *Juran's Quality Handbook, Fifth Edition* (edited by J. M. Juran, A. Blanton Godfrey, R. E. Hoogstoel, and E. G. Schilling), McGraw-Hill, New York. [5, 273]

Dudewicz, E. J., Chen, P., and Taneja, B. K. (1989). *Modern Elementary Probability and Statistics, with Statistical Programming in SAS, MINITAB, & BMDP* (Second Printing), American Sciences Press, Inc., Columbus, Ohio. [99, 215]

Dudewicz, E. J. and Karian, Z. A. (1985). *Modern Design and Analysis of Discrete-Event Computer Simulations*, IEEE Computer Society, New York. [109]

Dudewicz, E. J. and Karian, Z. A. (1996). "The Extended Generalized Lambda Distribution (EGLD) system for fitting distributions to data with moments, II: Tables," *American Journal of Mathematical and Management Sciences*, 16(3 & 4), 271–332. [63, 126, 309, 331]

Dudewicz, E. J. and Karian, Z. A. (1999a). "Fitting the Generalized Lambda Distribution (GLD) system by a method of percentiles, II: Tables," *American Journal of Mathematical and Management Sciences*, 19 (1 & 2), 1–73. [357]

Dudewicz, E. J. and Karian, Z. A. (1999b). "The role of statistics in IS/IT: Practical gains from mined data," *Information Systems Frontiers*, 1, 259–266. [274, 276, 277, 278, 279, 282]

Dudewicz, E. J., Levy, G. C., Lienhart, J. L. and Wehrli, F. (1989). "Statistical analysis of magnetic resonance imaging data in the normal brain dData, screening normality, discrimination, variability), and implications for expert statistical programming for ESS^{TM} (the Expert Statistical System)," *American Journal of Mathematical and Management Sciences*, 9, 299–359. [98, 218, 253]

Dudewicz, E. J. and Mishra, S. N. (1988). *Modern Mathematical Statistics*, John Wiley & Sons, New York. [5, 51, 113, 196, 201, 225, 274]

Dudewicz, E. J., Mommaerts, W., and van der Meulen, E. C. (1991). "Maximum entropy methods in modern spectroscopy: a review and an empiric entropy approach," *The Frontiers of Statistical Scientific Theory & Industrial Applications* (Chief Editors A. Öztürk and E. C. van der Meulen), American Sciences Press, Inc., Columbus, Ohio, 115–160. [32]

Dudewicz, E. J. and van der Meulen, E. C. (1981). "Entropy-based tests of uniformity," *Journal of the American Statistical Association*, 76, 967–974. [95]

Dudewicz, E. J. and van der Meulen, E. C. (1987). "The empiric entropy, a new approach to nonparametric entropy estimation," *New Perspectives in Theoretical and Applied Statistics* (eds. M. L. Puri, J. P. Vilaplana, and W. Wertz), John Wiley & Sons, Inc., New York, 207–227. [196]

Elinder, C. G., Jönsson, L., Piscator, M. and Rahnster, B (1981). "Histo-pathological changes in relation to cadmium concentration in horse kidneys," *Environmental Research*, 26, 1–21. [96]

Farlie, D. J. G. (1960). "The performance of some correlation coefficients for a general bivariate distribution," *Biometrika*, 47, 307–323. [221]

Filliben, J. J. (1969). *Simple and Robust Linear Estimation of the Location Parameter of a Symmetric Distribution*, Ph.D. Thesis, Princeton University, New Jersey. [4]

Freimer, M., Kollia, G., Mudholkar, G. S., and Lin, C. T. (1988). "A study of the Generalized Tukey Lambda family," *Communications in Statistics–Theory and Methods*, 17, 3547–3567. [35, 39, 83]

Galton, F. (1875). "Statistics by intercomparison, with remarks on the law of frequency of error," *Philosophical Magazine*, 49, 33–46. [153]

Gan, F. F., Koehler, K. J., and Thompson, J. C. (1991). "Probability plots and distribution curves for assessing the fit of probability models," *The American Statistician*, 45 (1), 14–21. [96, 112]

Gibbons, J. D. (1997). *Nonparametric Methods for Quantitative Analysis (Third Edition)*, American Sciences Press, Inc., Columbus, Ohio. [96, 201]

Gilchrist, W. (1997). "Modelling with quantile distribution functions," *Journal of Applied Statistics*, 24, 113–122. [40]

Golden, B. L., Assad, A. A., and Zanakis, S. H. (eds.) (1984). *Statistics and Optimization: The Interface*, American Sciences Press, Inc., Columbus, Ohio. [273]

Golden, B. L. and Eiselt, H. A. (eds.) (1992). *Location Modeling in Practice: Applications (Site Location, Oil Field Generators, Emergency Facilities, Postal Boxes), Theory, and History*, American Sciences Press, Inc., Columbus, Ohio. [273]

Govindarajulu, Z. (1987). *The Sequential Statistical Analysis of Hypothesis Testing, Point and Interval Estimation, and Decision Theory*, American Sciences Press, Inc., Columbus, Ohio. [81]

Györfi, L., Liese, F., Vajda, I., and van der Meulen, E. C. (1998). "Distribution estimates consistent in χ^2–divergence," *Statistics*, 32, 31–57. [196]

Hahn, G. J. and Shapiro, S. S. (1967). *Statistical Models in Engineering*, John Wiley & Sons, New York. [105]

Hald, A. (1998). *A History of Mathematical Statistics from 1750 to 1930*, Wiley, New York. [3, 153]

Harter, H. Leon (1993). *The Chronological Annotated Bibliography of Order Statistics, Volumes I-VIII*, American Sciences Press, Inc., Columbus, Ohio. [76]

Hastings, C., Mosteller, F., Tukey J. W., and Winsor, C. P. (1947). "Low moments for small samples: a comparative study of statistics," *Annals of Mathematical Statistics*, 18, 413–426. [3]

Hayakawa, T., Aoshima, M., Shimizu, K., and Taneja, V. S. (Editors) (1995, 1996, 1997, 1998). *MSI-2000: Multivariate Statistical Analysis in Honor of Professor Minoru Siotani, Vol. I, II, III, IV*, American Sciences Press, Inc., Columbus, Ohio. [271]

Hogben, D. (1963). *Some Properties of Tukey's Test for Non-Additivity*, Ph.D. Thesis, Rutgers-The State University, New Jersey. [4, 40]

Hogg, R. V. (1972). "More light on the kurtosis and related statistics," *Journal of the American Statistical Association*, 67, 422–424. [201]

Hogg, R. V. and Tanis, E. A. (1997). *Probability and Statistical Inference (Fifth Edition)*, Prentice Hall, Upper Saddle River, New Jersey. [154]

Horton, T. R. (1979). "A preliminary radiological assessment of radon exhalation from phosphate gypsum piles and inactive uranium mill tailings piles," *EPA-520/5-79-004*, Environmental Protection Agency, Washington, D.C. [206]

Hutchinson, T. P. and Lai, C. D. (1990). *Continuous Bivariate Distributions, Emphasizing Applications,* Rumsby Scientific Publishing, Adelaide, South Australia. [221, 232, 238, 244, 271]

Johnson, M. E. (1987). *Multivariate Statistical Simulation.* Wiley, New York. [233]

Johnson, M. E., Wang, C., and Ramberg, J. S. (1984). "Generation of continuous multivariate distributions for statistical applications," *American Journal of Mathematical and Management Sciences,* 4, 225–248. [233]

Johnson, N. L. and Kotz, S. (1973). "Extended and multivariate Tukey lambda distributions," *Biometrika,* 60, 655–661. [270]

Johnson, R. A. and Wichern, D. W. (1992). *Applied Multivariate Statistical Analysis.* Prentice Hall, Englewood Cliffs, New Jersey. [253]

Joiner, B. L. and Rosenblatt, J. R. (1971). "Some properties of the range in samples from Tukey's symmetric Lambda distributions," *Journal of the American Statistical Association,* 66, 384–399. [39]

Karian, Z. A. and Dudewicz, E. J. (1991). *Modern Statistical, Systems, and GPSS Simulation* (First Edition), Computer Sciences Press/W. H. Freeman and Company, Publishers, New York. (Now available from American Sciences Press, Inc., Columbus, Ohio. For the Second Edition, see Karian and Dudewicz (1999b)). [274]

Karian, Z. A. and Dudewicz, E. J. (1999a). "Fitting the Generalized Lambda Distribution to data: a method based on percentiles," *Communications in Statistics: Simulation and Computation,* 28(3), 793–819. [37, 38, 111, 153, 204, 217, 219]

Karian, Z. A. and Dudewicz, E. J. (1999b). *Modern Statistical, Systems, and GPSS Simulation,* Second Edition, CRC Press, Boca Raton, Florida. [4, 5, 109, 151, 277, 413]

Karian, Z. A., Dudewicz, E. J. and McDonald, P. (1996). "The extended generalized lambda distribution system for fitting distributions to data: history, completion of theory, tables, applications, the 'final word' on moment fits," *Communications in Statistics: Simulation and Computation,* 25(3), 611–642. [16, 98, 113, 140, 141]

Kendall, M. G. and Stuart, A. (1969). *The Advanced Theory of Statistics, Volume I, Distribution Theory* (Third Edition), Hafner Publishing Company, New York, and Charles Griffen & Company Limited, London. [50, 85]

Klugman, S. A., Panjer, H. H., and Willmot, G. E. (1998). *Loss Models, From Data to Decisions,* John Wiley & Sons, Inc., New York. [1]

Kroese, A. H. (1994). *Distributional Inference: A Loss Function Approach,* Ph.D. Thesis, University of Groningen, The Netherlands. [282]

Lin, Y. (1997). "Asymptotics of bootstrapping mean on some smoothed empirical distribution," *Statistics & Decisions,* 15, 301–306. [278, 282]

Mardia, K. V. (1967). "Some contributions to contingency-type distributions," *Biometrika,* 54, 235–249. (Corrections: 1968, 55, 597.) [222, 231]

Mardia, K. V. (1970). *Families of Bivariate Distributions,* Hafner, Darien, Connecticut. [222]

McClave, U. T., Dietrich, F. H., II, and Sincich, T. (1997). *Statistics,* Seventh Edition, Prentice Hall, Upper Saddle River, New Jersey. [206, 208]

McLachlan, G. J. (1992). "Cluster analysis and related techniques in medical research," *Statistical Methods in Medical Research,* 1, 27–48. [219]

Micceri, T. (1989), "The unicorn, the normal curve, and other improbable creatures," *Psych. Bull.,* 105, 156–166. [65, 90]

Mishra, S. N. and Carpenter, M. (1998). "Parameter estimation for fitting generalized beta distributions," Preprint, Department of Mathematics and Statistics, University of South Alabama, Mobile, Alabama 36688. Submitted to *Proceedings of the International Conference on Statistical Inference, Combinatorics and Related Areas,* Varanasi, India, December 1997. [123]

Mishra, S. N., Shah, A. K., and Lefante, J. J. (1986). "Overlapping coefficient: The generalized *t*–approach," *Communications in Statistics,* 15, 123–128. [196]

Moore, D. S. (1977). "Generalized inverses, Wald's method, and the construction of chi-squared tests of fit," *Journal of the American Statistical Association,* 72, 131–137. [95]

Mudholkar, G. D. and Phatak, M. V. (1984). "Quantile function models for quantal response analysis: an outline," *Topics in Applied Statistics-Proc. Stat. 81 Canada*, 621–627.[4]

Müller, H.–G. (1997). "Density estimation (update)," article in *Encyclopedia of Statistical Sciences, Update Volume I* (S. Kotz, Editor-in-Chief), pp. 185–200, Wiley, New York. [1]

Mykytka, E. F. (1979). "Fitting a distribution to data using an alternative to moments," *IEEE Proceedings of the 1979 Winter Simulation Conference*, 361–374. [153, 242]

National Bureau of Standards (1953). *Tables of Normal Probability Functions, Applied Mathematics Series 23*, Issued June 5, 1953 (a reissue of Mathematical Table 14, with corrections), U. S. Government Printing Office, Washington, D. C. (available from National Technical Information Service, Springfield, Virginia 22151, Document No. PB–175 967). [41]

Nelson, W. (1982). *Applied Life Data Analysis*, John Wiley & Sons, Inc., New York. [79, 83, 84]

Oegerle, W. R., Hill, J. M., and Fitchett, M. J. (1995). "Observations of high dispersion clusters of galaxies: Constraints on cold dark matter," *The Astronomical Journal*, 110(1). [208]

Olsson, D. M. and Nelson, L. S. (1975). "The Nelder-Mead simplex procedure for function minimization," *Technometrics*, 17, 45–51. [54]

Omey, E. (1998). Review of *Modern Digital Simulation, II*, American Sciences Press, Inc., Columbus, Ohio, 1997, which contains the papers Beckwith and Dudewicz (1996) and Dudewicz and Karian (1996). In *Kwantitatieve Methoden*, 19(58), 155–156. [v]

Ortega, J. M. and Rheinboldt, W. C. (1970). *Iterative Solution of Nonlinear Equations in Several Variables*, Academic Press, New York. [54]

Öztürk, A. and Dale, R. F. (1982). "A study of fitting the Generalized Lambda Distribution to solar radiation data," *Journal of Applied Meteorology*, 12, 995–1004. [4]

Öztürk, A. and Dale, R. F. (1985). "Least squares estimation of the parameters of the Generalized Lambda Distribution," *Technometrics*, 27, 81–84. [4]

Pearson, E. S. and Pease, N. W. (1975). "Relation between the shape of population distribution and the robustness of four simple statistics," *Biometrika*, 62, 223–241. [65, 90]

Pearson, K. (1894). "Contribution to the mathematical theory of evolution," *Philos. Trans. Royal Soc. London, Series A*, 185, 71–110. Reprinted in K. Pearson (1948). [3]

Pearson, K. (1895). "Contributions to the mathematical theory of evolution. II. Skew variation in homogeneous material," *Philos. Trans. Royal Soc. London, Series A*, 186, 343–414. Reprinted in K. Pearson (1948). [3]

Pearson, K. (1900). "On the theory of contingency and its relation to association and normal correlation," *Philosophy Magazine*, 50, 157–175. [95]

Pearson, K. (1913). "Note on the surface of constant association," *Biometrika*, 9, 534–537. [270]

Pearson, K. (1948). *Karl Pearson's Early Statistical Papers* (E. S. Pearson, Editor), Cambridge University Press, Cambridge, United Kingdom.

Pearson, K. and Heron, D. (1913). "On theories of association," *Biometrika*, 9, 159–315. [270]

Plackett, R. L. (1965). "A class of bivariate distributions," *Biometrika*, 60, 516–522. [217, 252, 270]

Plackett, R. L. (1981). *The Analysis of Categorical Data (Second Edition)*. Macmillan Publishing Co., Inc., New York. [270]

Pregibon, D. (1980). "Goodness of link tests for generalized linear models," *Applied Statistics*, 29, 15–24. [4]

Ramberg, J. S. (1975). "A probability distribution with applications to Monte Carlo simulation studies," *Statistical Distributions in Scientific Work, Volume 2–Model Building and Model Selection* (ed. G. P. Patil, S. Kotz, and J. K. Ord), D. Reidel Publishing Company, Dordrecht-Holland, 51–64. [112]

Ramberg, J. S. and Schmeiser, B. W. (1972). "An approximate method for generating symmetric random variables," *Comm. ACM*, 15, 987–990. [3]

Ramberg, J. S. and Schmeiser, B. W. (1974). "An approximate method for generating asymmetric random variables," *Comm. ACM*, 17, 78–82. [3, 63]

Ramberg, J. S., Tadikamalla, P. R., Dudewicz, E. J., and Mykytka, E. F. (1979). "A probability distribution and its uses in fitting data," *Technometrics*, 21, 201–214. [3, 10, 28, 35, 63, 104]

Ricer, T. L. (1980). *Accounting for Non–Normality and Sampling Error in Analysis of Variance of Construction Data*, M. S. Thesis (Adviser Richard E. Larew), Department of Civil Engineering, The Ohio State University, Columbus, Ohio, xi+183 pp. [4, 96]

Rocke, D. M. (1993). "On the beta transformation family," *Technometrics*, 35, 72–81. [119]

Schreuder, H. T. and Hafley, W. L. (1977). "A useful bivariate distribution for describing stand structure of tree heights and diameters," *Biometrics*, 33, 471–478. [146]

Scott, D. W. (1992). *Multivariate Density Estimation*, Wiley, New York. [1]

Shapiro, S. S. (1980). *How to Test Normality and Other Distributional Assumptions*, Vol. 3 of The ASQC Basic References in Quality Control: Statistical Techniques (Edward J. Dudewicz, Editor), American Society for Quality, Milwaukee, Wisconsin. [95]

Shapiro, S. S. and Wilk, M. B. (1965). "An analysis of variance test for normality (complete samples)," *Biometrika*, 52, 591–611. [4, 40]

Shapiro, S. S., Wilk, M. B., and Chen, Hwei J. (1968). "A comparative study of various tests for normality," *Journal of the American Statistical Association*, 63, 1343–1372. [4]

Shimizu, K. (1993). "A bivariate mixed lognormal distribution with an analysis of rainfall data," *Journal of Applied Meteorology*, 32, 161–171. [102, 103]

Silver, E. A. (1977). "A safety factor approximation based upon Tukey's Lambda distribution," *Operational Research Quarterly*, 28, 743–746. [4]

Simonoff, J. S. (1996). *Smoothing Methods in Statistics*, Springer, New York. [1]

Sun, L. and Müller-Schwarze, D. (1996). "Statistical resampling methods in biology: a case study of beaver dispersal patterns," *American Journal of Mathematical and Management Sciences*, 16, 463–502. [274, 278]

Syracuse University (1999). *Graduate Catalog 1999–2000*, Volume 12, No. 2 (March 1999) of *Syracuse University Bulletin*, Syracuse University, Syracuse, New York. [*v*]

Tadikamalla, P. R. (1984). *Modern Digital Simulation, Vol. I*, American Sciences Press, Inc., Columbus, Ohio. [151]

Todd, M. J. (1989). "On convergence properties of algorithms for unconstrained minimization," *IMA Journal of Numerical Analysis*, 9, 435–441. [54]

Tukey, J. W. (1960). *The Practical Relationship Between the Common Transformations of Percentages of Counts and of Amounts*, Technical Report 36, Statistical Techniques Research Group, Princeton University. [3, 4]

Van Dyke, J. (1961). "Numerical investigation of the random variable $y = c(u^\lambda - (1 - u)^\lambda)$," *Working Paper*, Statistical Engineering Laboratory, National Bureau of Standards, Washington, D. C. [4]

Weibull, W. (1951). "A statistical distribution function of wide applicability," *Journal of Applied Mechanics*, 18, 293–297. [84]

Wilcox, R. R. (1990), "Comparing the means of two independent groups," *Biometrical Journal*, 7, 771–780. [65, 90]

Williams, C. J. and Zhou, L. (1998). "A comparison of tests for detecting genetic variance from human twin data based on absolute intra-twin differences," *Communications in Statistics–Simulation*, 27, 51–65. [99]

Young, G. A. (1994). "Bootstrap: more than a stab in the dark?," *Statistical Science*, 9, 382–415. [282]

Subject Index

A

$\alpha_1, \alpha_2, \alpha_3, \alpha_4$
  of data, 53–54
  of the GBD, 117
  of the GLD, 45
(α_3^2, α_4)-space, 50, 53
$\alpha_4 > 1 + \alpha_3^2$, 50
AD data, 254
Advantages,
  use of *Maple*, 285
  use of percentiles, 153
Algorithms,
  examples of use, 64, 166–167,
  GB, generalized bootstrap,
    274–275
  GBD–M, fitting via moments, 126
  GLD–BV, bivariate fitting, 252
  GLD–M, fitting via moments, 63
  GLD–P, fitting via percentiles, 166
  $L_1 L_2$, L_1 and L_2 errors, 198
  PM, confidence intervals in GB,
    275–276
Analysis,
  re-doing, 107
  refining theoretical, 107
Anderson-Darling A^2, 95
Appropriateness of support, 94
Approximate solutions, 54
Approximations of distributions
    via GLD, 66
  first check sup $|\hat{f}(x) - f(x)|$, 66

second check sup $|\hat{F}(x) - F(x)|$,
  66–67
third check $|\hat{\alpha}_i - \alpha_i|$ $(i = 1, 2, 3, 4)$,
  66–67
Assessment of goodness-of-fit, 93–96
  approximations to true, 93
  characteristics of fit, 93–94
  chi-square test, 95
  eyeball test, 95
  KS test, 96
  overplot of histogram and fit, 94–95
  situation S–1, S–2, S–3, S–4, 93
  statistical tests, 95
Author Index, 417
Autoregressive processes, 32

B

Backwards Gamma marginals, 249
Beckwith, Nelson B., *vi*
Bessel function, 244
Beta distribution, 52, 79, 113, 137,
    183, 214
  GBD/EGLD fit, 137
  generalized, 119
  GLD fit, 79–81, 183–184
  moment-based fit, 80
  moments, 80, 117
  percentile-based fits, 183
  percentiles, 183
Beta function, 43
  properties of, 44